# FUNDAMENTALS OF THE THEORY OF OPERATOR ALGEBRAS

VOLUME I

*Elementary Theory*

This is a volume in
PURE AND APPLIED MATHEMATICS

A Series of Monographs and Textbooks

Editors: SAMUEL EILENBERG AND HYMAN BASS

A list of recent titles in this series appears at the end of this volume.

# FUNDAMENTALS OF THE THEORY OF OPERATOR ALGEBRAS

## VOLUME I
*Elementary Theory*

**Richard V. Kadison**

*Department of Mathematics*
*University of Pennsylvania*
*Philadelphia, Pennsylvania*

**John R. Ringrose**

*School of Mathematics*
*University of Newcastle*
*Newcastle upon Tyne, England*

1983

**ACADEMIC PRESS**
*A Subsidiary of Harcourt Brace Jovanovich, Publishers*
New York London
Paris San Diego San Francisco São Paulo Sydney Tokyo Toronto

ACADEMIC PRESS, INC.
111 Fifth Avenue, New York, New York 10003

*United Kingdom Edition published by*
ACADEMIC PRESS, INC. (LONDON) LTD.
24/28 Oval Road, London NW1 7DX

Library of Congress Cataloging in Publication Data

Kadison, Richard V., date
Fundamentals of the theory of operator algebras.

(Pure and applied mathematics)
Includes index.
1. Operator algebras. I. Ringrose, John R.
II. Title. III. Series: Pure and applied mathematics
(Academic Press)
QA3.P8 [QA326] 512'.55 82-13768
ISBN 0-12-393301-3 (v. 1)

PRINTED IN THE UNITED STATES OF AMERICA

83 84 85 86 9 8 7 6 5 4 3 2 1

# CONTENTS

## Chapter 3. **Banach Algebras**



## Chapter 4. **Elementary $C^*$-Algebra Theory**



## Chapter 5. **Elementary von Neumann Algebra Theory**

# PREFACE

These volumes deal with a subject, introduced half a century ago, that has become increasingly important and popular in recent years. While they cover the fundamental aspects of this subject, they make no attempt to be encyclopaedic. Their primary goal is to *teach* the subject and lead the reader to the point where the vast recent research literature, both in the subject proper and in its many applications, becomes accessible.

Although we have put major emphasis on making the material presented clear and understandable, the subject is not easy; no account, however lucid, can make it so. If it is possible to browse in this subject and acquire a significant amount of information, we hope that these volumes present that opportunity—but they have been written primarily for the reader, either starting at the beginning or with enough preparation to enter at some intermediate stage, who works through the text systematically. The study of this material is best approached with equal measures of patience and persistence.

Our starting point in Chapter 1 is finite-dimensional linear algebra. We assume that the reader is familiar with the results of that subject and begin by proving the infinite-dimensional algebraic results that we need from time to time. These volumes deal almost exclusively with infinite-dimensional phenomena. Much of the intuition that the reader may have developed from contact with finite-dimensional algebra and geometry must be abandoned in this study. It will mislead as often as it guides. In its place, a new intuition about infinite-dimensional constructs must be cultivated. Results that are apparent in finite dimensions may be false, or may be difficult and important principles whose application yields great rewards, in the infinite-dimensional case.

Almost as much as the subject matter of these volumes is infinite dimensional, it is *non-commutative real analysis*. Despite this description, the reader will find a very large number of references to the "abelian" or "commutative" case—an important part of this first volume is an analysis of the abelian case. This case, parallel to function theory and measure theory, provides us with a major tool and an important guide to our intuition. A good part of what we know comes from extending to the non-commutative case results that are known in the commutative case. The "extension" process is usually difficult. The main techniques include elaborate interlacing of "abelian" segments. The reference to "real analysis" involves the fact that while we consider complex-valued functions and, non-commutatively, non-self-adjoint operators, the structures we study make simultaneously available to us the complex conjugates of those functions and, non-commutatively, the adjoints of those operators. In essence, we are studying the algebraic interrelations of systems of real functions and, non-commutatively, systems of self-adjoint operators. At its most primitive level, the non-commutativity makes itself visible in the fact that the product of a function and its conjugate is the same in either order while this is not in general true of the product of an operator and its adjoint.

In the sense that we consider an operator and its adjoint on the same footing, the subject matter we treat is referred to as the "self-adjoint theory." There is an emerging and important development of non-self-adjoint operator algebras that serves as a non-commutative analogue of complex function theory—algebras of holomorphic functions. This area is not treated in these volumes. Many important developments in the self-adjoint theory—both past and current—are not treated. The type I $C^*$-algebras and $C^*$-algebra $K$-theory are examples of important subjects not dealt with. The aim of teaching the basics and preparing the reader for individual work in research areas seems best served by a close adherence to the "classical" fundamentals of the subject. For this same reason, we have not included material on the important application of the subject to the mathematical foundation of theoretical quantum physics. With one exception, applications to the theory of representations of topological groups are omitted. Accounts of these vast research areas, within the scope of this treatise, would be necessarily superficial. We have preferred instead to devote space to clear and leisurely expositions of the fundamentals. For several important topics, two approaches are included.

Our emphasis on instruction rather than comprehensive coverage has led us to settle on a very brief bibliography. We cite just three textbooks (listed as [H], [K], and [R]) for background information on general topology and measure theory, and for this first volume, include only 25 items from the literature of our subject. Several extensive and excellent bibliographies are available (see, for example, [2,24,25]), and there would be little purpose in reproducing a modified version of one of the existing lists. We have included in our references items specifically referred to in the text and others that might provide profitable additional reading. As a consequence, we have made no attempt, either in the text or in the exercises, to credit sources on which we have drawn or to trace the historical background of the ideas and results that have gone into the development of the subject.

Each of the chapters of this first volume has a final section devoted to a substantial list of exercises, arranged roughly in the order of the appearance of topics in the chapter. They were designed to serve two purposes: to illustrate and extend the results and examples of the earlier sections of the chapter, and to help the reader to develop working technique and facility with the subject matter of the chapter. For the reader interested in acquiring an ability to work with the subject, a certain amount of exercise solving is indispensable. We *do not* recommend a rigid adherence to order—each exercise being solved in sequence and no new material attempted until all the exercises of the preceding chapter are solved. Somewhere between that approach and total disregard of the exercises a line must be drawn congenial to the individual reader's needs and circumstances. In general, we *do* recommend that the greater proportion of the reader's time be spent on a thorough understanding of the main text than on the exercises. In any event, *all* the exercises have been designed to be solved. Most exercises are separated into several parts with each of the parts manageable and some of them provided with hints. Some are routine, requiring nothing more than a clear understanding of a definition or result for their solutions. Other exercises (and groups of exercises) constitute small (guided) research projects.

On a first reading, as an introduction to the subject, certain sections may well be left unread and consulted on a few occasions as needed. Section 2.6, *Tensor products and the Hilbert–Schmidt class* (this "subsection" is the largest part of Section 2.6) will not be needed seriously until Chapter 11 (in Volume II). All the material on unbounded operators (and the material related to Stone's theorem) will not be needed until Chapter 9 (in Volume II). Thus Section 2.7, Section 3.2, *The Banach algebra* $L_1(\mathbb{R})$ *and Fourier analysis*, the last few pages of Chapter 4 (including Theorem 4.5.9), and Section 5.6, can be deferred to a later reading. Some readers, more or less familiar with the elements of functional analysis, may want to enter the text after Chapter 1 with occasional back references for notation or precise definitions and statements of results. The reader with a good general knowledge of basic functional analysis may consider beginning at Section 3.4 or perhaps with Chapter 4.

The various possible styles of reading this volume, related to the levels of preparation of the reader, suggest several styles and levels of courses for which it can be used. For all of these, a good working knowledge of point-set (general) topology, such as may be found in [K], is assumed. Somewhat less vital, but useful, is a knowledge of general measure theory, such as may be found in [H] and parts of [R]. Of course, full command of the fundamentals of real and complex analysis (we refer to [R] for these) is needed; and, as noted earlier, the elements of finite-dimensional linear algebra are used. The first three chapters form the basis of a course in elementary functional analysis with a slant toward operator algebras and its allied fields of group representations, harmonic analysis, and mathematical (quantum) physics. These chapters provide material for a brisk one-semester course at the first- or second-year graduate level or for a more leisurely one-year course at the advanced undergraduate or beginning graduate level. Chapters 3, 4, and 5 provide an introduction to the theory of operator algebras and have material that would serve as a one-semester graduate course at the second- or third-year level (especially if Section 5.6 is omitted). In any event, the book has been designed for individual study as well as for courses, so that the problem of a wide spread of preparation in a class can be dealt with by encouraging the better prepared students to proceed at their own paces. Seminar and reading-course possibilities are also available.

When several (good) terms for a mathematical construct are in common use, we have made no effort to choose one and then to use that one term consistently. On the contrary, we have used such terms interchangeably after introducing them simultaneously. This seems the best preparation for further reading in the research literature. Some examples of such terms are weaker, coarser (for topologies on a space), unitary transformation, and Hilbert space isomorphism (for structure-preserving mappings between Hilbert spaces). In cases where there is conflicting use of a term in the research literature (for example, "purely infinite" in connection with von Neumann algebras), we have avoided all use of the term and employed accepted terminology for each of the constructs involved. Since the symbol * is used to denote the adjoint operations on operators and on sets of operators, we have preferred to use a different symbol in the context of Banach dual spaces. We denote the dual space of a Banach space $\mathfrak{X}$ by $\mathfrak{X}^{\#}$. However, we felt compelled by usage to retain the terminology "weak *" for the topology induced by elements of $\mathfrak{X}$ (as linear functionals on $\mathfrak{X}^{\#}$).

Results in the body of the text are italicized, titled Theorem, Proposition, Lemma, and Corollary (in decreasing order of "importance"—though, as usual, the "heart of the matter" may be dealt with in a lemma and its most usable aspect may appear in a corollary). In addition, there are Remarks and Examples that extend and illuminate the material of a section, and of course there are the (formal) Definitions. None of these items is italicized, though a crucial phrase or word frequently is. Each of these segments of the text is preceded by a number, the first digit of which indicates the chapter, the second the section, and the last one- or two-digit number the position of the item in the section. Thus, "Proposition 5.5.18" refers to the eighteenth numbered item in the fifth section of the fifth chapter. A back or forward reference to such an item will include the title ("Theorem," "Remark," etc.), though the number alone would serve to locate it. Occasionally a displayed equation, formula, inequality, etc., is assigned a number in parentheses at the left of the display—for example, the "convolution formula" of Fourier transform theory appears as the display numbered (4) in the proof of Theorem 3.2.26. In its own section, it is referred to as (4) and elsewhere as 3.2(4).

The lack of illustrative examples in much of Chapter 1 results from our wish to bring the reader more rapidly to the subject of operator algebras rather than to dwell on the basics of general functional analysis. As compensation for their lack, the exercises supply much of the illustrative material for this chapter. Although the tensor product development in Section 2.6 may appear somewhat formal and forbidding at first, it turns out that the trouble and care taken at that point simplify subsequent application. The same can be said (perhaps more strongly) about Section 5.6. The material on unbounded operators (their spectral theory and function calculus) is so vital when needed and so susceptible to incorrect and incomplete application that it seemed well worth a careful and thorough treatment. We have chosen a powerful approach that permits such a treatment, much in the spirit of the theory of operator algebras.

Another (general) aspect of the organization of material in a text is the way the material of the text proper relates to the exercises. As a matter of specific policy, we have not relegated to the exercises whole arguments or parts of arguments. Reference is occasionally made to an exercise as an illustration of some point—for example, the fact that the statement resulting from the omission of some hypothesis from a theorem is false.

During the course of the preparation of these volumes, we have enjoyed, jointly and separately, the hospitality and facilities of several universities, aside from our home institutions. Notable among these are the Mathematics Institutes of the Universities of Aarhus and Copenhagen and the Theoretical Physics Institute of Marseille–Luminy. The subject matter of these volumes and its style of development is inextricably interwoven with the individual research of the authors. As a consequence, the support of that research by the National Science Foundation (U.S.A.) and the Science Research Council (U.K.) has had an oblique but vital influence on the formation of these volumes. It is the authors' pleasure to express their gratitude for this support and for the hospitality of the host institutions noted.

# CONTENTS OF VOLUME II

## *Advanced Theory*

## Chapter 6. **Comparison Theory of Projections**



## Chapter 7. **Normal States and Unitary Equivalence of von Neumann Algebras**



## Chapter 8. **The Trace**

## Chapter 9. **Algebra and Commutant**



## Chapter 10. **Special Representations of $C^*$-Algebras**



## Chapter 11. **Tensor Products**



## Chapter 12. **Approximation by Matrix Algebras**

## Chapter 13. **Crossed Products**



## Chapter 14. **Direct Integrals and Decompositions**



## **Bibliography**

# CHAPTER 1

# LINEAR SPACES

This chapter contains an account of those basic aspects of linear functional analysis that are needed, later in the book, in the study of operator algebras. The main topics – continuous linear operators, continuous linear functionals, weak topologies, convexity – are studied first in the context of linear topological spaces, then in the more restricted setting of normed spaces and Banach spaces. In preparation for this, some related material is treated in the purely algebraic situation (that is, without topological considerations).

## 1.1. Algebraic results

In this section we shall consider linear spaces (that is, vector spaces) over a field $\mathbb{K}$, and it will be assumed throughout that $\mathbb{K}$ is either the real field $\mathbb{R}$ or the complex field $\mathbb{C}$. We sometimes distinguish between these two cases by referring to *real* vector spaces or *complex* vector spaces. Our main concern is with linear functionals, convex sets, and the separation of convex sets by hyperplanes.

Suppose that $\mathscr{V}$ is a linear space with scalar field $\mathbb{K}$. If $X$ and $Y$ are non-empty subsets of $\mathscr{V}$, and $a \in \mathbb{K}$, we define further subsets $aX$, $X \pm Y$ by

$$aX = \{ax : x \in X\}, \qquad X + Y = \{x + y : x \in X, \ y \in Y\},$$

and

$$X - Y = X + (-1)Y.$$

When $X$ consists of a single element $x$, we write $x \pm Y$ in place of $X \pm Y$. To avoid ambiguity in the use of the symbol $-$, the set theoretic difference $\{x \in \mathbb{A} : x \notin \mathbb{B}\}$ of two sets $\mathbb{A}$ and $\mathbb{B}$ will be denoted by $\mathbb{A} \backslash \mathbb{B}$. A vector of the form $a_1x_1 + \cdots + a_nx_n$, where $x_1, \ldots, x_n \in X$ and $a_1, \ldots, a_n \in \mathbb{K}$, is called a (*finite*) *linear combination* of elements of $X$. The zero vector is always of this form (in a trivial way), with $\{x_1, \ldots, x_n\}$ an arbitrary finite subset of $X$, and $a_j = 0$ for each $j$. If it can be expressed as a *non-trivial* linear combination of elements of $X$ (that is, with $x_1, \ldots, x_n$ distinct, and at least one $a_j$ non-zero), then $X$ is said to be *linearly dependent*; otherwise $X$ is *linearly independent*. The set of all finite linear combinations of elements of $X$ is a linear subspace of $\mathscr{V}$, the smallest containing $X$; we refer to it as the *linear subspace generated by X*.

If $\mathscr{V}_0$ is a linear subspace of $\mathscr{V}$, we denote by $\mathscr{V}/\mathscr{V}_0$ the set of all cosets $x + \mathscr{V}_0$ $(x \in \mathscr{V})$ in the additive group $\mathscr{V}$. Of course, $\mathscr{V}/\mathscr{V}_0$ is a group, with addition defined by $(x + \mathscr{V}_0) + (y + \mathscr{V}_0) = (x + y) + \mathscr{V}_0$. If $a \in \mathbb{K}$, and $x_1 + \mathscr{V}_0 = x_2 + \mathscr{V}_0$, we have $ax_1 - ax_2 = a(x_1 - x_2) \in \mathscr{V}_0$, so $ax_1 + \mathscr{V}_0 = ax_2 + \mathscr{V}_0$. From this it follows easily that $\mathscr{V}/\mathscr{V}_0$ becomes a linear space over $\mathbb{K}$, the *quotient* of $\mathscr{V}$ by $\mathscr{V}_0$, when multiplication by scalars is defined (unambiguously) by $a(x + \mathscr{V}_0) = ax + \mathscr{V}_0$. If $\mathscr{V}/\mathscr{V}_0$ has finite dimension $n$, we say that $\mathscr{V}_0$ has finite *codimension n* in $\mathscr{V}$.

Suppose that $\mathscr{V}$ and $\mathscr{W}$ are linear spaces over $\mathbb{K}$. By a *linear operator* (or *linear transformation*) from $\mathscr{V}$ into $\mathscr{W}$, we mean a mapping $T: \mathscr{V} \to \mathscr{W}$ such that

$$T(ax + by) = aTx + bTy$$

whenever $x, y \in \mathscr{V}$ and $a, b \in \mathbb{K}$ (the notation $T: \mathscr{V} \to \mathscr{W}$ indicates that $T$ is defined on $\mathscr{V}$ and takes values in $\mathscr{W}$; it can be read "$T$, from $\mathscr{V}$ into $\mathscr{W}$"). If $\mathscr{V}_0$ is a linear subspace of $\mathscr{V}$, the equation $Qx = x + \mathscr{V}_0$ defines a linear operator $Q$ from $\mathscr{V}$ onto $\mathscr{V}/\mathscr{V}_0$, the *quotient mapping*. When $T: \mathscr{V} \to \mathscr{W}$ is a linear operator, the *null space* of $T$ is the linear subspace $\{x \in \mathscr{V} : Tx = 0\}$ of $\mathscr{V}$, and the image (or range) $T(\mathscr{V}) = \{Tx : x \in \mathscr{V}\}$ is a linear subspace of $\mathscr{W}$. If $T(\mathscr{V}_0) = \{0\}$, the condition $x + \mathscr{V}_0 = y + \mathscr{V}_0$ entails $x - y \in \mathscr{V}_0$, and hence $Tx - Ty = 0$; moreover, if $\mathscr{V}_0$ is the null space of $T$, $Tx = 0$ entails $x \in \mathscr{V}_0$. From this, the equation $T_0(x + \mathscr{V}_0) = Tx$ defines (unambiguously) a linear operator $T_0$ from $\mathscr{V}/\mathscr{V}_0$ onto $T(\mathscr{V})$ $(\subseteq \mathscr{W})$, when $T(\mathscr{V}_0) = \{0\}$; and $T_0$ is one-to-one if $\mathscr{V}_0$ is the null space of $T$. Note that $T = T_0Q$, a fact sometimes described by saying that *T factors through* $\mathscr{V}/\mathscr{V}_0$ when $T(\mathscr{V}_0) = \{0\}$. Given any linear operators $S, T$: $\mathscr{V} \to \mathscr{W}$ and scalars $a$, $b$, the equation $(aS + bT)x = aSx + bTx$ $(x \in \mathscr{V})$ defines another such operator $aS + bT$, and in this way, the set of all linear operators from $\mathscr{V}$ into $\mathscr{W}$ becomes a linear space over $\mathbb{K}$.

By a *linear functional* on $\mathscr{V}$ we mean a linear operator $\rho: \mathscr{V} \to \mathbb{K}$ (of course, $\mathbb{K}$ is a one-dimensional linear space over $\mathbb{K}$). The set of all linear functionals on $\mathscr{V}$ is itself a linear space over $\mathbb{K}$, the *algebraic dual space of* $\mathscr{V}$. When $\rho$ is a *non-zero* linear functional on $\mathscr{V}$ (that is, $\rho$ does not vanish identically on $\mathscr{V}$) the image $\rho(\mathscr{V})$ is $\mathbb{K}$.

1.1.1. Proposition. *If $\rho$ is a linear functional on a linear space $\mathscr{V}$, then every linear functional on $\mathscr{V}$ that vanishes on the null space $\mathscr{V}_0$ of $\rho$ is a scalar multiple of $\rho$. If $\rho \neq 0$, $\mathscr{V}_0$ has codimension* 1 *in $\mathscr{V}$. Conversely each linear subspace of codimension* 1 *in $\mathscr{V}$ is the null space of a non-zero linear functional. If $\rho_1, \ldots, \rho_n$ are linear functionals on $\mathscr{V}$, then every linear functional on $\mathscr{V}$ that vanishes on the intersection of the null spaces of $\rho_1, \ldots, \rho_n$ is a linear combination of $\rho_1, \ldots, \rho_n$.*

*Proof.* We may suppose that $\rho \neq 0$. The equation $\rho_0(x + \mathscr{V}_0) = \rho(x)$ defines a one-to-one linear operator $\rho_0$ from $\mathscr{V}/\mathscr{V}_0$ onto the one-dimensional linear space $\mathbb{K}$; so $\mathscr{V}/\mathscr{V}_0$ is one dimensional. Of course, $\rho_0$ is a non-zero linear functional on $\mathscr{V}/\mathscr{V}_0$; and in the same way, if a linear functional $\sigma$ on $\mathscr{V}$ vanishes on $\mathscr{V}_0$, there is a linear functional $\sigma_0$ on $\mathscr{V}/\mathscr{V}_0$, defined by $\sigma_0(x + \mathscr{V}_0) = \sigma(x)$. Since $\mathscr{V}/\mathscr{V}_0$ is one dimensional, $\sigma_0 = a\rho_0$ for some scalar $a$, and $\sigma = \sigma_0 Q = a\rho_0 Q = a\rho$, where $Q$ is the quotient mapping from $\mathscr{V}$ onto $\mathscr{V}/\mathscr{V}_0$.

If $\mathscr{V}_1$ is a linear subspace with codimension 1 in $\mathscr{V}$, there is a non-zero linear functional $\tau_1$ on the one-dimensional linear space $\mathscr{V}/\mathscr{V}_1$, and $\tau_1 : \mathscr{V}/\mathscr{V}_1 \to \mathbb{K}$ is a one-to-one mapping. Accordingly, the equation $\tau(x) = \tau_1(x + \mathscr{V}_1)$ defines a non-zero linear functional $\tau$ on $\mathscr{V}$ whose null space is $\mathscr{V}_1$.

The final assertion of the proposition is proved by induction on $n$. We make the inductive assumption that it is valid when $n = k$ (the initial case, in which $n = 1$, reduces to the statement in the first sentence of the proposition, and has already been proved). Now suppose that $\sigma$, $\rho_1, \ldots, \rho_k$, $\rho_{k+1}$ are linear functionals on $\mathscr{V}$, and $\sigma$ vanishes on the intersection of the null spaces of $\rho_1, \ldots, \rho_k$, $\rho_{k+1}$. Let $\mathscr{V}_{k+1}$ denote the null space of $\rho_{k+1}$, and consider the restrictions $\sigma|\mathscr{V}_{k+1}$, $\rho_1|\mathscr{V}_{k+1}, \ldots, \rho_k|\mathscr{V}_{k+1}$, $\rho_{k+1}|\mathscr{V}_{k+1}$ $(= 0)$. Since $\sigma|\mathscr{V}_{k+1}$ vanishes on the intersection of the null spaces of the $k$ linear functionals $\rho_1|\mathscr{V}_{k+1}, \ldots, \rho_k|\mathscr{V}_{k+1}$, it is a linear combination $(a_1\rho_1 + \cdots + a_k\rho_k)|\mathscr{V}_{k+1}$ of those linear functionals by our inductive assumption. Thus $\sigma - a_1\rho_1 - \cdots - a_k\rho_k$ vanishes on the null space $\mathscr{V}_{k+1}$ of $\rho_{k+1}$, and is therefore a multiple $a_{k+1}\rho_{k+1}$ by the first assertion of the proposition; and $\sigma = a_1\rho_1 + \cdots + a_k\rho_k + a_{k+1}\rho_{k+1}$. ∎

Suppose that $\mathscr{V}$ is a linear space over $\mathbb{K}$ $(= \mathbb{R}$ or $\mathbb{C})$, and $X$, $Y \subseteq \mathscr{V}$. By a (*finite*) *convex combination* of elements of $X$, we mean a vector of the form $a_1x_1 + \cdots + a_nx_n$, where $x_1, \ldots, x_n \in X$ and $a_1, \ldots, a_n$ are real scalars satisfying $a_j > 0$ $(j = 1, \ldots, n)$ and $\sum a_j = 1$. It makes no difference in this definition if the condition $a_j > 0$ is relaxed to $a_j \geqslant 0$ (because zero terms can be deleted), but strict inequality is slightly more convenient for our present purposes. We say that $Y$ is *convex* if $b_1y_1 + b_2y_2 \in Y$ whenever $y_1$, $y_2 \in Y$ and $b_1$, $b_2$ are positive real numbers with sum 1 (that is, $Y$ contains each convex combination of just *two* elements of $Y$; geometrically, this means that each line segment with endpoints in $Y$ lies wholly in $Y$). A simple proof, by induction on $n$, shows that a convex set $Y$ contains every convex combination $a_1x_1 + \cdots + a_nx_n$ of elements $x_1, \ldots, x_n$ of $Y$; the "inductive step up," from $n - 1$ to $n$, depends on the observation that

$$a_1x_1 + \cdots + a_nx_n = b_1y_1 + b_2y_2,$$

where $b_1 = a_1$, $b_2 = a_2 + \cdots + a_n$, $y_1 = x_1$, and $y_2$ is the convex combination $b_2^{-1}(a_2x_2 + \cdots + a_nx_n)$ of $x_2, \ldots, x_n$.

It is sometimes useful to note that a subset $Y$ of $\mathscr{V}$ is convex if and only if $a_1Y + a_2Y = (a_1 + a_2)Y$ whenever $a_1$ and $a_2$ are non-negative scalars; for this is equivalent to $(a_1 + a_2)^{-1}(a_1 Y + a_2 Y) = Y$, when $a_1$ and $a_2$ are non-negative scalars, not both 0.

When $X \subseteq \mathscr{V}$, we denote by co $X$ the set of all finite convex combinations of elements of $X$. A straightforward calculation shows that if $y_1, \ldots, y_n \in \text{co } X$, then every convex combination of $y_1, \ldots, y_n$ lies in co $X$. Thus co $X$ is a convex set, the smallest one containing $X$; it is called the *convex hull* of $X$. By an *internal point* of $X$ we mean a vector $x$ in $X$ with the following property: given any $y$ in $\mathscr{V}$, there is a positive real number $c$ such that $x + ay \in X$ whenever $0 \leqslant a < c$.

Our next result is concerned with *real* vector spaces. By a *hyperplane*, in a linear space $\mathscr{V}$ over $\mathbb{R}$, we mean a set of the form $x_0 + \mathscr{V}_0$, where $x_0 \in \mathscr{V}$ and $\mathscr{V}_0$ is a linear subspace with codimension 1 in $\mathscr{V}$. From Proposition 1.1.1, a subset $H$ of $\mathscr{V}$ is a hyperplane if and only if it can be expressed in the form

$$H = \{x \in \mathscr{V} : \rho(x) = k\},$$

where $\rho$ is a non-zero linear functional on $\mathscr{V}$ and $k \in \mathbb{R}$; of course, $\rho$ and $k$ are not uniquely determined by $H$, but the only possible variation is to replace them by $a\rho$ and $ak$, respectively, where $a$ is a non-zero real number. With the hyperplane $H$ we can associate the two *closed half-spaces*, $\{x \in \mathscr{V} : \rho(x) \geqslant k\}$ and $\{x \in \mathscr{V} : \rho(x) \leqslant k\}$, and the two *open half-spaces*, which are defined similarly but with strict inequalities. We say that $H$ *separates* two subsets $Y$ and $Z$ of $\mathscr{V}$ if $Y$ is contained in one of the closed half-spaces determined by $H$ and $Z$ is contained in the other; *strict separation* is defined similarly in terms of the open half-spaces. If the hyperplane is described in terms of $a\rho$ and $ak$, the property of separation remains unchanged (although the two half-spaces are interchanged if $a < 0$).

1.1.2. THEOREM. *If $Y$ and $Z$ are non-empty disjoint convex subsets of a real vector space $\mathscr{V}$, at least one of which has an internal point, they are separated by a hyperplane $H$ in $\mathscr{V}$. If either $Y$ or $Z$ consists entirely of internal points, it is contained in one of the open half-spaces determined by $H$. If both $Y$ and $Z$ consist entirely of internal points, they are strictly separated by $H$.*

*Proof.* We may suppose that $Y$ has an internal point, and denote by $Y_i$ the set of all internal points of $Y$. It is easily verified that $Y_i$ is a convex subset of $Y$, and that $(1 - a)y_1 + ay \in Y_i$ whenever $y_1 \in Y_i$, $y \in Y$, and $0 \leqslant a < 1$.

We assert that every point of $Y_i$ is an internal point of $Y_i$, and that a hyperplane which separates $Y_i$ and $Z$ also separates $Y$ and $Z$. For this, suppose that $y_1 \in Y_i$ and $x \in \mathscr{V}$. Since $y_1$ is an internal point of $Y$, $y_1 + cx \in Y$ for some positive scalar $c$. From the preceding paragraph,

$$y_1 + acx = (1 - a)y_1 + a(y_1 + cx) \in Y_i$$

when $0 \leqslant a < 1$; so $y_1 + bx \in Y_i$ whenever $0 \leqslant b < c$, and thus $y_1$ is an internal point of $Y_i$. If $H$ is a hyperplane separating $Y_i$ and $Z$, there is a non-zero linear functional $\rho$ on $\mathscr{V}$ and a scalar $k$ such that $H = \{x \in \mathscr{V} : \rho(x) = k\}$ and

$$\rho(y_1) \geqslant k \geqslant \rho(z) \qquad (y_1 \in Y_i, \quad z \in Z).$$

Given $y$ in $Y$, choose any $y_1$ in $Y_i$. Since $(1-a)y_1 + ay \in Y_i$ when $0 \leqslant a < 1$, we have

$$(1-a)\rho(y_1) + a\rho(y) = \rho((1-a)y_1 + ay) \geqslant k;$$

and when $a \to 1$, we obtain $\rho(y) \geqslant k$. Thus $H$ separates $Y$ and $Z$.

Upon replacing $Y$ by $Y_i$, it now suffices to prove the theorem under the additional assumption that *each* point of $Y$ is an internal point of $Y$. In this case, $Y - Z$ is a convex subset of $\mathscr{V}$ consisting entirely of internal points and not containing 0. Let $\mathscr{C}$ be the family of all convex subsets $C$ of $\mathscr{V}$ for which $0 \notin C$, $Y - Z \subseteq C$, and each point of $C$ is an internal point of $C$. Then $\mathscr{C}$ is partially ordered by the inclusion relation $\subseteq$. If $\mathscr{C}_0$ is a totally ordered subfamily of $\mathscr{C}$, let $C_1$ be the union of all the sets in $\mathscr{C}_0$. It is apparent that $0 \notin C_1$, $Y - Z \subseteq C_1$, and $C_1$ consists entirely of internal points. Given $u$ and $v$ in $C_1$, there is a *single* set $C_0$ in $\mathscr{C}_0$ containing both $u$ and $v$ (because $\mathscr{C}_0$ is totally ordered by $\subseteq$). Thus $C_0$ (and hence, also $C_1$) contains every convex combination of $u$ and $v$, and so $C_1$ is convex. Accordingly, $C_1 \in \mathscr{C}$, and $C_1$ is an upper bound for $\mathscr{C}_0$. It now follows from Zorn's lemma that there is an element $C$ of $\mathscr{C}$ that is maximal with respect to inclusion.

It is immediately verified that the set

$$\{au : u \in C, \ a > 0\}$$

is an element of $\mathscr{C}$ and contains $C$. By maximality, it coincides with $C$; so $au \in C$ whenever $u \in C$ and $a > 0$. From this, and since $C$ is convex and $0 \notin C$, it now follows that $C \cap -C = \varnothing$ (the empty set), and

$$au \in C, \qquad au + bv \in C, \qquad bw \in \mathscr{V} \backslash C$$

whenever $u, v \in C$, $w \in \mathscr{V} \backslash C$, $a > 0$, and $b \geqslant 0$.

We assert next that $au + bv \in \mathscr{V} \backslash C$ whenever $u, v \in \mathscr{V} \backslash C$ and $a, b \geqslant 0$. For this, suppose the contrary, so that $au + bv \in C$ for some $a, b, u, v$ satisfying the stated conditions. From the preceding paragraph, $au, bv \in \mathscr{V} \backslash C$; so, upon replacing $u$ by $au$ and $v$ by $bv$, we may suppose that $u, v \in \mathscr{V} \backslash C$ and $u + v \in C$. When $r \geqslant 0$, we have

$$2rv \in \mathscr{V} \backslash C, \qquad 2rv = r(v+u) + r(v-u)$$

and $v + u \in C$; so, again from the preceding paragraph, $r(v-u) \notin C$. Accordingly, if

$$C_1 = \{x + r(u-v) : x \in C, \ r \geqslant 0\},$$

then $0 \notin C_1$, $C_1 \supseteq C$ $(\supseteq Y - Z)$, and $C_1$ is convex and consists entirely of internal points. Thus $C_1 \in \mathscr{C}$, by maximality $C_1 = C$, and hence

$$2u = (u + v) + (u - v) \in C_1 = C,$$

a contradiction (since $au \in \mathscr{V} \backslash C$ for all $a \geqslant 0$). This proves our assertion that $au + bv \in \mathscr{V} \backslash C$ whenever $u, v \in \mathscr{V} \backslash C$ and $a, b \geqslant 0$. It follows that the set

$$\mathscr{V}_0 = \{x \in \mathscr{V} : x, -x \in \mathscr{V} \backslash C\} = \mathscr{V} \backslash (C \cup -C)$$

is a linear subspace of $\mathscr{V}$ (and $\mathscr{V}_0 \cap C = \varnothing$, whence $\mathscr{V}_0 \neq \mathscr{V}$).

We prove next that $\mathscr{V}_0$ has codimension 1 in $\mathscr{V}$. To this end, we have to show that any two non-zero elements of $\mathscr{V}/\mathscr{V}_0$ are linearly dependent; that is, if $u, v \in \mathscr{V} \backslash \mathscr{V}_0$, then $au + bv \in \mathscr{V}_0$ for suitable non-zero scalars $a, b$. Since $\mathscr{V} \backslash \mathscr{V}_0 = C \cup -C$, we may suppose (upon replacing $u$ by $-u$, or $v$ by $-v$, if necessary) that $u \in C$ and $v \in -C$. Since $C$ consists entirely of internal points, the same is true of $-C$. From this, the disjoint subsets

$$S_0 = \{s \in [0, 1] : u + s(v - u) \in C\},$$

$$S_1 = \{s \in [0, 1] : u + s(v - u) \in -C\}$$

of the real interval $[0, 1]$ are both open; indeed, if $u + s_0(v - u) \in C$ (or $-C$), then $u + s(v - u) \in C$ (or $-C$) for all $s$ sufficiently close to $s_0$, since both the points $u + s_0(v - u) \pm t(v - u)$ lie in $C$ (or $-C$) for all sufficiently small non-negative $t$. Since $0 \in S_0$, $1 \in S_1$, and $[0, 1]$ is connected, $S_0 \cup S_1$ is not the whole of $[0, 1]$; so there is a real number $s$ such that $0 < s < 1$ and

$$(1 - s)u + sv = u + s(v - u) \in \mathscr{V} \backslash (C \cup -C) = \mathscr{V}_0.$$

This completes the proof that $\mathscr{V}_0$ has codimension 1 in $\mathscr{V}$.

Let $\rho$ be a (non-zero) linear functional on $\mathscr{V}$ whose null space is $\mathscr{V}_0$. Since $C$ is convex and $\mathscr{V}_0 \cap C = \varnothing$, the subset $\rho(C)$ of $\mathbb{R}$ is convex and does not contain 0; so either $\rho(C) \subseteq (0, \infty)$ or $\rho(C) \subseteq (-\infty, 0)$. Upon replacing $\rho$ by $-\rho$ if necessary, we may suppose that $\rho(u) > 0$ for all $u$ in $C$. Since $Y - Z \subseteq C$, it follows that $\rho(y) > \rho(z)$ whenever $y \in Y$ and $z \in Z$. From this, the subset $\rho(Z)$ of $\mathbb{R}$ is bounded above, and its least upper bound $k$ satisfies

$$\rho(y) \geqslant k \geqslant \rho(z) \qquad (y \in Y, \quad z \in Z).$$

Thus the hyperplane $\{x \in \mathscr{V} : \rho(x) = k\}$ $(= H)$ separates $Y$ and $Z$.

From the assumption that $Y$ consists entirely of internal points, we now deduce that it is contained in the *open* half-space $\{x \in \mathscr{V} : \rho(x) > k\}$. For this, suppose that $y \in Y$, and choose $x_0$ in $\mathscr{V}$ such that $\rho(x_0) > 0$. Then $y - ax_0 \in Y$ (and therefore $\rho(y) - a\rho(x_0) \geqslant k$) for all sufficiently small positive scalars $a$, and thus $\rho(y) > k$. If $Z$ (as well as $Y$) consists entirely of internal points, a similar argument shows that $\rho(z) < k$ for each $z$ in $Z$; so in this case $Y$ and $Z$ are strictly separated by $H$. ■

Theorem 1.1.2 is our first example from a group of related results, described loosely as Hahn–Banach theorems. These results, which occur both in the present algebraic setting and also in the context of linear topological spaces, can be divided broadly into two main types. The first group (*separation theorems*) is concerned with separation of convex sets; closely related to this, there is a second group (*extension theorems* for linear functionals).

A complex vector space $\mathscr{V}$ can be viewed also as a real vector space simply by restricting attention to real scalars. Occasionally, for emphasis, we shall denote the real vector space so obtained by $\mathscr{V}_r$. A linear functional on $\mathscr{V}_r$ is described as a *real-linear functional* on $\mathscr{V}$, and linear subspaces of $\mathscr{V}_r$ are called *real-linear subspaces* of $\mathscr{V}$. A set $X$ ($\subseteq \mathscr{V}$) is convex if and only if it is convex when viewed as a subset of $\mathscr{V}_r$, and the internal points of $X$ are the same in both cases, since the concepts of "convex set" and "internal point" depend only on real scalars. In proving Hahn–Banach theorems for complex vector spaces, we shall require the following simple result.

1.1.3. LEMMA. *If $\rho$ is a linear functional on a complex vector space $\mathscr{V}$, the equation $\rho_r(x) = \operatorname{Re}\rho(x)$ defines a real-linear functional $\rho_r$ on $\mathscr{V}$, and*

$$\rho(x) = \rho_r(x) - i\rho_r(ix) \qquad (x \in \mathscr{V}).$$

*Every real-linear functional on $\mathscr{V}$ arises, in this way, from a linear functional.*

*Proof.* It is apparent that, given a linear functional $\rho$ on $\mathscr{V}$, the stated equation defines a real-linear functional $\rho_r$. Moreover

$$\operatorname{Im}\rho(x) = -\operatorname{Re} i\rho(x) = -\operatorname{Re}\rho(ix) = -\rho_r(ix),$$

whence $\rho(x) = \operatorname{Re}\rho(x) + i\operatorname{Im}\rho(x) = \rho_r(x) - i\rho_r(ix)$.

Suppose next that $\sigma$ is a real-linear functional on $\mathscr{V}$, and define a function $\sigma_c : \mathscr{V} \to \mathbb{C}$ by

$$\sigma_c(x) = \sigma(x) - i\sigma(ix) \qquad (x \in \mathscr{V}).$$

It is clear that $\sigma(x) = \operatorname{Re}\sigma_c(x)$, $\sigma_c(x + y) = \sigma_c(x) + \sigma_c(y)$, and $\sigma_c(ax) = a\sigma_c(x)$, whenever $x, y \in \mathscr{V}$ and $a$ is a *real* scalar. Since, also,

$$\begin{aligned}\sigma_c(ix) &= \sigma(ix) - i\sigma(-x) = \sigma(ix) + i\sigma(x)\\ &= i[\sigma(x) - i\sigma(ix)] = i\sigma_c(x),\end{aligned}$$

it follows that $\sigma_c$ is a linear functional on $\mathscr{V}$. ∎

We now obtain the analogue, for complex vector spaces, of Theorem 1.1.2.

1.1.4. THEOREM. *If $Y$ and $Z$ are non-empty disjoint convex subsets of a complex vector space $\mathscr{V}$, at least one of which has an internal point, there is a non-zero linear functional $\rho$ on $\mathscr{V}$, and a real number $k$, such that*

$$\operatorname{Re}\rho(y) \geqslant k \geqslant \operatorname{Re}\rho(z) \qquad (y \in Y, \quad z \in Z).$$

*Moreover,* $\operatorname{Re}\rho(y) > k$ $(y \in Y)$ *if $Y$ consists entirely of internal points, and* $k > \operatorname{Re}\rho(z)$ $(z \in Z)$ *if $Z$ consists entirely of internal points.*

*Proof.* By considering $Y$ and $Z$ as subsets of the real vector space $\mathscr{V}_r$ obtained from $\mathscr{V}$, it follows from Theorem 1.1.2 that there is a non-zero real-linear functional $\sigma$ on $\mathscr{V}$ and a real number $k$ such that $\sigma(y) \geqslant k \geqslant \sigma(z)$ whenever $y \in Y$ and $z \in Z$. Moreover, if either $Y$ or $Z$ consists entirely of internal points, the corresponding one of the inequalities $\geqslant$ can be replaced by $>$. By Lemma 1.1.3, there is a linear functional $\rho$ on $\mathscr{V}$ such that $\sigma(x) = \operatorname{Re}\rho(x)$ for each $x$ in $\mathscr{V}$. ■

Let $\mathscr{V}$ be a linear space with scalar field $\mathbb{K}$ $(= \mathbb{R}$ or $\mathbb{C})$. By a *sublinear functional* on $\mathscr{V}$ we mean a function $p: \mathscr{V} \to \mathbb{R}$ such that

$$p(x + y) \leqslant p(x) + p(y), \qquad p(ax) = ap(x)$$

whenever $x, y \in \mathscr{V}$ and $a$ is a non-negative real number. If, further,

$$p(ax) = |a|p(x) \qquad (x \in \mathscr{V}, \quad a \in \mathbb{K}),$$

$p$ is described as a *semi-norm* on $\mathscr{V}$. If $p$ is a semi-norm, then

$$p(x) \geqslant 0, \qquad |p(x) - p(y)| \leqslant p(x - y) \qquad (x, y \in \mathscr{V}).$$

Indeed, $2p(x) = p(x) + p(-x) \geqslant p(x - x) = 0$; while

$$p(x) = p((x - y) + y) \leqslant p(x - y) + p(y),$$

whence $p(x) - p(y) \leqslant p(x - y)$, and similarly

$$p(y) - p(x) \leqslant p(y - x) = p(x - y).$$

By a *norm* on $\mathscr{V}$, we mean a semi-norm $p$ such that $p(x) > 0$ whenever $x \in \mathscr{V}$, $x \neq 0$.

As an example note that if $\mathbb{K}$ is $\mathbb{R}$ or $\mathbb{C}$ and $n$ is a positive integer, the set $\mathbb{K}^n$ consists of all ordered $n$-tuples $(a_1, \ldots, a_n)$ of elements of $\mathbb{K}$, and is an $n$-dimensional vector space when the algebraic structure is defined by

$$a(a_1, \ldots, a_n) + b(b_1, \ldots, b_n) = (aa_1 + bb_1, \ldots, aa_n + bb_n).$$

The equations

$$p_1((a_1, \ldots, a_n)) = |a_1| + \cdots + |a_n|,$$

$$p_\infty((a_1, \ldots, a_n)) = \max\{|a_1|, \ldots, |a_n|\}$$

define norms, $p_1$ and $p_\infty$, on $\mathbb{K}^n$. In particular, the modulus function is a norm on $\mathbb{K}$.

A subset $Y$ of $\mathscr{V}$ is said to be *balanced* if $ay \in Y$ whenever $y \in Y$, $a \in \mathbb{K}$, and $|a| \leqslant 1$. If $p$ is a sublinear functional on $\mathscr{V}$, it is immediately verified that the set $V_p = \{x \in \mathscr{V} : p(x) < 1\}$ is convex, contains 0, and consists entirely of internal points; $V_p$ is balanced if $p$ is a semi-norm.

1.1.5. PROPOSITION. *Suppose that $V$ is a convex subset of a linear space $\mathscr{V}$ over $\mathbb{K}$ (= $\mathbb{R}$ or $\mathbb{C}$), and 0 is an internal point of $V$. Then the equation*

$$p(x) = \inf\{c : c \in \mathbb{R},\ c > 0,\ x \in cV\} \qquad (x \in \mathscr{V})$$

*defines a sublinear functional $p$ on $\mathscr{V}$. If $V$ consists entirely of internal points, then $V = \{x \in \mathscr{V} : p(x) < 1\}$. If $V$ is balanced, $p$ is a semi-norm.*

*Proof.* Given $x$ in $\mathscr{V}$, $c^{-1}x \in V$ (and thus $x \in cV$) for all sufficiently large positive scalars $c$, since 0 is an internal point of $V$; so $p(x)$, as defined in the proposition, is a non-negative real number.

Suppose that $x, y \in \mathscr{V}$ and $a > 0$. Since $x \in cV$ if and only if $ax \in acV$, it follows that $p(ax) = ap(x)$ (and this remains true when $a = 0$, since it is apparent that $p(0) = 0$). Given any positive real number $\varepsilon$, we can choose real numbers $b$ and $c$ so that

$$0 < b < p(x) + \varepsilon, \qquad 0 < c < p(y) + \varepsilon, \qquad x \in bV, \quad y \in cV.$$

Since $V$ is convex, $x + y \in bV + cV = (b + c)V$, and

$$p(x + y) \leqslant b + c < p(x) + p(y) + 2\varepsilon.$$

Since $\varepsilon$ ($> 0$) is arbitrary, $p(x + y) \leqslant p(x) + p(y)$; so $p$ is a sublinear functional on $\mathscr{V}$.

If $V$ is balanced, the condition $ax \in cV$ is equivalent to $|a|x \in cV$, when $x \in \mathscr{V}$, $a \in \mathbb{K}$, and $c > 0$. Thus $p(ax) = p(|a|x) = |a|\,p(x)$, and $p$ is a semi-norm.

Suppose finally that $V$ consists entirely of internal points. Given $y$ in $V$, $y + \varepsilon y \in V$ (and hence $y \in (1 + \varepsilon)^{-1}V$) for all sufficiently small $\varepsilon$ ($> 0$); so $p(y) \leqslant (1 + \varepsilon)^{-1} < 1$. Conversely, if $z \in \mathscr{V}$ and $p(z) < 1$, there is a real number $c$ such that $0 < c < 1$ and $z \in cV$. Since $V$ is convex, $c^{-1} > 1$, and $0, c^{-1}z \in V$, it now follows that $z \in V$. Accordingly, $V = \{x \in \mathscr{V} : p(x) < 1\}$. ∎

The sublinear functional $p$ occurring in Proposition 1.1.5 is called the *support functional* of $V$.

Our next two results are Hahn–Banach theorems of the "extension" type.

1.1.6. THEOREM. *If $p$ is a sublinear functional on a real vector space $\mathscr{V}$, while $\rho_0$ is a linear functional on a linear subspace $\mathscr{V}_0$ of $\mathscr{V}$, and*

$$\rho_0(y) \leqslant p(y) \qquad (y \in \mathscr{V}_0),$$

*there is a linear functional $\rho$ on $\mathscr{V}$ such that*

$$\rho(x) \leqslant p(x) \quad (x \in \mathscr{V}), \qquad \rho(y) = \rho_0(y) \quad (y \in \mathscr{V}_0).$$

*Proof.* The product set $\mathbb{R} \times \mathscr{V}$ becomes a real vector space when addition and scalar multiplication are defined by

$$(r, x) + (s, y) = (r + s, x + y), \qquad a(r, x) = (ar, ax),$$

for $x$, $y$ in $\mathscr{V}$ and $a$, $r$, $s$ in $\mathbb{R}$. From the defining properties of sublinear functionals, it is immediately verified that the set

$$V = \{(r, x) \in \mathbb{R} \times \mathscr{V} : r > p(x)\} \ (\subseteq \mathbb{R} \times \mathscr{V})$$

is non-empty, convex, and consists entirely of internal points. The set

$$W = \{(\rho_0(y), y) : y \in \mathscr{V}_0\}$$

is a linear subspace of $\mathbb{R} \times \mathscr{V}$ (and is therefore convex), and $V \cap W = \emptyset$. From Theorem 1.1.2, there is a linear functional $\sigma$ on $\mathbb{R} \times \mathscr{V}$ and a real number $k$ such that

$$\sigma(v) > k \geqslant \sigma(w) \qquad (v \in V, \quad w \in W).$$

If $w \in W$, then $aw \in W$, and thus $a\sigma(w) = \sigma(aw) \leqslant k$, for every scalar $a$; so

$$\sigma(w) = 0 \qquad (w \in W),$$

and $k \geqslant 0$. From this, and since $(1, 0) \in V$, it follows that $\sigma((1, 0)) > k \geqslant 0$; upon replacing $\sigma$ by a suitable positive multiple of $\sigma$, we may assume that $\sigma((1, 0)) = 1$.

The equation $\rho(x) = -\sigma((0, x))$ defines a linear functional $\rho$ on $\mathscr{V}$, and

$$\sigma((r, x)) = \sigma(r(1, 0) + (0, x)) = r - \rho(x) \qquad (r \in \mathbb{R}, \quad x \in \mathscr{V}).$$

Given any $x$ in $\mathscr{V}$, we have $(r, x) \in V$, and therefore

$$r - \rho(x) = \sigma((r, x)) > k \geqslant 0,$$

whenever $r > p(x)$; so $\rho(x) \leqslant p(x)$. When $y \in \mathscr{V}_0$, $(\rho_0(y), y) \in W$, and thus

$$\rho_0(y) - \rho(y) = \sigma((\rho_0(y), y)) = 0. \quad \blacksquare$$

1.1.7. Theorem. *If $p$ is a semi-norm on a linear space $\mathscr{V}$ over $\mathbb{K}$ $(= \mathbb{R}$ or $\mathbb{C})$, while $\rho_0$ is a linear functional on a linear subspace $\mathscr{V}_0$ of $\mathscr{V}$, and*

$$|\rho_0(y)| \leqslant p(y) \qquad (y \in \mathscr{V}_0),$$

*there is a linear functional $\rho$ on $\mathscr{V}$ such that*

$$|\rho(x)| \leqslant p(x) \quad (x \in \mathscr{V}), \qquad \rho(y) = \rho_0(y) \quad (y \in \mathscr{V}_0).$$

*Proof.* If $\mathbb{K} = \mathbb{R}$, $p$ is a sublinear functional on $\mathscr{V}$, and $\rho_0(y) \leqslant p(y)$ for all $y$ in $\mathscr{V}_0$. By Theorem 1.1.6, there is a linear functional $\rho$ on $\mathscr{V}$ such that $\rho(y) = \rho_0(y)$ when $y \in \mathscr{V}_0$ and $\rho(x) \leqslant p(x)$ for each $x$ in $\mathscr{V}$. Since, also,

$$-\rho(x) = \rho(-x) \leqslant p(-x) = p(x),$$

it follows that $|\rho(x)| \leqslant p(x)$ when $x \in \mathscr{V}$.

Suppose now that $\mathbb{K} = \mathbb{C}$, and let $\mathscr{V}_r$ be the real vector space obtained from $\mathscr{V}$ by restricting the scalar field. Then $p$ is a sublinear functional on $\mathscr{V}_r$, the equation

$$\sigma_0(y) = \operatorname{Re} \rho_0(y)$$

defines a linear functional $\sigma_0$ on the linear subspace $\mathscr{V}_0$ of $\mathscr{V}_r$, and $\sigma_0(y) \leqslant p(y)$ for each $y$ in $\mathscr{V}_0$. By Theorem 1.1.6, there is a linear functional $\sigma$ on $\mathscr{V}_r$ such that

$$\sigma(y) = \sigma_0(y) \quad (y \in \mathscr{V}_0), \qquad \sigma(x) \leqslant p(x) \quad (x \in \mathscr{V}_r).$$

Thus $\sigma$ is a real-linear functional on $\mathscr{V}$. By Lemma 1.1.3, there is a linear functional $\rho$ on $\mathscr{V}$ such that $\sigma(x) = \operatorname{Re} \rho(x)$, and

$$\rho(y) = \sigma(y) - i\sigma(iy) = \sigma_0(y) - i\sigma_0(iy) = \rho_0(y) \qquad (y \in \mathscr{V}_0).$$

When $x \in \mathscr{V}$, we can choose a scalar $a$ so that $|a| = 1$, $|\rho(x)| = a\rho(x)$; and

$$\begin{aligned} |\rho(x)| &= \rho(ax) = \operatorname{Re} \rho(ax) \\ &= \sigma(ax) \leqslant p(ax) = |a|p(x) = p(x). \quad \blacksquare \end{aligned}$$

If $\mathscr{U}$, $\mathscr{V}$, $\mathscr{W}$ are vector spaces over the same scalar field $\mathbb{K}$, and $S: \mathscr{V} \to \mathscr{W}$, $T: \mathscr{U} \to \mathscr{V}$ are linear operators, the composition of $S$ and $T$ is a linear operator, which we write as a product $ST$, from $\mathscr{U}$ into $\mathscr{W}$. This applies, in particular, when $\mathscr{U} = \mathscr{V} = \mathscr{W}$; and with the multiplication so obtained, the linear space of all linear operators $T: \mathscr{V} \to \mathscr{V}$ becomes an associative linear algebra. It has a unit, the identity mapping $I$ on $\mathscr{V}$. We now identify the idempotents in this algebra (those elements $E$ such that $E^2 = E$).

If $E: \mathscr{V} \to \mathscr{V}$ is a linear operator and $E^2 = E$, the sets

$$Y = \{x \in \mathscr{V} : Ex = x\}, \qquad Z = \{x \in \mathscr{V} : Ex = 0\} \tag{1}$$

are linear subspaces of $\mathscr{V}$. Given $x$ in $\mathscr{V}$, let $y = Ex$, $z = x - Ex$; note that $x = y + z$, $y \in Y$ (because $Ey = E^2x = Ex = y$) and $z \in Z$ (because $Ez = Ex - E^2x = 0$). If $x$ has another expression as $x = y_1 + z_1$, with $y_1$ in $Y$ and $z_1$ in $Z$, then $Ey_1 = y_1$ and $Ez_1 = 0$ from (1), and thus

$$y_1 = E(y_1 + z_1) = Ex = y, \qquad z_1 = x - y_1 = x - Ex = z.$$

Accordingly, the subspaces $Y$ and $Z$ of $\mathscr{V}$ have the following property: each element $x$ of $\mathscr{V}$ can be expressed uniquely in the form $x = y + z$, with $y$ in $Y$ and $z$ in $Z$. Two linear subspaces of $\mathscr{V}$ with this property are described as *complementary subspaces* of $\mathscr{V}$.

We assert that two subspaces $Y$ and $Z$ of $\mathscr{V}$ are complementary if and only if $Y + Z = \mathscr{V}$ and $Y \cap Z = \{0\}$. Indeed, the first of these conditions asserts that each $x$ in $\mathscr{V}$ has *at least one* expression as $x = y + z$, with $y$ in $Y$ and $z$ in $Z$. If it has another such expression, $x = y_1 + z_1$ $(= y + z)$, then

$$y_1 - y = z - z_1 \in Y \cap Z;$$

so the most general expression for $x$, in the stated form, is $x = (y + u) + (z - u)$, where $u \in Y \cap Z$. Accordingly, $x$ is *uniquely* expressible in this form if and only if $Y \cap Z = \{0\}$.

Now suppose that $Y$ and $Z$ are complementary subspaces of $\mathscr{V}$; we shall show that they can be obtained as in (1) from an idempotent linear operator acting on $\mathscr{V}$. We may define a mapping $E: \mathscr{V} \to \mathscr{V}$ as follows: when $x \in \mathscr{V}$, take the unique expression for $x$ in the form $y + z$ (with $y$ in $Y$ and $z$ in $Z$), and let $Ex$ be $y$. It is then apparent that $Y$ and $Z$ are related to $E$ as in (1); moreover, $Ex \in Y$, and therefore $E(Ex) = Ex$, for each $x$ in $\mathscr{V}$. If, for $j = 1, 2$, we have $x_j = y_j + z_j$ (where $y_j \in Y$ and $z_j \in Z$) and $a_j \in \mathbb{K}$, then

$$a_1 x_1 + a_2 x_2 = (a_1 y_1 + a_2 y_2) + (a_1 z_1 + a_2 z_2)$$

and $a_1 y_1 + a_2 y_2 \in Y$, $a_1 z_1 + a_2 z_2 \in Z$. Thus

$$E(a_1 x_1 + a_2 x_2) = a_1 y_1 + a_2 y_2 = a_1 E x_1 + a_2 E x_2;$$

so $E: \mathscr{V} \to \mathscr{V}$ is a linear operator and $E^2 = E$.

The following theorem embodies the main results of the preceding discussion.

1.1.8. THEOREM. *Two linear subspaces $Y$ and $Z$ of a linear space $\mathscr{V}$ are complementary if and only if $Y + Z = \mathscr{V}$ and $Y \cap Z = \{0\}$. When these conditions are satisfied, the equation*

$$E(y + z) = y \qquad (y \in Y, \quad z \in Z)$$

*defines a linear operator $E: \mathscr{V} \to \mathscr{V}$; and $E^2 = E$,*

$$Y = \{x \in \mathscr{V} : Ex = x\}, \qquad Z = \{x \in \mathscr{V} : Ex = 0\}.$$

*Conversely, every linear operator $E: \mathscr{V} \to \mathscr{V}$ satisfying $E^2 = E$ arises in the above manner from a pair of complementary subspaces of $\mathscr{V}$.*

The operator $E$ occurring in Theorem 1.1.8 is described as the *projection from $\mathscr{V}$ onto $Y$, parallel to $Z$.*

## 1.2. Linear topological spaces

Suppose that a set $\mathscr{V}$ is both a linear space with scalar field $\mathbb{K}$ ($= \mathbb{R}$ or $\mathbb{C}$) and also a Hausdorff topological space. If the algebraic and topological structures are so related that the mappings

$$(x, y) \to x + y \ : \ \mathscr{V} \times \mathscr{V} \to \mathscr{V},$$

$$(a, x) \to ax \ : \ \mathbb{K} \times \mathscr{V} \to \mathscr{V}$$

are continuous (when $\mathscr{V} \times \mathscr{V}$ and $\mathbb{K} \times \mathscr{V}$ have their product topologies), then $\mathscr{V}$ is said to be a *linear topological space.*

The simplest examples of linear topological spaces over $\mathbb{K}$ are the sets $\mathbb{K}^n$ ($n = 1, 2, \ldots$), with their usual vector space structure and with the product topology.

It is apparent that a complex linear topological space can be viewed also as a real one, simply by restricting the scalar field. A linear subspace of a linear topological space is itself a linear topological space, with the relative topology.

The given topology, in a linear topological space $\mathscr{V}$, is sometimes described as the *initial topology* in order to distinguish it from other topologies that can naturally (and usefully) be introduced, such as the weak topologies described in Section 1.3. It is usual to develop the early parts of the theory without the assumption that the initial topology is Hausdorff. However, for most purposes, an easy quotient procedure permits an immediate reduction to the Hausdorff case; and the initial topology is Hausdorff in all the cases we shall encounter in later chapters. Accordingly, we have included this condition as part of our definition of linear topological spaces.

If $\mathscr{V}_1$ is a linear subspace of a linear topological space $\mathscr{V}$, the closure $\mathscr{V}_0$ of $\mathscr{V}_1$ is also a linear subspace. Indeed, suppose that $x_0, y_0 \in \mathscr{V}_0$ and $a, b$ are scalars. Then $(x_0, y_0)$ lies in the closure $\mathscr{V}_0 \times \mathscr{V}_0$ of $\mathscr{V}_1 \times \mathscr{V}_1$, and is therefore the limit of a net $\{(x_j, y_j)\}$ in $\mathscr{V}_1 \times \mathscr{V}_1$. Since $\mathscr{V}_1$ is a subspace of $\mathscr{V}$, and the mapping $(x, y) \to ax + by : \mathscr{V} \times \mathscr{V} \to \mathscr{V}$ is continuous, we have

$$ax_j + by_j \in \mathscr{V}_1, \qquad ax_j + by_j \to ax_0 + by_0,$$

and $ax_0 + by_0$ lies in the closure $\mathscr{V}_0$ of $\mathscr{V}_1$. Similar arguments show that if a subset of $\mathscr{V}$ is balanced (or convex), then the same is true of its closure. Note also that an open set $G$ in $\mathscr{V}$ consists entirely of internal points. For this, suppose that $x \in G$, $y \in \mathscr{V}$. Since the mapping $a \to x + ay : \mathbb{R} \to \mathscr{V}$ is continuous and takes 0 into the open set $G$, it carries some real interval $(-c, c)$ into $G$; and in particular, $x + ay \in G$ whenever $0 \leqslant a < c$.

Suppose that $\mathscr{V}$ is a linear topological space with scalar field $\mathbb{K}$, $x_0 \in \mathscr{V}$, and $V \subseteq \mathscr{V}$. Since the continuous mapping $x \to x + x_0 : \mathscr{V} \to \mathscr{V}$ has a continuous inverse mapping $x \to x - x_0$, it follows that $V$ is a neighborhood of 0 if and only if $x_0 + V$ is a neighborhood of $x_0$. Accordingly, the topology of $\mathscr{V}$ is determined once a base of neighborhoods of 0 has been specified. If $V$ is a neighborhood of 0, then so is $aV$ for each non-zero scalar $a$, since the one-to-one mapping $x \to ax$, from $\mathscr{V}$ onto $\mathscr{V}$, is bicontinuous; in particular, $-V$ is a neighborhood of 0. From continuity at $(0, 0)$ of the mapping $(x, y) \to x + y$, there is a neighborhood $V_0$ of 0 such that $V_0 + V_0 \subseteq V$. From continuity at $(0, 0)$ of the mapping $(a, x) \to ax$, there exist a neighborhood $V_1$ of 0, and a positive real number $\varepsilon$, such that $ax \in V$ whenever $x \in V_1$ and $|a| \leqslant \varepsilon$. From this, $\bigcup\{aV_1 : 0 < |a| \leqslant \varepsilon\}$ is a balanced open subset of $V$; so every neighborhood of 0 contains a balanced neighborhood of 0.

Let $\mathscr{N}$ denote the set of all neighborhoods of 0 in $\mathscr{V}$, and with each $V$ in $\mathscr{N}$ associate the subset

$$\mathscr{E}(V) = \{(x, y) : y - x \in V\}$$

of $\mathscr{V} \times \mathscr{V}$. If $V_0 + V_0 \subseteq V$, the conditions $(x, z) \in \mathscr{E}(V_0)$ and $(z, y) \in \mathscr{E}(V_0)$ entail $(x, y) \in \mathscr{E}(V)$. Since, also, $\mathscr{E}(V_1) \cap \mathscr{E}(V_2) = \mathscr{E}(V_1 \cap V_2)$, and $(x, y) \in \mathscr{E}(V)$ if and only if $(y, x) \in \mathscr{E}(-V)$, it is apparent that the sets $\mathscr{E}(V)$ $(V \in \mathscr{N})$ form a base of a uniform structure (uniformity) on $\mathscr{V}$ [K: p. 176 et seq.]. It is often convenient to use a *base* $\mathscr{N}_0$ of neighborhoods of 0, rather than the family $\mathscr{N}$ of *all* neighborhoods of 0; it is evident that the sets $\mathscr{E}(V)$ $(V \in \mathscr{N}_0)$ form a base of the same uniform structure. In this way, $\mathscr{V}$ becomes a Hausdorff uniform space, and the topology derived from the uniform structure coincides with the initial topology, since in both cases the sets

$$\{y \in \mathscr{V} : (x, y) \in \mathscr{E}(V)\} = \{y \in \mathscr{V} : y - x \in V\} = x + V$$

form a base of neighborhoods of $x$, when $V$ runs through $\mathscr{N}_0$.

When we refer to any "uniform" concept (such as uniform continuity, or completeness) in relation to a linear topological space $\mathscr{V}$, it is understood that the uniform structure just described is the one in question. From time to time we shall make use of the fact [K: p. 195] that a uniformly continuous mapping from a subset $X$ of a uniform space $\mathscr{V}$ into a complete Hausdorff uniform space $\mathscr{W}$ extends uniquely to a uniformly continuous mapping from the closure of $X$ into $\mathscr{W}$. Our use of this result will often be indicated by a reference to "extension by continuity."

1.2.1. Proposition. *Suppose that $\mathscr{V}$ and $\mathscr{W}$ are linear topological spaces with the same scalar field $\mathbb{K}$ $(= \mathbb{R}$ or $\mathbb{C})$, and $T: \mathscr{V} \to \mathscr{W}$ is a linear operator.*

(i) *If $x_0 \in \mathscr{V}$ and $T$ is continuous at $x_0$, then $T$ is uniformly continuous on $\mathscr{V}$.*

(ii) *If $C$ is a balanced convex subset of $\mathscr{V}$ and the restriction $T|C$ is continuous at* 0, *then $T|C$ is uniformly continuous on $C$.*

*Proof.* (i) Since $T$ is continuous at $x_0$, given any neighborhood $W$ of 0 in $\mathscr{W}$, there is a neighborhood $V$ of 0 in $\mathscr{V}$ such that $Tx \in Tx_0 + W$ whenever $x \in x_0 + V$. Let $\mathscr{E}(V)$ $(\subseteq \mathscr{V} \times \mathscr{V})$ be the set occurring in the above discussion of the uniform structure on $\mathscr{V}$, and let $\mathscr{E}(W)$ $(\subseteq \mathscr{W} \times \mathscr{W})$ be defined similarly. When $(x, y) \in \mathscr{E}(V)$, we have $y - x \in V$, $x_0 + y - x \in x_0 + V$, and therefore $Tx_0 + Ty - Tx \in Tx_0 + W$; so $Ty - Tx \in W$, and $(Tx, Ty) \in \mathscr{E}(W)$.

The above argument shows that, given any neighborhood $W$ of 0 in $\mathscr{W}$, there is a neighborhood $V$ of 0 in $\mathscr{V}$ such that $(Tx, Ty) \in \mathscr{E}(W)$ whenever $(x, y) \in \mathscr{E}(V)$; so $T$ is uniformly continuous on $\mathscr{V}$.

(ii) Since the restriction $T|C$ is continuous at 0, given any neighborhood $W$ of 0 in $\mathscr{W}$, there is a balanced neighborhood $V$ of 0 in $\mathscr{V}$ such that $Tx \in \frac{1}{2}W$ whenever $x \in V \cap C$.

Suppose that $x, y \in C$ and $y - x \in V$. Since $C$ is convex, and both $C$ and $V$ are balanced, we have $\frac{1}{2}y - \frac{1}{2}x \in V \cap C$; hence $\frac{1}{2}Ty - \frac{1}{2}Tx \in \frac{1}{2}W$, and $Ty - Tx \in W$. Accordingly, $(Tx, Ty) \in \mathscr{E}(W)$ whenever $x, y \in C$ and $(x, y) \in \mathscr{E}(V)$; so $T|C$ is uniformly continuous on $C$. ■

1.2.2. Remark. For complex linear topological spaces, Proposition 1.2.1 remains true under the weakened assumption that $T: \mathscr{V} \to \mathscr{W}$ is a *real*-linear operator, since $T$ can then be viewed as a linear operator between the real linear topological spaces obtained from $\mathscr{V}$ and $\mathscr{W}$ by restricting the scalar field. In particular, therefore, Proposition 1.2.1 applies to *conjugate-linear* operators (those mappings $T: \mathscr{V} \to \mathscr{W}$ satisfying $T(ax + by) = \bar{a}Tx + \bar{b}Ty$, where $\bar{a}$ denotes the complex conjugate of $a$). ■

1.2.3. Corollary. *If $\mathscr{V}$ and $\mathscr{W}$ are linear topological spaces, $\mathscr{W}$ is complete, $\mathscr{V}_0$ is an everywhere-dense subspace of $\mathscr{V}$, and $T_0: \mathscr{V}_0 \to \mathscr{W}$ is a continuous linear operator, then $T_0$ extends uniquely to a continuous linear operator $T: \mathscr{V} \to \mathscr{W}$.*

*Proof.* By Proposition 1.2.1, $T_0$ is uniformly continuous on $\mathscr{V}_0$, and so extends uniquely to a uniformly continuous mapping $T: \mathscr{V} \to \mathscr{W}$. Given any scalars $a$, $b$, the equation

$$g(x, y) = T(ax + by) - aTx - bTy$$

defines a continuous mapping $g: \mathscr{V} \times \mathscr{V} \to \mathscr{W}$; and $g$ vanishes on $\mathscr{V} \times \mathscr{V}$ since it vanishes on the everywhere-dense subset $\mathscr{V}_0 \times \mathscr{V}_0$. Thus $T$ is linear. ■

The lemma that follows, concerning continuity of linear functionals and semi-norms, is written in a form that applies to both real and complex linear topological spaces; the notation Re, occurring in part (i), is redundant in the real case.

1.2.4. Lemma. *Suppose that $\mathscr{V}$ is a linear topological space, $\rho$ is a linear functional on $\mathscr{V}$, and $p$ is a semi-norm on $\mathscr{V}$.*

(i) *If there is a non-empty open set $G$ in $\mathscr{V}$ and a real number $c$ such that* $\operatorname{Re} \rho(x) < c$ *whenever $x \in G$, then $\rho$ is (uniformly) continuous on $\mathscr{V}$.*

(ii) *If $p$ is bounded on some neighborhood of* 0 *in $\mathscr{V}$, then $p$ is (uniformly) continuous on $\mathscr{V}$.*

*Proof.* (i) If $G$ and $c$ have the stated properties, we can choose $x_0$ in $G$ and a balanced neighborhood $V_0$ of 0 in $\mathscr{V}$ such that $x_0 + V_0 \subseteq G$. Given $x$ in $V_0$, let $a$ be a scalar such that $|a| = 1$ and $|\rho(x)| = \rho(ax)$. Then $ax \in V_0$, $x_0 + ax \in x_0 + V_0 \subseteq G$, and therefore

$$c > \operatorname{Re} \rho(x_0 + ax) = \operatorname{Re} \rho(x_0) + |\rho(x)|.$$

Hence $|\rho(x)| < b$ for all $x$ in $V_0$, where $b = c - \operatorname{Re} \rho(x_0)$ $(> 0)$.

Given any positive $\varepsilon$, $|\rho(x)| < \varepsilon$ for all $x$ in the neighborhood $\varepsilon b^{-1} V_0$ of 0. Thus $\rho$ is continuous at 0, and is therefore uniformly continuous on $\mathscr{V}$ by Proposition 1.2.1.

(ii) Suppose that there is a neighborhood $V$ of 0 in $\mathscr{V}$ and a positive real number $b$ such that $p(x) < b$ whenever $x \in V$. Given any positive $\varepsilon$, the set $\varepsilon b^{-1} V$ is a neighborhood of 0 in $\mathscr{V}$, and

$$|p(y) - p(x)| \leqslant p(y - x) < \varepsilon$$

whenever $x, y \in \mathscr{V}$ and $y - x \in \varepsilon b^{-1} V$. Thus $p$ is uniformly continuous on $\mathscr{V}$. ■

1.2.5. Corollary. *A linear functional $\rho$ on a linear topological space $\mathscr{V}$ is continuous if and only if its null space $\rho^{-1}(0)$ is closed in $\mathscr{V}$.*

*Proof.* We may assume that $\rho \neq 0$; it is evident that $\rho^{-1}(0)$ is closed if $\rho$ is continuous.

Conversely, suppose that $\rho^{-1}(0)$ is closed. We can choose $x_0$ in $\mathscr{V}$ so that $\rho(x_0) = 1$. Since $x_0 \notin \rho^{-1}(0)$, there is a balanced neighborhood $V_0$ of 0 in $\mathscr{V}$ such that $x_0 + V_0$ does not meet $\rho^{-1}(0)$.

If $x \in V_0$ and $|\rho(x)| \geqslant 1$, we can choose a scalar $a$ so that $|a| \leqslant 1$ and $\rho(ax) = -1$. Then $ax \in V_0$, $x_0 + ax \in x_0 + V_0$, and $\rho(x_0 + ax) = 0$, contrary to our assumption that $x_0 + V_0$ does not meet $\rho^{-1}(0)$. From this, it follows that $|\rho(x)| < 1$ whenever $x \in V_0$, and $\rho$ is continuous by Lemma 1.2.4(i). ■

If $\rho$ is a non-zero continuous linear functional on a (real or complex) linear topological space $\mathscr{V}$, the set $A = \{x \in \mathscr{V} : \operatorname{Re} \rho(x) < 1\}$ is a non-empty convex open subset of $\mathscr{V}$, but is not the whole of $\mathscr{V}$. There are examples of linear topological spaces that have no subset $A$ with the properties just listed, and these spaces therefore have no non-zero continuous linear functionals. We now introduce a class of linear topological spaces, each of which (as we shall see in Corollary 1.2.11) has an abundance of continuous linear functionals.

A *locally convex space* is a linear topological space in which the topology has a base consisting of convex sets. By a *locally convex topology*, on a (real or complex) vector space $\mathscr{V}$, we mean a topology with which $\mathscr{V}$ becomes a locally convex space.

The vector spaces $\mathbb{R}^n$ and $\mathbb{C}^n$ provide the simplest examples of locally convex spaces, since the product topology has a base (for example, the open balls) consisting of convex sets.

In the theorem that follows we show that locally convex topologies, on a linear space $\mathscr{V}$ over $\mathbb{K}$ ($= \mathbb{R}$ or $\mathbb{C}$), are closely associated with certain families of semi-norms. A semi-norm $p$ on $\mathscr{V}$ can be regarded as an analogue of the "modulus" function on $\mathbb{K}$; the "triangle inequality," $p(x + y) \leqslant p(x) + p(y)$, and its consequences (for example, the inequality $|p(x) - p(y)| \leqslant p(x - y)$) are used in analysis in locally convex spaces, in much the same way that the corresponding properties of the modulus are used in elementary real or complex analysis.

1.2.6. THEOREM. *Suppose that $\mathscr{V}$ is a (real or complex) vector space, and $\Gamma$ is a family of semi-norms on $\mathscr{V}$ that separates the points of $\mathscr{V}$ in the following sense: if $x \in \mathscr{V}$ and $x \neq 0$, there is an element $p$ of $\Gamma$ for which $p(x) \neq 0$. Then there is a locally convex topology on $\mathscr{V}$ in which, for each $x_0$ in $\mathscr{V}$, the family of all sets*

$$V(x_0 : p_1, \ldots, p_m; \varepsilon) = \{x \in \mathscr{V} : p_j(x - x_0) < \varepsilon \ (j = 1, \ldots, m)\}$$

*(where $\varepsilon > 0$ and $p_1, \ldots, p_m \in \Gamma$) is a base of neighborhoods of $x_0$. With this topology, each of the semi-norms in $\Gamma$ is continuous. Moreover, every locally convex topology on $\mathscr{V}$ arises, in this way, from a suitable family of semi-norms.*

*Proof.* If

$$x_0 \in V(y_0 : p_1, \ldots, p_m; \delta) \cap V(z_0 : q_1, \ldots, q_n; \eta),$$

we can choose $\varepsilon$ ($> 0$) so that

$$p_j(x_0 - y_0) + \varepsilon < \delta, \qquad q_k(x_0 - z_0) + \varepsilon < \eta \qquad (j = i, \ldots, m, \quad k = 1, \ldots, n).$$

It then follows easily, from the triangle inequality, that

$$V(x_0 : p_1, \ldots, p_m, q_1, \ldots, q_n; \varepsilon) \subseteq V(y_0 : p_1, \ldots, p_m; \delta) \cap V(z_0 : q_1, \ldots, q_n; \eta).$$

In particular, if $V(y_0 : p_1, \ldots, p_m; \delta)$ contains $x_0$, it also contains a set of the form $V(x_0 : p_1, \ldots, p_m; \varepsilon)$. From these facts, and since $x_0 \in V(x_0 : p_1, \ldots, p_m; \varepsilon)$, it follows that the family

$$\{V(x_0 : p_1, \ldots, p_m; \varepsilon) : x_0 \in \mathscr{V}, p_1, \ldots, p_m \in \Gamma, \varepsilon > 0\}$$

is a base for a topology on $\mathscr{V}$ [K: p. 47], in which

$$\{V(x_0 : p_1, \ldots, p_m; \varepsilon) : p_1, \ldots, p_m \in \Gamma, \varepsilon > 0\}$$

is a base of neighborhoods of $x_0$.

If $x_0, y_0 \in \mathscr{V}$ and $x_0 \neq y_0$, we can choose $p$ in $\Gamma$ so that $p(x_0 - y_0) > 0$. When $0 < \varepsilon < \frac{1}{2}p(x_0 - y_0)$, the sets $V(x_0 : p; \varepsilon)$ and $V(y_0 : p; \varepsilon)$ are disjoint neighborhoods of $x_0$ and $y_0$, respectively; so the topology is Hausdorff. Since

$$p(x + y - x_0 - y_0) \leqslant p(x - x_0) + p(y - y_0),$$

$$\begin{aligned} p(ax - a_0x_0) &= p(a_0(x - x_0) + (a - a_0)x_0 + (a - a_0)(x - x_0)) \\ &\leqslant |a_0|\, p(x - x_0) + |a - a_0|\, p(x_0) + |a - a_0|\, p(x - x_0), \end{aligned}$$

for every semi-norm $p$, it follows that

$$x + y \in V(x_0 + y_0 : p_1, \ldots, p_m; \varepsilon),$$

$$ax \in V(a_0x_0 : p_1, \ldots, p_m; \varepsilon),$$

whenever

$$x \in V(x_0 \colon p_1, \ldots, p_m; \delta), \qquad y \in V(y_0 \colon p_1, \ldots, p_m; \delta)$$

and $|a - a_0| < \delta$, provided that the positive real number $\delta$ is sufficiently small to ensure that

$$2\delta < \varepsilon, \qquad |a_0|\delta + \delta p_j(x_0) + \delta^2 < \varepsilon \qquad (j = 1, \ldots, m).$$

This establishes the continuity of the mappings $(x, y) \to x + y, (a, x) \to ax$, and so shows that $\mathscr{V}$ is a linear topological space. It is locally convex, since each of the basic neighborhoods $V(x_0 \colon p_1, \ldots, p_m; \varepsilon)$ is convex. When $p \in \Gamma$, $p$ is bounded on the neighborhood $V(0 \colon p; 1)$, so that $p$ is (uniformly) continuous on $\mathscr{V}$ from Lemma 1.2.4(ii).

Conversely, suppose that $\tau$ is a locally convex topology on $\mathscr{V}$. We shall show that $\tau$ can be obtained, by the process described in the theorem, from a suitable family of semi-norms. To this end, let $V$ be a $\tau$-neighborhood of 0 in $\mathscr{V}$. Then $V$ contains a convex $\tau$-neighborhood $V_0$ of 0. Moreover, $V_0$ contains a balanced $\tau$-neighborhood $V_1$ of 0, and thus contains the convex hull $V_2$ of $V_1$; so $V_2 \subseteq V$. Since $V_2$ consists of all finite convex combinations $a_1x_1 + \cdots + a_nx_n$ of elements of $V_1$ (with each $a_j > 0$), it too is balanced. It is $\tau$-open, and so consists entirely of internal points, since it can be expressed as a union of sets of the form $a_1x_1 + \cdots + a_{n-1}x_{n-1} + a_nV_1$. By Proposition 1.1.5 there is a semi-norm $p$ on $\mathscr{V}$ such that

$$V_2 = \{x \in \mathscr{V} \colon p(x) < 1\},$$

and since $V_2$ is a $\tau$-neighborhood of 0, it results from Lemma 1.2.4(ii) that $p$ is $\tau$-continuous.

Let $\Gamma_0$ denote the set of all $\tau$-continuous semi-norms on $\mathscr{V}$. The preceding paragraph shows that every $\tau$-neighborhood $V$ of 0 contains a set of the form $\{x \in \mathscr{V} \colon p(x) < 1\}$, with $p$ in $\Gamma_0$. From this, if $x \in \mathscr{V}$ and $p(x) = 0$ for every $p$ in $\Gamma_0$, then $x$ lies in each $\tau$-neighborhood of 0, and hence $x = 0$; so $\Gamma_0$ separates the points of $\mathscr{V}$, in the sense stated in the theorem. Moreover, when $x_0 \in \mathscr{V}$, the $\tau$-neighborhood $x_0 + V$ contains the set

$$\begin{aligned} \{x_0 + x \colon x \in \mathscr{V}, p(x) < 1\} &= \{x \in \mathscr{V} \colon p(x - x_0) < 1\} \\ &= V(x_0 \colon p; 1); \end{aligned}$$

and $V(x_0 \colon p_1, \ldots, p_m; \varepsilon)$ is a $\tau$-neighborhood of $x_0$, whenever $p_1, \ldots, p_m \in \Gamma_0$ and $\varepsilon > 0$, since $p_1, \ldots, p_m$ are $\tau$-continuous. It follows that the sets $V(x_0 \colon p_1, \ldots, p_m; \varepsilon)$ form a base of $\tau$-neighborhoods of $x_0$; therefore $\tau$ coincides with the topology obtained from $\Gamma_0$ by the process described in the theorem. ■

1.2.7. Corollary. *In a locally convex space there is a base of neighborhoods of* 0 *consisting of balanced convex sets.*

*Proof.* The sets $V(0: p_1, \ldots, p_m; \varepsilon)$ are balanced and convex. ■

Observe that the topology on the locally convex space $\mathbb{K}^n$ (where $\mathbb{K} = \mathbb{R}$ or $\mathbb{C}$) can be obtained, as in Theorem 1.2.6, from a family $\Gamma$ consisting of just one norm. For example, either of the norms $p_1$ and $p_\infty$, defined in the discussion preceding Proposition 1.1.5, will suffice for this purpose.

We now give criteria for the continuity of semi-norms and linear operators on locally convex spaces, in terms of families of semi-norms defining the locally convex topologies.

1.2.8. PROPOSITION. *Suppose that $\mathscr{V}_1$ and $\mathscr{V}_2$ are locally convex spaces with the same scalar field $\mathbb{K}$ ($= \mathbb{R}$ or $\mathbb{C}$); and for $j = 1, 2$, let $\Gamma_j$ be a separating family of semi-norms on $\mathscr{V}_j$ that gives rise to the topology of $\mathscr{V}_j$.*

(i) *A semi-norm $p$ on $\mathscr{V}_1$ is continuous if and only if there is a positive real number $C$ and a finite set $p_1, \ldots, p_m$ of elements of $\Gamma_1$ such that*

$$p(x) \leqslant C \max\{p_1(x), \ldots, p_m(x)\} \qquad (x \in \mathscr{V}_1).$$

(ii) *A linear operator $T: \mathscr{V}_1 \to \mathscr{V}_2$ is continuous if and only if, given any $q$ in $\Gamma_2$, there is a positive real number $C$ and a finite set $p_1, \ldots, p_m$ of elements of $\Gamma_1$ such that*

$$q(Tx) \leqslant C \max\{p_1(x), \ldots, p_m(x)\} \qquad (x \in \mathscr{V}_1).$$

(iii) *A linear functional $\rho$ on $\mathscr{V}_1$ is continuous if and only if there is a positive real number $C$ and a finite set $p_1, \ldots, p_m$ of elements of $\Gamma_1$ such that*

$$|\rho(x)| \leqslant C \max\{p_1(x), \ldots, p_m(x)\} \qquad (x \in \mathscr{V}_1).$$

*Proof.* (i) If $p$ is continuous, the set $\{x \in \mathscr{V}_1 : p(x) < 1\}$ is a neighborhood of 0 in $\mathscr{V}_1$; so it contains one of the basic neighborhoods $V(0: p_1, \ldots, p_m; \varepsilon)$, where $\varepsilon > 0$, and $p_1, \ldots, p_m \in \Gamma_1$. If $x \in \mathscr{V}_1$ and $p(x) > \varepsilon^{-1} \max\{p_1(x), \ldots, p_m(x)\}$, we may assume (upon replacing $x$ by $cx$, for some positive $c$) that $p(x) = 1$ and $p_j(x) < \varepsilon$ ($j = 1, \ldots, m$); that is, $p(x) = 1$ and $x \in V(0: p_1, \ldots, p_m; \varepsilon)$, contradicting our assumption concerning this basic neighborhood. Accordingly, the stated condition

$$p(x) \leqslant C \max\{p_1(x), \ldots, p_m(x)\} \qquad (x \in \mathscr{V}_1),$$

is satisfied (with $C = \varepsilon^{-1}$). Conversely, if this condition is satisfied, then $p$ is bounded on $V(0: p_1, \ldots, p_m; 1)$; by Lemma 1.2.4(ii), $p$ is continuous on $\mathscr{V}_1$.

(ii) When $q$ is a semi-norm on $\mathscr{V}_2$, the composite mapping $q \circ T$ is a semi-norm on $\mathscr{V}_1$. In view of this, and taking into account part (i) of the proposition, we have to show that $T$ is continuous if and only if $q \circ T$ is continuous whenever $q \in \Gamma_2$. By Theorem 1.2.6, each $q$ in $\Gamma_2$ is continuous on $\mathscr{V}_2$; so continuity of $T$ entails continuity of $q \circ T$.

Conversely, suppose that $q \circ T$ is continuous, for each $q$ in $\Gamma_2$. Every neighborhood $V$ of 0 in $\mathscr{V}_2$ contains a basic neighborhood $V(0:q_1,\ldots,q_m;\varepsilon)$, where $\varepsilon > 0$ and $q_1,\ldots,q_m \in \Gamma_2$. Since $q_j \circ T$ is continuous ($j = 1,\ldots,m$), the set

$$W = \{x \in \mathscr{V}_1 : q_j(Tx) < \varepsilon \ (j = 1,\ldots,m)\}$$

is a neighborhood of 0 in $\mathscr{V}_1$, and it is apparent that

$$T(W) \subseteq V(0:q_1,\ldots,q_m;\varepsilon) \subseteq V.$$

Thus $T$ is continuous (at 0, and therefore throughout $\mathscr{V}_1$).

(iii) The scalar field $\mathbb{K}$ is a locally convex space, its (usual) topology being obtained from a single norm, the modulus function; and $\rho: \mathscr{V}_1 \to \mathbb{K}$ is a linear operator. Thus (iii) is a special case of (ii). ■

Our next two results are Hahn–Banach (separation) theorems. They are formulated so as to apply to both real and complex linear topological spaces, the notation Re being redundant in the real case.

1.2.9. Theorem. *If $Y$ and $Z$ are disjoint non-empty convex subsets of a linear topological space $\mathscr{V}$, and $Y$ is open, there is a continuous linear functional $\rho$ on $\mathscr{V}$ and a real number $k$ such that*

$$\operatorname{Re}\rho(y) > k \geqslant \operatorname{Re}\rho(z) \qquad (y \in Y, \quad z \in Z).$$

*If, further, $Z$ is open, then $k > \operatorname{Re}\rho(z)$ for each $z$ in $Z$.*

*Proof.* In view of the fact that an open set consists entirely of internal points, the assumptions of Theorem 1.1.2 are fulfilled in the real case, while those of Theorem 1.1.4 obtain in the complex case. From those theorems, there is a linear functional $\rho$ on $\mathscr{V}$ that satisfies the stated inequalities; and by applying Lemma 1.2.4(i), with $-\rho$ and $Y$ in place of $\rho$ and $G$, it follows that $\rho$ is continuous. ■

1.2.10. Theorem. *If $Y$ and $Z$ are disjoint non-empty closed convex subsets of a locally convex space $\mathscr{V}$, at least one of which is compact, there are real numbers $a, b$ and a continuous linear functional $\rho$ on $\mathscr{V}$ such that*

$$\operatorname{Re}\rho(y) \geqslant a > b \geqslant \operatorname{Re}\rho(z) \qquad (y \in Y, \quad z \in Z).$$

*Proof.* We may suppose that $Y$ is compact. For each $y$ in $Y$, there is a balanced convex neighborhood $V_y$ of 0 such that $(y + V_y) \cap Z = \varnothing$, since $Z$ is closed and $y \notin Z$. The open covering $\{y + \frac{1}{3}V_y : y \in Y\}$ of $Y$ has a finite subcovering $\{y(j) + \frac{1}{3}V_{y(j)} : j = 1,\ldots,m\}$, and the set $V = \bigcap_{j=1}^{m} \frac{1}{3}V_{y(j)}$ is a balanced convex neighborhood of 0.

The convex sets $Y + V$ and $Z + V$ are open since, for example, $Y + V = \bigcup\{y + V : y \in Y\}$; we assert also that they are disjoint. For this, suppose the contrary, and choose $y$ in $Y$, $z$ in $Z$, and $v_1$, $v_2$ in $V$, so that $y + v_1 = z + v_2$. For some $j$ $(1 \leqslant j \leqslant m)$, $y \in y(j) + \frac{1}{3}V_{y(j)}$; moreover $v_1, -v_2 \in V \subseteq \frac{1}{3}V_{y(j)}$. Thus

$$z = y + v_1 - v_2 = y(j) + (y - y(j)) + v_1 - v_2$$
$$\in y(j) + \tfrac{1}{3}V_{y(j)} + \tfrac{1}{3}V_{y(j)} + \tfrac{1}{3}V_{y(j)} = y(j) + V_{y(j)},$$

contradicting our assumption that $(y(j) + V_{y(j)}) \cap Z = \varnothing$. This proves our assertion that $Y + V$, $Z + V$ are disjoint.

From Theorem 1.2.9, we can choose a continuous linear functional $\rho$ on $\mathcal{V}$ and a real number $k$ such that

$$\operatorname{Re} \rho(y) > k > \operatorname{Re} \rho(z) \qquad (y \in Y + V, \quad z \in Z + V);$$

in particular, these inequalities are satisfied when $y \in Y$ and $z \in Z$. Since $Y$ is compact, the continuous function $g\colon Y \to \mathbb{R}$, defined by $g(y) = \operatorname{Re} \rho(y)$, attains its lower bound at a point $y_0$ of $Y$. Hence

$$\operatorname{Re} \rho(y) \geqslant \operatorname{Re} \rho(y_0) > k > \operatorname{Re} \rho(z) \qquad (y \in Y, \quad z \in Z),$$

and it suffices to take $a = \operatorname{Re} \rho(y_0)$ and $b = k$. ■

1.2.11. COROLLARY. *If $x$ is a non-zero vector in a locally convex space $\mathcal{V}$, there is a continuous linear functional $\rho$ on $\mathcal{V}$ such that $\rho(x) \neq 0$.*

*Proof.* This follows from Theorem 1.2.10, with $Y = \{x\}$ and $Z = \{0\}$. ■

1.2.12. COROLLARY. *If $Z$ is a closed convex subset of a locally convex space $\mathcal{V}$, and $y \in \mathcal{V} \backslash Z$, there is a continuous linear functional $\rho$ on $\mathcal{V}$ and a real number $b$ such that* $\operatorname{Re} \rho(y) > b$, $\operatorname{Re} \rho(z) \leqslant b$ $(z \in Z)$.

*Proof.* This follows from Theorem 1.2.10, with $Y = \{y\}$. ■

1.2.13. COROLLARY. *If $Z$ is a closed subspace of a locally convex space $\mathcal{V}$, and $y \in \mathcal{V} \backslash Z$, there is a continuous linear functional $\rho$ on $\mathcal{V}$ such that $\rho(y) \neq 0$, $\rho(z) = 0$ $(z \in Z)$.*

*Proof.* From Corollary 1.2.12 there is a continuous linear functional $\rho$ on $\mathcal{V}$ and a real number $b$ such that

$$\operatorname{Re} \rho(y) > b, \qquad \operatorname{Re} \rho(z) \leqslant b \qquad (z \in Z).$$

The latter inequality implies that the range of values assumed by the linear functional $\rho | Z$ on $Z$ is not the whole of the scalar field. Hence $\rho | Z = 0$, $b \geqslant 0$, $\operatorname{Re} \rho(y) > b \geqslant 0$, and thus $\rho(y) \neq 0$. ■

We now prove a Hahn–Banach extension theorem.

1.2.14. THEOREM. *If $\rho_0$ is a continuous linear functional on a subspace $\mathscr{V}_0$ of a locally convex space $\mathscr{V}$, there is a continuous linear functional $\rho$ on $\mathscr{V}$ such that $\rho|\mathscr{V}_0 = \rho_0$.*

*Proof.* Let $\Gamma$ be a family of semi-norms that gives rise, as in Theorem 1.2.6, to the topology on $\mathscr{V}$. By restricting each member of $\Gamma$ to $\mathscr{V}_0$, we obtain a family of semi-norms that defines the relative topology on $\mathscr{V}_0$. Since $\rho_0$ is continuous, it follows from Proposition 1.2.8(iii) that there is a positive real number $C$ and a finite set $p_1, \ldots, p_m$ of elements of $\Gamma$ such that

$$|\rho_0(y)| \leqslant C \max\{p_1(y), \ldots, p_m(y)\} \qquad (y \in \mathscr{V}_0).$$

By Theorem 1.1.7, $\rho_0$ extends to a linear functional $\rho$ on $\mathscr{V}$, such that

$$|\rho(x)| \leqslant C \max\{p_1(x), \ldots, p_m(x)\} \qquad (x \in \mathscr{V}),$$

since the equation

$$p(x) = C \max\{p_1(x), \ldots, p_m(x)\}$$

defines a semi-norm $p$ on $\mathscr{V}$. A further application of Proposition 1.2.8(iii) shows that $\rho$ is continuous. ■

If $\mathscr{V}$ is a locally convex space and $\mathfrak{X} \subseteq \mathscr{V}$, we denote by $[\mathfrak{X}]$ the closure of the set of all finite linear combinations of elements of $\mathfrak{X}$. It is apparent that $[\mathfrak{X}]$ is the smallest closed subspace of $\mathscr{V}$ that contains $\mathfrak{X}$; we describe it as the *closed subspace generated by* $\mathfrak{X}$. When $\mathfrak{X}$ is a finite set, $\{x_1, \ldots, x_n\}$, we write $[x_1, \ldots, x_n]$ in place of $[\mathfrak{X}]$. In fact, $[x_1, \ldots, x_n]$ is the set of all linear combinations of $x_1, \ldots, x_n$, since it results from Theorem 1.2.17 that this set is closed in $\mathscr{V}$.

We now consider some properties of finite-dimensional subspaces of locally convex spaces. In fact, the main results obtained remain valid in all linear topological spaces, without the restriction of local convexity. However, we shall not need that degree of generality, and have preferred to derive these results as simple consequences of the Hahn–Banach theorems in the locally convex case.

We have already noted that, when $\mathbb{K}$ is $\mathbb{R}$ or $\mathbb{C}$ and $n$ is a positive integer, the finite-dimensional vector space $\mathbb{K}^n$, with its usual (product) topology, is a locally convex space. Each linear functional $\rho$ on $\mathbb{K}^n$ is continuous, since it is given by a formula

$$\rho((a_1, \ldots, a_n)) = a_1 c_1 + \cdots + a_n c_n,$$

where $c_1, \ldots, c_n$ are fixed elements of $\mathbb{K}$. From elementary linear algebra, if $\mathscr{M}$ is a proper subspace of $\mathbb{K}^n$, there is a non-zero linear functional on $\mathbb{K}^n$ that vanishes on $\mathscr{M}$.

1.2.15. LEMMA. *Suppose that* $\mathscr{V}$ *is a locally convex space with scalar field* $\mathbb{K}$ $(= \mathbb{R}$ or $\mathbb{C})$ *and* $\{x_1, \ldots, x_n\}$ *is a basis of a finite-dimensional subspace* $\mathscr{V}_0$ *of* $\mathscr{V}$. *Then there are continuous linear functionals* $\rho_1, \ldots, \rho_n$ *on* $\mathscr{V}$ *such that* $\rho_j(x_j) = 1$ *and* $\rho_j(x_k) = 0$ *when* $j \neq k$. *The equation*

$$Tx = (\rho_1(x), \ldots, \rho_n(x)) \qquad (x \in \mathscr{V})$$

*defines a continuous linear operator* $T: \mathscr{V} \to \mathbb{K}^n$, *and the restriction* $T|\mathscr{V}_0$ *is a one-to-one bicontinuous linear operator from* $\mathscr{V}_0$ *onto* $\mathbb{K}^n$.

*Proof.* The set $\mathscr{V}^\sharp$ of all continuous linear functionals on $\mathscr{V}$ is a linear subspace of the algebraic dual space of $\mathscr{V}$. The equation

$$S\rho = (\rho(x_1), \ldots, \rho(x_n)) \qquad (\rho \in \mathscr{V}^\sharp)$$

defines a linear operator $S: \mathscr{V}^\sharp \to \mathbb{K}^n$.

If the subspace $S(\mathscr{V}^\sharp)$ is not the whole of $\mathbb{K}^n$, there is a non-zero linear functional on $\mathbb{K}^n$ that vanishes on $S(\mathscr{V}^\sharp)$; in other words, we can choose $c_1, \ldots, c_n$ in $\mathbb{K}$, not all 0, so that

$$c_1\rho(x_1) + \cdots + c_n\rho(x_n) = 0 \qquad (\rho \in \mathscr{V}^\sharp).$$

Thus every continuous linear functional $\rho$ on $\mathscr{V}$ vanishes at the non-zero vector $c_1x_1 + \cdots + c_nx_n$, contradicting the conclusion of Corollary 1.2.11. It follows that $S(\mathscr{V}^\sharp) = \mathbb{K}^n$. Accordingly, we can find $\rho_1, \ldots, \rho_n$ in $\mathscr{V}^\sharp$ such that $S\rho_j = (0, \ldots, 0, 1, 0, \ldots, 0)$ (with the 1 in the $j$th place); that is, $\rho_1, \ldots, \rho_n$ are continuous linear functionals on $\mathscr{V}$, and $\rho_j(x_k)$ is 1 or 0 according as $j = k$ or $j \neq k$.

It is apparent that $T$, as defined in the lemma, is a continuous linear operator from $\mathscr{V}$ into $\mathbb{K}^n$. Moreover, since

$$\rho_j(c_1x_1 + \cdots + c_nx_n) = c_j \qquad (j = 1, \ldots, n),$$

if follows that

$$T(c_1x_1 + \cdots + c_nx_n) = (c_1, \ldots, c_n).$$

Thus $T$ carries $\mathscr{V}_0$ onto $\mathbb{K}^n$, and the restriction $T|\mathscr{V}_0$ has a continuous inverse mapping

$$(c_1, \ldots, c_n) \to c_1x_1 + \cdots + c_nx_n: \quad \mathbb{K}^n \to \mathscr{V}_0. \quad \blacksquare$$

1.2.16. PROPOSITION. *If* $\mathscr{V}$ *is a finite-dimensional linear space, with scalar field* $\mathbb{K}$ $(= \mathbb{R}$ *or* $\mathbb{C})$, *there is a unique locally convex topology on* $\mathscr{V}$.

*Proof.* Let $\{x_1, \ldots, x_n\}$ be a basis of $\mathscr{V}$, and define a one-to-one linear mapping $T$, from $\mathscr{V}$ onto $\mathbb{K}^n$, by

$$T(c_1x_1 + \cdots + c_nx_n) = (c_1, \ldots, c_n).$$

There is a unique topology $\tau$ on $\mathscr{V}$, which makes $T$ a homeomorphism; $\tau$ is locally convex, since $\mathbb{K}^n$ is a locally convex space.

If $\mathscr{V}$ has another locally convex topology $\tau_0$, we can apply Lemma 1.2.15 (with $\mathscr{V}_0 = \mathscr{V}$). Since the mapping $T$ just defined is the same as the one occurring in the lemma, $\tau_0$ makes $T$ a homeomorphism, and thus coincides with $\tau$. ■

1.2.17. THEOREM. *If $\mathscr{V}_0$ is a finite-dimensional subspace of a locally convex space $\mathscr{V}$, then $\mathscr{V}_0$ is closed in $\mathscr{V}$; moreover, there is a closed subspace $\mathscr{V}_1$ of $\mathscr{V}$ such that $\mathscr{V}_0$ and $\mathscr{V}_1$ are complementary subspaces of $\mathscr{V}$ and the projection from $\mathscr{V}$ onto $\mathscr{V}_0$ parallel to $\mathscr{V}_1$ is continuous.*

*Proof.* Let $\{x_1, \ldots, x_n\}$ be a basis of $\mathscr{V}_0$. By Lemma 1.2.15, there are continuous linear functionals $\rho_1, \ldots, \rho_n$ on $\mathscr{V}$ such that $\rho_j(x_k)$ is 1 or 0 according as $j = k$ or $j \neq k$. Each element $y$ of $\mathscr{V}_0$ is a linear combination $c_1x_1 + \cdots + c_nx_n$, and $\rho_j(y) = c_j$; so

$$y = \sum_{j=1}^{n} \rho_j(y)x_j \qquad (y \in \mathscr{V}_0).$$

The equation

$$Ex = \sum_{j=1}^{n} \rho_j(x)x_j \qquad (x \in \mathscr{V})$$

defines a continuous linear operator $E: \mathscr{V} \to \mathscr{V}$. From the preceding paragraph, $Ey = y$ when $y \in \mathscr{V}_0$, and it is apparent that $E(\mathscr{V}) \subseteq \mathscr{V}_0$. It follows easily that

$$\mathscr{V}_0 = \{x \in \mathscr{V} : Ex = x\},$$

that $E^2x = Ex$ for each $x$ in $\mathscr{V}$, and hence that $E^2 = E$. By Theorem 1.1.8, $\mathscr{V}_0$ has a complementary subspace,

$$\mathscr{V}_1 = \{x \in \mathscr{V} : Ex = 0\},$$

and $E$ is the projection from $\mathscr{V}$ onto $\mathscr{V}_0$, parallel to $\mathscr{V}_1$. Moreover, since $E$ is continuous, the above descriptions of $\mathscr{V}_0$ and $\mathscr{V}_1$, in terms of $E$, show that both these subspaces are closed. ■

1.2.18. THEOREM. *A locally convex space $\mathscr{V}$ is locally compact if and only if it is finite dimensional.*

*Proof.* If $\mathscr{V}$ has finite dimension $n$, it is homeomorphic to $\mathbb{K}^n$, where $\mathbb{K}$ is the scalar field, by Lemma 1.2.15 (with $\mathscr{V}_0 = \mathscr{V}$); so $\mathscr{V}$ is locally compact.

Conversely, suppose that $\mathscr{V}$ is locally compact, and let $V$ be a neighborhood of 0 in $\mathscr{V}$ whose closure $V^-$ is compact. If $x_0 \in \mathscr{V}$ and $x_0 \neq 0$, the one-dimensional subspace $[x_0]$ is closed in $\mathscr{V}$ (Theorem 1.2.17), but is not compact since it is homeomorphic to $\mathbb{K}$ (Lemma 1.2.15). Accordingly, $[x_0] \nsubseteq V^-$; and since, also, $[x_0]$ meets $V$ (at 0, for example), there is an element $ax_0$ of $[x_0]$ in the boundary $V^-\backslash V$ of $V$.

For each $y$ in the compact set $V^-\backslash V$, since $y \neq 0$, there is a continuous linear functional $\rho_y$ on $\mathscr{V}$ such that $\rho_y(y) \neq 0$ (Corollary 1.2.11). We may suppose that $|\rho_y(y)| > 1$, and define a neighborhood $G_y$ of $y$ by $G_y = \{x \in \mathscr{V}: |\rho_y(x)| > 1\}$. The open covering $\{G_y: y \in V^-\backslash V\}$ of $V^-\backslash V$ has a finite subcovering; and if the linear functionals $\rho_y$, corresponding to the sets $G_y$ in this subcovering, are enumerated as $\rho_1, \ldots, \rho_n$, then for each $x$ in $V^-\backslash V$ there is at least one integer $j$ such that $1 \leqslant j \leqslant n$ and $|\rho_j(x)| > 1$.

The equation $Tx = (\rho_1(x), \ldots, \rho_n(x))$ defines a linear mapping $T$ from $\mathscr{V}$ into $\mathbb{K}^n$. In order to prove that $\mathscr{V}$ is finite dimensional, it suffices to show that $T$ is one-to-one, and this follows easily from the two preceding paragraphs. Indeed, if $x_0 \in \mathscr{V}\backslash\{0\}$, then $ax_0 \in V^-\backslash V$ for some scalar $a$, and $|\rho_j(ax_0)| > 1$ for some integer $j$ with $1 \leqslant j \leqslant n$; so that $\rho_j(x_0) \neq 0$, and hence $Tx_0 \neq 0$. ■

We conclude this section with a discussion of nets and (unordered) infinite sums in a linear topological space $\mathscr{V}$. Suppose that $v \in \mathscr{V}$, and $(v_j, j \in J, \geqslant)$ (or, more briefly, $\{v_j\}$) is a net in $\mathscr{V}$, the index set $J$ being directed by the binary relation $\geqslant$. Then $\{v_j\}$ converges to $v$ if and only if, given any neighborhood $V$ of 0, there is an index $j_0$ such that $v_j \in v + V$ (equivalently, $v_j - v \in V$) whenever $j \geqslant j_0$. From the definition of the uniform structure on $\mathscr{V}$, as set out in the discussion preceding Proposition 1.2.1, $\{v_j\}$ is a Cauchy net if and only if the following condition is satisfied: given any neighborhood $V$ of 0, there is an index $j_0$ such that $v_j - v_k \in V$ whenever $j, k \geqslant j_0$. If $\mathscr{V}$ is a locally convex space and $\Gamma$ is a separating family of semi-norms that determines its topology, each neighborhood $V$ of 0 contains a basic neighborhood

$$V(0: p_1, \ldots, p_m; \varepsilon) = \bigcap_{j=1}^{m} \{x \in \mathscr{V}: p_j(x) < \varepsilon\}.$$

In this case, $\{v_j\}$ converges to $v$ if and only if, given any $p$ in $\Gamma$ and any positive $\varepsilon$, there is an index $j_0$ such that $p(v_j - v) < \varepsilon$ whenever $j \geqslant j_0$; and $\{v_j\}$ is a Cauchy net if and only if, given any such $p$ and $\varepsilon$, there is an index $j_0$ such that $p(v_j - v_k) < \varepsilon$ when $j, k \geqslant j_0$.

Now suppose that $\{x_a: a \in \mathbb{A}\}$ is a family of vectors in a linear topological space $\mathscr{V}$, and denote by $\mathscr{F}$ the family of all finite subsets of the index set $\mathbb{A}$. Then $\mathscr{F}$ is directed by the inclusion relation $\supseteq$, and for each $\mathbb{F}$ in $\mathscr{F}$ we can form the finite sum

$$s(\mathbb{F}) = \sum_{a \in \mathbb{F}} x_a.$$

In this way we obtain a net $(s(\mathbb{F}), \mathbb{F} \in \mathscr{F}, \supseteq)$ (briefly, $\{s(\mathbb{F})\}$) of elements of $\mathscr{V}$. If it converges to a vector $x$ in $\mathscr{V}$, we say that the family $\{x_a: a \in \mathbb{A}\}$ is *summable*, or that *the sum* $\sum_{a \in \mathbb{A}} x_a$ *exists*, and we write $\sum_{a \in \mathbb{A}} x_a = x$.

The sums considered here are *unordered*, in that we do not assume any order structure on the index set $\mathbb{A}$. We shall show that, at least in certain elementary respects, such sums can be handled in much the same way as convergent series; moreover, subject to suitable completeness properties in $\mathscr{V}$, they have some of the properties associated, in elementary analysis, with absolute convergence.

Suppose that $\mathscr{V}$ and $\mathscr{W}$ are linear topological spaces and $\{x_a : a \in \mathbb{A}\}$, $\{y_a : a \in \mathbb{A}\}$ are summable families of elements of $\mathscr{V}$, with sums $x$ and $y$, respectively. When $\mathbb{F} \in \mathscr{F}$, define $s(\mathbb{F})$ as above, and let $t(\mathbb{F}) = \sum_{a \in \mathbb{F}} y_a$. If $c$ is a scalar and $T: \mathscr{V} \to \mathscr{W}$ is a continuous linear operator, we have

$$\sum_{a \in \mathbb{F}} cx_a = cs(\mathbb{F}),$$

$$\sum_{a \in \mathbb{F}} (x_a + y_a) = s(\mathbb{F}) + t(\mathbb{F}),$$

$$\sum_{a \in \mathbb{F}} Tx_a = Ts(\mathbb{F}).$$

From continuity, of $T$ and of the algebraic operations in $\mathscr{V}$, it follows that the nets $\{cs(\mathbb{F})\}$, $\{s(\mathbb{F}) + t(\mathbb{F})\}$, $\{Ts(\mathbb{F})\}$ converge to $cx$, $x + y$, $Tx$, respectively. Thus the families $\{cx_a\}$, $\{x_a + y_a\}$, $\{Tx_a\}$ are summable, and

$$\sum cx_a = c\left(\sum x_a\right),$$

$$\sum (x_a + y_a) = \sum x_a + \sum y_a,$$

$$\sum Tx_a = T\left(\sum x_a\right).$$

If $\{x_a : a \in \mathbb{A}\}$ is a family of vectors in a linear topological space $\mathscr{V}$, the corresponding net $\{s(\mathbb{F})\}$ of finite sums is a Cauchy net if and only if the following condition is satisfied: given any neighborhood $V$ of 0, there is a finite subset $\mathbb{F}_0$ of $\mathbb{A}$ such that $s(\mathbb{F}_1) - s(\mathbb{F}_2) \in V$ whenever $\mathbb{F}_1$, $\mathbb{F}_2$ are finite subsets of $\mathbb{A}$ and $\mathbb{F}_0 \subseteq \mathbb{F}_1 \cap \mathbb{F}_2$. Now

$$s(\mathbb{F}_1) - s(\mathbb{F}_2) = s(\mathbb{F}_1 \backslash \mathbb{F}_0) - s(\mathbb{F}_2 \backslash \mathbb{F}_0),$$

and given any neighborhood $V_1$ of 0, there is a neighborhood $V_2$ of 0 such that $V_2 - V_2 \subseteq V_1$. From this, it follows easily that $\{s(\mathbb{F})\}$ is a Cauchy net if and only if it satisfies the following "Cauchy criterion": given any neighborhood $V$ of 0, there is a finite subset $\mathbb{F}_0$ of $\mathbb{A}$ such that $s(\mathbb{F}) \in V$ for every finite subset $\mathbb{F}$ of $\mathbb{A} \backslash \mathbb{F}_0$. If $\mathscr{V}$ is a locally convex space and $\Gamma$ is a separating family of semi-norms that determines its topology, then $\{s(\mathbb{F})\}$ is a Cauchy net if and only if, given any $p$ in $\Gamma$ and any positive $\varepsilon$, there is a finite subset $\mathbb{F}_0$ of $\mathbb{A}$ such that $p(s(\mathbb{F})) < \varepsilon$ for every finite subset $\mathbb{F}$ of $\mathbb{A} \backslash \mathbb{F}_0$.

1.2.19. PROPOSITION. *Suppose that $\mathscr{V}$ is a linear topological space, $X \subseteq \mathscr{V}$, and $X$ is complete (in its relative uniform structure). Let $\{x_d : d \in \mathbb{D}\}$ be a summable family of elements of $\mathscr{V}$, and suppose that, for each finite subset $\mathbb{F}$ of $\mathbb{D}$, the sum $s(\mathbb{F}) = \sum_{d \in \mathbb{F}} x_d$ lies in $X$.*

(i) *If $\mathbb{A} \subseteq \mathbb{D}$, the family $\{x_d : d \in \mathbb{A}\}$ is summable to an element of $X$.*
(ii) *If $\mathbb{D}$ is a disjoint union, $\mathbb{D} = \bigcup\{\mathbb{A}_b : b \in \mathbb{B}\}$, and*

$$y_b = \sum_{d \in \mathbb{A}_b} x_d \qquad (b \in \mathbb{B}),$$

*then the family $\{y_b : b \in \mathbb{B}\}$ is summable and $\sum_{b \in \mathbb{B}} y_b = \sum_{d \in \mathbb{D}} x_d$.*

*Proof.* (i) Since $\{x_d : d \in \mathbb{D}\}$ is summable, the corresponding net of finite sums converges, and is therefore a Cauchy net. Given any neighborhood $V$ of 0, let $\mathbb{F}_0$ be a finite subset of $\mathbb{D}$ such that $s(\mathbb{F}) \in V$ whenever $\mathbb{F}$ is a finite subset of $\mathbb{D} \backslash \mathbb{F}_0$. If $\mathbb{F}_1 = \mathbb{F}_0 \cap \mathbb{A}$, we have $\mathbb{A} \backslash \mathbb{F}_1 \subseteq \mathbb{D} \backslash \mathbb{F}_0$, and thus $s(\mathbb{F}) \in V$ for every finite subset $\mathbb{F}$ of $\mathbb{A} \backslash \mathbb{F}_1$. Accordingly, the net of finite sums $s(\mathbb{F})$, with $\mathbb{F} \subseteq \mathbb{A}$, is a Cauchy net in the complete set $X$, and therefore converges (and $\{x_d : d \in \mathbb{A}\}$ is summable) to an element of $X$.

(ii) Let $x = \sum_{d \in \mathbb{D}} x_d$. Given any neighborhood $V$ of 0, let $V_1$ be a neighborhood of 0 such that $V_1 + V_1 \subseteq V$, and let $\mathbb{F}_0$ be a finite subset of $\mathbb{D}$ such that

$$\sum_{d \in \mathbb{F}} x_d - x \in V_1$$

whenever $\mathbb{F}$ is a finite subset of $\mathbb{D}$, and $\mathbb{F}_0 \subseteq \mathbb{F}$. Note that

$$\mathbb{F}_0 = \bigcup_{b \in \mathbb{B}} \mathbb{A}_b \cap \mathbb{F}_0 = \bigcup_{b \in \mathbb{F}_1} \mathbb{A}_b \cap \mathbb{F}_0,$$

where $\mathbb{F}_1$ is the finite set $\{b \in \mathbb{B} : \mathbb{A}_b \cap \mathbb{F}_0 \neq \varnothing\}$.

It now suffices to prove that

$$\sum_{b \in \mathbb{F}} y_b - x \in V$$

whenever $\mathbb{F}$ is a finite set satisfying $\mathbb{F}_1 \subseteq \mathbb{F} \subseteq \mathbb{B}$. To this end, suppose that $\mathbb{F}$ has $m$ members, and choose a neighborhood $V_2$ of 0 for which $V_1 \supseteq V_2 + V_2 + \cdots + V_2$ ($m$ terms). For each $b$ in $\mathbb{F}$, there is a finite set $\mathbb{F}_b$ such that

$$\mathbb{A}_b \cap \mathbb{F}_0 \subseteq \mathbb{F}_b \subseteq \mathbb{A}_b, \qquad y_b - \sum_{d \in \mathbb{F}_b} x_d \in V_2.$$

Since

$$\bigcup_{b \in \mathbb{F}} \mathbb{F}_b \supseteq \bigcup_{b \in \mathbb{F}_1} \mathbb{A}_b \cap \mathbb{F}_0 = \mathbb{F}_0,$$

we have

$$\sum_{b\in\mathbb{F}} y_b - x = \sum_{b\in\mathbb{F}}\left(y_b - \sum_{d\in\mathbb{F}_b} x_d\right) + \left(\sum_{b\in\mathbb{F}}\sum_{d\in\mathbb{F}_b} x_d - x\right)$$

$$\in V_2 + V_2 + \cdots + V_2 + V_1 \subseteq V_1 + V_1 \subseteq V. \quad \blacksquare$$

A family $\{z_a : a\in\mathbb{A}\}$ of complex numbers is summable if and only if $\{|z_a| : a\in\mathbb{A}\}$ is summable. To prove this, we use the Cauchy criterion set out in the paragraph preceding Proposition 1.2.19. For every finite subset $\mathbb{F}$ of $\mathbb{A}$, if $z_a = x_a + iy_a$,

$$\left|\sum_{a\in\mathbb{F}} z_a\right| \leqslant \sum_{a\in\mathbb{F}} |z_a| \leqslant \sum_{a\in\mathbb{F}} |x_a| + \sum_{a\in\mathbb{F}} |y_a|$$

$$= \sum_{a\in\mathbb{F}_1} x_a + \sum_{a\in\mathbb{F}_2} -x_a + \sum_{a\in\mathbb{F}_3} y_a + \sum_{a\in\mathbb{F}_4} -y_a$$

$$\leqslant 4\max\left\{\sum_{a\in\mathbb{F}_j}\right\} \leqslant 4\max\left|\sum_{a\in\mathbb{F}_j} z_a\right|,$$

where $\mathbb{F}_1, \ldots, \mathbb{F}_4$ are suitable subsets of $\mathbb{F}$. It follows easily that, if either of the families $\{z_a\}$, $\{|z_a|\}$ satisfies the Cauchy criterion (which is necessary and sufficient for summability, since $\mathbb{C}$ is complete), then so does the other.

If $\{x_a : a\in\mathbb{A}\}$ is a family of non-negative real numbers, the corresponding net $\{s(\mathbb{F})\}$ of finite sums is monotonic increasing; so the family is summable if and only if the net is bounded above. When this is so, $\sum_{a\in\mathbb{A}} x_a$ is the least upper bound of $\{s(\mathbb{F})\}$, and the set of non-zero terms $x_a$ is at most countably infinite (otherwise, for some positive integer $n$, infinitely many of the terms $x_a$ exceed $1/n$).

## 1.3. Weak topologies

Suppose that $\mathscr{V}$ is a linear space with scalar field $\mathbb{K}$ ($= \mathbb{R}$ or $\mathbb{C}$), and $\mathscr{F}$ is a family of linear functionals on $\mathscr{V}$, which separates the points of $\mathscr{V}$ in the following sense: if $x$ is a non-zero vector in $\mathscr{V}$ then, for some $\rho$ in $\mathscr{F}$, $\rho(x) \neq 0$. When $\rho\in\mathscr{F}$, the equation $p_\rho(x) = |\rho(x)|$ defines a semi-norm $p_\rho$ on $\mathscr{V}$. The separating family $\{p_\rho : \rho\in\mathscr{F}\}$ of semi-norms gives rise, as in Theorem 1.2.6, to a locally convex topology on $\mathscr{V}$, the *weak topology induced on $\mathscr{V}$ by $\mathscr{F}$*. In this topology, which we denote by $\sigma(\mathscr{V}, \mathscr{F})$, each point $x_0$ of $\mathscr{V}$ has a base of neighborhoods that consists of all sets of the form

$$V(x_0 : \rho_1, \ldots, \rho_m; \varepsilon) = \{x\in\mathscr{V} : |\rho_j(x) - \rho_j(x_0)| < \varepsilon \ (j = 1, \ldots, m)\},$$

where $\varepsilon > 0$ and $\rho_1, \ldots, \rho_m \in \mathscr{F}$.

Since $|\rho(x)-\rho(x_0)|<\varepsilon$ when $x\in V(x_0:\rho;\varepsilon)$, each of the linear functionals $\rho$ in $\mathscr{F}$ is continuous relative to the topology $\sigma(\mathscr{V},\mathscr{F})$. However, if $\tau$ is a topology on $\mathscr{V}$, and each linear functional in $\mathscr{F}$ is $\tau$-continuous, then all the sets $V(x_0:\rho_1,\ldots,\rho_m;\varepsilon)$ are $\tau$-open; from this, $\sigma(\mathscr{V},\mathscr{F})$ is coarser (weaker) than $\tau$. Accordingly, $\sigma(\mathscr{V},\mathscr{F})$ is the coarsest (weakest) topology on $\mathscr{V}$ relative to which each element of $\mathscr{F}$ is a continuous mapping from $\mathscr{V}$ into $\mathbb{K}$.

It is apparent that $\sigma(\mathscr{V},\mathscr{F})$ is coarser than $\sigma(\mathscr{V},\mathscr{G})$, when $\mathscr{F}\subseteq\mathscr{G}$.

Suppose that $v\in\mathscr{V}$ and $\{v_j\}$ is a net of elements of $\mathscr{V}$. From the discussion following Theorem 1.2.18, $\{v_j\}$ converges to $v$, in the topology $\sigma(\mathscr{V},\mathscr{F})$, if and only if, given any $\rho$ in $\mathscr{F}$ and any positive $\varepsilon$, there is an index $j_0$ such that $p_\rho(v_j-v)<\varepsilon$ (that is, $|\rho(v_j)-\rho(v)|<\varepsilon$) whenever $j\geqslant j_0$. Thus $\{v_j\}$ is $\sigma(\mathscr{V},\mathscr{F})$-convergent to $v$ if and only if, for each $\rho$ in $\mathscr{F}$, the net $\{\rho(v_j)\}$ of scalars converges to $\rho(v)$. Similarly, $\{v_j\}$ is a Cauchy net, in the uniform structure associated with $\sigma(\mathscr{V},\mathscr{F})$, if and only if $\{\rho(v_j)\}$ is a Cauchy (and hence, convergent) net of scalars for each $\rho$ in $\mathscr{F}$.

1.3.1. THEOREM. *Suppose that $\mathscr{V}$ is a linear space with scalar field $\mathbb{K}$ ($=\mathbb{R}$ or $\mathbb{C}$), $\mathscr{F}$ is a separating family of linear functionals on $\mathscr{V}$, and $\mathscr{L}$ is the set of all finite linear combinations of elements of $\mathscr{F}$. Then $\sigma(\mathscr{V},\mathscr{L})$ coincides with $\sigma(\mathscr{V},\mathscr{F})$, and $\mathscr{L}$ is the set of all $\sigma(\mathscr{V},\mathscr{F})$-continuous linear functionals on $\mathscr{V}$.*

*Proof.* Since $\mathscr{F}\subseteq\mathscr{L}$, $\sigma(\mathscr{V},\mathscr{F})$ is coarser than $\sigma(\mathscr{V},\mathscr{L})$. Every linear functional $\rho$ in $\mathscr{L}$ is a linear combination $a_1\rho_1+\cdots+a_m\rho_m$ of elements $\rho_1,\ldots,\rho_m$ of $\mathscr{F}$; moreover, each $\rho_j$ is $\sigma(\mathscr{V},\mathscr{F})$-continuous, so the same is true of $\rho$. Since $\sigma(\mathscr{V},\mathscr{L})$ is the coarsest topology that makes every element of $\mathscr{L}$ continuous, it now follows that $\sigma(\mathscr{V},\mathscr{L})$ is coarser than $\sigma(\mathscr{V},\mathscr{F})$; so these two topologies coincide.

It remains to prove that each $\sigma(\mathscr{V},\mathscr{F})$-continuous linear functional $\rho_0$ lies in $\mathscr{L}$. Given such $\rho_0$, we can apply Proposition 1.2.8(iii), bearing in mind the form of the semi-norms $p_\rho$ (with $\rho$ in $\mathscr{F}$), which give rise to the topology $\sigma(\mathscr{V},\mathscr{F})$. It follows that there are elements $\rho_1,\ldots,\rho_m$ of $\mathscr{F}$ and a positive real number $C$ such that

$$|\rho_0(x)|\leqslant C\max\{|\rho_1(x)|,\ldots,|\rho_m(x)|\}$$

for each $x$ in $\mathscr{V}$. In particular, $\rho_0(x)=0$ if $\rho_j(x)=0$ for $j=1,\ldots,m$. From Proposition 1.1.1, $\rho_0$ is a linear combination of $\rho_1,\ldots,\rho_m$, and thus $\rho_0\in\mathscr{L}$. ■

1.3.2. PROPOSITION. *Suppose that $\mathscr{V}$ is a linear space with scalar field $\mathbb{K}$ ($=\mathbb{R}$ or $\mathbb{C}$), $\mathscr{F}$ is a separating family of linear functionals on $\mathscr{V}$, and $\varphi$ is a mapping from a topological space $\mathscr{S}$ into $\mathscr{V}$. Then $\varphi$ is continuous, relative to the topology $\sigma(\mathscr{V},\mathscr{F})$ on $\mathscr{V}$, if and only if each of the composite mappings $\rho\circ\varphi:\mathscr{S}\to\mathbb{K}$ ($\rho\in\mathscr{F}$) is continuous.*

*Proof.* Since each $\rho$ in $\mathscr{F}$ is $\sigma(\mathscr{V}, \mathscr{F})$-continuous, it is evident that continuity of $\varphi$ entails continuity of the composite mapping $\rho \circ \varphi$.

Conversely, suppose that $\rho \circ \varphi$ is continuous for each $\rho$ in $\mathscr{F}$. In order to establish the continuity of $\varphi$, it suffices to show that $\varphi^{-1}(V)$ is an open subset of $\mathscr{S}$ whenever $V$ is one of the basic $\sigma(\mathscr{V}, \mathscr{F})$-open sets $V(x_0 : \rho_1, \ldots, \rho_m; \varepsilon)$. Of course, $\rho_1, \ldots, \rho_m \in \mathscr{F}$, so $\rho_j \circ \varphi : \mathscr{S} \to \mathbb{K}$ is continuous for $j = 1, \ldots, m$. From this, and since

$$\varphi^{-1}(V) = \{s \in \mathscr{S} : \varphi(s) \in V\}$$
$$= \{s \in \mathscr{S} : |\rho_j(\varphi(s)) - \rho_j(x_0)| < \varepsilon \; (j = 1, \ldots, m)\},$$

$\varphi^{-1}(V)$ is open, as required. ■

Suppose that $\mathscr{V}$ is a locally convex space. The set of all continuous linear functionals on $\mathscr{V}$ is a subspace of the algebraic dual space of $\mathscr{V}$; it is denoted by $\mathscr{V}^\sharp$, and is described as the *continuous dual space* of $\mathscr{V}$. By Corollary 1.2.11, it separates the points of $\mathscr{V}$. The topology $\sigma(\mathscr{V}, \mathscr{V}^\sharp)$ is called the *weak topology on* $\mathscr{V}$. It is the coarsest topology on $\mathscr{V}$ that makes all the linear functionals in $\mathscr{V}^\sharp$ continuous; in particular, therefore, it is coarser than the initial topology on $\mathscr{V}$. When using topological concepts, such as continuity or compactness, in relation to the weak topology, we refer to weak continuity, weak compactness, etc. By Theorem 1.3.1, with $\mathscr{F} = \mathscr{L} = \mathscr{V}^\sharp$, the weakly continuous linear functionals on $\mathscr{V}$ are precisely the members of $\mathscr{V}^\sharp$; in other words, a linear functional on $\mathscr{V}$ is weakly continuous if and only if it is continuous relative to the initial topology on $\mathscr{V}$.

1.3.3. PROPOSITION. *If $\mathscr{V}$ and $\mathscr{W}$ are locally convex spaces and $T : \mathscr{W} \to \mathscr{V}$ is a continuous linear operator, then $T$ is continuous also relative to the weak topologies on $\mathscr{V}$ and $\mathscr{W}$.*

*Proof.* When $\rho \in \mathscr{V}^\sharp$, the linear functional $\rho \circ T$ on $\mathscr{W}$ is continuous relative to the initial topology on $\mathscr{W}$, and is therefore continuous also relative to the weak topology $\sigma(\mathscr{W}, \mathscr{W}^\sharp)$. If we now apply Proposition 1.3.2, taking for $\mathscr{S}$ the space $\mathscr{W}$ with its weak topology, it follows that $T$ is continuous relative to the topologies $\sigma(\mathscr{W}, \mathscr{W}^\sharp)$ and $\sigma(\mathscr{V}, \mathscr{V}^\sharp)$. ■

1.3.4. THEOREM. *A convex subset $Z$ of a locally convex space $\mathscr{V}$ has the same closures, relative to the initial and weak topologies on $\mathscr{V}$.*

*Proof.* Since the weak topology on $\mathscr{V}$ is coarser than the initial topology $\tau$, it suffices to prove the following assertion: if $y \in \mathscr{V}$ and $y$ is not in the $\tau$-closure $Z^-$ of $Z$, then $y$ is not in the weak closure of $Z$. Now $Z^-$ is convex, and it follows from Corollary 1.2.12 that there is a continuous linear functional $\rho$ on $\mathscr{V}$ and a real number $b$ such that

$$\operatorname{Re} \rho(y) > b, \qquad \operatorname{Re} \rho(z) \leqslant b \qquad (z \in Z^-).$$

The set $S = \{x \in \mathscr{V} : \mathrm{Re}\, \rho(x) \leqslant b\}$ is weakly closed since $\rho$ is weakly continuous; moreover, $y \notin S$, and $S$ contains $Z$ and so contains the weak closure of $Z$. ■

When $\mathscr{V}$ is a locally convex space and $x \in \mathscr{V}$, the equation

$$\hat{x}(\rho) = \rho(x) \qquad (\rho \in \mathscr{V}^{\#})$$

defines a linear functional $\hat{x}$ on the continuous dual space $\mathscr{V}^{\#}$. The set

$$\hat{\mathscr{V}} = \{\hat{x} : x \in \mathscr{V}\}$$

is a linear subspace of the algebraic dual space of $\mathscr{V}^{\#}$. It separates the points of $\mathscr{V}^{\#}$ since, if $\rho \in \mathscr{V}^{\#}$ and $\hat{x}(\rho) = 0$ for all $x$ in $\mathscr{V}$, then $\rho(x) = 0$ $(x \in \mathscr{V})$, and thus $\rho = 0$. We often write $\sigma(\mathscr{V}^{\#}, \mathscr{V})$, rather than $\sigma(\mathscr{V}^{\#}, \hat{\mathscr{V}})$, for the weak topology induced on $\mathscr{V}^{\#}$ by $\hat{\mathscr{V}}$; and we refer to $\sigma(\mathscr{V}^{\#}, \mathscr{V})$ as the *weak* topology* (sometimes called the $w^*$-topology) on $\mathscr{V}^{\#}$. Note that each $\rho_0$ in $\mathscr{V}^{\#}$ has a base of neighborhoods consisting of sets of the form

$$\{\rho \in \mathscr{V}^{\#} : |\rho(x_j) - \rho_0(x_j)| < \varepsilon \ (j = 1, \ldots, m)\},$$

where $\varepsilon > 0$ and $x_1, \ldots, x_m \in \mathscr{V}$. From Theorem 1.3.1 (with $\mathscr{V}$ replaced by $\mathscr{V}^{\#}$, and $\mathscr{F} = \mathscr{L} = \hat{\mathscr{V}}$), the weak* continuous linear functionals on $\mathscr{V}^{\#}$ are precisely the elements of $\hat{\mathscr{V}}$; so we have the following result.

1.3.5. PROPOSITION. *A linear functional $\omega$, on the continuous dual space $\mathscr{V}^{\#}$ of a locally convex space $\mathscr{V}$, is weak* continuous if and only if there is an element $x$ of $\mathscr{V}$ such that $\omega(\rho) = \rho(x)$ for each $\rho$ in $\mathscr{V}^{\#}$.*

## 1.4. Extreme points

Suppose that $\mathscr{V}$ is a locally convex space. By the *closed convex hull* of a subset $Y$ of $\mathscr{V}$ we mean the closure $\overline{\mathrm{co}}\, Y$ of the convex hull $\mathrm{co}\, Y$; it is clear that this is the smallest closed convex set that contains $Y$. An element $x_0$ of a convex set $X$ in $\mathscr{V}$ is described as an *extreme point* of $X$ if the only way in which it can be expressed as a convex combination $x_0 = (1 - a)x_1 + ax_2$, with $0 < a < 1$ and $x_1$, $x_2$ in $X$, is by taking $x_1 = x_2 = x_0$. We shall prove (Theorem 1.4.3) that every compact convex subset of $\mathscr{V}$ has extreme points and is the closed convex hull of the set of all its extreme points.

In the locally convex space $\mathbb{R}^2$ (that is, in the plane), a closed triangle is a convex set that has just three extreme points, its vertices. For a closed disk in $\mathbb{R}^2$, the extreme points are precisely the boundary points. In each of these examples, it is apparent that the set specified is the convex hull of its extreme points. In the case of a triangle, one might expect that the sides, as well as the vertices, have some significance in terms of convexity structure; in fact, each side is a "face," in a sense now to be defined.

By a *face* of a convex set $X$ in $\mathscr{V}$ we mean a non-empty convex subset $F$ of $X$, such that the conditions

$$0 < a < 1, \qquad x_1, x_2 \in X, \qquad (1-a)x_1 + ax_2 \in F,$$

imply that $x_1, x_2 \in F$. Note that $x_0$ is an extreme point of $X$ if and only if the one-point set $\{x_0\}$ is a face of $X$. It is apparent that, if a family of faces of $X$ has non-empty intersection, then this intersection is itself a face of $X$. Moreover, if $Y$ is a face of $X$ and $Z$ is a face of $Y$, then $Z$ is a face of $X$. In particular, therefore, an extreme point of a face of $X$ is also an extreme point of $X$.

The following lemma provides a slight strengthening of the defining property of a face; and upon specializing to "one-point" faces, it reduces to a similar assertion about extreme points.

1.4.1. LEMMA. *If $F$ is a face of a convex set $X$ and $a_1x_1 + \cdots + a_nx_n \in F$, where $x_1, \ldots, x_n \in X$ and $a_1, \ldots, a_n$ are non-negative real numbers with sum 1, then $x_j \in F$ when $1 \leqslant j \leqslant n$ and $a_j > 0$.*

*Proof.* It suffices to show that $x_1 \in F$ if $a_1 > 0$. If $a_1 = 1$, then $a_2 = a_3 = \cdots = a_n = 0$ and $x_1 = a_1x_1 + \cdots + a_nx_n \in F$. If $0 < a_1 < 1$, let $a = 1 - a_1$ and let $y$ be the convex combination $a^{-1}(a_2x_2 + \cdots + a_nx_n)$ of $x_2, \ldots, x_n$. Then $x_1, y \in X$, $0 < a < 1$, and

$$(1-a)x_1 + ay = a_1x_1 + a_2x_2 + \cdots + a_nx_n \in F.$$

Since $F$ is a face of $X$, it follows that $x_1 \in F$. ■

The results that follow are formulated so as to apply to both real and complex locally convex spaces, the notation Re being redundant in the real case.

1.4.2. LEMMA. *If $X$ is a non-empty compact convex set in a locally convex space $\mathscr{V}$, $\rho$ is a continuous linear functional on $\mathscr{V}$, and*

$$c = \sup\{\operatorname{Re}\rho(x) : x \in X\},$$

*then the set $F = \{x \in X : \operatorname{Re}\rho(x) = c\}$ is a compact face of $X$.*

*Proof.* Since a continuous real-valued function on a compact set attains its supremum, $F$ is not empty; and it is evident that $F$ is compact and convex. If $x_1, x_2 \in X$, $0 < a < 1$, and $(1-a)x_1 + ax_2 \in F$, we have $\operatorname{Re}\rho(x_1) \leqslant c$, $\operatorname{Re}\rho(x_2) \leqslant c$, and

$$(1-a)\operatorname{Re}\rho(x_1) + a\operatorname{Re}\rho(x_2) = \operatorname{Re}\rho((1-a)x_1 + ax_2) = c;$$

so $\operatorname{Re}\rho(x_1) = \operatorname{Re}\rho(x_2) = c$, and $x_1, x_2 \in F$. Thus $F$ is a face of $X$. ■

1.4.3. THEOREM (Krein–Milman). *If $X$ is a non-empty compact convex set in a locally convex space $\mathscr{V}$, then $X$ has an extreme point. Moreover, $X = \overline{\operatorname{co}}\, E$, where $E$ is the set of all extreme points of $X$.*

*Proof.* The family $\mathscr{F}$ of all compact faces of $X$ is non-empty since $X \in \mathscr{F}$, and is partially ordered by the inclusion relation $\subseteq$. Let $\mathscr{F}_0$ be a subfamily of $\mathscr{F}$ that is totally ordered by inclusion. It is evident that $\mathscr{F}_0$ has the finite-intersection property, so by compactness the set $F_0 = \bigcap\{F : F \in \mathscr{F}_0\}$ is non-empty. Thus $F_0$ is a compact face of $X$, and is a lower bound of $\mathscr{F}_0$ in $\mathscr{F}$.

Since every totally ordered subset of $\mathscr{F}$ has a lower bound in $\mathscr{F}$, it follows from Zorn's lemma that $\mathscr{F}$ has an element $F$ that is minimal with respect to inclusion. We shall show that $F$ consists of a single point $x$, and since $F$ is a (compact) face of $X$, it then follows that $x$ is an extreme point of $X$. To this end, suppose the contrary, and let $x_1$, $x_2$ be distinct elements of $F$. By the Hahn–Banach theorem, there is a continuous linear functional $\rho$ on $\mathscr{V}$ such that $\operatorname{Re}\rho(x_1) \neq \operatorname{Re}\rho(x_2)$. From Lemma 1.4.2 we can choose a real number $c$ so that the set

$$F_0 = \{x \in F : \operatorname{Re}\rho(x) = c\}$$

is a compact face of $F$. Accordingly, $F_0$ is a compact face of $X$; that is, $F_0 \in \mathscr{F}$. Since $\operatorname{Re}\rho(x_1) \neq \operatorname{Re}\rho(x_2)$, at least one of $x_1$, $x_2$ lies outside $F_0$; so $F_0$ is a proper subset of $F$, contrary to our minimality assumption. Hence $F$ consists of a single point.

So far, we have shown that each non-empty compact convex subset of $\mathscr{V}$ has an extreme point. If $E$ denotes the set of all extreme points of $X$, it is clear that $\overline{\text{co}}\, E \subseteq X$, and we have to show that equality occurs. Suppose the contrary, and let $x_0 \in X \backslash \overline{\text{co}}\, E$; we shall obtain a contradiction. From the Hahn–Banach theorem, we can find a continuous linear functional $\rho$ on $\mathscr{V}$ and a real number $a$ such that

$$\text{(1)} \qquad \operatorname{Re}\rho(x_0) > a \geqslant \operatorname{Re}\rho(y) \qquad (y \in \overline{\text{co}}\, E).$$

If $c_1 = \sup\{\operatorname{Re}\rho(x) : x \in X\}$, then $c_1 > a$, and the set

$$F_1 = \{x \in X : \operatorname{Re}\rho(x) = c_1\}$$

is a compact face of $X$ by Lemma 1.4.2. In particular, $F_1$ is a non-empty compact convex subset of $\mathscr{V}$, and so has an extreme point $x_1$. Since $x_1$ is an extreme point of a face of $X$, it is an extreme point of $X$; that is, $x_1 \in E$. However, $\operatorname{Re}\rho(x_1) = c_1 > a$, contradicting (1). ∎

1.4.4. Corollary. *If $X$ is a non-empty compact convex set in a locally convex space $\mathscr{V}$ and $\rho$ is a continuous linear functional on $\mathscr{V}$, there is an extreme point $x_0$ of $X$ such that* $\operatorname{Re}\rho(x) \leqslant \operatorname{Re}\rho(x_0)$ *for each $x$ in $X$.*

*Proof.* Let

$$c = \sup\{\operatorname{Re}\rho(x) : x \in X\}.$$

By Lemma 1.4.2, the set $\{x \in X : \operatorname{Re}\rho(x) = c\}$ is a compact face of $X$. In particular, it is a non-empty compact convex set in $\mathscr{V}$, and so has an extreme point $x_0$. Since $x_0$ is an extreme point of a face of $X$, it is an extreme point of $X$; and

$$\operatorname{Re} \rho(x_0) = c \geqslant \operatorname{Re} \rho(x) \qquad (x \in X). \quad \blacksquare$$

1.4.5. THEOREM. *If $X$ is a non-empty compact convex set in a locally convex space $\mathscr{V}$ and $Y$ is a closed subset of $X$ such that $\overline{\text{co}}\, Y = X$, then $Y$ contains the extreme points of $X$.*

*Proof.* Suppose that $x_0$ is an extreme point of $X$. In order to show that $x_0 \in Y$, it suffices to prove that $x_0 + V$ meets $Y$ whenever $V$ is a balanced convex neighborhood of 0 in $\mathscr{V}$; for $Y$ is closed, and sets of the form $x_0 + V$ constitute a base of neighborhoods of $x_0$. Given $V$ as above, the family $\{y + \frac{1}{2}V : y \in Y\}$ is an open covering of the compact set $Y$, and so has a finite subcovering $\{y_j + \frac{1}{2}V : j = 1, \ldots, n\}$, with $y_1, \ldots, y_n$ in $Y$. Let $V^-$ denote the closure of $V$, and for $j = 1, \ldots, n$, let $X_j$ be the non-empty compact convex set $(y_j + \frac{1}{2}V^-) \cap X$. Then

$$Y = Y \cap X \subseteq \left[ \bigcup_{j=1}^{n} (y_j + \tfrac{1}{2}V) \right] \cap X \subseteq \bigcup_{j=1}^{n} X_j.$$

Let $S$ be the set of all vectors of the form $a_1x_1 + \cdots + a_nx_n$, where $x_j \in X_j$ $(j = 1, \ldots, n)$ and the coefficients $a_1, \ldots, a_n$ are non-negative real numbers with sum 1. Then $S$ contains each $X_j$, and so contains $Y$; from the convexity of $X_1, \ldots, X_n$, it is readily verified that $S$ is convex. We assert also that $S$ is compact. To prove this, let $A$ be the compact subset

$$\{(a_1, \ldots, a_n) : a_1 \geqslant 0, \ldots, a_n \geqslant 0,\ a_1 + \cdots + a_n = 1\}$$

of $\mathbb{R}^n$, and write $\mathbf{a}$ for the element $(a_1, \ldots, a_n)$ of $A$. By Tychonoff's theorem, the set $A \times X_1 \times \cdots \times X_n$ is compact in the product topology; and hence $S$ is compact, since it is the image of the product set under the continuous mapping

$$(\mathbf{a}, x_1, \ldots, x_n) \to a_1x_1 + \cdots + a_nx_n.$$

From the preceding paragraph, $S$ is a compact convex set containing $Y$. Thus $S$ contains the closed convex hull $X$ of $Y$; in particular, $x_0 \in S$. Accordingly, $x_0$ can be expressed as a convex combination $a_1x_1 + \cdots + a_nx_n$, where $x_j \in X_j$ $(\subseteq X)$. Since $x_0$ is an extreme point of $X$, $x_0 = x_j$ for some $j$ in $\{1, \ldots, n\}$ (take any $j$ with $a_j > 0$), and

$$x_0 = x_j \in X_j \subseteq y_j + \tfrac{1}{2}V^-.$$

Thus $x_0 - y_j$ lies in the closure of $\frac{1}{2}V$; so the neighborhood $x_0 - y_j + \frac{1}{2}V$ meets $\frac{1}{2}V$. Since $V$ is balanced and convex, it now follows that $y_j \in x_0 + V$, whence $x_0 + V$ meets $Y$. $\blacksquare$

## 1.5. Normed spaces

In the discussion preceding Proposition 1.1.5, we introduced the concepts of "semi-norm" and "norm" on a (real or complex) linear space. In Theorem 1.2.6, we showed how a separating family $\Gamma$ of semi-norms on such a space gives rise to a locally convex topology. The present section is concerned with the case in which $\Gamma$ consists of a single norm.

By a *normed space* we mean a pair $(\mathfrak{X}, p)$ in which $\mathfrak{X}$ is a linear space whose scalar field $\mathbb{K}$ is either $\mathbb{R}$ or $\mathbb{C}$ and $p$ is a norm on $\mathfrak{X}$. When $x \in \mathfrak{X}$, we usually write $\|x\|$ rather than $p(x)$, and refer to $\|x\|$ as "the norm of $x$." With this notation, the defining properties of a norm can be set out as follows: whenever $x, y \in \mathfrak{X}$ and $a \in \mathbb{K}$,

(a) $\|x\| \geqslant 0$, with equality only when $x = 0$;
(b) $\|ax\| = |a| \, \|x\|$;
(c) $\|x + y\| \leqslant \|x\| + \|y\|$ (the triangle inequality).

We recall also another property, $|\|x\| - \|y\|| \leqslant \|x - y\|$, which is an easy consequence of the triangle inequality.

Suppose that $(\mathfrak{X}, \| \ \|)$ is a normed space. From the properties (a), (b), and (c), just noted, it is apparent that the equation

$$d(x, y) = \|x - y\| \qquad (x, y \in \mathfrak{X})$$

defines a metric $d$ on $\mathfrak{X}$. Also, $\mathfrak{X}$ has a locally convex topology, the *norm topology*, derived as in Theorem 1.2.6 from the family $\Gamma$ consisting of the single norm $\| \ \|$ on $\mathfrak{X}$; and this topology gives rise to a uniform structure on $\mathfrak{X}$, described in the discussion preceding Proposition 1.2.1. In the norm topology, each $x_0$ in $\mathfrak{X}$ has a base of neighborhoods consisting of the sets $V(x_0 : \| \ \|; \varepsilon)$ $(\varepsilon > 0)$, where

$$V(x_0 : \| \ \|; \varepsilon) = \{x \in \mathfrak{X} : \|x - x_0\| < \varepsilon\} = \{x \in \mathfrak{X} : d(x, x_0) < \varepsilon\},$$

the "open ball" with center $x_0$ and radius $\varepsilon$. The corresponding uniform structure has a base, consisting of the sets

$$\begin{aligned}\{(x, y) \in \mathfrak{X} \times \mathfrak{X} : x - y \in V(0 : \| \ \|; \varepsilon)\} &= \{(x, y) \in \mathfrak{X} \times \mathfrak{X} : \|x - y\| < \varepsilon\} \\ &= \{(x, y) \in \mathfrak{X} \times \mathfrak{X} : d(x, y) < \varepsilon\}.\end{aligned}$$

Accordingly, the norm topology, and the associated uniform structure, are precisely the topology and uniform structure derived in the usual way [K: pp. 119, 184] from the metric $d$ on $\mathfrak{X}$. Since $\mathfrak{X}$, with the norm topology, is a linear topological space, the mappings $(x, y) \to x + y$ and $(a, x) \to ax$ are continuous; and so is the mapping $x \to \|x\|$, by Theorem 1.2.6. Of course, these continuity assertions can easily be verified directly, using the defining properties (a), (b), and (c) of the norm.

A normed space $(\mathfrak{X}, \| \ \|)$ is described as a *Banach space* if it is complete relative to the uniform structure just mentioned; that is, if it becomes a complete metric space, with the metric $d$ defined above. The scalar field $\mathbb{K}$ is a one-dimensional Banach space, the modulus function being a norm which gives rise to the usual topology on $\mathbb{K}$. With either of the norms $p_1, p_\infty$, defined in the discussion preceding Proposition 1.1.5, $\mathbb{K}^n$ becomes a Banach space (in either case, the norm topology being the usual product topology).

When $\mathfrak{X}$ is a normed space and $r > 0$, we denote by $(\mathfrak{X})_r$ the closed ball $\{x \in \mathfrak{X} : \|x\| \leqslant r\}$; we refer to $(\mathfrak{X})_1$ as the *unit ball* of $\mathfrak{X}$. Since $(\mathfrak{X})_r$ is convex and is closed in the initial (that is, the norm) topology, it is weakly closed, by Theorem 1.3.4. A subset $\mathscr{C}$ of $\mathfrak{X}$ is *bounded* if $\mathscr{C} \subseteq (\mathfrak{X})_r$ for some positive real number $r$.

A linear mapping $T$ from a normed space $\mathfrak{X}$ into another such space $\mathscr{Y}$ is said to be *norm preserving* if $\|Tx\| = \|x\|$ for each $x$ in $\mathfrak{X}$. Such a mapping is necessarily isometric (that is, distance preserving) and hence one-to-one, since

$$d(Tx_1, Tx_2) = \|Tx_1 - Tx_2\| = \|T(x_1 - x_2)\| = \|x_1 - x_2\| = d(x_1, x_2);$$

and conversely, an isometric linear mapping is norm preserving. A norm-preserving linear mapping from a normed space $\mathfrak{X}$ onto another such space $\mathscr{Y}$ is sometimes described as an *isometric isomorphism* from $\mathfrak{X}$ onto $\mathscr{Y}$.

A subspace $\mathfrak{X}$ of a normed space $\mathscr{Y}$ is itself a normed space since the norm on $\mathscr{Y}$ restricts to a norm on $\mathfrak{X}$. The following theorem shows that every normed space can be viewed as an everywhere-dense subspace of an (essentially unique) Banach space.

1.5.1. THEOREM. *If $\mathfrak{X}$ is a normed space, there is a Banach space $\mathscr{Y}$ that contains $\mathfrak{X}$ as an everywhere-dense subspace (and such that the norm on $\mathfrak{X}$ is the restriction of the norm on $\mathscr{Y}$). If $\mathscr{Y}_1$ is another Banach space with these properties, the identity mapping on $\mathfrak{X}$ extends to an isometric isomorphism from $\mathscr{Y}$ onto $\mathscr{Y}_1$.*

*Proof.* Let $d$ be the metric on $\mathfrak{X}$ derived from the norm; let $\hat{\mathfrak{X}}$ be the completion of the metric space $\mathfrak{X}$ so obtained, and let $\hat{d}$ denote the metric on $\hat{\mathfrak{X}}$. Thus $\hat{\mathfrak{X}}$ is a complete metric space, $\mathfrak{X}$ is an everywhere-dense subset of $\hat{\mathfrak{X}}$, and

$$\hat{d}(u, v) = d(u, v) = \|u - v\| \qquad (u, v \in \mathfrak{X}).$$

We shall show that $\hat{\mathfrak{X}}$ can be made into a Banach space, with addition, scalar multiplication, and norm, extending those of $\mathfrak{X}$.

When $u, u', v, v' \in \mathfrak{X}$ and $a \in \mathbb{K}$ (the scalar field),

$$\|(u + v) - (u' + v')\| = \|(u - u') + (v - v')\| \leqslant \|u - u'\| + \|v - v'\|,$$

$$\|au - au'\| = |a| \, \|u - u'\|, \qquad |\|u\| - \|u'\|| \leqslant \|u - u'\|.$$

Accordingly, the equations

$$f(u, v) = u + v, \qquad g_a(u) = au, \qquad h(u) = \|u\|$$

define uniformly continuous mappings

$$f: \mathfrak{X} \times \mathfrak{X} \to \mathfrak{X} \ (\subseteq \hat{\mathfrak{X}}), \qquad g_a: \mathfrak{X} \to \mathfrak{X} \ (\subseteq \hat{\mathfrak{X}}), \qquad h: \mathfrak{X} \to \mathbb{R}.$$

Since $\hat{\mathfrak{X}}$ and $\mathbb{R}$ are complete, and $\mathfrak{X}$ is everywhere dense in $\hat{\mathfrak{X}}$ (so that $\mathfrak{X} \times \mathfrak{X}$ is everywhere dense in $\hat{\mathfrak{X}} \times \hat{\mathfrak{X}}$), it follows that $f$, $g_a$, $h$ extend by continuity to uniformly continuous mappings

$$\hat{f}: \hat{\mathfrak{X}} \times \hat{\mathfrak{X}} \to \hat{\mathfrak{X}}, \qquad \hat{g}_a: \hat{\mathfrak{X}} \to \hat{\mathfrak{X}}, \qquad \hat{h}: \hat{\mathfrak{X}} \to \mathbb{R},$$

respectively.

The addition and scalar multiplication, already defined for elements of $\mathfrak{X}$, can now be extended to $\hat{\mathfrak{X}}$ by the equations

$$u + v = \hat{f}(u, v), \qquad au = \hat{g}_a(u) \qquad (u, v \in \hat{\mathfrak{X}}, \quad a \in \mathbb{K});$$

and we assert that, in this way, $\hat{\mathfrak{X}}$ becomes a linear space over $\mathbb{K}$. To prove this, it suffices to verify the relations

$$u + v = v + u, \qquad u + (v + w) = (u + v) + w, \qquad u + 0 = u,$$

$$u + (-1)u = 0, \qquad (a + b)u = au + bu, \qquad a(u + v) = au + av,$$

$$a(bu) = (ab)u, \qquad 1u = u,$$

for all $u, v, w$ in $\hat{\mathfrak{X}}$ and $a, b$ in $\mathbb{K}$. Of course, all these relations are satisfied when $u, v, w \in \mathfrak{X}$, and simple continuity arguments show that they remain valid for elements of $\hat{\mathfrak{X}}$. For example, the relation $a(u + v) = au + av$ can be rewritten in the form

$$\hat{g}_a(\hat{f}(u, v)) - \hat{f}(\hat{g}_a(u), \hat{g}_a(v)) = 0,$$

and is satisfied when $u, v \in \mathfrak{X}$. The left-hand side of this last equation is a continuous function of $(u, v)$ on $\hat{\mathfrak{X}} \times \hat{\mathfrak{X}}$, which vanishes on the everywhere-dense subset $\mathfrak{X} \times \mathfrak{X}$, and so vanishes throughout $\hat{\mathfrak{X}} \times \hat{\mathfrak{X}}$ (as required). Similar arguments establish the other relations; so $\hat{\mathfrak{X}}$ is a linear space over $\mathbb{K}$.

We prove next that $\hat{h}$ is a norm on $\hat{\mathfrak{X}}$. The relations

$$\hat{h}(u - v) - \hat{d}(u, v) = 0, \qquad \hat{h}(au) = |a| \, \hat{h}(u) \qquad (a \in \mathbb{K})$$

are satisfied when $u, v \in \mathfrak{X}$, since $\hat{h}$ extends the norm on $\mathfrak{X}$. By continuity, they remain valid for all $u, v$ in $\hat{\mathfrak{X}}$. Accordingly,

$$\hat{h}(u + v) = \hat{d}(u + v, 0) \leqslant \hat{d}(u + v, v) + \hat{d}(v, 0) = \hat{h}(u) + \hat{h}(v),$$

and $\hat{h}(u) = \hat{d}(u, 0) \geqslant 0$ (with equality only when $u = 0$), when $u, v \in \hat{\mathfrak{X}}$. It now follows that $\hat{h}$ is a norm on $\hat{\mathfrak{X}}$, which extends the norm on $\mathfrak{X}$, and gives rise to the complete metric $\hat{d}$ on $\hat{\mathfrak{X}}$. Hence $\hat{\mathfrak{X}}$ is a Banach space and contains $\mathfrak{X}$ as an everywhere-dense subspace.

Finally, suppose that $\mathscr{Y}$ and $\mathscr{Y}_1$ are Banach spaces, each containing $\mathfrak{X}$ as an everywhere-dense subspace and each with its norm extending the norm on $\mathfrak{X}$.

The identity mapping on $\mathfrak{X}$ can be viewed as a continuous linear operator, from an everywhere-dense subspace of the Banach space $\mathscr{Y}$ into the Banach space $\mathscr{Y}_1$. By Corollary 1.2.3, it extends to a continuous linear operator $T$ from $\mathscr{Y}$ into $\mathscr{Y}_1$. The equation $f(u) = \|Tu\| - \|u\|$ defines a continuous mapping $f: \mathscr{Y} \to \mathbb{R}$, and $f$ vanishes on the everywhere-dense subset $\mathfrak{X}$ of $\mathscr{Y}$. Hence $f$ vanishes throughout $\mathscr{Y}$, and $T$ is norm preserving, and therefore isometric. Since $\mathscr{Y}$ is complete and $T$ is isometric, the range $T(\mathscr{Y})$ of $T$ is complete, and is therefore closed in $\mathscr{Y}_1$. Thus $T(\mathscr{Y}) = \mathscr{Y}_1$, since $T(\mathscr{Y})$ contains the everywhere-dense subset $\mathfrak{X}$ of $\mathscr{Y}_1$; and $T$ is an isometric isomorphism from $\mathscr{Y}$ onto $\mathscr{Y}_1$, which extends the identity mapping on $\mathfrak{X}$. ■

The Banach space $\mathscr{Y}$ occurring in the statement of Theorem 1.5.1 is called the *completion* of the normed space $\mathfrak{X}$.

The following lemma provides a useful criterion for the completeness of a normed space. It is couched in terms of the convergence of certain infinite series of vectors in the space; by the convergence of such a series, $\sum_1^\infty x_n$, we mean convergence of the sequence of partial sums $s_n = x_1 + \cdots + x_n$. The result could easily be reformulated in terms of unordered sums of the type considered at the end of section 1.2.

1.5.2. LEMMA. *If $\mathfrak{X}$ is a normed space, the following two conditions are equivalent*:

(i) *$\mathfrak{X}$ is a Banach space.*

(ii) *If $x_1, x_2, \ldots \in \mathfrak{X}$ and $\sum \|x_n\| < \infty$, the series $\sum x_n$ converges, in the metric of $\mathfrak{X}$, to an element of $\mathfrak{X}$.*

*Proof.* Suppose first that $\mathfrak{X}$ is a Banach space. Let $\{x_n\}$ be a sequence of elements of $\mathfrak{X}$, such that $\sum \|x_n\| < \infty$, and write $s_n$ for the partial sum $x_1 + \cdots + x_n$. Given any positive $\varepsilon$, there is a positive integer $N$ such that $\sum_{n+1}^m \|x_r\| < \varepsilon$ (and hence $\|s_m - s_n\| = \|\sum_{n+1}^m x_r\| < \varepsilon$) whenever $m > n \geqslant N$. Hence $\{s_n\}$ is a Cauchy sequence, in the complete space $\mathfrak{X}$, and so converges (that is, $\sum x_n$ converges).

Conversely, suppose that condition (ii) is satisfied. If $\{y_n\}$ is a Cauchy sequence in $\mathfrak{X}$, there is a strictly increasing sequence $\{n(1), n(2), \ldots\}$ of positive integers such that

$$\|y_m - y_n\| < 2^{-k} \qquad (m > n \geqslant n(k)).$$

In particular, $\|y_{n(k+1)} - y_{n(k)}\| < 2^{-k}$, and therefore

$$\|y_{n(1)}\| + \sum_{k=1}^{\infty} \|y_{n(k+1)} - y_{n(k)}\| < \infty.$$

From condition (ii), the series $y_{n(1)} + \sum_1^\infty (y_{n(k+1)} - y_{n(k)})$ converges to an element $y$ of $\mathfrak{X}$; that is, the sequence $\{y_{n(k)}\}$ converges to $y$ (because $y_{n(k)}$ is the

$k$th partial sum of the series). Finally, when $n > n(k)$,

$$\|y_n - y\| \leqslant \|y_n - y_{n(k)}\| + \|y_{n(k)} - y\| < 2^{-k} + \|y_{n(k)} - y\|;$$

and since the right-hand side tends to 0 when $k \to \infty$, $\{y_n\}$ converges to $y$. Hence $\mathfrak{X}$ is a Banach space. ■

1.5.3. THEOREM. *If $\mathscr{Y}$ is a closed subspace of a normed space $\mathfrak{X}$, the equation*

$$\|x + \mathscr{Y}\|_0 = \inf\{\|x + y\| : y \in \mathscr{Y}\} \qquad (x \in \mathfrak{X})$$

*defines a norm $\|\ \|_0$ on the quotient space $\mathfrak{X}/\mathscr{Y}$. With this norm on $\mathfrak{X}/\mathscr{Y}$, the quotient mapping*

$$Q : x \to x + \mathscr{Y} : \quad \mathfrak{X} \to \mathfrak{X}/\mathscr{Y}$$

*is a continuous linear operator, and $\|Qx\|_0 \leqslant \|x\|$; and $\mathfrak{X}/\mathscr{Y}$ is a Banach space if $\mathfrak{X}$ is a Banach space.*

*Proof.* Suppose that $x$, $x_1$, $x_2 \in \mathfrak{X}$ and $a$ is a scalar. Since

$$\inf\{\|x + y\| : y \in \mathscr{Y}\} = \inf\{\|u\| : u \in x + \mathscr{Y}\},$$

the definition of $\|x + \mathscr{Y}\|_0$ is unambiguous (that is, it depends only on the coset $x + \mathscr{Y}$, not on the choice of $x$ within that coset); moreover, $\|x + \mathscr{Y}\|_0 \geqslant 0$. For all $y$, $y_1$, and $y_2$ in $\mathscr{Y}$,

$$\|x_1 + x_2 + y_1 + y_2\| \leqslant \|x_1 + y_1\| + \|x_2 + y_2\|, \qquad \|ax + ay\| = |a|\,\|x + y\|.$$

Thus

$$\|x_1 + x_2 + \mathscr{Y}\|_0 \leqslant \|x_1 + \mathscr{Y}\|_0 + \|x_2 + \mathscr{Y}\|_0, \qquad \|ax + \mathscr{Y}\|_0 = |a|\,\|x + \mathscr{Y}\|_0.$$

If $\|x + \mathscr{Y}\|_0 = 0$, there is a sequence $\{y_n\}$ in $\mathscr{Y}$ such that $\|x + y_n\| < 1/n$ $(n = 1, 2, \ldots)$; since $\mathscr{Y}$ is closed, while $-y_n \in \mathscr{Y}$ and $\lim(-y_n) = x$, it follows that $x \in \mathscr{Y}$, whence $x + \mathscr{Y}$ is the zero vector in $\mathfrak{X}/\mathscr{Y}$. Hence $\|\ \|_0$ is a norm on $\mathfrak{X}/\mathscr{Y}$.

The quotient mapping $Q$ is a linear operator, and is continuous since

$$\begin{aligned}\|Qx_1 - Qx_2\|_0 &= \|x_1 - x_2 + \mathscr{Y}\|_0 \\ &= \inf\{\|x_1 - x_2 + y\| : y \in \mathscr{Y}\} \leqslant \|x_1 - x_2\|\end{aligned}$$

for all $x_1$ and $x_2$ in $\mathfrak{X}$.

Suppose that $\mathfrak{X}$ is a Banach space. If $\{x_n + \mathscr{Y}\}$ is a sequence of elements of $\mathfrak{X}/\mathscr{Y}$, and $\sum\|x_n + \mathscr{Y}\|_0 < \infty$, we can choose $y_1, y_2, \ldots$ in $\mathscr{Y}$ so that $\|x_n + y_n\| < \|x_n + \mathscr{Y}\|_0 + 2^{-n}$. Thus $\sum\|x_n + y_n\| < \infty$, and by Lemma 1.5.2, the series $\sum(x_n + y_n)$ converges to an element $z$ of $\mathfrak{X}$. Since $Q$ is a continuous linear operator, and $Q(x_n + y_n) = x_n + y_n + \mathscr{Y} = x_n + \mathscr{Y}$, it follows that

$$\sum_{n=1}^{m} (x_n + \mathscr{Y}) = Q\left( \sum_{n=1}^{m} (x_n + y_n) \right) \to Qz$$

as $m \to \infty$; that is, $\sum(x_n + \mathscr{Y})$ converges to $Qz$ ($\in \mathfrak{X}/\mathscr{Y}$). Again by Lemma 1.5.2, $\mathfrak{X}/\mathscr{Y}$ is a Banach space. ■

1.5.4. Corollary. *If $\mathscr{Y}$ and $\mathscr{Z}$ are subspaces of a normed space $\mathfrak{X}$, with $\mathscr{Y}$ closed and $\mathscr{Z}$ finite-dimensional, then $\mathscr{Y} + \mathscr{Z}$ is closed in $\mathfrak{X}$.*

*Proof.* The quotient mapping $Q: \mathfrak{X} \to \mathfrak{X}/\mathscr{Y}$ is a continuous linear operator; the subspace $Q(\mathscr{Z})$ of $\mathfrak{X}/\mathscr{Y}$ is finite-dimensional and is therefore closed in $\mathfrak{X}/\mathscr{Y}$ (Theorem 1.2.17); and $\mathscr{Y} + \mathscr{Z}$ is the inverse image $Q^{-1}(Q(\mathscr{Z}))$. ■

We now consider some elementary properties of linear operators acting on normed spaces.

1.5.5. Theorem. *If $\mathfrak{X}$ and $\mathscr{Y}$ are normed spaces and $T: \mathfrak{X} \to \mathscr{Y}$ is a linear operator, the following four conditions are equivalent.*

(i) *$T$ is continuous.*

(ii) *There is a non-negative real number $C$ such that $\|Tx\| \leqslant C\|x\|$ for each $x$ in $\mathfrak{X}$.*

(iii) $\sup\{\|Tx\|/\|x\| : x \in \mathfrak{X},\ x \neq 0\} < \infty$.

(iv) $\sup\{\|Tx\| : x \in \mathfrak{X},\ \|x\| = 1\} < \infty$.

*When these conditions are satisfied, the suprema occurring in* (iii) *and* (iv) *are both equal to the smallest real number $C$ with the property set out in* (ii).

*Proof.* The equivalence of (i) and (ii) is a special case of Proposition 1.2.8(ii), with both $\Gamma_1$ and $\Gamma_2$ consisting of a single norm.

A real number $C$ has the property set out in (ii) if and only if

$$\|Tx\|/\|x\| \leqslant C \qquad (x \in \mathfrak{X}, \quad x \neq 0).$$

Upon taking $x_1 = \|x\|^{-1}x$, it follows that this last condition is satisfied if and only if

$$\|Tx_1\| \leqslant C \qquad (x_1 \in \mathfrak{X}, \quad \|x_1\| = 1).$$

This proves the equivalence of (ii), (iii), and (iv), and shows that the suprema in (iii) and (iv) coincide with the smallest possible value for $C$ in (ii). ■

When $\mathfrak{X}, \mathscr{Y}$ are normed spaces and $T: \mathfrak{X} \to \mathscr{Y}$ is a linear operator, we denote by $\|T\|$ the (equal, and possibly infinite) suprema occurring in parts (iii) and (iv) of Theorem 1.5.5; we refer to $\|T\|$ as the (*operator*) *bound* of $T$. Thus $T$ is continuous if and only if $\|T\| < \infty$; then

$$\|Tx\| \leqslant \|T\|\ \|x\| \qquad (x \in \mathfrak{X}),$$

and $\|T\|$ is the smallest real number with this property. It is clear that $\|T\| = 0$ only when $T = 0$. Since continuity is equivalent to the existence of a finite bound, continuous linear operators between normed spaces are often described as *bounded* linear operators.

The set $\mathscr{B}(\mathfrak{X}, \mathscr{Y})$ of all bounded linear operators from $\mathfrak{X}$ into $\mathscr{Y}$ is a linear space (with the same scalar field $\mathbb{K}$ as $\mathfrak{X}$ and $\mathscr{Y}$). If $S$, $T \in \mathscr{B}(\mathfrak{X}, \mathscr{Y})$ and $a \in \mathbb{K}$,

$$\|(S + T)x\| = \|Sx + Tx\| \leqslant \|Sx\| + \|Tx\| \leqslant \|S\| + \|T\|,$$

$$\|(aT)x\| = \|a(Tx)\| = |a| \, \|Tx\|$$

whenever $x \in \mathfrak{X}$ and $\|x\| = 1$. By taking suprema, as $x$ varies subject to the conditions just stated, we obtain

$$\|S + T\| \leqslant \|S\| + \|T\|, \qquad \|aT\| = |a| \, \|T\|.$$

Since, also, $\|T\| \geqslant 0$, with equality only when $T = 0$, it follows that $\mathscr{B}(\mathfrak{X}, \mathscr{Y})$ is a normed space, with the operator bound as norm. When $\mathscr{Y} = \mathfrak{X}$, we usually write $\mathscr{B}(\mathfrak{X})$ rather than $\mathscr{B}(\mathfrak{X}, \mathfrak{X})$.

If $\mathfrak{X}$, $\mathscr{Y}$, $\mathscr{Z}$ are normed spaces, and $S: \mathscr{Y} \to \mathscr{Z}$, $T: \mathfrak{X} \to \mathscr{Y}$ are continuous linear operators, then $ST: \mathfrak{X} \to \mathscr{Z}$ is continuous. Since

$$\|STx\| \leqslant \|S\| \, \|Tx\| \leqslant \|S\| \, \|T\| \qquad (x \in \mathfrak{X}, \quad \|x\| = 1),$$

it follows that $\|ST\| \leqslant \|S\| \, \|T\|$. This applies, in particular, when $\mathfrak{X} = \mathscr{Y} = \mathscr{Z}$. The set $\mathscr{B}(\mathfrak{X})$ is an associative linear algebra with a unit element $I$ (the identity mapping on $\mathfrak{X}$); it is also a normed space, and its norm (the operator bound) satisfies $\|I\| = 1$, $\|ST\| \leqslant \|S\| \, \|T\|$. These properties of $\mathscr{B}(\mathfrak{X})$ are characteristic of Banach algebras, which are studied in Chapter 3.

1.5.6. Theorem. *If $\mathfrak{X}$ is a normed space and $\mathscr{Y}$ is a Banach space, both having the same scalar field $\mathbb{K}$, then the set $\mathscr{B}(\mathfrak{X}, \mathscr{Y})$ of all bounded linear operators from $\mathfrak{X}$ into $\mathscr{Y}$ is a Banach space over $\mathbb{K}$, with the operator bound as norm.*

*Proof.* We have already seen that $\mathscr{B}(\mathfrak{X}, \mathscr{Y})$ is a normed space, so it remains to prove that it is complete. Let $\{T_n\}$ be a Cauchy sequence in $\mathscr{B}(\mathfrak{X}, \mathscr{Y})$. Given any positive $\varepsilon$, there is a positive integer $N(\varepsilon)$ such that

$$\|T_m - T_n\| \leqslant \varepsilon \qquad (m > n \geqslant N(\varepsilon)).$$

When $x \in \mathfrak{X}$,

$$\|T_m x - T_n x\| \leqslant \|T_m - T_n\| \, \|x\| \leqslant \varepsilon \|x\| \qquad (m > n \geqslant N(\varepsilon)). \tag{1}$$

Thus $\{T_n x\}$ is a Cauchy sequence in the Banach space $\mathscr{Y}$, and so converges to an element of $\mathscr{Y}$. Accordingly, the equation

$$Tx = \lim_{n \to \infty} T_n x \qquad (x \in \mathfrak{X})$$

defines a mapping $T: \mathfrak{X} \to \mathscr{Y}$, and it is apparent that $T$ is a linear operator. Upon taking limits as $m \to \infty$ in (1), we obtain

$$\|Tx - T_n x\| \leqslant \varepsilon \|x\| \qquad (n \geqslant N(\varepsilon), \quad x \in \mathfrak{X}).$$

This shows that $T - T_n$ is a bounded linear operator (whence, so is $T$) and that $\|T - T_n\| \leqslant \varepsilon$ whenever $n \geqslant N(\varepsilon)$. It follows that $\{T_n\}$ converges to the element $T$ of $\mathscr{B}(\mathfrak{X}, \mathscr{Y})$, and hence $\mathscr{B}(\mathfrak{X}, \mathscr{Y})$ is complete. ■

1.5.7. THEOREM. *If $\mathfrak{X}$ is a normed space and $\mathscr{Y}$ is a Banach space, both having the same scalar field, then every bounded linear operator $T: \mathfrak{X} \to \mathscr{Y}$ extends uniquely to a bounded linear operator $\hat{T}: \hat{\mathfrak{X}} \to \mathscr{Y}$, where $\hat{\mathfrak{X}}$ is the completion of $\mathfrak{X}$. The mapping $T \to \hat{T}$ is an isometric isomorphism from $\mathscr{B}(\mathfrak{X}, \mathscr{Y})$ onto $\mathscr{B}(\hat{\mathfrak{X}}, \mathscr{Y})$.*

*Proof.* By Corollary 1.2.3, each continuous linear operator $T: \mathfrak{X} \to \mathscr{Y}$ extends uniquely to a continuous linear operator $\hat{T}: \hat{\mathfrak{X}} \to \mathscr{Y}$. The inequality $\|T\| \|x\| - \|\hat{T}x\| \geqslant 0$ is satisfied when $x \in \mathfrak{X}$, and by continuity it remains valid for all $x$ in $\hat{\mathfrak{X}}$. Thus $\|\hat{T}\| \leqslant \|T\|$; the reverse inequality is evident since $\hat{T}$ extends $T$, so $\|\hat{T}\| = \|T\|$.

It is apparent that the norm-preserving mapping

$$T \to \hat{T}: \quad \mathscr{B}(\mathfrak{X}, \mathscr{Y}) \to \mathscr{B}(\hat{\mathfrak{X}}, \mathscr{Y})$$

is linear. Its range is the whole of $\mathscr{B}(\hat{\mathfrak{X}}, \mathscr{Y})$ since, when $S_0 \in \mathscr{B}(\hat{\mathfrak{X}}, \mathscr{Y})$, we have $S_0 = \hat{T}_0$, where $T_0$ $(\in \mathscr{B}(\mathfrak{X}, \mathscr{Y}))$ is the restriction $S_0 | \mathfrak{X}$. ■

1.5.8. THEOREM. *Suppose that $\mathfrak{X}$ and $\mathscr{Y}$ are normed spaces, $T: \mathfrak{X} \to \mathscr{Y}$ is a bounded linear operator, $\mathfrak{X}_0$ is a closed subspace of $\mathfrak{X}$ such that $T(\mathfrak{X}_0) = \{0\}$, and $Q: \mathfrak{X} \to \mathfrak{X}/\mathfrak{X}_0$ is the quotient mapping. Then there is a bounded linear operator $T_0: \mathfrak{X}/\mathfrak{X}_0 \to \mathscr{Y}$ such that $T = T_0 Q$; moreover, $\|T_0\| = \|T\|$, and $T_0$ is one-to-one if $\mathfrak{X}_0$ is the null space of $T$.*

*Proof.* From purely algebraic considerations $T$ has a factorization $T_0 Q$, where $T_0: \mathfrak{X}/\mathfrak{X}_0 \to \mathscr{Y}$ is a linear operator, and is one-to-one if $\mathfrak{X}_0$ is the null space of $T$. When $x \in \mathfrak{X}$, $\|Qx\| \leqslant \|x\|$ by Theorem 1.5.3; so $\|Q\| \leqslant 1$. Moreover,

$$\|T_0 Q x\| = \|Tx\| = \|T(x + x_0)\| \leqslant \|T\| \|x + x_0\| \qquad (x_0 \in \mathfrak{X}_0),$$

so

$$\|T_0 Q x\| \leqslant \|T\| \inf\{\|x + x_0\| : x_0 \in \mathfrak{X}_0\} = \|T\| \|Qx\|.$$

Since $Q(\mathfrak{X}) = \mathfrak{X}/\mathfrak{X}_0$, it now follows that $T_0$ is bounded, and

$$\|T_0\| \leqslant \|T\| = \|T_0 Q\| \leqslant \|T_0\| \|Q\| \leqslant \|T_0\|;$$

so $\|T_0\| = \|T\|$. ■

1.5.9. LEMMA. *If $\mathfrak{X}$ and $\mathscr{Y}$ are normed spaces, $S: \mathfrak{X} \to \mathscr{Y}$ and $T: \mathscr{Y} \to \mathfrak{X}$ are linear operators such that $ST = I$ (the identity operator on $\mathscr{Y}$), and*

$$a = \inf\{\|Sx\| : x \in T(\mathscr{Y}),\ \|x\| = 1\},$$

*then $\|T\| = a^{-1}$ (where $0^{-1}$ is to be interpreted as $\infty$). In particular, $T$ is bounded if and only if $a > 0$.*

*Proof.* Since $ST = I$, $T$ can be viewed as a one-to-one linear mapping from $\mathscr{Y}$ onto $T(\mathscr{Y})$, and as such, it has an inverse mapping, the restriction $S|T(\mathscr{Y})$. Also, when $0 \neq z \in T(\mathscr{Y})$, $\|z\|^{-1}z$ is a unit vector $x$ in $T(\mathscr{Y})$. Thus

$$\begin{aligned}\|T\| &= \sup\left\{\frac{\|Ty\|}{\|y\|} : y \in \mathscr{Y},\ y \neq 0\right\}\\ &= \sup\left\{\frac{\|z\|}{\|Sz\|} : z \in T(\mathscr{Y}),\ z \neq 0\right\}\\ &= \sup\{\|Sx\|^{-1} : x \in T(\mathscr{Y}),\ \|x\| = 1\} = a^{-1}. \quad \blacksquare\end{aligned}$$

1.5.10. COROLLARY. *Suppose that $\mathfrak{X}$ and $\mathscr{Y}$ are normed spaces, $S$ is a one-to-one linear operator from $\mathfrak{X}$ onto $\mathscr{Y}$, $T: \mathscr{Y} \to \mathfrak{X}$ is its inverse operator, and*

$$a = \inf\{\|Sx\| : x \in \mathfrak{X},\ \|x\| = 1\}.$$

(i) $\|T\| = a^{-1}$ *(where $0^{-1}$ is to be interpreted as $\infty$); in particular, $T$ is bounded if and only if $a > 0$.*

(ii) *If $\mathfrak{X}$ is a Banach space, $S$ is bounded, and $a > 0$, then $\mathscr{Y}$ is a Banach space.*

*Proof.* (i) This follows from Lemma 1.5.9, since $T(\mathscr{Y}) = \mathfrak{X}$.

(ii) Since $a > 0$, $T$ (as well as $S$) is bounded; so both $S: \mathfrak{X} \to \mathscr{Y}$ and its inverse $T: \mathscr{Y} \to \mathfrak{X}$ are uniformly continuous. Hence the completeness of $\mathfrak{X}$ entails completeness of $\mathscr{Y}$. $\blacksquare$

## 1.6. Linear functionals on normed spaces

In this section we shall be concerned with the continuous dual space $\mathfrak{X}^\sharp$ of a normed space $\mathfrak{X}$ and with the properties of the weak topology $\sigma(\mathfrak{X}, \mathfrak{X}^\sharp)$ on $\mathfrak{X}$ and the weak* topology $\sigma(\mathfrak{X}^\sharp, \mathfrak{X})$ on $\mathfrak{X}^\sharp$. It turns out that, in a natural way, $\mathfrak{X}^\sharp$ becomes a Banach space and $\mathfrak{X}$ is isometrically isomorphic to a subspace of the second dual space $\mathfrak{X}^{\sharp\sharp}$ $(= (\mathfrak{X}^\sharp)^\sharp)$. We describe a necessary and sufficient condition for this subspace to be the whole of $\mathfrak{X}^{\sharp\sharp}$.

In Section 1.5 we considered linear operators from a normed space $\mathfrak{X}$ into another such space $\mathscr{Y}$ and introduced the normed space $\mathscr{B}(\mathfrak{X}, \mathscr{Y})$ of continuous linear operators. By taking, for $\mathscr{Y}$, the scalar field $\mathbb{K}$, we obtain results concerning linear functionals. From Theorem 1.5.5, a linear functional $\rho$ on $\mathfrak{X}$ is continuous if and only if it is bounded, in the sense that there is a non-negative real number $C$ such that

$$|\rho(x)| \leqslant C\|x\| \qquad (x \in \mathfrak{X}).$$

When $\rho$ is bounded, the least possible value of $C$ is the bound $\|\rho\|$ of $\rho$, defined by

$$\text{(1)} \qquad \|\rho\| = \sup\left\{\frac{|\rho(x)|}{\|x\|} : x \in \mathfrak{X},\ x \neq 0\right\} = \sup\{|\rho(x)| : x \in \mathfrak{X},\ \|x\| = 1\}.$$

The continuous dual space $\mathfrak{X}^{\#}$ of $\mathfrak{X}$, defined (as in Section 1.3) as the linear space of all continuous linear functionals on $\mathfrak{X}$, coincides with $\mathscr{B}(\mathfrak{X}, \mathbb{K})$; by Theorem 1.5.6, it is a Banach space, its norm being given by (1). We refer to $\mathfrak{X}^{\#}$, with this norm, as the *Banach dual space* of $\mathfrak{X}$.

For normed spaces, we have the following Hahn–Banach extension theorem.

1.6.1. THEOREM. *If $\mathfrak{X}_0$ is a subspace of a normed space $\mathfrak{X}$ and $\rho_0$ is a bounded linear functional on $\mathfrak{X}_0$, there is a bounded linear functional $\rho$ on $\mathfrak{X}$ such that $\|\rho\| = \|\rho_0\|$ and $\rho(x) = \rho_0(x)$ when $x \in \mathfrak{X}_0$.*

*Proof.* This follows at once from Theorem 1.1.7, with

$$p(x) = \|\rho_0\|\,\|x\| \qquad (x \in \mathfrak{X}). \quad ■$$

1.6.2. COROLLARY. *If $x_0$ is a non-zero vector in a normed space $\mathfrak{X}$, there is a bounded linear functional $\rho$ on $\mathfrak{X}$ such that $\|\rho\| = 1$ and $\rho(x_0) = \|x_0\|$.*

*Proof.* The equation $\rho_0(cx_0) = c\|x_0\|$ defines a bounded linear functional $\rho_0$ on the one-dimensional subspace $\mathfrak{X}_0$ generated by $x_0$; moreover, $\rho_0(x_0) = \|x_0\|$, and $\|\rho_0\| = 1$. By Theorem 1.6.1, $\rho_0$ extends (still with norm 1) to a bounded linear functional $\rho$ on $\mathfrak{X}$. ■

1.6.3. COROLLARY. *If $\mathscr{Y}$ is a closed subspace of a normed space $\mathfrak{X}$ and $x_0 \in \mathfrak{X} \backslash \mathscr{Y}$, there is a bounded linear functional $\rho$ on $\mathfrak{X}$ such that $\|\rho\| = 1$, $\rho(y) = 0$ for each $y$ in $\mathscr{Y}$, and $\rho(x_0) = d$ $(> 0)$, where*

$$d = \inf\{\|x_0 + y\| : y \in \mathscr{Y}\},$$

*the distance from $x_0$ to $\mathscr{Y}$.*

*Proof.* The quotient mapping $Q : \mathfrak{X} \to \mathfrak{X}/\mathscr{Y}$ is a bounded linear operator, and $d = \|Qx_0\| > 0$. By Corollary 1.6.2, there is a bounded linear functional $\rho_0$ on $\mathfrak{X}/\mathscr{Y}$ such that $\|\rho_0\| = 1$ and $\rho_0(Qx_0) = d$. The equation $\rho(x) = \rho_0(Qx)$ defines a bounded linear functional $\rho$ on $\mathfrak{X}$; $\rho(x_0) = d$, and $\rho(y) = 0$ for each $y$

in $\mathscr{Y}$. Since $\rho$ has the factorization $\rho_0 Q$ through $\mathfrak{X}/\mathscr{Y}$, it follows from Theorem 1.5.8 that $\|\rho\| = \|\rho_0\| = 1$. ■

1.6.4. THEOREM. *If $\mathfrak{X}$ is a normed space and $x \in \mathfrak{X}$, the equation*

$$\hat{x}(\rho) = \rho(x) \qquad (\rho \in \mathfrak{X}^{\#})$$

*defines a bounded linear functional $\hat{x}$ on the Banach dual space $\mathfrak{X}^{\#}$. The mapping $x \to \hat{x}$ is an isometric isomorphism from $\mathfrak{X}$ onto the subspace $\hat{\mathfrak{X}} = \{\hat{x} : x \in \mathfrak{X}\}$ of the second dual space $\mathfrak{X}^{\#\#}$.*

*Proof.* It is evident that $\hat{x}$, as defined in the theorem, is a linear functional on $\mathfrak{X}^{\#}$, and that the mapping $x \to \hat{x}$ is a linear operator from $\mathfrak{X}$ into the algebraic dual space of $\mathfrak{X}^{\#}$. When $x_0 \in \mathfrak{X}$,

$$|\hat{x}_0(\rho)| = |\rho(x_0)| \leqslant \|\rho\| \, \|x_0\| \qquad (\rho \in \mathfrak{X}^{\#}).$$

When $\rho$ is chosen as in Corollary 1.6.2,

$$|\hat{x}_0(\rho)| = \|x_0\| = \|\rho\| \, \|x_0\|.$$

Thus $\hat{x}_0$ is a bounded linear functional, and $\|\hat{x}_0\| = \|x_0\|$; so the mapping $x \to \hat{x}$ is an isometric isomorphism from $\mathfrak{X}$ onto a subspace $\hat{\mathfrak{X}}$ of $\mathfrak{X}^{\#\#}$. ■

When $\mathfrak{X}$ is a normed space, the mapping $x \to \hat{x}$ occurring in Theorem 1.6.4 is called the *natural isometric isomorphism from $\mathfrak{X}$ into $\mathfrak{X}^{\#\#}$*, and $\hat{\mathfrak{X}}$ is described as the *natural image of $\mathfrak{X}$ in $\mathfrak{X}^{\#\#}$*. The weak* topology $\sigma(\mathfrak{X}^{\#}, \mathfrak{X})$ (as defined, in the discussion preceding Proposition 1.3.5, for locally convex spaces) is the weak topology induced on $\mathfrak{X}^{\#}$ by $\hat{\mathfrak{X}}$.

If $\hat{\mathfrak{X}} = \mathfrak{X}^{\#\#}$, the normed space $\mathfrak{X}$ is said to be *reflexive*. A reflexive normed space $\mathfrak{X}$ is necessarily a Banach space since it is isometrically isomorphic to the Banach dual space $\mathfrak{X}^{\#\#}$. However, many Banach spaces are not reflexive (see, for example, Exercise 1.9.24).

Part (i) of the following result is known as the Alaoglu–Bourbaki theorem.

1.6.5. THEOREM. *Suppose that $\mathfrak{X}$ is a normed space and $\hat{\mathfrak{X}}$ is the natural image of $\mathfrak{X}$ in $\mathfrak{X}^{\#\#}$.*

(i) *The unit ball $(\mathfrak{X}^{\#})_1$ is compact in the weak* topology $\sigma(\mathfrak{X}^{\#}, \mathfrak{X})$ on $\mathfrak{X}^{\#}$.*
(ii) *The weak* closure in $\mathfrak{X}^{\#\#}$ of the unit ball $(\hat{\mathfrak{X}})_1$ of $\hat{\mathfrak{X}}$ is the unit ball $(\mathfrak{X}^{\#\#})_1$ of $\mathfrak{X}^{\#\#}$.*

*Proof.* (i) For each $x$ in $\mathfrak{X}$, let $D_x$ denote the compact subset $\{a : |a| \leqslant \|x\|\}$ of the scalar field $\mathbb{K}$. The product topological space

$$P = \prod_{x \in \mathfrak{X}} D_x$$

is compact, by Tychonoff's theorem. It consists of all functions $p : \mathfrak{X} \to \mathbb{K}$ such that $p(x) \in D_x$ $(x \in \mathfrak{X})$; each element $p_0$ of $P$ has a base of neighborhoods consisting of all sets of the form

$$\{p \in P : |p(x_j) - p_0(x_j)| < \varepsilon \ (j = 1, \ldots, m)\},$$

where $\varepsilon > 0$ and $x_1, \ldots, x_m \in \mathfrak{X}$. For each $x$ in $\mathfrak{X}$, the mapping $p \to p(x) : P \to \mathbb{K}$ is continuous.

The unit ball $(\mathfrak{X}^\#)_1$ consists of all *linear* mappings $\rho : \mathfrak{X} \to \mathbb{K}$ such that $\rho(x) \in D_x$ (that is, $|\rho(x)| \leqslant \|x\|$) for each $x$ in $\mathfrak{X}$. Thus

$$(\mathfrak{X}^\#)_1 = \{p \in P : p(ax + by) - ap(x) - bp(y) = 0 \ (x, y \in \mathfrak{X};\ a, b \in \mathbb{K})\}.$$

From the final sentence of the preceding paragraph, it now follows that $(\mathfrak{X}^\#)_1$ is a closed subset of $P$ and is therefore compact in the relative topology.

In view of the form of the basic neighborhoods of points in $P$ (as described above) and the definition and discussion of the weak* topology (preceding Proposition 1.3.5), it is clear that the relative topology on $(\mathfrak{X}^\#)_1$ as a subset of $P$ coincides with its relative weak* topology as a subset of $\mathfrak{X}^\#$. Thus $(\mathfrak{X}^\#)_1$ is compact in the latter topology.

(ii) From (i), $(\mathfrak{X}^{\#\#})_1$ is compact in the weak* topology $\sigma(\mathfrak{X}^{\#\#}, \mathfrak{X}^\#)$, and so contains the weak* closure $\mathscr{C}$ of its convex subset $(\hat{\mathfrak{X}})_1$; we have to show that $\mathscr{C} = (\mathfrak{X}^{\#\#})_1$. Suppose the contrary, and choose $\sigma_0$ in $(\mathfrak{X}^{\#\#})_1 \backslash \mathscr{C}$; we shall obtain a contradiction. By the Hahn–Banach theorem, there is a weak* continuous linear functional $\omega_0$ on $\mathfrak{X}^{\#\#}$ and a real number $a$ such that

$$\operatorname{Re} \omega_0(\sigma_0) > a, \qquad \operatorname{Re} \omega_0(\sigma) \leqslant a \qquad (\sigma \in \mathscr{C}).$$

By Proposition 1.3.5, there is an element $\rho_0$ of $\mathfrak{X}^\#$ such that $\omega_0(\sigma) = \sigma(\rho_0)$ for all $\sigma$ in $\mathfrak{X}^{\#\#}$.

When $x \in \mathfrak{X}$ and $\|x\| = 1$, we can choose a scalar $b$ such that $|b| = 1$ and

$$|\rho_0(x)| = b\rho_0(x) = b\hat{x}(\rho_0) = \omega_0(b\hat{x}) = \operatorname{Re} \omega_0(b\hat{x});$$

and since $b\hat{x} \in (\hat{\mathfrak{X}})_1 \subseteq \mathscr{C}$, it follows that $|\rho_0(x)| \leqslant a$. Thus $\|\rho_0\| \leqslant a$. However, since $\|\sigma_0\| \leqslant 1$, we have

$$\|\rho_0\| \geqslant |\sigma_0(\rho_0)| \geqslant \operatorname{Re} \sigma_0(\rho_0) = \operatorname{Re} \omega_0(\sigma_0) > a,$$

contradicting the previous inequality. ∎

1.6.6. Corollary. *If $\mathscr{S}$ is a bounded weak* closed subset of the Banach dual space $\mathfrak{X}^\#$ of a normed space $\mathfrak{X}$, then $\mathscr{S}$ is weak* compact. If, in addition, $\mathscr{S}$ is convex, it is the weak* closed convex hull of its extreme points.*

*Proof.* For some positive $r$, $\mathscr{S}$ is a (weak* closed) subset of the ball $(\mathfrak{X}^\#)_r$ $(= r(\mathfrak{X}^\#)_1)$, and this ball is weak* compact by Theorem 1.6.5(i); so $\mathscr{S}$ is weak* compact. The final assertion in the corollary now follows from the Krein–Milman theorem (1.4.3), since $\mathfrak{X}^\#$, with the weak* topology, is a locally convex space. ∎

1.6.7. THEOREM. *A normed space $\mathfrak{X}$ is reflexive if and only if its unit ball $(\mathfrak{X})_1$ is compact in the weak topology.*

*Proof.* It is clear that $\mathfrak{X}$ is reflexive (that is, $\hat{\mathfrak{X}} = \mathfrak{X}^{\#\#}$) if and only if $(\hat{\mathfrak{X}})_1 = (\mathfrak{X}^{\#\#})_1$. By Theorem 1.6.5, $(\mathfrak{X}^{\#\#})_1$ is weak* compact, and is the weak* closure of $(\hat{\mathfrak{X}})_1$. Thus $(\hat{\mathfrak{X}})_1 = (\mathfrak{X}^{\#\#})_1$ if and only if $(\hat{\mathfrak{X}})_1$ is weak* compact.

The natural isometric isomorphism $x \to \hat{x} : \mathfrak{X} \to \hat{\mathfrak{X}}\ (\subseteq \mathfrak{X}^{\#\#})$ carries $(\mathfrak{X})_1$ onto $(\hat{\mathfrak{X}})_1$. When $x_0 \in \mathfrak{X}$, $\rho_1, \ldots, \rho_m \in \mathfrak{X}^{\#}$, and $\varepsilon > 0$, it carries the basic neighborhood

$$\{x \in \mathfrak{X} : |\rho_j(x) - \rho_j(x_0)| < \varepsilon\ (j = 1, \ldots, m)\}$$

of $x_0$ (in the weak topology on $\mathfrak{X}$) onto the basic neighborhood

$$\{\hat{x} \in \hat{\mathfrak{X}} : |\hat{x}(\rho_j) - \hat{x}_0(\rho_j)| < \varepsilon\ (j = 1, \ldots, m)\}$$

of $\hat{x}_0$ (in the relative weak* topology on $\hat{\mathfrak{X}}$, as a subset of $\mathfrak{X}^{\#\#}$). It is therefore a homeomorphism between $\mathfrak{X}$ and $\hat{\mathfrak{X}}$, with the topologies just mentioned. Thus $(\hat{\mathfrak{X}})_1$ is weak* compact if and only if $(\mathfrak{X})_1$ is weakly compact. ■

Suppose that $\mathfrak{X}$ and $\mathscr{Y}$ are normed spaces and $T : \mathfrak{X} \to \mathscr{Y}$ is a bounded linear operator. If $\rho$ is a continuous linear functional on $\mathscr{Y}$, the composite mapping $\rho \circ T$ is a continuous linear functional on $\mathfrak{X}$. Accordingly, we can define a mapping $T^{\#} : \mathscr{Y}^{\#} \to \mathfrak{X}^{\#}$ by

$$T^{\#}\rho = \rho \circ T \qquad (\rho \in \mathscr{Y}^{\#}).$$

We assert that $T^{\#}$ is a bounded linear operator and that $\|T^{\#}\| = \|T\|$. The linearity of $T^{\#}$ follows from the fact that

$$(a_1\rho_1 + a_2\rho_2) \circ T = a_1(\rho_1 \circ T) + a_2(\rho_2 \circ T),$$

when $\rho_1, \rho_2 \in \mathscr{Y}^{\#}$ and $a_1, a_2$ are scalars. For each $\rho$ in $\mathscr{Y}^{\#}$,

$$|(T^{\#}\rho)(x)| = |\rho(Tx)| \leqslant \|\rho\|\,\|Tx\| \leqslant \|\rho\|\,\|T\|\,\|x\| \qquad (x \in \mathfrak{X}),$$

and thus $\|T^{\#}\rho\| \leqslant \|T\|\,\|\rho\|$; so $T^{\#}$ is bounded, and $\|T^{\#}\| \leqslant \|T\|$. To prove the reverse inequality, it suffices to show that $\|Tx\| \leqslant \|T^{\#}\|\,\|x\|$ for each $x$ in $\mathfrak{X}$. Given such $x$, it follows from Corollary 1.6.2 that we can choose $\rho$ in $\mathscr{Y}^{\#}$ so that $\|\rho\| = 1$ and $\rho(Tx) = \|Tx\|$; and

$$\begin{aligned} \|Tx\| &= |\rho(Tx)| = |(T^{\#}\rho)(x)| \\ &\leqslant \|T^{\#}\rho\|\,\|x\| \leqslant \|T^{\#}\|\,\|\rho\|\,\|x\| = \|T^{\#}\|\,\|x\|, \end{aligned}$$

as required.

When $T_1, T_2 : \mathfrak{X} \to \mathscr{Y}$ are bounded linear operators and $\rho \in \mathscr{Y}^{\#}$, it results from the linearity of $\rho$ that

$$\begin{aligned} (a_1T_1 + a_2T_2)^{\#}\rho &= \rho \circ (a_1T_1 + a_2T_2) \\ &= a_1(\rho \circ T_1) + a_2(\rho \circ T_2) = a_1T_1^{\#}\rho + a_2T_2^{\#}\rho \end{aligned}$$

for all scalars $a_1, a_2$. Thus

$$(a_1T_1 + a_2T_2)^\# = a_1T_1^\# + a_2T_2^\#,$$

and the mapping $T \to T^\#$ is a norm-preserving linear operator from $\mathscr{B}(\mathfrak{X}, \mathscr{Y})$ into $\mathscr{B}(\mathscr{Y}^\#, \mathfrak{X}^\#)$.

If $\mathfrak{X}$, $\mathscr{Y}$, $\mathscr{Z}$ are normed spaces, and $S \in \mathscr{B}(\mathscr{Y}, \mathscr{Z})$, $T \in \mathscr{B}(\mathfrak{X}, \mathscr{Y})$, then

$$(ST)^\#\rho = \rho \circ (ST) = \rho \circ (S \circ T) = (\rho \circ S) \circ T = T^\#(S^\#\rho)$$

for each $\rho$ in $\mathscr{Z}^\#$; so $(ST)^\# = T^\#S^\#$.

The operator $T^\#: \mathscr{Y}^\# \to \mathfrak{X}^\#$ is called the (*Banach*) *adjoint* of the bounded linear operator $T: \mathfrak{X} \to \mathscr{Y}$. When $\mathfrak{X}$ and $\mathscr{Y}$ are Hilbert spaces, there is another (and, in that case, more important) adjoint operator, the (*Hilbert*) *adjoint* $T^*: \mathscr{Y} \to \mathfrak{X}$, which will be described in Section 2.4.

1.6.8. PROPOSITION. *If $T$ is a bounded linear operator from a normed space $\mathfrak{X}$ into another such space $\mathscr{Y}$, then $T^\#$ is continuous relative to the weak* topologies on $\mathscr{Y}^\#$ and $\mathfrak{X}^\#$.*

*Proof.* We use the criterion set out in Proposition 1.3.2. The weak* topology on $\mathfrak{X}^\#$ is $\sigma(\mathfrak{X}^\#, \hat{\mathfrak{X}})$, where $\hat{\mathfrak{X}}$ is the natural image of $\mathfrak{X}$ in $\mathfrak{X}^{\#\#}$. Accordingly, it suffices to show that the linear functionals $\hat{x} \circ T^\#$ $(x \in \mathfrak{X})$ on $\mathscr{Y}^\#$ are weak* continuous. Suppose $x \in \mathfrak{X}$, and let $y = Tx$; for each $\rho$ in $\mathscr{Y}^\#$,

$$\begin{aligned}(\hat{x} \circ T^\#)(\rho) &= \hat{x}(T^\#\rho) = (T^\#\rho)(x) \\ &= \rho(Tx) = \rho(y) = \hat{y}(\rho).\end{aligned}$$

Thus $\hat{x} \circ T^\# = \hat{y} \in \hat{\mathscr{Y}}$, and therefore $\hat{x} \circ T^\#$ is continuous in the weak* topology $\sigma(\mathscr{Y}^\#, \hat{\mathscr{Y}})$. ■

## 1.7. Some examples of Banach spaces

In this section we describe some of the Banach spaces that will be used in later chapters. In some cases, we shall first indicate a general process by which, from a given Banach space $\mathfrak{X}$, another such space can be constructed; we then obtain specific examples by taking $\mathfrak{X} = \mathbb{R}$ or $\mathbb{C}$.

When $f$, $g$ are mappings from a set $\mathbb{A}$ into a Banach space $\mathfrak{X}$, and $c$ is a scalar, $f + g$ and $cf$ will always denote the mappings defined by

$$(f + g)(a) = f(a) + g(a), \qquad (cf)(a) = cf(a) \qquad (a \in \mathbb{A}).$$

1.7.1. EXAMPLE. *$l_\infty$ spaces.* If $\mathbb{A}$ is a set and $\mathfrak{X}$ is a Banach space with scalar field $\mathbb{K}$, we denote by $l_\infty(\mathbb{A}, \mathfrak{X})$ the set of all functions $f: \mathbb{A} \to \mathfrak{X}$ such that $\sup\{\|f(a)\| : a \in \mathbb{A}\} < \infty$. Given two such functions, $f$ and $g$, and a scalar $c$,

$f+g$ and $cf$ are functions of the same type. Thus $l_\infty(\mathbb{A},\mathfrak{X})$ is a linear space over $\mathbb{K}$, and it has a norm defined by

$$\|f\| = \sup\{\|f(a)\| : a \in \mathbb{A}\}.$$

*We shall show that, with this norm,* $l_\infty(\mathbb{A},\mathfrak{X})$ *is a Banach space.*

To this end, suppose that $\{f_n\}$ is a Cauchy sequence in $l_\infty(\mathbb{A},\mathfrak{X})$; we have to show that it converges to some element $f$ of $l_\infty(\mathbb{A},\mathfrak{X})$. Given any positive $\varepsilon$, there is a positive integer $N(\varepsilon)$ such that $\|f_m - f_n\| \leqslant \varepsilon$ whenever $m > n \geqslant N(\varepsilon)$. Hence

$$(1) \qquad \|f_m(a) - f_n(a)\| \leqslant \|f_m - f_n\| \leqslant \varepsilon \qquad (a \in \mathbb{A}, \quad m > n \geqslant N(\varepsilon)).$$

Accordingly, for each $a$ in $\mathbb{A}$, $\{f_n(a)\}$ is a Cauchy sequence in the Banach space $\mathfrak{X}$, and so converges; we can define a mapping $f : \mathbb{A} \to \mathfrak{X}$ by $f(a) = \lim f_n(a)$. When $m \to \infty$ in (1), we obtain

$$(2) \qquad \|f(a) - f_n(a)\| \leqslant \varepsilon \qquad (a \in \mathbb{A}, \quad n \geqslant N(\varepsilon)).$$

Thus $\|f(a)\| \leqslant \varepsilon + \|f_n(a)\| \leqslant \varepsilon + \|f_n\|$, when $a \in \mathbb{A}$ and $n \geqslant N(\varepsilon)$; so $\sup\{\|f(a)\| : a \in \mathbb{A}\} < \infty$, and $f \in l_\infty(\mathbb{A},\mathfrak{X})$. Since $\|f - f_n\| \leqslant \varepsilon$ when $n \geqslant N(\varepsilon)$, by (2), $\{f_n\}$ converges to $f$; and $l_\infty(\mathbb{A},\mathfrak{X})$ is a Banach space.

If a sequence (or net) $\{f_n\}$ converges in $l_\infty(\mathbb{A},\mathfrak{X})$, with limit $f$, then

$$\sup_{a \in \mathbb{A}} \|f(a) - f_n(a)\| = \|f - f_n\| \to 0;$$

that is, $f_n(a) \to f(a)$ uniformly on $\mathbb{A}$. Convergence in the Banach space $l_\infty(\mathbb{A},\mathfrak{X})$ is *uniform convergence on* $\mathbb{A}$.

By taking $\mathfrak{X}$ to be $\mathbb{R}$ or $\mathbb{C}$, the construction just described gives rise to Banach spaces $l_\infty(\mathbb{A},\mathbb{R})$ and $l_\infty(\mathbb{A},\mathbb{C})$; the latter is usually denoted by $l_\infty(\mathbb{A})$. ■

1.7.2. Example. *Spaces of continuous functions.* If $S$ is a topological space and $\mathfrak{X}$ is a Banach space with scalar field $\mathbb{K}$, we denote by $C(S,\mathfrak{X})$ the set of all continuous functions $f : S \to \mathfrak{X}$ such that $\sup\{\|f(s)\| : s \in S\} < \infty$. It is apparent that $C(S,\mathfrak{X})$ is a subspace of the Banach space $l_\infty(S,\mathfrak{X})$ defined in Example 1.7.1; we shall show that it is a *closed* subspace. From this it follows that $C(S,\mathfrak{X})$ *becomes a Banach space when the norm is defined by*

$$\|f\| = \sup\{\|f(s)\| : s \in S\}.$$

Suppose, then, that $f \in l_\infty(S,\mathfrak{X})$, and $f$ is the limit of a net $\{f_n\}$ of elements of the subspace $C(S,\mathfrak{X})$. Then $f_n(s) \to f(s)$ uniformly on $S$, and since $f$ is a uniform limit of continuous functions, it too is continuous on $S$. Thus $f \in C(S,\mathfrak{X})$, and $C(S,\mathfrak{X})$ is closed in $l_\infty(S,\mathfrak{X})$.

We shall usually be interested in the case in which $S$ is a compact Hausdorff space. In this case, $C(S,\mathfrak{X})$ consists of *all* continuous functions $f : S \to \mathfrak{X}$.

Indeed, given any such $f$, the mapping $s \to \|f(s)\|: S \to \mathbb{R}$ is continuous, and is therefore bounded on $S$; so $\sup\{\|f(s)\|: s \in S\} < \infty$.

By taking $\mathfrak{X} = \mathbb{R}$ and $\mathbb{C}$, we obtain Banach spaces $C(S, \mathbb{R})$ and $C(S, \mathbb{C})$; the latter is usually denoted by $C(S)$. ■

In discussing the next example, we shall make use of *Minkowski's inequality* [R: p. 62, Theorem 3.5]: if $1 \leqslant p < \infty$, and $x_1, \ldots, x_n, y_1, \ldots, y_n \in \mathbb{C}$, then

$$\left\{\sum_{j=1}^{n} |x_j + y_j|^p\right\}^{1/p} \leqslant \left\{\sum_{j=1}^{n} |x_j|^p\right\}^{1/p} + \left\{\sum_{j=1}^{n} |y_j|^p\right\}^{1/p}. \tag{3}$$

The inequality extends at once to infinite sums; if $f$ and $g$ are complex-valued functions defined on a set $\mathbb{A}$ and the sums $\sum |f(a)|^p$, $\sum |g(a)|^p$ converge, then so does $\sum |f(a) + g(a)|^p$, and

$$\left\{\sum_{a \in \mathbb{A}} |f(a) + g(a)|^p\right\}^{1/p} \leqslant \left\{\sum_{a \in \mathbb{A}} |f(a)|^p\right\}^{1/p} + \left\{\sum_{a \in \mathbb{A}} |g(a)|^p\right\}^{1/p}. \tag{4}$$

For this, it suffices to observe that (by Minkowski's inequality for finite sums), the net of finite subsums of $\sum |f(a) + g(a)|^p$ is bounded above by

$$\left[\left\{\sum_{a \in \mathbb{A}} |f(a)|^p\right\}^{1/p} + \left\{\sum_{a \in \mathbb{A}} |g(a)|^p\right\}^{1/p}\right]^p.$$

1.7.3. Example. *$l_p$ spaces.* If $\mathbb{A}$ is a set, $\mathfrak{X}$ is a Banach space with scalar field $\mathbb{K}$, and $1 \leqslant p < \infty$, we denote by $l_p(\mathbb{A}, \mathfrak{X})$ the set of all functions $f: \mathbb{A} \to \mathfrak{X}$ such that $\sum_{a \in \mathbb{A}} \|f(a)\|^p < \infty$. Given two such functions, $f$ and $g$, it follows from Minkowski's inequality that

$$\begin{aligned}\left\{\sum_{a \in \mathbb{A}} \|f(a) + g(a)\|^p\right\}^{1/p} &\leqslant \left\{\sum_{a \in \mathbb{A}} [\|f(a)\| + \|g(a)\|]^p\right\}^{1/p} \\ &\leqslant \left\{\sum_{a \in \mathbb{A}} \|f(a)\|^p\right\}^{1/p} + \left\{\sum_{a \in \mathbb{A}} \|g(a)\|^p\right\}^{1/p} < \infty;\end{aligned}$$

so

$$f + g \in l_p(\mathbb{A}, \mathfrak{X}).$$

Also, $cf \in l_p(\mathbb{A}, \mathfrak{X})$, and

$$\left\{\sum_{a \in \mathbb{A}} \|cf(a)\|^p\right\}^{1/p} = |c| \left\{\sum_{a \in \mathbb{A}} \|f(a)\|^p\right\}^{1/p}$$

for each scalar $c$. Accordingly, $l_p(\mathbb{A}, \mathfrak{X})$ is a linear space over $\mathbb{K}$, and has a norm defined by

$$\|f\| = \left\{ \sum_{a \in \mathbb{A}} \|f(a)\|^p \right\}^{1/p}.$$

*We shall show that, with this norm, $l_p(\mathbb{A}, \mathfrak{X})$ is a Banach space.*

To this end, suppose that $\{f_n\}$ is a Cauchy sequence in $l_p(\mathbb{A}, \mathfrak{X})$. Given any positive $\varepsilon$, there is a positive integer $N(\varepsilon)$ such that $\|f_m - f_n\| \leqslant \varepsilon$ whenever $m > n \geqslant N(\varepsilon)$; that is,

$$\sum_{a \in \mathbb{A}} \|f_m(a) - f_n(a)\|^p \leqslant \varepsilon^p \qquad (m > n \geqslant N(\varepsilon)). \tag{5}$$

It follows that $\|f_m(a) - f_n(a)\| \leqslant \varepsilon$ when $m > n \geqslant N(\varepsilon)$ and $a \in \mathbb{A}$; so, for each $a$ in $\mathbb{A}$, $\{f_n(a)\}$ is a Cauchy sequence in the Banach space $\mathfrak{X}$, and therefore converges. Accordingly, we can define a mapping $f\colon \mathbb{A} \to \mathfrak{X}$ by $f(a) = \lim f_n(a)$. For each finite subset $\mathbb{F}$ of $\mathbb{A}$, it results from (5) that

$$\sum_{a \in \mathbb{F}} \|f_m(a) - f_n(a)\|^p \leqslant \varepsilon^p \qquad (m > n \geqslant N(\varepsilon)).$$

When $m \to \infty$, we obtain

$$\sum_{a \in \mathbb{F}} \|f(a) - f_n(a)\|^p \leqslant \varepsilon^p \qquad (n \geqslant N(\varepsilon));$$

and since this last inequality is satisfied for every finite subset $\mathbb{F}$ of $\mathbb{A}$,

$$\sum_{a \in \mathbb{A}} \|f(a) - f_n(a)\|^p \leqslant \varepsilon^p \qquad (n \geqslant N(\varepsilon)). \tag{6}$$

Thus

$$f - f_n \in l_p(\mathbb{A}, \mathfrak{X})$$

when $n \geqslant N(\varepsilon)$, and therefore $f = (f - f_n) + f_n \in l_p(\mathbb{A}, \mathfrak{X})$. By (6), $\|f - f_n\| \leqslant \varepsilon$ whenever $n \geqslant N(\varepsilon)$, so $\{f_n\}$ converges to $f$. This proves that $l_p(\mathbb{A}, \mathfrak{X})$ is a Banach space.

The construction just described gives rise, in particular, to Banach spaces $l_p(\mathbb{A}, \mathbb{R})$ and $l_p(\mathbb{A}, \mathbb{C})$; the latter is usually denoted by $l_p(\mathbb{A})$. ■

By taking for $\mathbb{A}$ the set $\{1, 2, \ldots, n\}$ in Examples 1.7.1 and 1.7.3, it follows that there are norms $\|\ \|_p$ $(1 \leqslant p \leqslant \infty)$ on the linear space $\mathbb{K}^n$ (where $\mathbb{K}$ is $\mathbb{R}$ or $\mathbb{C}$), defined by the equations

$$\|(c_1, \ldots, c_n)\|_p = [|c_1|^p + \cdots + |c_n|^p]^{1/p} \qquad (1 \leqslant p < \infty),$$

$$\|(c_1, \ldots, c_n)\|_\infty = \max\{|c_1|, \ldots, |c_n|\}.$$

(In the case of $\|\ \|_1$ and $\|\ \|_\infty$, this has already been noted near the beginning of Section 1.5, the two norms being denoted there by $p_1, p_\infty$.) It is easily verified that each of these norms gives rise to the usual product topology on $\mathbb{K}^n$ (necessarily so, by Proposition 1.2.16).

Our next two examples are drawn from measure theory, and we refer to [H, R] as standard sources of information on this subject. For the sake of simplicity we confine attention to $\sigma$-finite measures, which suffice for our purposes. Accordingly, we shall assume throughout the remainder of this section that *$m$ is a $\sigma$-finite measure defined on a $\sigma$-algebra $\mathscr{S}$ of subsets of a set $S$.*

1.7.4. EXAMPLE. *$L_\infty$ spaces.* We denote by $L_\infty$ $(= L_\infty(S, \mathscr{S}, m))$ the set of all measurable complex-valued functions $f$ on $S$ that are *essentially bounded* in the following sense: there is a positive real number $c$ such that $|f(s)| \leqslant c$ for almost all $s$ in $S$. It is evident that $L_\infty$ is a complex vector space, containing as a subspace the set $N$ of all null functions (those that vanish almost everywhere) on $S$. We shall observe that, *with a suitable norm* (*defined in* (7) *below*), *the quotient space $L_\infty/N$ is a Banach space.*

It is easily verified that, when $f \in L_\infty$, there is a *smallest* constant $c$ such that $|f(s)| \leqslant c$ almost everywhere on $S$ [R: p. 64]. It is denoted by $\operatorname{ess\,sup}\{|f(s)| : s \in S\}$, the *essential supremum* of $|f|$; and it is 0 only when $f$ is a null function. There is an equivalence relation $\sim$ on $L_\infty$, in which $f \sim g$ if and only if $f(s) = g(s)$ for almost all $s$ in $S$; and the equivalence class $[f]$ of $f$ is the coset $f + N$. The equation

$$\|[f]\| = \operatorname{ess\,sup}\{|f(s)| : s \in S\} \qquad (f \in L_\infty) \tag{7}$$

defines a norm on $L_\infty/N$. It is a straightforward result in measure theory that, with this norm, $L_\infty/N$ is complete, and is therefore a Banach space [R: p. 66, Theorem 3.11].

There is a common convention, which we shall usually follow, that the Banach space just defined is denoted by $L_\infty$ (rather than $L_\infty/N$), and that its elements are described as functions (although, strictly speaking, they are *equivalence classes* of functions, modulo null functions). This convention is convenient, and should not lead to confusion, provided it is remembered that two essentially bounded functions must be regarded as the same member of $L_\infty$ if they are equal almost everywhere.

It is not difficult to verify that a sequence $\{f_n\}$ in $L_\infty$ converges, in the norm topology, if and only if there is a null set $Z$ such that the functions $\{f_n(s)\}$ converge uniformly on $S \backslash Z$ [R: p. 67]. If $m$ is a regular Borel measure on a compact Hausdorff space $S$ and $f \in L_\infty$, there is a sequence $\{f_n\}$ of continuous functions on $S$, such that $f(s) = \lim f_n(s)$ almost everywhere on $S$ and $|f_n(s)| \leqslant \|f\|$ for all $s$ in $S$ and all $n = 1, 2, \ldots$ [R: p. 54, Corollary]. ∎

1.7.5. EXAMPLE. *$L_p$ spaces.* Suppose that $1 \leqslant p < \infty$, and denote by $L_p$ $(= L_p(S, \mathscr{S}, m))$ the set of all measurable complex-valued functions $f$ on $S$ for which $\int_S |f(s)|^p \, dm(s)$ is finite. When $f$, $g \in L_p$, it follows from Minkowski's inequality for integrals [R: p. 62, Theorem 3.5] that $f+g \in L_p$ and

$$\left\{\int_S |f(s)+g(s)|^p\,dm(s)\right\}^{1/p} \leqslant \left\{\int_S |f(s)|^p\,dm(s)\right\}^{1/p} + \left\{\int_S |g(s)|^p\,dm(s)\right\}^{1/p}.$$

From this, it is evident that $L_p$ is a complex vector space, which contains as a subspace the set $N$ of all null functions. The quotient space $L_p/N$ has a norm defined by

$$\|[f]\| = \left\{\int_S |f(s)|^p\,dm(s)\right\}^{1/p} \qquad (f \in L_p), \tag{8}$$

where $[f]$ denotes the coset $f+N$. It is a result in measure theory [R: p. 66, Theorem 3.11] that, *with this norm, $L_p/N$ is complete and is therefore a Banach space.*

Just as in the preceding example, we shall adopt a convenient (though not *strictly* accurate) convention, by referring to the Banach space $L_p$ (rather than $L_p/N$) and describing its elements as functions (rather than equivalence classes of functions).

We note three further results from measure theory [R: pp. 67, 68; Theorems 3.12, 3.13, 3.14] concerning properties of $L_p$ spaces. First, a sequence $\{f_n\}$ in $L_p$ that converges in the norm topology to an element $f$ has a subsequence that converges almost everywhere to $f$. Second, the set of all simple functions in $L_p$ is an everywhere-dense subset of $L_p$. Finally, if $m$ is a regular Borel measure on a compact Hausdorff space $S$, the continuous functions form an everywhere-dense subset of $L_p$. ∎

1.7.6. Remark. Several of the Banach spaces described above will be used frequently in later chapters. Our primary concern in this book is with a certain class of Banach spaces (namely, Hilbert spaces), and with certain algebras ($C^*$-algebras) that can be represented as algebras of bounded linear operators acting on Hilbert spaces. Now $l_2$ spaces and $L_2$ spaces are the simplest examples of Hilbert spaces (see Examples 2.1.12 and 2.1.14). Moreover, Banach spaces of the types $l_\infty$, $L_\infty$, $C(S)$ can be provided with additional algebraic structure (multiplication and involution), and then become abelian $C^*$-algebras.

As one would expect, our use of $L_2$ and $L_\infty$ spaces involves occasional appeal to results from measure theory. In fact, measure theory will sometimes be needed also in connection with $C(S)$, where $S$ is a compact Hausdorff space. One of the main tools in the study of $C^*$-algebras is the theory of positive linear functionals; and the *Riesz representation theorem* [R: p. 40, Theorem 2.14] asserts that there is a one-to-one correspondence between positive linear functionals $\rho$ on $C(S)$ and regular Borel measures $m$ on $S$, determined by the equation

$$\rho(f) = \int_S f(s)\, dm(s) \qquad (f \in C(S)).$$

(We shall always assume, as one of the defining properties of a regular Borel measure $m$ on a topological space $S$, that $m(K) < \infty$ for each compact subset $K$ of $S$. In the present case, $S$ itself is compact, and thus $m(S) < \infty$.) ■

Our next objective, achieved in Theorem 1.7.8 below, is to show that $L_\infty$ can be identified with the Banach dual space of $L_1$. For this purpose we shall require the following simple result concerning $\sigma$-finite measures.

1.7.7. Lemma. *Suppose that $c > 0$ and $g$ is a measurable complex-valued function on $S$. If, for every measurable set $X$ $(\subseteq S)$ of finite measure, $g$ is integrable over $X$ and*

$$\left| \int_X g(s)\, dm(s) \right| \leqslant cm(X),$$

*then $|g(s)| \leqslant c$ for almost all $s$ in $S$.*

*Proof.* We have to show that the set $Y = \{s \in S: |g(s)| > c\}$ is null. Let $\{z_1, z_2, \ldots\}$ be a countable everywhere-dense subset of the unit circle

$$\{z \in \mathbb{C}: |z| = 1\},$$

and note that

$$Y = \bigcup_{j,k=1}^{\infty} Y_{jk},$$

where $Y_{jk} = \{s \in S: \operatorname{Re} z_j g(s) \geqslant c + 1/k\}$. Thus it suffices to show that each of the sets $Y_{jk}$ is null.

If $Y_{jk}$ is not null, it has a measurable subset $X$ such that $0 < m(X) < \infty$ (since $m$ is $\sigma$-finite). Then,

$$\begin{aligned}
\left| \int_X g(s)\, dm(s) \right| &= \left| \int_X z_j g(s)\, dm(s) \right| \\
&\geqslant \int_X \operatorname{Re} z_j g(s)\, dm(s) \\
&\geqslant \int_X \left( c + \frac{1}{k} \right) dm(s) = \left( c + \frac{1}{k} \right) m(X) > cm(X),
\end{aligned}$$

a contradiction; so each $Y_{jk}$ is a null set. ■

1.7.8. THEOREM. *Suppose that $m$ is a $\sigma$-finite measure defined on a $\sigma$-algebra $\mathscr{S}$ of subsets of a set $S$. For each $g$ in $L_\infty$, the equation*

$$\rho_g(f) = \int_S f(s)g(s)\,dm(s) \qquad (f \in L_1)$$

*defines a bounded linear functional $\rho_g$ on the Banach space $L_1$, and the mapping $g \to \rho_g$ is an isometric isomorphism from $L_\infty$ onto the Banach dual space $(L_1)^\sharp$.*

*Proof.* The usual norms on $L_1$ and $L_\infty$ will be denoted by $\|\ \|_1$ and $\|\ \|_\infty$ respectively. When $f \in L_1$ and $g \in L_\infty$, the function $fg$ is measurable, and

$$|f(s)g(s)| \leqslant \|g\|_\infty |f(s)|$$

for almost all $s$ in $S$. Thus $fg \in L_1$ and

$$\left| \int_S f(s)g(s)\,dm(s) \right| \leqslant \|g\|_\infty \int_S |f(s)|\,dm(s) = \|g\|_\infty \|f\|_1 .$$

From this, it follows easily that $\rho_g$, as defined in the theorem, is a bounded linear functional on $L_1$, with $\|\rho_g\| \leqslant \|g\|_\infty$. When $X (\subseteq S)$ is a measurable set of finite measure, its characteristic function $\chi_X$ is an element of $L_1$, and

$$\begin{aligned} \left| \int_X g(s)\,dm(s) \right| &= \left| \int_S \chi_X(s)g(s)\,dm(s) \right| = |\rho_g(\chi_X)| \\ &\leqslant \|\rho_g\|\,\|\chi_X\|_1 = \|\rho_g\| m(X). \end{aligned}$$

From Lemma 1.7.7, $|g(s)| \leqslant \|\rho_g\|$ for almost all $s$ in $S$; so $\|g\|_\infty \leqslant \|\rho_g\|$, and therefore $\|\rho_g\| = \|g\|_\infty$.

The mapping $g \to \rho_g$ is linear, and from the preceding paragraph it is an isometric isomorphism from $L_\infty$ onto a subspace of $(L_1)^\sharp$. It remains to prove that its range is the whole of $(L_1)^\sharp$.

Suppose that $\rho \in (L_1)^\sharp$; we shall show in due course that $\rho = \rho_g$ for some $g$ in $L_\infty$. Observe first that, if $\{Y_n\}$ is an increasing sequence of measurable sets whose union $Y$ has finite measure, then

$$\|\chi_Y - \chi_{Y_n}\|_1 = m(Y \setminus Y_n) \to 0;$$

and, since $\rho$ is continuous, it follows that

$$\rho(\chi_Y) = \lim_{n \to \infty} \rho(\chi_{Y_n}).$$

We now choose (and, for the moment, fix) a measurable set $X$ of finite measure, and we define a complex-valued function $\mu$ on $\mathscr{S}$ by

$$\mu(Y) = \rho(\chi_{X \cap Y}) \qquad (Y \in \mathscr{S}). \tag{9}$$

It is apparent that $\mu$ is a finitely additive set function, and

$$\mu(Y) = \mu(Y \cap X), \qquad \mu(Z) = 0 \qquad (Y, Z \in \mathscr{S}, \quad m(Z) = 0). \tag{10}$$

Moreover, if $\{Y_n\}$ is an increasing sequence of measurable sets with union $Y$, it results from the preceding paragraph that

$$\mu(Y) = \rho(\chi_{X \cap Y}) = \lim_{n\to\infty} \rho(\chi_{X \cap Y_n}) = \lim_{n\to\infty} \mu(Y_n).$$

Thus $\mu$ is a complex measure on $\mathscr{S}$. Since it vanishes on the null sets of $m$, it follows from the Radon–Nikodým theorem that there is an element $h$ of $L_1$ such that

$$\mu(Y) = \int_Y h(s)\, dm(s) \qquad (Y \in \mathscr{S}). \tag{11}$$

Now

$$\left| \int_Y h(s)\, dm(s) \right| = |\mu(Y)| = |\rho(\chi_{X \cap Y})|$$

$$\leqslant \|\rho\|\, \|\chi_{X\cap Y}\|_1 = \|\rho\| m(X \cap Y) \leqslant \|\rho\| m(Y) \qquad (Y \in \mathscr{S}).$$

By Lemma 1.7.7, $|h(s)| \leqslant \|\rho\|$ for almost all $s$ in $S$; moreover, $\int_Y h(s)\, dm(s) = 0$ whenever $Y$ is a measurable subset of $S \backslash X$, and thus $h(s) = 0$ almost everywhere on $S \backslash X$. When the values of $h$ are suitably adjusted on a null set, (11) remains valid, and in addition

$$h \in L_1, \qquad |h(s)| \leqslant \|\rho\| \quad (s \in S), \qquad h(t) = 0 \quad (t \in S \backslash X). \tag{12}$$

Since $m$ is $\sigma$-finite, $S$ is the disjoint union of a sequence $\{X_1, X_2, \ldots\}$ of measurable sets of finite measure. For each $X_n$, we can use the process described in the preceding paragraph to obtain a complex measure $\mu_n$ and the corresponding Radon–Nikodým derivative $h_n$; and $X_n$, $h_n$, $\mu_n$ satisfy conditions analogous to (9), (11), and (12). Each $s$ in $S$ lies in exactly one of the sets $X_n$; so the sequence $\{h_n(s)\}$ has at most one non-zero term, and

$$\sum_{n=1}^{\infty} |h_n(s)| \leqslant \|\rho\|.$$

The equation

$$g(s) = \sum_{n=1}^{\infty} h_n(s)$$

defines a bounded measurable function $g$ on $S$ (so $g \in L_\infty$).

We shall prove that $\rho_g = \rho$. To this end, it suffices to show that $\rho_g(f) = \rho(f)$ whenever $f$ is the characteristic function of a measurable set $Y$ ($\subseteq S$) of finite measure; for $\rho$ and $\rho_g$ are continuous linear functionals, and linear combinations of such functions $f$ form an everywhere-dense subset of $L_1$

(the integrable simple functions). Now $Y = \bigcup Y_n$, where

$$Y_n = Y \cap (X_1 \cup X_2 \cup \cdots \cup X_n) \qquad (n = 1, 2, \ldots).$$

Thus

$$\begin{aligned}\rho(\chi_Y) &= \lim_{n\to\infty} \rho(\chi_{Y_n}) = \lim_{n\to\infty} \rho\left(\sum_{j=1}^{n} \chi_{Y\cap X_j}\right) \\ &= \lim_{n\to\infty} \sum_{j=1}^{n} \rho(\chi_{Y\cap X_j}) = \sum_{j=1}^{\infty} \mu_j(Y) \\ &= \sum_{j=1}^{\infty} \int_Y h_j(s)\, dm(s) = \int_Y g(s)\, dm(s);\end{aligned}$$

the last step follows from the dominated convergence theorem, since $m(Y) < \infty$. Hence $\rho(\chi_Y) = \rho_g(\chi_Y)$, as required. ■

A topological space is said to be *separable* if it has a countable everywhere-dense subset. It is not difficult to verify that a metric space is separable if and only if its topology has a countable base [K: p. 120, Theorem 11]; and from this, it follows that a subset of a separable metric space is itself separable in the relative topology.

In proving the two following results, we shall use the term *rational complex number* to describe a complex number whose real and imaginary parts are both rational.

1.7.9. PROPOSITION. *If $\mathbb{A}$ is a set and $1 \leqslant p < \infty$, the Banach space $l_p(\mathbb{A})$ is separable if and only if $\mathbb{A}$ is countable.*

*Proof.* If $\mathbb{A}$ is countable, we may enumerate it as a (possibly finite) sequence $\{a_1, a_2, \ldots\}$. For each positive integer $r$, let $Y_r$ be the set of all functions $y$ on $\mathbb{A}$ such that $y(a_n)$ is a rational complex number when $1 \leqslant n \leqslant r$ and $y(a_n) = 0$ if $n > r$. Since each $Y_r$ is a countable subset of $l_p(\mathbb{A})$, so is the set $Y = \bigcup Y_r$.

If $x \in l_p(\mathbb{A})$ and $\varepsilon > 0$, we can choose first a positive integer $r$ and then rational complex numbers $c_1, \ldots, c_r$ so that

$$\sum_{n>r} |x(a_n)|^p < \tfrac{1}{2}\varepsilon^p, \qquad \sum_{n=1}^{r} |x(a_n) - c_n|^p < \tfrac{1}{2}\varepsilon^p.$$

If $y(a_n) = c_n$ $(1 \leqslant n \leqslant r)$ and $y(a_n) = 0$ $(n > r)$, then $y \in Y_r$ $(\subseteq Y)$ and $\|x - y\| < \varepsilon$. Hence $Y$ is an everywhere-dense countable subset of $l_p(\mathbb{A})$.

Conversely, if $l_p(\mathbb{A})$ is separable, the same is true of its subset $X = \{x_a : a \in \mathbb{A}\}$, where $x_a$ denotes the function whose value is 1 at $a$ and 0 elsewhere on $\mathbb{A}$. Thus $X$ has an everywhere-dense countable subset, which must be the whole of $X$, since $\|x_a - x_b\| = 2^{1/p}$ when $a \neq b$. In other words, $X$ (and therefore, also, $\mathbb{A}$) is countable. ■

1.7.10. Proposition. *Suppose that $1 \leqslant p < \infty$ and $m$ is a $\sigma$-finite measure defined on a $\sigma$-algebra $\mathscr{S}$ of subsets of a set $S$. Then the Banach space $L_p(S, \mathscr{S}, m)$ is separable if and only if there is a sequence $\{X_1, X_2, \ldots\}$ of measurable sets of finite measure with the following property: given any measurable set $X$ of finite measure and any positive $\varepsilon$, there is an integer $j$ for which*

$$m((X_j \backslash X) \cup (X \backslash X_j)) < \varepsilon.$$

*Proof.* Let $\mathscr{S}_0$ be the family of all measurable sets of finite measure. When $X, Y \in \mathscr{S}_0$, the characteristic functions $\chi_X$, $\chi_Y$ lie in $L_p$, and

$$\|\chi_X - \chi_Y\|^p = m((X \backslash Y) \cup (Y \backslash X)).$$

Accordingly, the existence of a sequence $\{X_1, X_2, \ldots\}$ with the properties set out in the proposition is equivalent to the existence of a countable everywhere-dense subset of the set $C = \{\chi_X : X \in \mathscr{S}_0\}$ $(\subseteq L_p)$.

If $L_p$ is separable, so is $C$. Conversely, suppose that $C$ has a countable everywhere-dense subset $\{g_1, g_2, \ldots\}$. Let $R$ be the countable subset of $L_p$ that consists of all finite linear combinations $z_1 g_1 + \cdots + z_n g_n$, in which the coefficients $z_1, \ldots, z_n$ are rational complex numbers. Given any $f$ in $L_p$ and any positive $\varepsilon$, there is a simple function $f_1$ in $L_p$ such that $\|f - f_1\| < \frac{1}{2}\varepsilon$. Since $f_1$ is a finite linear combination (with complex coefficients) of elements of $C$, there is a finite linear combination $f_2$ (with rational complex coefficients) of elements of the everywhere-dense subset $\{g_1, g_2, \ldots\}$ of $C$ such that $\|f_1 - f_2\| < \frac{1}{2}\varepsilon$. Then $f_2 \in R$ and $\|f - f_2\| < \varepsilon$; so $R$ is a countable everywhere-dense subset of $L_p$. ■

1.7.11. Remark. It follows from Proposition 1.7.10 that, when $1 \leqslant p < \infty$, the $L_p$ space, associated with Lebesgue measure on a measurable subset $E$ of $\mathbb{R}^n$, is separable. To prove this, we consider the set of all "rational cells"

$$\{(x_1, \ldots, x_n) : a_j \leqslant x_j \leqslant b_j \ (j = 1, \ldots, n)\}$$

in $\mathbb{R}^n$, where the $a_j$'s and $b_j$'s are specified rational numbers. We can list, as a sequence $\{Y_1, Y_2, \ldots\}$, the set of all finite unions of rational cells, and take $X_j = E \cap Y_j$. The sequence $\{X_j\}$ then has the properties set out in Proposition 1.7.10.

We shall show, in Exercise 2.8.7, that certain $\sigma$-finite measures give rise to inseparable $L_2$ spaces.

We shall see later that $C(S)$ is separable when $S$ is a compact metric space (see Remark 3.4.15), while infinite-dimensional $l_\infty$ and $L_\infty$ spaces are not separable (Exercises 1.9.31, 1.9.32). ■

## 1.8. Linear operators acting on Banach spaces

In this section we prove four basic results concerning linear operators acting on Banach spaces, namely, the open mapping theorem, the Banach inversion theorem, the closed graph theorem, and the principle of uniform boundedness. The first three are so closely related as to be more or less equivalent, and the fourth is easily deduced from the third.

We recall that a mapping $\varphi$ from a topological space $X$ into another such space $Y$ is said to be *open* if $\varphi(G)$ is open in $Y$ whenever $G$ is an open subset of $X$.

1.8.1. PROPOSITION. *Suppose that $\mathfrak{X}$, $\mathscr{Y}$ are normed spaces and $T: \mathfrak{X} \to \mathscr{Y}$ is a linear operator. Then $T$ is open if and only if the image $\{Tx : x \in (\mathfrak{X})_1\}$ of the unit ball $(\mathfrak{X})_1$ contains the ball $(\mathscr{Y})_r$, for some $r$ $(> 0)$. If $T$ is open, $T(\mathfrak{X}) = \mathscr{Y}$.*

*Proof.* If $T$ is open, it carries the open unit ball $\{x \in \mathfrak{X} : \|x\| < 1\}$ onto an open subset of $\mathscr{Y}$ that contains 0 and so contains $(\mathscr{Y})_r$ for some $r$ $(> 0)$. Thus

$$(\mathscr{Y})_r \subseteq \{Tx : x \in \mathfrak{X},\ \|x\| < 1\} \subseteq T((\mathfrak{X})_1).$$

Moreover, the subspace $T(\mathfrak{X})$ of $\mathscr{Y}$ contains $(\mathscr{Y})_r$, and is therefore the whole of $\mathscr{Y}$.

Conversely, suppose that $r > 0$ and $(\mathscr{Y})_r \subseteq T((\mathfrak{X})_1)$. We have to show that $T(G)$ is open in $\mathscr{Y}$ when $G$ is an open subset of $\mathfrak{X}$. If $x_0 \in G$, then $G$ contains a ball $x_0 + c(\mathfrak{X})_1$ (where $c > 0$), and

$$T(G) \supseteq Tx_0 + cT((\mathfrak{X})_1) \supseteq Tx_0 + c(\mathscr{Y})_r.$$

Hence each point $Tx_0$ of $T(G)$ is an interior point, and $T(G)$ is open, as required. ∎

1.8.2. PROPOSITION. *If $T$ is a bounded linear operator from a Banach space $\mathfrak{X}$ into a normed space $\mathscr{Y}$, and the closure $C^-$ of the set $C = \{Tx : x \in (\mathfrak{X})_1\}$ contains the ball $(\mathscr{Y})_r$, for some $r$ $(> 0)$, then $T$ is open.*

*Proof.* From Proposition 1.8.1, it suffices to show that $C$ contains the ball $(\mathscr{Y})_{r/2}$. To this end, suppose that $y \in \mathscr{Y}$ and $\|y\| \leqslant \frac{1}{2}r$. Since $2y \in (\mathscr{Y})_r \subseteq C^-$, we can choose $y_1$ in $C$ so that

$$\|2y - y_1\| < \tfrac{1}{2}r.$$

Since $2^2y - 2y_1 \in (\mathscr{Y})_r \subseteq C^-$, we can choose $y_2$ in $C$ so that

$$\|2^2y - 2y_1 - y_2\| < \tfrac{1}{2}r.$$

Since $2^3y - 2^2y_1 - 2y_2 \in (\mathscr{Y})_r \subseteq C^-$, we can choose $y_3$ in $C$ so that

$$\|2^3y - 2^2y_1 - 2y_2 - y_3\| < \tfrac{1}{2}r.$$

By continuing in this way, we obtain a sequence $\{y_1, y_2, \ldots\}$ in $C$ such that

$$\|2^n y - 2^{n-1} y_1 - 2^{n-2} y_2 - \cdots - y_n\| < \tfrac{1}{2} r \qquad (n = 1, 2, \ldots).$$

Thus

$$\|y - \sum_{j=1}^{n} 2^{-j} y_j\| < 2^{-n-1} r \qquad (n = 1, 2, \ldots),$$

and $y = \sum_{j=1}^{\infty} 2^{-j} y_j$.

Now $y_j \in C$, so $y_j = Tx_j$ for some $x_j$ in $(\mathfrak{X})_1$. Since $\mathfrak{X}$ is a Banach space and

$$\sum_{j=1}^{\infty} \|2^{-j} x_j\| \leqslant \sum_{j=1}^{\infty} 2^{-j} = 1,$$

the series $\sum_{j=1}^{\infty} 2^{-j} x_j$ converges to an element $x$ of $(\mathfrak{X})_1$. Moreover

$$Tx = \sum_{j=1}^{\infty} 2^{-j} Tx_j = \sum_{j=1}^{\infty} 2^{-j} y_j = y,$$

and since $x \in (\mathfrak{X})_1$, it follows that $y \in C$, as required. ∎

When $\mathfrak{X}$ is reflexive, Proposition 1.8.2 reduces immediately to Proposition 1.8.1, since in this case the set $C$ is closed (see Exercise 1.9.17).

1.8.3. LEMMA. *If $T$ is a linear operator from a normed space $\mathfrak{X}$ onto a Banach space $\mathscr{Y}$, there exist positive real numbers $r$, $s$, and an element $y_0$ of $\mathscr{Y}$, such that the closure in $\mathscr{Y}$ of the set $\{Tx : x \in (\mathfrak{X})_s\}$ contains the ball $y_0 + (\mathscr{Y})_r$.*

*Proof.* For each positive integer $n$, let

$$C_n = \{Tx : x \in (\mathfrak{X})_n\}.$$

When $y \in \mathscr{Y}$ and $r > 0$, let $B(y, r)$ denote the closed ball $y + (\mathscr{Y})_r$. It suffices to show that, for some $n$, the closure $C_n^-$ of $C_n$ contains such a ball. We assume the contrary, and in due course obtain a contradiction. (There is a short cut available here for the reader who is familiar with the following result, known as the Baire category theorem: *a complete metric space $X$ cannot be expressed as the union of a sequence $\{X_n\}$ of subsets, each of which is nowhere dense in $X$.* Indeed, $\bigcup_{n=1}^{\infty} C_n$ is the complete metric space $\mathscr{Y}$ since $T$ has range $\mathscr{Y}$, and the category theorem implies the required conclusion that at least one of the sets $C_n$ is not nowhere dense in $\mathscr{Y}$ (that is, for at least one value of $n$, the closure $C_n^-$ has non-empty interior [K: p. 201]). The argument that follows is in fact a proof of the category theorem, within the particular context now under consideration, but in a form easily adapted so as to apply to the general case.)

From our assumption, $C_1^-$ does not contain the unit ball $(\mathscr{Y})_1$; so we can choose $y_1$ ($\in \mathscr{Y}$) and $r_1$ ($> 0$) so that

$$y_1 \in (\mathscr{Y})_1 \backslash C_1^-, \qquad B(y_1, r_1) \cap C_1 = \varnothing.$$

Since $C_2^-$ does not contain the ball $B(y_1, \frac{1}{2}r_1)$, we can choose $y_2$ ($\in \mathscr{Y}$) and $r_2$ ($> 0$) so that

$$y_2 \in B(y_1, \tfrac{1}{2}r_1) \backslash C_2^-, \qquad B(y_2, r_2) \cap C_2 = \varnothing, \qquad r_2 < \tfrac{1}{2}r_1.$$

Since $C_3^-$ does not contain the ball $B(y_2, \frac{1}{2}r_2)$, we can choose $y_3$ ($\in \mathscr{Y}$) and $r_3$ ($> 0$) so that

$$y_3 \in B(y_2, \tfrac{1}{2}r_2) \backslash C_3^-, \qquad B(y_3, r_3) \cap C_3 = \varnothing, \qquad r_3 < \tfrac{1}{2}r_2.$$

By continuing in this way, we obtain sequences $\{y_n\}$ in $\mathscr{Y}$ and $\{r_n\}$ in $\mathbb{R}$ such that, for $n = 2, 3, \ldots,$

$$y_n \in B(y_{n-1}, \tfrac{1}{2}r_{n-1}) \backslash C_n^-, \qquad B(y_n, r_n) \cap C_n = \varnothing, \qquad 0 < r_n < \tfrac{1}{2}r_{n-1}.$$

These conditions imply that

$$B(y_n, r_n) \subseteq B(y_{n-1}, r_{n-1}) \quad (n \geqslant 2), \qquad \lim_{n \to \infty} r_n = 0,$$

and from this,

$$y_m \in B(y_m, r_m) \subseteq B(y_n, r_n), \qquad \|y_m - y_n\| \leqslant r_n \qquad (1 \leqslant n < m).$$

It follows that $\{y_n\}$ is a Cauchy sequence in the Banach space $\mathscr{Y}$, and so converges to an element $y$ of $\mathscr{Y}$. The closed ball $B(y_n, r_n)$ contains the sequence $\{y_n, y_{n+1}, y_{n+2}, \ldots\}$, so $y \in B(y_n, r_n)$ for all $n = 1, 2, \ldots$ .

Since $T(\mathfrak{X}) = \mathscr{Y}$, we have $y = Tx$ for some $x$ in $\mathfrak{X}$. If $n > \|x\|$, then

$$y \in B(y_n, r_n) \cap C_n,$$

contradicting the earlier assertion that $B(y_n, r_n) \cap C_n = \varnothing$. ■

1.8.4. Theorem (Open mapping theorem). *A bounded linear operator $T$ from a Banach space $\mathfrak{X}$ onto a Banach space $\mathscr{Y}$ is open.*

*Proof.* By Lemma 1.8.3 we can choose positive real numbers $r$, $s$, and an element $y_0$ of $\mathscr{Y}$ so that the closure $C^-$ of the set $C = \{Tx : x \in (\mathfrak{X})_s\}$ contains the ball $y_0 + (\mathscr{Y})_r$. Upon replacing $y_0$ by $s^{-1}y_0$ and $r$ by $s^{-1}r$, we may assume that $s = 1$.

Suppose that $y \in (\mathscr{Y})_r$. Since $C^-$ is balanced and convex, and

$$y_0 \pm y \in y_0 + (\mathscr{Y})_r \subseteq C^-,$$

it follows that

$$y = \tfrac{1}{2}[(y_0 + y) - (y_0 - y)] \in C^-.$$

Hence $(\mathscr{Y})_r \subseteq C^-$, and $T$ is open by Proposition 1.8.2. ■

1.8.5. Theorem (Banach inversion theorem). *If $T$ is a one-to-one bounded linear operator from a Banach space $\mathfrak{X}$ onto a Banach space $\mathscr{Y}$, the inverse $T^{-1} : \mathscr{Y} \to \mathfrak{X}$ is a bounded linear operator.*

*Proof.* It is apparent that $T^{-1}$ is a linear operator. We have to show that it is continuous; that is, we must prove that the inverse image, under $T^{-1}$, of any open subset $G$ of $\mathfrak{X}$, is open in $\mathcal{Y}$. Since this inverse image is $T(G)$, the result follows at once from Theorem 1.8.4. ■

Suppose that $\mathfrak{X}$ and $\mathcal{Y}$ are normed spaces with the same scalar field $\mathbb{K}$. The product set $\mathfrak{X} \times \mathcal{Y}$ is a normed space when the algebraic structure and norm are defined as follows:

$$(x_1, y_1) + (x_2, y_2) = (x_1 + x_2, y_1 + y_2),$$
$$a(x, y) = (ax, ay), \qquad \|(x, y)\| = \|x\| + \|y\|.$$

When $\mathfrak{X}$ and $\mathcal{Y}$ are both Banach spaces, so is $\mathfrak{X} \times \mathcal{Y}$.

If $T: \mathfrak{X} \to \mathcal{Y}$ is a linear operator, the *graph* of $T$ is the subspace $\mathcal{G}(T)$ of $\mathfrak{X} \times \mathcal{Y}$ defined by

$$\mathcal{G}(T) = \{(x, Tx): x \in \mathfrak{X}\}.$$

As a linear subspace of a normed space, $\mathcal{G}(T)$ is itself a normed space.

Note that $\mathcal{G}(T)$ is closed (in $\mathfrak{X} \times \mathcal{Y}$) if and only if the following condition is satisfied: if a sequence $\{x_n\}$ in $\mathfrak{X}$ converges to an element $x$ of $\mathfrak{X}$, while $\{Tx_n\}$ converges to an element $y$ of $\mathcal{Y}$, then $Tx = y$. It is clear that a bounded linear operator has a closed graph; for operators acting on Banach spaces, there is a converse.

1.8.6. THEOREM (Closed graph theorem). *If $\mathfrak{X}$ and $\mathcal{Y}$ are Banach spaces, and $T$ is a linear operator from $\mathfrak{X}$ into $\mathcal{Y}$, then the graph of $T$ is closed if and only if $T$ is bounded.*

*Proof.* In view of the preceding discussion, it suffices to show that $T$ is bounded when its graph $\mathcal{G}(T)$ is closed. In this case, $\mathcal{G}(T)$ is complete, and is therefore a Banach space, since it is a closed subset of the complete metric space $\mathfrak{X} \times \mathcal{Y}$. The equation

$$H(x, Tx) = x \qquad (x \in \mathfrak{X})$$

defines a one-to-one linear operator $H$ from $\mathcal{G}(T)$ onto $\mathfrak{X}$; and $H$ is bounded with $\|H\| \leqslant 1$, since

$$\|H(x, Tx)\| = \|x\| \leqslant \|x\| + \|Tx\| = \|(x, Tx)\|.$$

By the Banach inversion theorem, $H^{-1}$ is bounded; and the same is true of $T$ since

$$\|Tx\| \leqslant \|(x, Tx)\| = \|H^{-1}x\| \leqslant \|H^{-1}\| \, \|x\| \qquad (x \in \mathfrak{X}). \quad ■$$

If $E$ is an idempotent bounded linear operator acting on a normed space $\mathfrak{X}$, the corresponding complementary subspaces

$$Y = \{x \in \mathfrak{X}: Ex = x\}, \qquad Z = \{x \in \mathfrak{X}: Ex = 0\}$$

of $\mathfrak{X}$ are closed, being the null spaces of the continuous linear operators $I - E$ and $E$, respectively. For Banach spaces, there is a converse.

1.8.7. THEOREM. *If $Y$ and $Z$ are closed complementary subspaces of a Banach space $\mathfrak{X}$, then the projection $E$ from $\mathfrak{X}$ onto $Y$ parallel to $Z$ is bounded.*

*Proof.* The graph of $E$ can be expressed in the form

$$\{(x, y) \in \mathfrak{X} \times \mathfrak{X} : y \in Y,\ x - y \in Z\}$$

and is therefore closed in $\mathfrak{X} \times \mathfrak{X}$; so the result follows from the closed graph theorem. ∎

1.8.8. COROLLARY. *If $Y$ and $Z$ are closed subspaces of a Banach space $\mathfrak{X}$ and $Y \cap Z = \{0\}$, then $Y + Z$ is closed in $\mathfrak{X}$ if and only if there is a positive real number $C$ such that*

$$\|y\| \leqslant C\|y + z\| \qquad (y \in Y, \quad z \in Z).$$

*Proof.* Suppose that there is such a constant $C$. If a sequence $\{x_n\}$ in $Y + Z$ converges to an element $x$ of $\mathfrak{X}$, let $x_n = y_n + z_n$, where $y_n \in Y$ and $z_n \in Z$. Since

$$\|y_m - y_n\| \leqslant C\|(y_m - y_n) + (z_m - z_n)\| = C\|x_m - x_n\|,$$

$\{y_n\}$ is a Cauchy sequence in the closed (and hence complete) subspace $Y$ of $\mathfrak{X}$, and so converges to an element $y$ of $Y$. Since $Z$ is closed, and

$$x - y = \lim(x_n - y_n) = \lim z_n,$$

it follows that

$$x - y \in Z, \qquad x = y + (x - y) \in Y + Z.$$

Hence $Y + Z$ is closed in $\mathfrak{X}$.

Conversely, if $Y + Z$ is closed in $\mathfrak{X}$, it is a Banach space $\mathfrak{X}_0$ containing $Y$ and $Z$ as complementary subspaces. From Theorem 1.8.7, the projection $E$ from $\mathfrak{X}_0$ onto $Y$ parallel to $Z$ is bounded; and the stated condition is satisfied, with $C = \|E\|$. ∎

If $y$ and $z$ are unit vectors in a Banach space $\mathfrak{X}$, it is reasonable to consider the "angle" between $y$ and $z$ to be large, or small, according as $\|y - z\|$ is large or small. Accordingly, when $Y$ and $Z$ are closed subspaces of $\mathfrak{X}$, the lower bound

$$\inf\{\|y - z\| : y \in Y,\ z \in Z,\ \|y\| = \|z\| = 1\}$$

can be regarded as indicating the (minimum) angle between $Y$ and $Z$. It is not difficult to show that this lower bound is strictly positive if and only if there is a constant $C$ with the property set out in Corollary 1.8.8 (see Exercise 1.9.5).

Thus the corollary can be interpreted in the following geometrical form: if $Y \cap Z = \{0\}$, then $Y + Z$ is closed if and only if the angle between $Y$ and $Z$ is strictly positive.

We conclude this section with various forms of the principle of uniform boundedness.

1.8.9. THEOREM. *Suppose that $\{T_a : a \in \mathbb{A}\}$ is a family of bounded linear operators from a Banach space $\mathfrak{X}$ into a normed space $\mathscr{Y}$ and*

$$\sup\{\|T_a x\| : a \in \mathbb{A}\} < \infty$$

*for each $x$ in $\mathfrak{X}$. Then* $\sup\{\|T_a\| : a \in \mathbb{A}\} < \infty$.

*Proof.* Since $T_a$ can be viewed as a bounded linear operator from $\mathfrak{X}$ into the completion $\hat{\mathscr{Y}}$ of $\mathscr{Y}$, we may suppose that $\mathscr{Y}$ is a Banach space. For each $x$ in $\mathfrak{X}$, the equation

$$(Sx)(a) = T_a x \qquad (a \in \mathbb{A})$$

defines a mapping $Sx : \mathbb{A} \to \mathscr{Y}$, and $Sx$ is an element of the Banach space $l_\infty(\mathbb{A}, \mathscr{Y})$ (Example 1.7.1). It is evident that the mapping $x \to Sx$ is a linear operator $S$ from $\mathfrak{X}$ into $l_\infty(\mathbb{A}, \mathscr{Y})$.

We assert that the graph of $S$ is closed. For this, suppose that a sequence $\{x_n\}$ in $\mathfrak{X}$ converges to $x$ $(\in \mathfrak{X})$, while $\{Sx_n\}$ converges to an element $f$ of $l_\infty(\mathbb{A}, \mathscr{Y})$; we have to show that $Sx = f$. For each $a$ in $\mathbb{A}$,

$$\|f(a) - (Sx_n)(a)\| \leqslant \|f - Sx_n\| \to 0.$$

From this, and since $T_a$ is continuous,

$$f(a) = \lim(Sx_n)(a) = \lim T_a x_n = T_a x = (Sx)(a);$$

so $f = Sx$, and the graph of $S$ is closed.

From the closed graph theorem, $S$ is bounded. For each $x$ in $\mathfrak{X}$ and $a$ in $\mathbb{A}$,

$$\|T_a x\| = \|(Sx)(a)\| \leqslant \|Sx\| \leqslant \|S\| \, \|x\|;$$

so $\|T_a\| \leqslant \|S\|$ $(a \in \mathbb{A})$. ∎

1.8.10. THEOREM. *Suppose that $\{\rho_a : a \in \mathbb{A}\}$ is a family of bounded linear functionals on a Banach space $\mathfrak{X}$ and* $\sup\{|\rho_a(x)| : a \in \mathbb{A}\} < \infty$ *for each $x$ in $\mathfrak{X}$. Then* $\sup\{\|\rho_a\| : a \in \mathbb{A}\} < \infty$.

*Proof.* This follows from Theorem 1.8.9, with $\mathscr{Y}$ the scalar field. ∎

1.8.11. COROLLARY. *Suppose that $\{x_a : a \in \mathbb{A}\}$ is a family of elements of a normed space $\mathfrak{X}$ and* $\sup\{|\rho(x_a)| : a \in \mathbb{A}\} < \infty$ *for each bounded linear functional $\rho$ on $\mathfrak{X}$. Then* $\sup\{\|x_a\| : a \in \mathbb{A}\} < \infty$.

*Proof.* For each $\rho$ in the Banach dual space $\mathfrak{X}^\#$,

$$\sup\{|\hat{x}_a(\rho)| : a \in \mathbb{A}\} = \sup\{|\rho(x_a)| : a \in \mathbb{A}\} < \infty$$

(where $x \to \hat{x}$ is the natural isometric isomorphism from $\mathfrak{X}$ into $\mathfrak{X}^{\#\#}$). From Theorem 1.8.10 (with $\mathfrak{X}^\#$ in place of $\mathfrak{X}$ and $\hat{x}_a$ in place of $\rho_a$), and since $\|\hat{x}_a\| = \|x_a\|$, we have $\sup\{\|x_a\| : a \in \mathbb{A}\} < \infty$. ∎

1.8.12. THEOREM. *Suppose that* $\{T_a : a \in \mathbb{A}\}$ *is a family of bounded linear operators from a Banach space* $\mathfrak{X}$ *into a normed space* $\mathcal{Y}$ *and*

$$\sup\{|\rho(T_a x)| : a \in \mathbb{A}\} < \infty$$

*for each* $x$ *in* $\mathfrak{X}$ *and* $\rho$ *in* $\mathcal{Y}^\#$. *Then* $\sup\{\|T_a\| : a \in \mathbb{A}\} < \infty$.

*Proof.* By Corollary 1.8.11, $\sup\{\|T_a x\| : a \in \mathbb{A}\} < \infty$ for each $x$ in $\mathfrak{X}$; so the result follows from Theorem 1.8.9. ∎

1.8.13. REMARK. In the context of complex Banach spaces, the open mapping, Banach inversion, and closed graph theorems, and the various forms of the principle of uniform boundedness, apply also to *conjugate-linear* operators; for such operators can be viewed as linear mappings between the corresponding real Banach spaces. ∎

## 1.9. Exercises

1.9.1. Show that, if $q$ is a sublinear functional on a (real or complex) vector space $\mathcal{V}$, and $V$ is the convex set $\{x \in \mathcal{V} : q(x) < 1\}$, then the support functional $p$ of $V$ is given by

$$p(x) = \max\{q(x), 0\} \qquad (x \in \mathcal{V}).$$

1.9.2. Suppose that $q_1$ and $q_2$ are semi-norms on a (real or complex) vector space $\mathcal{V}$, $\rho$ is a linear functional on $\mathcal{V}$, and

$$|\rho(x)| \leqslant q_1(x) + q_2(x) \qquad (x \in \mathcal{V}).$$

Show that $\rho$ can be expressed in the form $\rho_1 + \rho_2$, where $\rho_1$, $\rho_2$ are linear functionals on $\mathcal{V}$, and

$$|\rho_j(x)| \leqslant q_j(x) \qquad (x \in \mathcal{V},\ j = 1, 2).$$

[*Hint.* Define a semi-norm $p$ on the vector space $\mathcal{V} \times \mathcal{V}$ and a linear functional $\sigma_0$ on the "diagonal" subspace $\{(x, x) : x \in \mathcal{V}\}$ of $\mathcal{V} \times \mathcal{V}$ by

$$p((x, y)) = q_1(x) + q_2(y), \qquad \sigma_0((x, x)) = \rho(x).]$$

1.9.3. Suppose that $\mathscr{V}$ is a (real or complex) vector space on which two locally convex topologies, $\tau_1$ and $\tau_2$, are specified. Let $\Gamma_j$ be the set of all $\tau_j$-continuous semi-norms on $\mathscr{V}$, for $j = 1, 2$; and let

$$\Gamma = \{q_1 + q_2 : q_1 \in \Gamma_1, q_2 \in \Gamma_2\}.$$

Prove that $\Gamma$ is a separating family of semi-norms on $\mathscr{V}$, and that the corresponding locally convex topology $\tau$ is the coarsest locally convex topology on $\mathscr{V}$ that is finer than both $\tau_1$ and $\tau_2$. Show also that a linear functional $\rho$ on $\mathscr{V}$ is $\tau$-continuous if and only if it has the form $\rho_1 + \rho_2$, where $\rho_j$ is a $\tau_j$-continuous linear functional on $\mathscr{V}$ for $j = 1, 2$.

1.9.4. Show that, if $\{x_1, \ldots, x_n\}$ are elements of a normed space $\mathfrak{X}$ such that 0 is in the closure of

$$\left\{a_1 x_1 + \cdots + a_n x_n : a_j \text{ scalars}, \prod_{j=1}^{n} (a_j - 1) = 0\right\},$$

then $\{x_1, \ldots, x_n\}$ are linearly dependent.

1.9.5. Suppose that $Y$ and $Z$ are closed subspaces of a Banach space $\mathfrak{X}$ such that $Y \cap Z = \{0\}$, and let

$$k = \inf\{\|y - z\| : y \in Y, z \in Z, \|y\| = \|z\| = 1\}.$$

Show that the subspace $Y + Z$ of $\mathfrak{X}$ is closed if and only if $k > 0$. [See Corollary 1.8.8 and the discussion following it.]

1.9.6. Show that, if $\mathfrak{X}$ is a real Banach space, then $\mathfrak{X} \times \mathfrak{X}$ becomes a complex Banach space $\mathfrak{X}_{\mathbb{C}}$ when its linear structure and norm are defined by

$$(x, y) + (u, v) = (x + u, y + v),$$

$$(a + ib)(x, y) = (ax - by, bx + ay),$$

$$\|(x, y)\| = \sup\{\|(\cos\theta)x + (\sin\theta)y\| : 0 \leqslant \theta \leqslant 2\pi\}$$

for all $x, y, u, v$ in $\mathfrak{X}$ and $a, b$ in $\mathbb{R}$.

Prove also that the set $\{(x, 0) : x \in \mathfrak{X}\}$ is a closed real-linear subspace $\mathfrak{X}_{\mathbb{R}}$ of $\mathfrak{X}_{\mathbb{C}}$, that $\mathfrak{X}_{\mathbb{C}} = \{h + ik : h, k \in \mathfrak{X}_{\mathbb{R}}\}$, and that the mapping $x \to (x, 0)$ is an isometric isomorphism from $\mathfrak{X}$ onto (the real Banach space) $\mathfrak{X}_{\mathbb{R}}$.

1.9.7. Suppose that $\mathfrak{X}$ is an infinite-dimensional normed space, and $V$ is a neighborhood of 0 in the weak topology on $\mathfrak{X}$. Show that $V$ contains a closed subspace of finite codimension in $\mathfrak{X}$. Deduce that the weak topology on $\mathfrak{X}$ is strictly coarser than the norm topology.

1.9.8. Show that, if $\mathfrak{X}$ is a separable Banach space, each bounded sequence $\{\rho_n\}$ in $\mathfrak{X}^\#$ has a subsequence that is weak* convergent to an element of $\mathfrak{X}^\#$.

1.9.9. Prove that, if $\mathfrak{X}$ and $\mathscr{Y}$ are normed spaces, $\mathfrak{X} \neq \{0\}$, and $\mathscr{B}(\mathfrak{X}, \mathscr{Y})$ is complete, then $\mathscr{Y}$ is complete.

1.9.10. Suppose that $\mathfrak{X}_0$ is a closed subspace of a Banach space $\mathfrak{X}$, $Q: \mathfrak{X} \to \mathfrak{X}/\mathfrak{X}_0$ is the quotient mapping, and $\mathfrak{X}_0^\perp$ $(\subseteq \mathfrak{X}^\#)$ is the closed subspace consisting of all bounded linear functionals on $\mathfrak{X}$ that vanish on $\mathfrak{X}_0$.

(i) Show that the (Banach) adjoint operator $Q^\#$ is an isometric linear mapping from $(\mathfrak{X}/\mathfrak{X}_0)^\#$ onto $\mathfrak{X}_0^\perp$.
(ii) Show that the mapping $T: \rho + \mathfrak{X}_0^\perp \to \rho | \mathfrak{X}_0$ is an isometric isomorphism from $\mathfrak{X}^\#/\mathfrak{X}_0^\perp$ onto $\mathfrak{X}_0^\#$.

1.9.11. Show that, if $\mathfrak{X}_0$ is a closed subspace of a Banach space $\mathfrak{X}$, then $\mathfrak{X}$ is reflexive if and only if both $\mathfrak{X}_0$ and $\mathfrak{X}/\mathfrak{X}_0$ are reflexive.

1.9.12. Show that a Banach space $\mathfrak{X}$ is reflexive if and only if its dual space $\mathfrak{X}^\#$ is reflexive.

1.9.13. Suppose that $\mathfrak{X}$ is a Banach space with the following property: given any positive real number $\varepsilon$, there is a positive real number $\delta(\varepsilon)$ such that $\|x - y\| < \varepsilon$ whenever $x$ and $y$ lie in the unit ball $(\mathfrak{X})_1$ and $\|\frac{1}{2}(x + y)\| > 1 - \delta(\varepsilon)$ (such a Banach space is said to be *uniformly convex*). Prove that $\mathfrak{X}$ is reflexive. [*Hint*. Suppose that $\Omega_0 \in \mathfrak{X}^{\#\#}$ and $\|\Omega_0\| = 1$. Choose $\rho_0$ in $\mathfrak{X}^\#$ so that $\|\rho_0\| = 1$ and $|\Omega_0(\rho_0) - 1| < \delta(\varepsilon)$; and let

$$\mathscr{K} = \{\hat{x}: x \in (\mathfrak{X})_1, \ |\rho_0(x) - 1| < \delta(\varepsilon)\},$$

where $x \to \hat{x}$ is the natural isometric isomorphism from $\mathfrak{X}$ into $\mathfrak{X}^{\#\#}$. Prove that $\Omega_0$ lies in the weak* closure of $\mathscr{K}$, and that $\|\hat{x} - \hat{y}\| < \varepsilon$ whenever $\hat{x}, \hat{y} \in \mathscr{K}$. Deduce that $\|\Omega_0 - \hat{x}\| \leqslant \varepsilon$ for each $\hat{x}$ in $\mathscr{K}$.]

1.9.14. Suppose that $\mathfrak{X}$ is a normed space, $\mathscr{M}$ is a linear subspace of the dual space $\mathfrak{X}^\#$, and $\mathscr{M}$ separates the points of $\mathfrak{X}$. Let $\rho$ be a linear functional on $\mathfrak{X}$.

(i) Suppose that the restriction $\rho | (\mathfrak{X})_1$ of $\rho$ to the unit ball $(\mathfrak{X})_1$ is continuous in the weak topology $\sigma(\mathfrak{X}, \mathscr{M})$. Show that $\rho \in \mathfrak{X}^\#$. Prove also that, if $\varepsilon > 0$, there is a finite subset $\{\omega_1, \ldots, \omega_n\}$ of $\mathscr{M}$ such that

$$|\rho(x)| < \varepsilon \quad \text{whenever} \quad x \in (\mathfrak{X})_1 \quad \text{and} \quad \sum_{j=1}^{n} |\omega_j(x)| < 1.$$

Deduce that

$$|\rho(x)| \leqslant \varepsilon\|x\| + \|\rho\| \sum_{j=1}^{n} |\omega_j(x)| \qquad (x \in \mathfrak{X}).$$

From this inequality, together with the result of Exercise 1.9.2, deduce that $\rho$ has the form $\rho_1 + \rho_2$, where $\rho_1 \in \mathfrak{X}^\sharp$, $\|\rho_1\| \leqslant \varepsilon$, and $\rho_2 \in \mathscr{M}$.

(ii) Prove that $\rho|(\mathfrak{X})_1$ is $\sigma(\mathfrak{X}, \mathscr{M})$-continuous if and only if $\rho$ lies in the norm closure $\mathscr{M}^=$ of $\mathscr{M}$ in $\mathfrak{X}^\sharp$.

1.9.15. Suppose that $\mathfrak{X}$ is a Banach space, $x \to \hat{x}$ is the natural isometric isomorphism from $\mathfrak{X}$ into $\mathfrak{X}^{\sharp\sharp}$, and $\omega \in \mathfrak{X}^{\sharp\sharp}$. Show that, if the restriction $\omega|(\mathfrak{X}^\sharp)_1$ of $\omega$ to the unit ball of $\mathfrak{X}^\sharp$ is continuous in the weak* topology $\sigma(\mathfrak{X}^\sharp, \mathfrak{X})$, then $\omega = \hat{x}$ for some $x$ in $\mathfrak{X}$. [*Hint*. Use the result of Exercise 1.9.14(ii).]

1.9.16. Suppose that $\mathfrak{X}$ and $\mathscr{Y}$ are Banach spaces, and $S \in \mathscr{B}(\mathscr{Y}^\sharp, \mathfrak{X}^\sharp)$.

(i) Prove that, if $S$ is continuous relative to the weak* topologies on $\mathscr{Y}^\sharp$ and $\mathfrak{X}^\sharp$, then $S = T^\sharp$ for some $T$ in $\mathscr{B}(\mathfrak{X}, \mathscr{Y})$.

(ii) By using the result of Exercise 1.9.15, show that (i) remains valid when the weak* continuity of $S$ is replaced by weak* continuity of $S|(\mathscr{Y}^\sharp)_1$.

1.9.17. Suppose that $\mathfrak{X}$, $\mathscr{Y}$ are Banach spaces and $T \in \mathscr{B}(\mathfrak{X}, \mathscr{Y})$. Prove that

(i) the set $\{T^\sharp\rho : \rho \in (\mathscr{Y}^\sharp)_1\}$ is weak* compact, and hence norm closed in $\mathfrak{X}^\sharp$;

(ii) if $\mathfrak{X}$ is reflexive, the set $\{Tx : x \in (\mathfrak{X})_1\}$ is weakly compact, and hence norm closed in $\mathscr{Y}$.

1.9.18. Suppose that $\mathfrak{X}$ and $\mathscr{Y}$ are Banach spaces, $T \in \mathscr{B}(\mathfrak{X}, \mathscr{Y})$, and the image $T(\mathfrak{X})$ $(= \{Tx : x \in \mathfrak{X}\})$ is a closed subspace of $\mathscr{Y}$. Prove that $T^\sharp(\mathscr{Y}^\sharp)$ is the closed subspace of $\mathfrak{X}^\sharp$ consisting of all bounded linear functionals on $\mathfrak{X}$ that vanish on the null space of $T$. [*Hint*. If $\rho \in \mathfrak{X}^\sharp$ and $\rho$ vanishes on the null space of $T$, the equation $\omega_0(Tx) = \rho(x)$ defines (unambiguously) a linear functional $\omega_0$ on $T(\mathfrak{X})$. By applying the open mapping theorem to $T$, as an operator from $\mathfrak{X}$ onto $T(\mathfrak{X})$, prove that $\omega_0$ is bounded. Deduce that $\rho = T^\sharp\omega$, for some $\omega$ in $\mathscr{Y}^\sharp$.]

1.9.19. Let $l_\infty$ denote the Banach space $l_\infty(\mathbb{N}, \mathbb{C})$ of Example 1.7.1, where $\mathbb{N}$ is the set of positive integers; an element of $l_\infty$ is a bounded complex sequence $\{x_1, x_2, \ldots\}$, and $\|\{x_n\}\| = \sup\{|x_n| : n \in \mathbb{N}\}$. Let $c$ and $c_0$ be the linear subspaces of $l_\infty$ defined by

$$c = \left\{\{x_n\} \in l_\infty : \lim_{n \to \infty} x_n \text{ exists}\right\},$$

$$c_0 = \left\{\{x_n\} \in l_\infty : \lim_{n \to \infty} x_n = 0\right\}.$$

Prove that

(i) $c$ and $c_0$ are closed subspaces of $l_\infty$;

(ii) the sequence $\{1, 1, 1, \ldots\}$ is an extreme point in the closed unit ball of $l_\infty$, and also in the closed unit ball of $c$;

(iii) the closed unit ball of $c_0$ has no extreme point.

Deduce that $c_0$ is isometrically isomorphic neither to $c$, nor to any Banach dual space.

1.9.20. With the notation of Exercise 1.9.19, let $U$ be the element $\{1, 1, 1, \ldots\}$ of $c$; and, for $k = 1, 2, \ldots$, let $E_k$ (in $c_0$) be the sequence that has 1 in the $k$th position and zeros elsewhere.

(i) Prove that, if $X = \{x_n\} \in c_0$, then $X = \sum_{k=1}^{\infty} x_k E_k$, the series converging in the norm topology of $c_0$ (that is, $\|X - \sum_{k=1}^{m} x_k E_k\| \to 0$ as $m \to \infty$).

(ii) Prove that, if $X = \{x_n\} \in c$ and $x = \lim_{n\to\infty} x_n$, then $X - xU \in c_0$ and $X = xU + \sum_{k=1}^{\infty} (x_k - x)E_k$, the series converging in the norm topology of $c$.

1.9.21. Adopt the notation of Exercises 1.9.19 and 1.9.20; in addition, let $l_1$ denote the Banach space $l_1(\mathbb{N}, \mathbb{C})$ of Example 1.7.3, so that an element of $l_1$ is a complex sequence $Y = \{y_1, y_2, \ldots\}$ such that $(\|Y\| =)\ \sum_{n=1}^{\infty} |y_n| < \infty$.

(i) Show that, if $Y = \{y_1, y_2, \ldots\} \in l_1$, the equation

$$\rho(X) = \sum_{n=1}^{\infty} y_n x_n \qquad (X = \{x_n\} \in c_0)$$

defines a bounded linear functional $\rho$ on $c_0$, and $\|\rho\| = \|Y\|$. Prove also that $y_n = \rho(E_n)$.

(ii) Show that each bounded linear functional on $c_0$ arises, as in (i), from an element $Y = \{y_1, y_2, \ldots\}$ of $l_1$.

(iii) Deduce that the Banach dual space $c_0^\sharp$ is isometrically isomorphic to $l_1$.

1.9.22. Let $c$ and $l_1$ be the Banach spaces defined in Exercises 1.9.19 and 1.9.21.

(i) Show that, if $\{y_0, y_1, y_2, \ldots\}$ is a complex sequence such that $\sum_{n=0}^{\infty} |y_n| < \infty$, the equation

$$\rho(X) = y_0 \lim_{n\to\infty} x_n + \sum_{n=1}^{\infty} y_n x_n \qquad (X = \{x_n\} \in c)$$

defines a bounded linear functional $\rho$ on $c$, and $\|\rho\| = \sum_{n=0}^{\infty} |y_n|$.

(ii) Prove that every bounded linear functional on $c$ arises, as in (i), from such a sequence $\{y_0, y_1, y_2, \ldots\}$. [*Hint.* Use the results of Exercises 1.9.20(ii) and 1.9.21(ii).]

(iii) Deduce that the Banach dual space $c^\sharp$ is isometrically isomorphic to $l_1$.

(iv) Deduce that $c_0^\sharp$ and $c^\sharp$ are isometrically isomorphic, while $c_0$ and $c$ are not.

1.9.23. Let $l_\infty$ and $l_1$ be the Banach spaces defined in Exercises 1.9.19 and 1.9.21, and, for each positive integer $k$, let $e_k$ (in $l_1$) be the sequence that has 1 in the $k$th position and zeros elsewhere. Without using Theorem 1.7.8:

(i) Prove that, if $Y = \{y_1, y_2, \ldots\} \in l_1$, then $Y = \sum_{k=1}^{\infty} y_k e_k$, the series converging in the norm topology on $l_1$.

(ii) Show that, if $X = \{x_1, x_2, \ldots\} \in l_\infty$, the equation

$$\rho(Y) = \sum_{n=1}^{\infty} x_n y_n \qquad (Y = \{y_n\} \in l_1)$$

defines a bounded linear functional $\rho$ on $l_1$, and $\|\rho\| = \|X\|$.

(iii) Prove that each bounded linear functional on $l_1$ arises, as in (ii), from an element $X$ of $l_\infty$.

(iv) Deduce that the Banach dual space $l_1^\sharp$ is isometrically isomorphic to $l_\infty$.

1.9.24. By using the results of Exercises 1.9.21, 1.9.22, and 1.9.23, show that neither of the Banach spaces $c_0$, $c$ is reflexive. Deduce that neither of the Banach spaces $l_1$, $l_\infty$ is reflexive.

1.9.25. Give a second proof that none of the Banach spaces $c_0$, $c$, $l_\infty$ is reflexive, by using the results of Exercises 1.9.19 and 1.9.11.

1.9.26. (i) Suppose that $\varepsilon > 0$ and $\{x_{m,n} \colon m, n = 1, 2, \ldots\}$ is a double sequence of complex numbers that satisfies the conditions

$$4\varepsilon < \sum_{n=1}^{\infty} |x_{m,n}| < \infty \qquad (m = 1, 2, \ldots),$$

$$\lim_{m \to \infty} x_{m,n} = 0 \qquad (n = 1, 2, \ldots).$$

Show that there exist integers $0 = n(0) < n(1) < n(2) < \cdots$ and $1 = m(1) < m(2) < m(3) < \cdots$ such that, for $j = 1, 2, \ldots$,

$$\sum_{n=1}^{n(j-1)} |x_{m(j),n}| < \varepsilon,$$

$$\sum_{n=1+n(j-1)}^{n(j)} |x_{m(j),n}| > 3\varepsilon,$$

$$\sum_{n=1+n(j)}^{\infty} |x_{m(j),n}| < \varepsilon.$$

Prove also that there is a sequence $\{y_n\}$ of complex numbers of modulus 1 such that

$$\left|\sum_{n=1}^{\infty} y_n x_{m(j),n}\right| > \varepsilon \qquad (j = 1, 2, \ldots).$$

(ii) Prove that, if a sequence $\{X_m\}$ of elements of the Banach space $l_1$ is weakly convergent to 0, then it converges to 0 in the norm topology.

1.9.27. Suppose that $1 < p < \infty$, $q = p/(p-1)$, and $l_p$ is the Banach space $l_p(\mathbb{N}, \mathbb{C})$ of Example 1.7.3, so that an element of $l_p$ is a complex sequence $X = \{x_1, x_2, \ldots\}$ such that $\sum_{n=1}^{\infty} |x_n|^p \ (= \|X\|^p) < \infty$.

(i) Show that, for each $Y = \{y_1, y_2, \ldots\}$ in $l_q$, the equation

$$\rho_Y(X) = \sum_{n=1}^{\infty} y_n x_n \qquad (X = \{x_n\} \in l_p)$$

defines a bounded linear functional $\rho_Y$ on $l_p$, and $\|\rho_Y\| \leqslant \|Y\|_q$ (the norm of $Y$ in $l_q$). By considering the sequence $X_0 = \{t_n |y_n|^{q/p}\}$, where $t_1, t_2, \ldots$ are suitable complex numbers of modulus 1, show that $\|\rho_Y\| = \|Y\|_q$.

(ii) Prove that the mapping $Y \to \rho_Y$ is an isometric isomorphism from $l_q$ onto the Banach dual space $l_p^{\sharp}$, and deduce that $l_p$ is reflexive (that is, that the *natural* isomorphism of $l_p$ into $l_p^{\sharp\sharp}$ is onto).

1.9.28. Suppose that $p > 1$, $q = p/(p-1)$, and $Y = \{y_1, y_2, \ldots\}$ is a complex sequence with the following property: whenever $X = \{x_1, x_2, \ldots\} \in l_p$, the sequence $\{y_n x_n\}$ is an element of $l_1$. Prove that $Y \in l_q$.

1.9.29. Suppose that $1 < p < \infty$, $q = p/(p-1)$, and $L_p$ is the Banach space associated, as in Example 1.7.5, with a $\sigma$-finite measure space $(S, \mathcal{S}, m)$.

(i) Prove that, for each $f$ in $L_q$, the equation

$$\rho_f(g) = \int_S f(s) g(s)\, dm(s) \qquad (g \in L_p)$$

defines a bounded linear functional $\rho_f$ on $L_p$, and $\|\rho_f\| \leqslant \|f\|_q$ (the norm of $f$ in $L_q$). By considering a suitable function $g_0$, of the form $g_0(s) = |f(s)|^{q/p} t(s)$, where $t(s) f(s) = |f(s)|$, show that $\|\rho_f\| = \|f\|_q$.

(ii) Suppose that $m(S) < \infty$, and let $\rho$ be a bounded linear functional on $L_p$. Show that, if $X\ (\subseteq S)$ is measurable, the characteristic function $\chi_X$ lies in $L_p$. Prove that there is an element $f$ of $L_1$ such that

$$\rho(\chi_X) = \int_S f(s) \chi_X(s)\, dm(s)$$

for every measurable subset $X$ of $S$. Show that $f \in L_q$, and $\rho = \rho_f$.

(iii) In the general case (with $m$ $\sigma$-finite, but not necessarily finite) prove that the mapping $f \to \rho_f$ is an isometric isomorphism from $L_q$ onto the Banach dual space $L_p^\sharp$, and deduce that $L_p$ is reflexive.

1.9.30. Suppose that $1 < p < \infty$, $q = p/(p-1)$, and $L_p$, $L_q$ are the corresponding Banach spaces associated with a $\sigma$-finite measure space $(S, \mathscr{S}, m)$. Show that, if $f$ is a complex-valued function on $S$, and $fg \in L_1$ for each $g$ in $L_p$, then $f \in L_q$. [*Hint*. Show that $f$ is the limit, almost everywhere, of a sequence $\{f_n\}$ of functions in $L_q$ such that $|f_n(s)| \leqslant |f(s)|$, and consider the corresponding sequence $\{\rho_n\}$ of bounded linear functionals on $L_p$.]

1.9.31. Show that the Banach space $l_\infty(\mathbb{A})$ is separable if and only if the set $\mathbb{A}$ is finite. Deduce that a separable Banach space may have a non-separable dual.

1.9.32. Show that the Banach space $L_\infty$, associated with a $\sigma$-finite measure space $(S, \mathscr{S}, m)$, is separable if and only if $S$ can be expressed as the disjoint union of a finite family of "atoms." (An *atom* is a measurable subset $S_0$ of $S$ such that $m(S_0) > 0$ and each measurable subset of $S_0$ has measure 0 or $m(S_0)$.)

1.9.33. Consider the Banach spaces $L_1$ and $L_\infty$ associated with a $\sigma$-finite measure space $(S, \mathscr{S}, m)$, and the isometric isomorphism (or Theorem 1.7.8) from $L_\infty$ onto $L_1^\sharp$.

(i) Suppose that $g, g_1, g_2, \ldots \in L_\infty$, and let $\rho, \rho_1, \rho_2, \ldots$ be the corresponding bounded linear functionals on $L_1$. Show that, if

$$\sup\{\|g_n\|_\infty : n \in \mathbb{N}\} < \infty$$

and $g(s) = \lim_{n\to\infty} g_n(s)$ for almost all $s$, then the sequence $\{\rho_n\}$ is weak* convergent to $\rho$.

(ii) Show that, if $m$ is Lebesgue measure on the interval $[0, 1]$, and $g_1, g_2, g_3, \ldots$ (in $L_\infty$) are defined by

$$g_n(s) = (-1)^r \qquad (2^{-n}r \leqslant s < 2^{-n}(r+1), \quad r = 0, 1, \ldots, 2^n - 1),$$

then the corresponding sequence $\{\rho_n\}$ of bounded linear functionals on $L_1$ is weak* convergent to 0.

1.9.34. Suppose that $\sum_{n=1}^\infty x_n$ is a series of elements of a complex Banach space $\mathfrak{X}$ such that, for every strictly increasing sequence $\{n(1), n(2), \ldots\}$ of positive integers, the subseries $\sum_{j=1}^\infty x_{n(j)}$ converges in the weak topology to an element of $\mathfrak{X}$.

(i) Prove that $\sum_{n=1}^{\infty} |\rho(x_n)| < \infty$ for each $\rho$ in $\mathfrak{X}^\#$.

(ii) Show that the equation $S\rho = \{\rho(x_1), \rho(x_2), \ldots\}$ defines a bounded linear operator $S$ from $\mathfrak{X}^\#$ into the Banach space $l_1$.

(iii) Prove that, if $A = \{a_1, a_2, \ldots\} \in l_\infty$, the equation

$$\Omega_A(\rho) = \sum_{n=1}^{\infty} \rho(a_n x_n) \qquad (\rho \in \mathfrak{X}^\#)$$

defines a bounded linear functional $\Omega_A$ on $\mathfrak{X}^\#$, and $\|\Omega_A\| \leqslant \|S\| \, \|A\|_\infty$.

(iv) Show that $\Omega_A$ lies in $\hat{\mathfrak{X}}$ (the natural image of $\mathfrak{X}$ in $\mathfrak{X}^{\#\#}$) whenever $A$ (in $l_\infty$) is a sequence that takes only finitely many distinct values. Deduce that $\Omega_A \in \hat{\mathfrak{X}}$ for all $A$ in $l_\infty$.

(v) Prove that, for every $A = \{a_1, a_2, \ldots\}$ in $l_\infty$, the series $\sum_{n=1}^{\infty} a_n x_n$ is weakly convergent to an element of $\mathfrak{X}$.

(vi) Show that, if a sequence $\{\rho_n\}$ in $\mathfrak{X}^\#$ is weak* convergent to an element $\rho$ of $\mathfrak{X}^\#$, then $\{S\rho_n\}$ is norm convergent to $S\rho$. [*Hint*. Use (v) and the result of Exercise 1.9.26.]

(vii) Prove that, if $\varepsilon > 0$, there is a positive integer $n(\varepsilon)$, such that $\sum_{n=n(\varepsilon)}^{\infty} |\rho(x_n)| \leqslant \varepsilon \|\rho\|$ for each $\rho$ in $\mathfrak{X}^\#$. [*Hint*. Upon replacing $\mathfrak{X}$ by the closed linear span of $\{x_n\}$, reduce to the case in which $\mathfrak{X}$ is separable. If the result were false, we could choose $\rho_1, \rho_2, \ldots$ in the unit ball of $\mathfrak{X}^\#$, satisfying $\sum_{n=k}^{\infty} |\rho_k(x_n)| \geqslant \varepsilon$. Obtain a contradiction by using Exercise 1.9.8 and (vi).]

(viii) Prove that, for each bounded complex sequence $\{a_1, a_2, \ldots\}$, the series $\sum_{n=1}^{\infty} a_n x_n$ converges in the norm topology on $\mathfrak{X}$. [The assertion, that weak convergence of every subseries entails norm convergence, is known as the Banach–Orlicz theorem.]

1.9.35. (i) Prove that, if the dual space $\mathfrak{X}^\#$ of a Banach space $\mathfrak{X}$ is separable, then $\mathfrak{X}$ is separable. [*Hint*. Let $\{\rho_n\}$ be a countable dense subset of the surface $\{\rho \in \mathfrak{X}^\# : \|\rho\| = 1\}$ of the unit ball $(\mathfrak{X}^\#)_1$; for each $n = 1, 2, \ldots,$ choose $x_n$ in $(\mathfrak{X})_1$ such that $|\rho_n(x_n)| > \frac{1}{2}$. Show that $\mathfrak{X}$ is the closed linear span of $\{x_n\}$.]

(ii) Show that a reflexive Banach space is separable if and only if its dual space is separable.

(iii) Give an example of a Banach space that is not reflexive but has a separable dual space (and is, therefore, separable).

1.9.36. Show that a bounded sequence $\{x_n\}$ of elements of a reflexive Banach space $\mathfrak{X}$ has a subsequence that is weakly convergent to an element of $\mathfrak{X}$. [*Hint*. Show that it is sufficient to consider the case in which $\mathfrak{X}$ is separable, by replacing $\mathfrak{X}$ by the closed linear span of $\{x_n\}$. In the separable case show, by use of Exercise 1.9.35(ii), that the required result can be deduced from Exercise 1.9.8.]

1.9.37. Suppose $\mathfrak{X}$ is a separable normed space and $\{x_n : n \in \mathbb{N}\}$ is a (norm-)dense subset of $(\mathfrak{X})_1$. Define

$$d(\rho, \rho') = \sum_{n=1}^{\infty} 2^{-n}|(\rho - \rho')(x_n)| \qquad (\rho, \rho' \in \mathfrak{X}^{\#}).$$

(i) Show that $d$ is a metric on $\mathfrak{X}^{\#}$.

(ii) Show that the metric topology induced by $d$ on $(\mathfrak{X}^{\#})_1$ is the weak* topology on $(\mathfrak{X}^{\#})_1$.

(iii) Use the fact that each sequence in a compact metric space has a convergent subsequence to solve Exercise 1.9.8 (and 1.9.36) again.

1.9.38. Use the Baire category theorem (see the proof of Lemma 1.8.3) to give another proof (direct) of the uniform boundedness principle (Theorem 1.8.9).

1.9.39. If $\{T_n\}$ is a sequence of bounded linear transformations of one Banach space $\mathfrak{X}$ into another Banach space $\mathcal{Y}$ and $\{T_n x\}$ converges for each $x$ in $\mathfrak{X}$, show that $T$, defined by $Tx = \lim_n T_n x$, is a bounded linear transformation of $\mathfrak{X}$ into $\mathcal{Y}$.

1.9.40. Suppose $\mathfrak{X}$ and $\mathcal{Y}$ are Banach spaces and $T$ is a linear transformation of $\mathfrak{X}$ into $\mathcal{Y}$. With $\eta$ in the algebraic dual of $\mathcal{Y}$, let $(T'\eta)(x)$ be $\eta(Tx)$ for each $x$ in $\mathfrak{X}$; and suppose that $T'\rho \in \mathfrak{X}^{\#}$ for each $\rho$ in $\mathcal{Y}^{\#}$. Show that $T$ is bounded and that $T'|\mathcal{Y}^{\#} = T^{\#}$. [*Hint*. Consider the graph of $T$.]

# CHAPTER 2

# BASICS OF HILBERT SPACE AND LINEAR OPERATORS

This chapter deals with the elementary geometry of Hilbert spaces and with the simplest properties of Hilbert space operators. Section 2.1 is concerned with inner products, and the corresponding norms, on linear spaces with complex (or, occasionally, real) scalars. It introduces the concept of Hilbert space and provides a number of examples. Section 2.2 is devoted to the notion of orthogonality in a Hilbert space. In it we deal with orthogonal complements of closed subspaces, orthogonal sets, orthonormal bases, dimension, and the classification of Hilbert spaces up to isomorphism. This is followed, in Section 2.3, by Riesz's representation theorem concerning the form of bounded linear functionals on a Hilbert space, and some corollaries concerning the weak topology of such a space. Section 2.4 is devoted to bounded linear operators acting on Hilbert spaces, with primary emphasis on elementary properties of the "Hilbert adjoint" of such an operator. Special classes of operators (normal, self-adjoint, positive, unitary) are considered briefly, and illustrative examples are given. Section 2.5 is concerned with orthogonal projections, corresponding to the decomposition of a Hilbert space as the direct sum of a closed subspace and its orthogonal complement. It includes an account of the order structure of projections, and its relation to the strong-operator topology. In Section 2.6 we deal with elementary constructions with Hilbert spaces, such as direct sums and tensor products, together with related aspects of operator theory. Section 2.7 is concerned with unbounded linear operators on Hilbert spaces.

## 2.1. Inner products on linear spaces

By an *inner product* on a complex vector space $\mathscr{H}$, we mean a mapping $(x, y) \to \langle x, y \rangle$, from $\mathscr{H} \times \mathscr{H}$ into the scalar field $\mathbb{C}$, such that

(i) $\langle ax + by, z \rangle = a\langle x, z \rangle + b\langle y, z \rangle$,

(ii) $\langle y, x \rangle = \overline{\langle x, y \rangle}$,

(iii) $\langle x, x \rangle \geqslant 0$,

whenever $x, y, z \in \mathscr{H}$ and $a, b \in \mathbb{C}$. If, in addition,

(iv) $\langle x, x \rangle = 0$ only when $x = 0$,

the inner product is said to be *definite* (sometimes, *positive definite* is used). In (ii) we adopt the convention that $\bar{c}$ denotes the complex conjugate of an element $c$ of $\mathbb{C}$. From (i) and (ii), an inner product satisfies the further condition (conjugate linearity in its second variable)

(v) $\langle z, ax + by \rangle = \bar{a}\langle z, x \rangle + \bar{b}\langle z, y \rangle$.

When $\langle\ ,\ \rangle$ is an inner product on a complex vector space $\mathscr{H}$, the pair $(\mathscr{H}, \langle\ ,\ \rangle)$ is called a (complex) *inner product space*, and we refer to the complex number $\langle x, y \rangle$ as the inner product of the vectors $x$ and $y$ in $\mathscr{H}$.

For real vector spaces, the definition of inner products is the same as the one given above, except that scalars and the values $\langle x, y \rangle$ are required to be real, so that the "bars" denoting complex conjugation no longer appear in (ii) and (v). As regards elementary geometrical properties of inner product spaces, there is very little difference between the real and complex cases. In the main, we shall restrict attention to the complex case, making only occasional comments on the modifications needed to deal with real spaces. For the theory of linear operators on inner product spaces and algebras of such operators, the complex case has significant advantage over the real one.

The finite-dimensional linear spaces $\mathbb{C}^n$ and $\mathbb{R}^n$ provide the simplest examples of inner product spaces, with the inner product defined by

$$\langle (a_1, \ldots, a_n), (b_1, \ldots, b_n) \rangle = a_1\bar{b}_1 + \cdots + a_n\bar{b}_n$$

(complex conjugation being redundant in the case of $\mathbb{R}^n$). Just as the real space $\mathbb{R}^n$ can be viewed as a real-linear subspace of the complex space $\mathbb{C}^n$, it can be shown that every real vector (or normed, or inner product) space can naturally be imbedded in a complex space of the same type (see Exercises 1.9.6 and 2.8.3).

2.1.1. Proposition. *Suppose that $\langle\ ,\ \rangle$ is an inner product on a complex vector space $\mathscr{H}$.*

(i) $|\langle x, y \rangle|^2 \leqslant \langle x, x \rangle\langle y, y \rangle$, *for all $x$ and $y$ in $\mathscr{H}$.*

(ii) *The set $\mathscr{L} = \{z \in \mathscr{H} : \langle z, z \rangle = 0\}$ is a linear subspace of $\mathscr{H}$, and the equation*

$$\langle x + \mathscr{L}, y + \mathscr{L} \rangle_1 = \langle x, y \rangle \qquad (x, y \in \mathscr{H})$$

*defines a definite inner product $\langle\ ,\ \rangle_1$ on the quotient space $\mathscr{H}/\mathscr{L}$.*

*Proof.* (i) When $x, y \in \mathscr{H}$ and $a, b \in \mathbb{C}$,

$$\begin{aligned}\langle ax + by, ax + by \rangle &= a\langle x, ax + by \rangle + b\langle y, ax + by \rangle \\ &= a\bar{a}\langle x, x \rangle + a\bar{b}\langle x, y \rangle + b\bar{a}\langle y, x \rangle + b\bar{b}\langle y, y \rangle,\end{aligned}$$

so

$$(1) \qquad \langle ax + by, ax + by \rangle = |a|^2\langle x, x \rangle + 2\ \mathrm{Re}\ a\bar{b}\langle x, y \rangle + |b|^2\langle y, y \rangle.$$

By taking $a = t\langle y, x\rangle$, where $t$ is real, and $b = 1$, we obtain

$$\begin{aligned} 0 &\leqslant \langle ax + by, ax + by\rangle \\ &= t^2|\langle x, y\rangle|^2\langle x, x\rangle + 2t|\langle x, y\rangle|^2 + \langle y, y\rangle \qquad (t \in \mathbb{R}). \end{aligned}$$

If $\langle x, x\rangle = 0$, it follows, by considering large negative $t$, that $|\langle x, y\rangle| = 0$, so $|\langle x, y\rangle|^2 = \langle x, x\rangle\langle y, y\rangle = 0$ in this case. If $\langle x, x\rangle > 0$, we can take $t = -1/\langle x, x\rangle$, to obtain

$$\begin{aligned} 0 &\leqslant |\langle x, y\rangle|^2/\langle x, x\rangle - 2|\langle x, y\rangle|^2/\langle x, x\rangle + \langle y, y\rangle \\ &= \langle y, y\rangle - |\langle x, y\rangle|^2/\langle x, x\rangle, \end{aligned}$$

whence $|\langle x, y\rangle|^2 \leqslant \langle x, x\rangle\langle y, y\rangle$.

(ii) Let

$$\mathscr{L}_1 = \{z \in \mathscr{H} : \langle z, y\rangle = 0 \text{ for each } y \text{ in } \mathscr{H}\}.$$

It is evident that $\mathscr{L}_1$ is a linear subspace of $\mathscr{H}$, contained in the set $\mathscr{L}$ defined in the proposition, and that

$$\mathscr{L}_1 = \{z \in \mathscr{H} : \langle y, z\rangle = 0 \text{ for each } y \text{ in } \mathscr{H}\}.$$

With $z$ in $\mathscr{L}$, it follows from (i) that

$$|\langle z, y\rangle|^2 \leqslant \langle z, z\rangle\langle y, y\rangle = 0, \qquad \langle z, y\rangle = 0 \qquad (y \in \mathscr{H}),$$

so $z \in \mathscr{L}_1$. Hence $\mathscr{L} = \mathscr{L}_1$, and $\mathscr{L}$ is a linear subspace of $\mathscr{H}$.

If $x, y \in \mathscr{H}$ and $z_1, z_2 \in \mathscr{L}$ $(= \mathscr{L}_1)$, we have

$$\langle x + z_1, y + z_2\rangle = \langle x, y\rangle + \langle x, z_2\rangle + \langle z_1, y\rangle + \langle z_1, z_2\rangle = \langle x, y\rangle.$$

It follows that the equation

$$\langle x + \mathscr{L}, y + \mathscr{L}\rangle_1 = \langle x, y\rangle \qquad (x, y \in \mathscr{H})$$

defines (unambiguously) a mapping $(u, v) \to \langle u, v\rangle_1$ from $(\mathscr{H}/\mathscr{L}) \times (\mathscr{H}/\mathscr{L})$ into $\mathbb{C}$. It is clear that $\langle\ ,\ \rangle_1$ inherits from $\langle\ ,\ \rangle$ the three defining properties of an inner product. If $0 = \langle x + \mathscr{L}, x + \mathscr{L}\rangle_1$ $(= \langle x, x\rangle)$, then $x \in \mathscr{L}$, and $x + \mathscr{L}$ is the zero element of $\mathscr{H}/\mathscr{L}$. Thus $\langle\ ,\ \rangle_1$ is a definite inner product on $\mathscr{H}/\mathscr{L}$. ∎

The inequality stated in Proposition 2.1.1(i) is known as the *Cauchy–Schwarz inequality*.

2.1.2. PROPOSITION. *If $\langle\ ,\ \rangle$ is an inner product on a complex vector space $\mathscr{H}$, the equation*

$$\|x\| = \langle x, x\rangle^{1/2} \qquad (x \in \mathscr{H}) \tag{2}$$

*defines a semi-norm $\|\ \|$ on $\mathscr{H}$. If the inner product is definite, $\|\ \|$ is a norm on $\mathscr{H}$.*

*Proof.* With $\|\ \|$ defined by (2), it is apparent that $\|x\| \geqslant 0$ and $\|ax\| = |a|\,\|x\|$ whenever $x \in \mathscr{H}$ and $a \in \mathbb{C}$. Moreover, if the inner product is definite,

$\|x\| = 0$ only when $x = 0$. The Cauchy–Schwarz inequality can be written in the form

$$|\langle x, y\rangle| \leqslant \langle x, x\rangle^{1/2}\langle y, y\rangle^{1/2} = \|x\|\,\|y\| \qquad (x, y \in \mathscr{H}).$$

From (1), with $a = b = 1$,

$$\begin{aligned}\|x + y\|^2 &= \|x\|^2 + 2\,\mathrm{Re}\langle x, y\rangle + \|y\|^2\\ &\leqslant \|x\|^2 + 2|\langle x, y\rangle| + \|y\|^2\\ &\leqslant \|x\|^2 + 2\|x\|\,\|y\| + \|y\|^2 = (\|x\| + \|y\|)^2.\end{aligned}$$

Hence $\|x + y\| \leqslant \|x\| + \|y\|$ for each $x$ and $y$ in $\mathscr{H}$. ■

When referring to the norm on a (definite) inner product space, it is understood, in the absence of an explicit statement to the contrary, that the norm intended is the one constructed as in Proposition 2.1.2 from the inner product. For such spaces, we have proved the triangle inequality

$$\|x + y\| \leqslant \|x\| + \|y\|$$

and the Cauchy–Schwarz inequality

$$|\langle x, y\rangle| \leqslant \|x\|\,\|y\|.$$

For each of these results, we now determine the conditions under which equality occurs.

2.1.3. Proposition. *If $\langle\ ,\ \rangle$ is a definite inner product on a complex vector space $\mathscr{H}$ and $x, y \in \mathscr{H}$, the following three conditions are equivalent:*

(i) $\|x + y\| = \|x\| + \|y\|$;
(ii) $\langle x, y\rangle = \|x\|\,\|y\|$;
(iii) *one of $x$ and $y$ is a non-negative scalar multiple of the other.*

*Proof.* For any scalars $a$ and $b$, it follows from (1) that

$$\|ax + by\|^2 = |a|^2\|x\|^2 + 2\,\mathrm{Re}\, a\bar{b}\langle x, y\rangle + |b|^2\|y\|^2. \tag{3}$$

Thus

$$(\|x\| + \|y\|)^2 - \|x + y\|^2 = 2(\|x\|\,\|y\| - \mathrm{Re}\langle x, y\rangle).$$

If (i) is satisfied, the last equation and the Cauchy–Schwarz inequality give

$$\mathrm{Re}\langle x, y\rangle = \|x\|\,\|y\| \geqslant |\langle x, y\rangle|,$$

and therefore $\langle x, y\rangle = \mathrm{Re}\langle x, y\rangle = \|x\|\,\|y\|$. Thus (i) implies (ii).

If (ii) is satisfied and $a, b$ are real, (3) gives

$$\|ax + by\|^2 = (a\|x\| + b\|y\|)^2.$$

With $a = \|y\|$ and $b = -\|x\|$, it follows that $\|y\|x - \|x\|y = 0$. Hence either $x = 0\ (= 0 \cdot y)$ or $y = \|x\|^{-1}\|y\|x$, and so, (ii) implies (iii).

If (iii) is satisfied, we may suppose that $x = ay$, where $a \geqslant 0$. Then

$$\begin{aligned}\|x + y\| &= \|(a + 1)y\| = |a + 1|\ \|y\| \\ &= (a + 1)\|y\| = \|ay\| + \|y\| = \|x\| + \|y\|;\end{aligned}$$

so (iii) implies (i). ■

2.1.4. Corollary. *If $\langle\ ,\ \rangle$ is a definite inner product on a complex vector space $\mathscr{H}$ and $x, y \in \mathscr{H}$, then $|\langle x, y\rangle| = \|x\|\ \|y\|$ if and only if $x$ and $y$ are linearly dependent.*

*Proof.* If $|\langle x, y\rangle| = \|x\|\,\|y\|$, we can choose a scalar $a$ so that $|a| = 1$ and $a\langle x, y\rangle = \|x\|\,\|y\|$; that is, $\langle ax, y\rangle = \|ax\|\,\|y\|$. By Proposition 2.1.3, one of $ax$ and $y$ is a non-negative scalar multiple of the other; so $x$ and $y$ are linearly dependent.

Conversely, suppose that $x$ and $y$ are linearly dependent. We may assume that $x = ay$ for some scalar $a$, and then

$$|\langle x, y\rangle| = |a\langle y, y\rangle| = |a|\,\|y\|^2 = \|x\|\,\|y\|. \quad ■$$

A complex normed space $\mathscr{H}$ is said to be a *pre-Hilbert space* if its norm $\|\ \|$ can be obtained, as in Proposition 2.1.2, from a (necessarily) definite inner product on $\mathscr{H}$. If, in addition, $\mathscr{H}$ is complete relative to $\|\ \|$, then $\mathscr{H}$ is described as a *Hilbert space* (of course, one can consider *real Hilbert spaces* — complete real inner product spaces). Accordingly, Hilbert spaces form a particular class of Banach spaces, and the theory developed in Chapter 1 for linear topological spaces, normed spaces, and Banach spaces is available in the case of Hilbert spaces. The geometry of Hilbert spaces is in many respects analogous to elementary euclidean geometry, and is simpler and more extensive than any corresponding theory for general Banach spaces. In consequence, the analysis of Hilbert space operators is more fully developed than its Banach space counterpart. The main objects studied in this book are certain algebras of linear operators acting on Hilbert spaces.

2.1.5. Proposition. *The inner product on a pre-Hilbert space $\mathscr{H}$ is a continuous mapping from $\mathscr{H} \times \mathscr{H}$ into $\mathbb{C}$.*

*Proof.* When $x, y, x_0, y_0 \in \mathscr{H}$,

$$\begin{aligned}\langle x, y\rangle &= \langle x_0 + (x - x_0), y_0 + (y - y_0)\rangle \\ &= \langle x_0, y_0\rangle + \langle x_0, y - y_0\rangle + \langle x - x_0, y_0\rangle + \langle x - x_0, y - y_0\rangle.\end{aligned}$$

From this and the Cauchy–Schwarz inequality,

$$(4)\qquad |\langle x, y\rangle - \langle x_0, y_0\rangle| \leqslant \|x_0\|\,\|y - y_0\| + \|x - x_0\|\,\|y_0\| + \|x - x_0\|\,\|y - y_0\|,$$

and the right-hand side is small when $x$ is close to $x_0$ and $y$ is close to $y_0$. ■

In Theorem 1.5.1, we showed that a normed space $\mathfrak{X}$ can be embedded, essentially uniquely, as an everywhere-dense subspace of a Banach space $\hat{\mathfrak{X}}$, the completion of $\mathfrak{X}$. We now consider the case in which $\mathfrak{X}$ is a pre-Hilbert space.

2.1.6. Proposition. *If $\mathscr{H}$ is a pre-Hilbert space, its completion $\hat{\mathscr{H}}$ is a Hilbert space.*

*Proof.* For $n = 1, 2, \ldots$, let

$$S_n = \{(x, y): x, y \in \mathscr{H},\ \|x\| < n,\ \|y\| < n\},$$
$$\hat{S}_n = \{(x, y): x, y \in \hat{\mathscr{H}},\ \|x\| < n,\ \|y\| < n\},$$

and note that $S_n$ is everywhere dense in $\hat{S}_n$ in the topology on $\hat{\mathscr{H}} \times \hat{\mathscr{H}}$. From (4),

$$|\langle x, y\rangle - \langle x_0, y_0\rangle| \leqslant n\|x - x_0\| + n\|y - y_0\| + \|x - x_0\|\,\|y - y_0\|,$$

when $(x, y)$, $(x_0, y_0) \in S_n$. From this, the mapping $f_n: S_n \to \mathbb{C}$, defined by

$$(5)\qquad f_n(x, y) = \langle x, y\rangle,$$

is uniformly continuous on $S_n$, and so extends uniquely to a continuous mapping $\hat{f}_n: \hat{S}_n \to \mathbb{C}$. When $m \geqslant n$, the restriction $\hat{f}_m | \hat{S}_n$ is another continuous extension of $f_n$, so $\hat{f}_m | \hat{S}_n = \hat{f}_n$. It follows that there is a mapping $f: \hat{\mathscr{H}} \times \hat{\mathscr{H}} \to \mathbb{C}$ such that $f | \hat{S}_n = \hat{f}_n$ $(n = 1, 2, \ldots)$.

We assert that $f$ is an inner product on $\hat{\mathscr{H}}$, and gives rise to its norm, whence $\hat{\mathscr{H}}$ is a Hilbert space. For this, we have to show that

$$f(ax + by, z) = af(x, z) + bf(y, z),$$
$$f(y, x) = \overline{f(x, y)},$$
$$f(x, x) = \|x\|^2,$$

whenever $x, y, z \in \hat{\mathscr{H}}$ and $a, b \in \mathbb{C}$. We can choose an integer $n$ that exceeds the norm of each of the vectors $x, y, z, ax + by$, and by continuity of $f | \hat{S}_n$ $(= \hat{f}_n)$, it then suffices to prove the three required equations under the additional assumption that $x, y, z \in \mathscr{H}$. However, in this case, these three equations follow at once from (5), since $f | S_n = f_n$. ■

Next, we prove two identities, both of which are frequently useful, concerning vectors in a pre-Hilbert space. The first of these is known as the *parallelogram law*.

2.1.7. PROPOSITION. *If $u, v, x$, and $y$ are vectors in a pre-Hilbert space,*

$$\|x + y\|^2 + \|x - y\|^2 = 2\|x\|^2 + 2\|y\|^2,$$

*and*

$$4\langle u, y\rangle = \langle u + v, x + y\rangle - \langle u - v, x - y\rangle + i\langle u + iv, x + iy\rangle - i\langle u - iv, x - iy\rangle.$$

*Proof.* The first identity is an immediate consequence of the equations

$$\|x \pm y\|^2 = \|x\|^2 \pm 2\,\mathrm{Re}\langle x, y\rangle + \|y\|^2,$$

which are particular cases of (3). For the second identity, note first that

$$\langle u \pm v, x \pm y\rangle = \langle u, x\rangle + \langle v, y\rangle \pm (\langle u, y\rangle + \langle v, x\rangle)$$

(with the same choice of the ambiguous sign throughout). Thus

$$\langle u + v, x + y\rangle - \langle u - v, x - y\rangle = 2\langle u, y\rangle + 2\langle v, x\rangle.$$

Upon replacing $v$ by $iv$ and $y$ by $iy$, we obtain

$$\langle u + iv, x + iy\rangle - \langle u - iv, x - iy\rangle = -2i\langle u, y\rangle + 2i\langle v, x\rangle.$$

From the last two equations,

$$\langle u + v, x + y\rangle - \langle u - v, x - y\rangle + i\langle u + iv, x + iy\rangle - i\langle u - iv, x - iy\rangle = 4\langle u, y\rangle. \quad \blacksquare$$

We illustrate the use of the two identities just established, in obtaining the following characterization of pre-Hilbert spaces within the class of normed spaces.

2.1.8. PROPOSITION. *A complex normed space $\mathscr{H}$ is a pre-Hilbert space if and only if*

$$\|x + y\|^2 + \|x - y\|^2 = 2\|x\|^2 + 2\|y\|^2 \qquad (x, y \in \mathscr{H}). \tag{6}$$

*When this condition is satisfied, there is a unique inner product on $\mathscr{H}$ that defines its norm, and this is given by*

$$4\langle x, y\rangle = \|x + y\|^2 - \|x - y\|^2 + i\|x + iy\|^2 - i\|x - iy\|^2. \tag{7}$$

*Proof.* When $\mathscr{H}$ is a pre-Hilbert space, it follows from Proposition 2.1.7, with $u = x$ and $v = y$, that the norm satisfies (6) and the inner product is determined by (7).

Conversely, suppose that $\mathscr{H}$ is a normed space whose norm satisfies (6); and when $x, y \in \mathscr{H}$, define a scalar $\langle x, y\rangle$ by (7). Then

$$\langle x, x\rangle = \tfrac{1}{4}(\|2x\|^2 + i\|(1 + i)x\|^2 - i\|(1 - i)x\|^2) = \tfrac{1}{4}\|x\|^2(4 + i|1 + i|^2 - i|1 - i|^2) = \|x\|^2,$$

and

$$\begin{aligned}\langle y,x\rangle &= \tfrac{1}{4}(\|y+x\|^2 - \|y-x\|^2 + i\|y+ix\|^2 - i\|y-ix\|^2)\\ &= \tfrac{1}{4}(\|x+y\|^2 - \|x-y\|^2 + i\|i(x-iy)\|^2 - i\|i(x+iy)\|^2)\\ &= \overline{\langle x,y\rangle}.\end{aligned}$$

In order to complete the proof that $\langle\ ,\ \rangle$ is an inner product on $\mathscr{H}$, which defines the norm on $\mathscr{H}$, it remains only to show that, for each fixed $y$ in $\mathscr{H}$, the equation

$$f(x) = \|x+y\|^2 - \|x-y\|^2 + i\|x+iy\|^2 - i\|x-iy\|^2 \tag{8}$$

defines a linear functional $f$ on $\mathscr{H}$. We begin by proving that

$$f(ix) = if(x), \qquad f(x_1+x_2) = f(x_1) + f(x_2), \tag{9}$$

for all $x, x_1, x_2$ in $\mathscr{H}$. For this, note that

$$\begin{aligned}f(ix) &= \|ix+y\|^2 - \|ix-y\|^2 + i\|i(x+y)\|^2 - i\|i(x-y)\|^2\\ &= i(-i\|i(x-iy)\|^2 + i\|i(x+iy)\|^2 + \|x+y\|^2 - \|x-y\|^2)\\ &= if(x).\end{aligned}$$

Since the norm satisfies the parallelogram law (6),

$$\begin{aligned}\|x_1+y\|^2 + \|x_2+y\|^2 &= \tfrac{1}{2}\|x_1+x_2+2y\|^2 + \tfrac{1}{2}\|x_1-x_2\|^2\\ &= 2\|\tfrac{1}{2}(x_1+x_2)+y\|^2 + \tfrac{1}{2}\|x_1-x_2\|^2.\end{aligned}$$

From this and the three similar equations obtained when $y$ is replaced by $-y, iy, -iy$, it follows that

$$f(x_1) + f(x_2) = 2f(\tfrac{1}{2}(x_1+x_2)). \tag{10}$$

With $x_1 = x$ and $x_2 = 0$, (10) gives

$$f(x) = 2f(\tfrac{1}{2}x) \qquad (x \in \mathscr{H}),$$

since it is apparent from (8) that $f(0) = 0$. Hence (10) can be rewritten in the form

$$f(x_1) + f(x_2) = f(x_1+x_2),$$

and (9) is proved.

It now remains to show that $f(ax) = af(x)$ whenever $x \in \mathscr{H}$ and $a \in \mathbb{C}$. Equivalently, we must prove that $\mathbb{F} = \mathbb{C}$ where

$$\mathbb{F} = \{a \in \mathbb{C} : f(ax) = af(x) \text{ for each } x \text{ in } \mathscr{H}\}.$$

From (8), $f$ is continuous, so $\mathbb{F}$ is closed. It is evident that $1 \in \mathbb{F}$, and that $ab$, $c^{-1} \in \mathbb{F}$ whenever $a, b, c\ (\neq 0) \in \mathbb{F}$. From (9), $i \in \mathbb{F}$, and $a + b \in \mathbb{F}$ whenever

$a, b \in \mathbb{F}$. The properties just listed imply that $s + it \in \mathbb{F}$ whenever $s$ and $t$ are rational, and so $\mathbb{F} = \mathbb{C}$, since $\mathbb{F}$ is closed. ■

2.1.9. REMARK. Most of the theory developed above for complex inner product spaces is valid also in the real case. For real inner product spaces, the second relation in Proposition 2.1.7 is omitted. In Proposition 2.1.8, the relation (7) between the inner product and norm is modified by the deletion of the last two terms on the right-hand side and is easily proved by direct computation. The remaining proofs require only minor alterations. ■

2.1.10. REMARK. Equation (7) gives an immediate alternative proof of the continuity of the inner product on a pre-Hilbert space. Moreover, if $\mathscr{H}$ is a normed space that satisfies the parallelogram law (6), it follows by continuity that the completion $\hat{\mathscr{H}}$ has the same property. This, together with Proposition 2.1.8, provides an alternative proof that the completion of a pre-Hilbert space is a Hilbert space. ■

2.1.11. EXAMPLE. With $n$ a positive integer, the complex vector space $\mathbb{C}^n$, consisting of all $n$-tuples $\mathbf{x} = (x_1, \ldots, x_n)$, $\mathbf{y} = (y_1, \ldots, y_n)$ of complex numbers, has a definite inner product defined by

$$\langle \mathbf{x}, \mathbf{y} \rangle = x_1\bar{y}_1 + \cdots + x_n\bar{y}_n.$$

The associated norm is given by

$$\|\mathbf{x}\| = (|x_1|^2 + \cdots + |x_n|^2)^{1/2}.$$

Since $\mathbb{C}^n$ is complete, relative to the metric $d(\mathbf{x}, \mathbf{y}) = \|\mathbf{x} - \mathbf{y}\|$, it is a Hilbert space. In this example, the Cauchy–Schwarz and triangle inequalities reduce to

$$\left|\sum_{j=1}^{n} x_j\bar{y}_j\right| \leqslant \left(\sum_{j=1}^{n} |x_j|^2\right)^{1/2} \left(\sum_{j=1}^{n} |y_j|^2\right)^{1/2},$$

$$\left(\sum_{j=1}^{n} |x_j + y_j|^2\right)^{1/2} \leqslant \left(\sum_{j=1}^{n} |x_j|^2\right)^{1/2} + \left(\sum_{j=1}^{n} |y_j|^2\right)^{1/2},$$

for all complex numbers $x_1, \ldots, x_n, y_1, \ldots, y_n$.

In the same way, the equation

$$\langle \mathbf{x}, \mathbf{y} \rangle = x_1y_1 + \cdots + x_ny_n \qquad (\mathbf{x}, \mathbf{y} \in \mathbb{R}^n)$$

defines a definite inner product on the real vector space $\mathbb{R}^n$.

The equation

$$\langle \mathbf{x}, \mathbf{y} \rangle_1 = x_1\bar{y}_1 \qquad (\mathbf{x}, \mathbf{y} \in \mathbb{C}^n)$$

defines an inner product on $\mathbb{C}^n$; when $n \geqslant 1$, $\langle\ ,\ \rangle_1$ is not definite, for $\mathscr{L}$ (see Proposition 2.1.1) consists of all vectors whose first component is zero. In this case, $\mathbb{C}^n/\mathscr{L}$ is a one-dimensional Hilbert space (isomorphic to $\mathbb{C}$). ■

2.1.12. Example. Given a set $\mathbb{A}$, the Banach space $l_2(\mathbb{A})$ described in Example 1.7.3 consists of all complex-valued functions $x$ on $\mathbb{A}$ for which the (unordered) sum $\sum_{a\in\mathbb{A}} |x(a)|^2$ is finite, and its norm is given by

$$\|x\| = \left( \sum_{a\in\mathbb{A}} |x(a)|^2 \right)^{1/2}. \tag{11}$$

When $x, y \in l_2(\mathbb{A})$, the sum $\sum_{a\in\mathbb{A}} x(a)\overline{y(a)}$ converges, since

$$|x(a)\overline{y(a)}| \leqslant \tfrac{1}{2}(|x(a)|^2 + |y(a)|^2), \qquad \sum_{a\in\mathbb{A}} (|x(a)|^2 + |y(a)|^2) < \infty.$$

From this, it follows easily that $l_2(\mathbb{A})$ has a definite inner product, defined by

$$\langle x, y\rangle = \sum_{a\in\mathbb{A}} x(a)\overline{y(a)},$$

which gives rise to the norm in (11). Hence $l_2(\mathbb{A})$ is a Hilbert space. In this example, the Cauchy–Schwarz and triangle inequalities assert that

$$\Big|\sum_{a\in\mathbb{A}} x(a)\overline{y(a)}\Big| \leqslant \left( \sum_{a\in\mathbb{A}} |x(a)|^2 \right)^{1/2} \left( \sum_{a\in\mathbb{A}} |y(a)|^2 \right)^{1/2},$$

$$\left( \sum_{a\in\mathbb{A}} |x(a) + y(a)|^2 \right)^{1/2} \leqslant \left( \sum_{a\in\mathbb{A}} |x(a)|^2 \right)^{1/2} + \left( \sum_{a\in\mathbb{A}} |y(a)|^2 \right)^{1/2}$$

for all $x$ and $y$ in $l_2(\mathbb{A})$.

When $\mathbb{A} = \{1, 2, \ldots, n\}$, $l_2(\mathbb{A})$ is the Hilbert space $\mathbb{C}^n$ considered in the preceding example. When $\mathbb{A}$ is the set $\{1, 2, 3, \ldots\}$ of all positive integers, we write $l_2$ in place of $l_2(\mathbb{A})$, and sometimes denote an element of this space as a sequence $\{x_n\}$. ■

2.1.13. Example. Let $l_2^{(0)}$ be the class of all complex-valued functions defined on the set $\mathbb{A} = \{1, 2, 3, \ldots\}$ that take non-zero values at only finitely many points of $\mathbb{A}$. Thus $l_2^{(0)}$ is a linear subspace of $l_2$, and so inherits from $l_2$, by restriction, a definite inner product and the associated norm. Hence $l_2^{(0)}$ is a pre-Hilbert space; we assert that it is *not* a Hilbert space, that is, it is incomplete. For this, we show that $l_2^{(0)}$ is everywhere dense in $l_2$, from which it follows (since $l_2 \neq l_2^{(0)}$) that $l_2^{(0)}$ is not closed in $l_2$, and therefore not complete.

With $x$ in $l_2$, define $x_1, x_2, x_3, \ldots$ in $l_2^{(0)}$ by

$$x_j(k) = \begin{cases} x(k) & \text{if } k \leqslant j \\ 0 & \text{if } k > j. \end{cases}$$

Then

$$\|x - x_j\| = \left( \sum_{k=1}^{\infty} |x(k) - x_j(k)|^2 \right)^{1/2} = \left( \sum_{k=j+1}^{\infty} |x(k)|^2 \right)^{1/2} \to 0$$

as $j \to \infty$. This shows that each element of $l_2$ is the limit of a sequence in $l_2^{(0)}$, and so proves our assertion that $l_2^{(0)}$ is everywhere dense in $l_2$. ■

2.1.14. EXAMPLE. Suppose that $m$ is a $\sigma$-finite measure defined on a $\sigma$-algebra $\mathcal{S}$ of subsets of a set $S$. The Banach space $L_2$ $(= L_2(S, \mathcal{S}, m))$, described in Example 1.7.5, consists of all (equivalence classes modulo null functions of) complex-valued measurable functions $x$ on $S$ for which

$$\int_S |x(s)|^2 \, dm(s) < \infty,$$

with the norm defined by

$$\|x\| = \left(\int_S |x(s)|^2 \, dm(s)\right)^{1/2}. \tag{12}$$

When $x, y \in L_2$, the function $x(s)\overline{y(s)}$ is integrable, since it is measurable and its absolute value is dominated by the integrable function $\frac{1}{2}(|x(s)|^2 + |y(s)|^2)$. From this, it follows easily that $L_2$ has a definite inner product, defined by

$$\langle x, y \rangle = \int_S x(s)\overline{y(s)} \, dm(s),$$

which gives rise to the norm in (12). Hence $L_2$ is a Hilbert space.

The Cauchy–Schwarz and triangle inequalities reduce, in this example, to

$$\left|\int_S x(s)\overline{y(s)} \, dm(s)\right| \leqslant \left(\int_S |x(s)|^2 \, dm(s)\right)^{1/2} \left(\int_S |y(s)|^2 \, dm(s)\right)^{1/2},$$

$$\left(\int_S |x(s) + y(s)|^2 \, dm(s)\right)^{1/2} \leqslant \left(\int_S |x(s)|^2 \, dm(s)\right)^{1/2} + \left(\int_S |y(s)|^2 \, dm(s)\right)^{1/2},$$

for all $x$ and $y$ in $L_2$. ■

## 2.2. Orthogonality

The theory of Hilbert spaces and Hilbert space operators is more tractable than its Banach space counterpart, largely because the presence of an inner product permits the introduction of a satisfactory concept of orthogonality. In the present section we study this concept, after obtaining some preliminary results.

We show first that, in a Hilbert space, the minimal distance from a point to a closed convex set is attained.

2.2.1. PROPOSITION. *If $Y$ is a closed convex subset of a Hilbert space $\mathscr{H}$, and $x_0 \in \mathscr{H}$, there is a unique element $y_0$ of $Y$ such that*

$$\|x_0 - y_0\| \leqslant \|x_0 - y\| \qquad (y \in Y). \tag{1}$$

*Moreover,*

$$\operatorname{Re}\langle y_0, x_0 - y_0\rangle \geqslant \operatorname{Re}\langle y, x_0 - y_0\rangle \qquad (y \in Y). \tag{2}$$

*Proof.* With

$$d = \inf\{\|x_0 - y\| : y \in Y\},$$

there is a sequence $\{y_n\}$ of elements of $Y$ such that $\|x_0 - y_n\| \to d$. By the parallelogram law,

$$2\|x_0 - y_m\|^2 + 2\|x_0 - y_n\|^2 = \|2x_0 - y_m - y_n\|^2 + \|y_n - y_m\|^2$$

for all positive integers $m$ and $n$. Since $\frac{1}{2}(y_m + y_n) \in Y$, we have

$$\|2x_0 - y_m - y_n\| = 2\|x_0 - \tfrac{1}{2}(y_m + y_n)\| \geqslant 2d,$$

and therefore

$$\begin{aligned}\|y_n - y_m\|^2 &= 2\|x_0 - y_m\|^2 + 2\|x_0 - y_n\|^2 - \|2x_0 - y_m - y_n\|^2\\ &\leqslant 2\|x_0 - y_m\|^2 + 2\|x_0 - y_n\|^2 - 4d^2 \to 0\end{aligned}$$

as $\min(m, n) \to \infty$. Hence $\{y_n\}$ is a Cauchy sequence, and so converges to an element $y_0$ of $\mathscr{H}$. Moreover, $y_0 \in Y$, since $Y$ is closed; and $y_0$ satisfies (1), since

$$\|x_0 - y_0\| = \lim_{n\to\infty} \|x_0 - y_n\| = d = \inf\{\|x_0 - y\| : y \in Y\}.$$

If $y_0'$ is another element of $Y$ that satisfies (1), then $\|x_0 - y_0'\| = \|x_0 - y_0\| = d$. We can apply the preceding reasoning, with $y_0$ and $y_0'$ in place of $y_m$ and $y_n$, to obtain

$$\begin{aligned}\|y_0' - y_0\|^2 &= 2\|x_0 - y_0\|^2 + 2\|x_0 - y_0'\|^2 - 4\|x_0 - \tfrac{1}{2}(y_0 + y_0')\|^2\\ &\leqslant 2d^2 + 2d^2 - 4d^2 = 0.\end{aligned}$$

Hence $y_0' = y_0$, and $y_0$ is uniquely determined by (1).

For each $y$ in $Y$ and $t$ in $(0, 1)$, $y_0 + t(y - y_0) \in Y$, and (1) gives

$$\begin{aligned}\|x_0 - y_0\|^2 &\leqslant \|x_0 - y_0 - t(y - y_0)\|^2\\ &= \|x_0 - y_0\|^2 - 2t\operatorname{Re}\langle y - y_0, x_0 - y_0\rangle + t^2\|y - y_0\|^2.\end{aligned}$$

Hence

$$-2\operatorname{Re}\langle y - y_0, x_0 - y_0\rangle + t\|y - y_0\|^2 \geqslant 0 \qquad (0 < t < 1),$$

and this gives (2) when $t \to 0$. ■

2.2.2. REMARK. We have proved, in Theorem 1.3.4, that a closed convex subset $Y$ of a locally convex space is closed also in the weak topology. For a Hilbert space $\mathscr{H}$, Proposition 2.2.1 permits an alternative proof of this result. For this, suppose that $x_0 \in \mathscr{H} \setminus Y$, and let $y_0$ be the element of $Y$ that satisfies (1) and (2). Then $\|x_0 - y_0\| > 0$, and

$$\begin{aligned}\operatorname{Re}\langle y, x_0 - y_0\rangle &\leqslant \operatorname{Re}\langle y_0, x_0 - y_0\rangle \\ &= \operatorname{Re}\langle x_0, x_0 - y_0\rangle - \langle x_0 - y_0, x_0 - y_0\rangle \\ &= \operatorname{Re}\langle x_0, x_0 - y_0\rangle - \|x_0 - y_0\|^2\end{aligned}$$

for each $y$ in $Y$. Thus

$$\mathscr{H} \setminus Y \supseteq V,$$

where

$$V = \{x \in \mathscr{H} : \operatorname{Re}\langle x, x_0 - y_0\rangle > \operatorname{Re}\langle x_0, x_0 - y_0\rangle - \|x_0 - y_0\|^2\}.$$

The equation $\rho(x) = \langle x, x_0 - y_0\rangle$ defines a linear functional $\rho$ on $\mathscr{H}$; and from continuity of the inner product, $\rho$ is bounded, and is therefore weakly continuous. Since

$$V = \{x \in \mathscr{H} : \operatorname{Re}\rho(x) > \operatorname{Re}\rho(x_0) - \|x_0 - y_0\|^2\},$$

it follows that $V(\subseteq \mathscr{H} \setminus Y)$ is a neighborhood of $x_0$ in the weak topology on $\mathscr{H}$. Hence $\mathscr{H} \setminus Y$ is weakly open, and $Y$ is weakly closed. ■

Suppose that $\mathscr{H}$ is a Hilbert space, $u, v \in \mathscr{H}$, and $X$, $Y$ are subsets of $\mathscr{H}$. We say that *$u$ is orthogonal to $v$* if $\langle u, v\rangle = 0$, that *$u$ is orthogonal to $Y$* if $\langle u, y\rangle = 0$ for each $y$ in $Y$, and that *$X$ is orthogonal to $Y$* if $\langle x, y\rangle = 0$ whenever $x \in X$ and $y \in Y$. The set of all vectors, in $\mathscr{H}$ and orthogonal to $Y$, is denoted by $Y^\perp$. When $Y$ is a closed subspace of $\mathscr{H}$, we sometimes write $\mathscr{H} \ominus Y$ in place of $Y^\perp$.

If $u$ is orthogonal to $v$, then also $v$ is orthogonal to $u$, and by expanding the inner product $\langle u + v, u + v\rangle$ we obtain

$$\|u + v\|^2 = \|u\|^2 + \|v\|^2.$$

From continuity and linearity of the inner product in its first variable, $Y^\perp$ is a closed subspace of $\mathscr{H}$. It is apparent that $X \subseteq Y^\perp$ if and only if $Y \subseteq X^\perp$, and that $X^\perp \subseteq Y^\perp$ if $X \supseteq Y$. If $u \in Y^\perp$, then $Y \subseteq \{u\}^\perp$. Moreover, since $\{u\}^\perp$ is a closed subspace containing $Y$, it contains the closed subspace $[Y]$ generated by $Y$, and so $u \in [Y]^\perp$. This shows that $Y^\perp \subseteq [Y]^\perp$, and the reverse inclusion is apparent since $Y \subseteq [Y]$; so

$$Y^\perp = [Y]^\perp.$$

If $y \in Y \cap Y^\perp$, then $\langle y, y\rangle = 0$, whence $y = 0$; so

$$Y \cap Y^\perp = \{0\}.$$

In particular, $\mathscr{H}^\perp = \mathscr{H} \cap \mathscr{H}^\perp = \{0\}$.

The following theorem includes the assertion that, if $Y$ is a closed subspace of a Hilbert space $\mathscr{H}$, then $Y$ and $Y^\perp$ are complementary subspaces in the sense discussed in Section 1.1 (preceding Theorem 1.1.8). For this reason, $Y^\perp$ is called the *orthogonal complement* of $Y$.

2.2.3. THEOREM. *If $Y$ is a closed subspace of a Hilbert space $\mathscr{H}$, each element $x_0$ of $\mathscr{H}$ can be expressed uniquely in the form $y_0 + z_0$, with $y_0$ in $Y$ and $z_0$ in $Y^\perp$. Moreover, $y_0$ is the unique point in $Y$ that is closest to $x_0$.*

*Proof.* Since $Y$ is a closed convex subset of $\mathscr{H}$, we can choose $y_0$ as in Proposition 2.2.1, and define $z_0 = x_0 - y_0$. From (1) and (2), $y_0$ is the (unique) point in $Y$ that is closest to $x_0$, and $\operatorname{Re}\langle y, z_0\rangle \leqslant \operatorname{Re}\langle y_0, z_0\rangle$ for each $y$ in $Y$. By writing $ay$ in place of $y$, we obtain

$$\operatorname{Re} a\langle y, z_0\rangle \leqslant \operatorname{Re}\langle y_0, z_0\rangle \qquad (y \in Y, \quad a \in \mathbb{C}).$$

Hence $\langle y, z_0\rangle = 0$ for each $y$ in $Y$, and $z_0 \in Y^\perp$. This proves the existence of a decomposition $x_0 = y_0 + z_0$, with $y_0$ in $Y$ and $z_0$ in $Y^\perp$. If, also, $x_0 = y_1 + z_1$, with $y_1$ in $Y$ and $z_1$ in $Y^\perp$, then

$$y_0 + z_0 = y_1 + z_1, \qquad y_0 - y_1 = z_1 - z_0 \in Y \cap Y^\perp = \{0\};$$

and so $y_0 = y_1$, $z_0 = z_1$. ∎

2.2.4. COROLLARY. *If $Y$ is a closed subspace of a Hilbert space $\mathscr{H}$ and $X \subseteq \mathscr{H}$, then*

$$(Y^\perp)^\perp = Y, \qquad (X^\perp)^\perp = [X].$$

*Moreover $Y = \mathscr{H}$ if and only if $Y^\perp = \{0\}$.*

*Proof.* Since $[X]$ is a closed subspace of $\mathscr{H}$, and $(X^\perp)^\perp = ([X]^\perp)^\perp$, it suffices to prove only the results concerning $Y$.

If $y \in Y$, then $y$ is orthogonal to each element of $Y^\perp$, and so $y \in (Y^\perp)^\perp$. This shows that $Y \subseteq (Y^\perp)^\perp$, and we have to prove the reverse inclusion. With $x_0$ in $(Y^\perp)^\perp$, we can choose $y_0$ in $Y$ and $z_0$ in $Y^\perp$ so that $x_0 = y_0 + z_0$, by Theorem 2.2.3. Then $x_0 \in (Y^\perp)^\perp$, $y_0 \in Y \subseteq (Y^\perp)^\perp$, and therefore $z_0 = x_0 - y_0 \in (Y^\perp)^\perp$. Hence

$$z_0 \in Y^\perp \cap (Y^\perp)^\perp = \{0\},$$

and $x_0 = y_0 \in Y$. This gives the required inclusion $(Y^\perp)^\perp \subseteq Y$, so $(Y^\perp)^\perp = Y$.

If $Y = \mathscr{H}$, then $Y^\perp = \mathscr{H}^\perp = \{0\}$; conversely, if $Y^\perp = \{0\}$, then $Y = (Y^\perp)^\perp = \{0\}^\perp = \mathscr{H}$. ∎

A subset $Y$ of a Hilbert space $\mathscr{H}$ is described as an *orthogonal set* if any two distinct elements of $Y$ are mutually orthogonal. By an *orthonormal set* we mean an orthogonal set of unit vectors. An orthonormal set $Y$ is linearly independent (by which we mean that every finite subset of $Y$ is linearly independent); for if $y_1, \ldots, y_n$ are distinct elements of $Y$, $a_1, \ldots, a_n \in \mathbb{C}$, and $\sum_{j=1}^{n} a_j y_j = 0$, then

$$a_k = \langle \sum_{j=1}^{n} a_j y_j, y_k \rangle = 0 \qquad (k = 1, \ldots, n).$$

In developing the theory of orthogonal expansions in a Hilbert space, we make use of the concept of unordered summation introduced in Section 1.2 (following Theorem 1.2.18).

2.2.5. Proposition. *If $Y$ is an orthogonal set in a Hilbert space $\mathscr{H}$, the sum $\sum_{y \in Y} y$ converges if and only if $\sum_{y \in Y} \|y\|^2 < \infty$. When this condition is satisfied,*

$$\|\sum_{y \in Y} y\|^2 = \sum_{y \in Y} \|y\|^2. \tag{3}$$

*Proof.* With $F$ a finite subset of $Y$, expansion of the inner product expression for $\|\sum_{y \in F} y\|^2$ gives

$$\|\sum_{y \in F} y\|^2 = \sum_{y \in F} \|y\|^2. \tag{4}$$

From this, and the Cauchy criterion for unordered sums, it follows that convergence of either of the sums in (3) implies convergence of the other. When these sums converge, they are limits of the finite subsums occurring in (4); since the norm is continuous, (3) is an immediate consequence of (4). ■

2.2.6. Corollary. *If $Y$ is an orthonormal set in a Hilbert space $\mathscr{H}$, and $f$ is a complex-valued function defined on $Y$, the sum $\sum_{y \in Y} f(y)y$ converges if and only if $\sum_{y \in Y} |f(y)|^2 < \infty$. When this condition is satisfied,*

$$\|\sum_{y \in Y} f(y)y\|^2 = \sum_{y \in Y} |f(y)|^2.$$

*Proof.* It suffices to apply Proposition 2.2.5 to the orthogonal set $\{f(y)y : y \in Y\}$. ■

2.2.7. Proposition. *If $Y$ is an orthonormal set in a Hilbert space $\mathscr{H}$ and $u \in \mathscr{H}$, then*

(i) $\sum_{y \in Y} |\langle u, y \rangle|^2 \leqslant \|u\|^2$;

(ii) *the sum* $\sum_{y \in Y} \langle u, y \rangle y$ *converges, and*

$$u - \sum_{y \in Y} \langle u, y \rangle y \in Y^{\perp};$$

(iii) $\|u - \sum_{y \in Y} \langle u, y \rangle y\|^2 = \|u\|^2 - \sum_{y \in Y} |\langle u, y \rangle|^2$.

*Proof.* With $F$ a finite subset of $Y$ and $\rho(y) = \langle u, y \rangle$ for each $y$ in $Y$,

$$\begin{aligned}
\|u - \sum_{y \in F} \langle u, y \rangle y\|^2 &= \langle u - \sum_{y \in F} \rho(y) y, u - \sum_{z \in F} \rho(z) z \rangle \\
&= \langle u, u \rangle - \sum_{y \in F} \rho(y) \langle y, u \rangle - \sum_{z \in F} \overline{\rho(z)} \langle u, z \rangle + \sum_{y, z \in F} \rho(y) \overline{\rho(z)} \langle y, z \rangle \\
&= \|u\|^2 - \sum_{y \in F} |\rho(y)|^2 - \sum_{z \in F} |\rho(z)|^2 + \sum_{z \in F} |\rho(z)|^2 \\
&= \|u\|^2 - \sum_{y \in F} |\langle u, y \rangle|^2.
\end{aligned}$$

Hence

$$\sum_{y \in F} |\langle u, y \rangle|^2 = \|u\|^2 - \|u - \sum_{y \in F} \langle u, y \rangle y\|^2 \leqslant \|u\|^2 \tag{5}$$

for each finite subset $F$ of $Y$. This proves (i), and the convergence of $\sum_{y \in Y} \langle u, y \rangle y$ now follows from Corollary 2.2.6. For each $y_0$ in $Y$, continuity and linearity of the inner product in its first variable entail

$$\begin{aligned}
\langle u - \sum_{y \in Y} \langle u, y \rangle y, y_0 \rangle &= \langle u, y_0 \rangle - \sum_{y \in Y} \langle u, y \rangle \langle y, y_0 \rangle \\
&= \langle u, y_0 \rangle - \langle u, y_0 \rangle = 0,
\end{aligned}$$

so $u - \sum_{y \in Y} \langle u, y \rangle y \in Y^{\perp}$. This proves (ii), and (iii) is an immediate consequence of (5) since the norm is continuous. ∎

The inequality in Proposition 2.2.7(i) is usually known as *Bessel's inequality*.

2.2.8. COROLLARY. *If $Y$ is an orthonormal set in a Hilbert space $\mathscr{H}$ and $u \in \mathscr{H}$, then $\sum_{y \in Y} \langle u, y \rangle y$ is the unique vector closest to $u$ in the closed subspace $[Y]$ generated by $Y$. Moreover, the following three conditions are equivalent*:

(i) $u \in [Y]$;
(ii) $u = \sum_{y \in Y} \langle u, y \rangle y$;
(iii) $\|u\|^2 = \sum_{y \in Y} |\langle u, y \rangle|^2$.

*Proof.* With $v = \sum_{y \in Y} \langle u, y \rangle y$, it is evident that $v \in [Y]$, and Proposition 2.2.7(ii) asserts that $u - v \in Y^{\perp} = [Y]^{\perp}$. Since, also, $u = v + (u - v)$, Theorem 2.2.3 now implies that $v$ is the unique point closest to $u$ in $[Y]$. From the last statement, it follows that $v = u$ if $u \in [Y]$, so (i) implies (ii). The reverse implication is apparent, and the equivalence of (ii) and (iii) follows from Proposition 2.2.7(iii). ∎

2.2.9. THEOREM. *If $Y$ is an orthonormal set in a Hilbert space $\mathscr{H}$, the following six conditions are equivalent*:

(i) *for each $u$ in $\mathscr{H}$, $u = \sum_{y \in Y} \langle u, y \rangle y$;*
(ii) *for each $u$ and $v$ in $\mathscr{H}$, $\langle u, v \rangle = \sum_{y \in Y} \langle u, y \rangle \langle y, v \rangle$;*
(iii) *for each $u$ in $\mathscr{H}$, $\|u\|^2 = \sum_{y \in Y} |\langle u, y \rangle|^2$;*
(iv) *$Y$ is not contained in any strictly larger orthonormal set $X$;*
(v) *$Y^{\perp} = \{0\}$;*
(vi) *$[Y] = \mathscr{H}$.*

*Proof.* By the linearity and continuity of the inner product in its first variable, (i) implies (ii). It is apparent, by taking $v = u$, that (ii) implies (iii).

If $Y$ is contained in a strictly larger orthonormal set $X$, (iii) fails since, when $x \in X \setminus Y$,

$$\sum_{y \in Y} |\langle x, y \rangle|^2 = 0 \neq 1 = \|x\|^2.$$

It follows that (iii) implies (iv).

If $Y^{\perp}$ has a non-zero element $x$, $Y$ is contained in a strictly larger orthonormal set $Y \cup \{\|x\|^{-1}x\}$; so (iv) implies (v).

By Corollary 2.2.4, and since $Y^{\perp} = [Y]^{\perp}$, (v) implies (vi). It follows, from the equivalence of the first two conditions stated in Corollary 2.2.8, that (vi) implies (i). ■

An orthonormal set $Y$ in a Hilbert space $\mathscr{H}$ that satisfies (any one, and hence all six, of) the equivalent conditions set out in Theorem 2.2.9 is called an *orthonormal basis* of $\mathscr{H}$. When $Y$ is an orthonormal basis, the equation in condition (ii) is known as *Parseval's equation*.

2.2.10. THEOREM. *Each Hilbert space $\mathscr{H}$ has an orthonormal basis, and every orthonormal set in $\mathscr{H}$ is contained in an orthonormal basis. Moreover, all orthonormal bases of $\mathscr{H}$ have the same cardinality.*

*Proof.* The class of all orthonormal sets in $\mathscr{H}$ is partially ordered by inclusion. If a family $\{Y_a\}$ of orthonormal sets is *totally* ordered by inclusion, then $\cup Y_a$ is an orthonormal set that contains each $Y_a$; for any two distinct elements $y, z$ of $\cup Y_a$ are contained in the union $Y_b \cup Y_c$ of two sets in the family, $Y_b \cup Y_c$ coincides with $Y_b$ or $Y_c$ and is therefore orthonormal, and so $\langle y, z \rangle = 0$. In view of this, it follows from Zorn's lemma that there is a maximal orthonormal set $Y_0$; since $Y_0$ is not contained in a strictly larger orthonormal set, it is an orthonormal basis.

If $Y$ is a given orthonormal set, we can repeat the above argument, restricting attention throughout to orthonormal sets containing $Y$. In this way, we prove that there is an orthonormal basis containing $Y$.

Suppose that $X$ and $Y$ are two orthonormal bases of $\mathscr{H}$, and that the cardinal numbers corresponding to these sets are $m$ and $n$, respectively. In proving that $m = n$, we consider separately two cases.

If $\mathscr{H}$ is finite dimensional, the (linearly independent) sets $X$ and $Y$ are necessarily finite. Since both satisfy condition (i) in Theorem 2.2.9, each has linear span $\mathscr{H}$, and is therefore a basis of $\mathscr{H}$ in the elementary algebraic sense. The basis theorem for finite-dimensional vector spaces now implies that $m = n$. We note also an alternative proof, that $m = n$ when $m$ and $n$ are known to be finite, which is more akin to the argument needed in the infinite-dimensional case. By condition (iii) in Theorem 2.2.9

$$m = \sum_{x \in X} \|x\|^2 = \sum_{x \in X} \sum_{y \in Y} |\langle x, y \rangle|^2$$

$$= \sum_{y \in Y} \sum_{x \in X} |\langle y, x \rangle|^2 = \sum_{y \in Y} \|y\|^2 = n.$$

If $\mathscr{H}$ is infinite dimensional, both $X$ and $Y$ are infinite sets, since $[X] = [Y] = \mathscr{H}$. If $m \neq n$, we may assume that $m < n$; we show, in due course, that this assumption leads to a contradiction. For each $x$ in $X$,

$$\sum_{y \in Y} |\langle x, y \rangle|^2 = \|x\|^2 = 1,$$

so the set $Y_x = \{y \in Y : \langle x, y \rangle \neq 0\}$ is countable. Moreover, $Y = \bigcup_{x \in X} Y_x$; for, if $y \in Y$,

$$\sum_{x \in X} |\langle x, y \rangle|^2 = \|y\|^2 = 1,$$

whence $\langle x, y \rangle \neq 0$ and $y \in Y_x$, for at least one element $x$ of $X$.

The remainder of the argument consists of elementary cardinal arithmetic, amounting, essentially, to the observation that $n \leqslant m\aleph_0 = m$ (where, as usual, $\aleph_0$ denotes the cardinal of the set of natural numbers). In order to prove that $n \leqslant m$, and so obtain the desired contradiction, we need only show that there is a mapping from $X$ onto $Y$. For this, it suffices to prove that $X$ can be expressed as a disjoint union $\bigcup_{x \in X} X_x$ of a family (indexed by $X$) of countably infinite subsets; for then each $X_x$ can be mapped onto the corresponding $Y_x$, and $X$ $(= \cup X_x)$ can be mapped onto $Y$ $(= \cup Y_x)$. To prove the existence of such a family $(X_x)_{x \in X}$, it is enough to show that $X \times Z$ has the same cardinality as $X$, when $Z$ is a countably infinite set. Since $Z \times Z$ is countably infinite, it now suffices to prove that $X$ has the same cardinality as $\mathbb{A} \times Z$, for some set $\mathbb{A}$. This, in turn, amounts to showing that $X$ can be expressed as the disjoint union $\bigcup_{a \in \mathbb{A}} Z_a$ of a family (with arbitrary index set $\mathbb{A}$) of countably infinite subsets of $X$. For this, observe that a simple argument using Zorn's lemma proves the existence of a maximal disjoint family $\{Z_a\}$ of countably infinite subsets of $X$. The maximality of this family implies that $X \backslash \cup Z_a$ has only a finite number of elements; by adding these to any one $Z_a$, we obtain the required partition $X = \cup Z_a$. ■

By the *dimension* of a Hilbert space $\mathscr{H}$ we mean the cardinal number $\dim \mathscr{H}$ corresponding to an orthonormal basis $Y$ of $\mathscr{H}$. From the preceding theorem, this does not depend on the choice of $Y$; moreover, it coincides with the elementary algebraic concept of dimension when $\mathscr{H}$ is finite dimensional.

Suppose that $\mathscr{H}_1$ and $\mathscr{H}_2$ are Hilbert spaces, and $U$ is a linear mapping from $\mathscr{H}_1$ onto $\mathscr{H}_2$. By expressing inner products in terms of norms, as in Proposition 2.1.8, it follows that $U$ preserves inner products if and only if it preserves norms. Accordingly, the concept of isomorphism, from $\mathscr{H}_1$ onto $\mathscr{H}_2$, is the same whether $\mathscr{H}_1$ and $\mathscr{H}_2$ are regarded as Banach spaces or as Hilbert spaces. It is evident that isomorphic Hilbert spaces have the same dimension.

2.2.11. Example. With $\mathbb{A}$ any set, the Hilbert space $l_2(\mathbb{A})$ has an orthonormal set $Y = \{y_a : a \in \mathbb{A}\}$, in which $y_a$ is the function taking the value 1 at $a$ and 0 elsewhere on $\mathbb{A}$. Since $\langle x, y_a \rangle = x(a)$ for each $x$ in $l_2(\mathbb{A})$ and $a$ in $\mathbb{A}$, it follows that $Y^\perp = \{0\}$. Hence $Y$ is an orthonormal basis of $l_2(\mathbb{A})$, and the dimension of $l_2(\mathbb{A})$ is the cardinal number corresponding to the set $\mathbb{A}$.

A necessary and sufficient condition for two spaces $l_2(\mathbb{A})$ and $l_2(\mathbb{B})$ to be isomorphic is that the sets $\mathbb{A}$ and $\mathbb{B}$ have the same cardinality. The condition is necessary because isomorphism preserves dimension; it is sufficient since, if $f$ is a one-to-one mapping from $\mathbb{A}$ onto $\mathbb{B}$, the equation $(Ux)(a) = x(f(a))$ defines an isomorphism $U$ from $l_2(\mathbb{B})$ onto $l_2(\mathbb{A})$. ■

2.2.12. Theorem. *Two Hilbert spaces are isomorphic if and only if they have the same dimension.*

*Proof.* In view of Example 2.2.11, it suffices to prove that a Hilbert space $\mathscr{H}$ with an orthonormal basis $Y$ is isomorphic to $l_2(Y)$. For each $x$ in $\mathscr{H}$, we can define a complex-valued function $Ux$ on $Y$ by $(Ux)(y) = \langle x, y \rangle$. From condition (iii) in Theorem 2.2.9,

$$\|x\|^2 = \sum_{y \in Y} |\langle x, y \rangle|^2 = \sum_{y \in Y} |(Ux)(y)|^2,$$

so $U$ is a norm-preserving linear mapping from $\mathscr{H}$ into $l_2(Y)$. With $f$ in $l_2(Y)$, it follows from Corollary 2.2.6 that the sum $\sum_{y \in Y} f(y)y$ converges to an element $x$ of $\mathscr{H}$. Moreover, for each $y_0$ in $Y$,

$$(Ux)(y_0) = \langle x, y_0 \rangle = \sum_{y \in Y} f(y)\langle y, y_0 \rangle = f(y_0),$$

so $Ux = f$. Hence $U$ is an isomorphism from $\mathscr{H}$ onto $l_2(Y)$. ■

2.2.13. Corollary. *Every Hilbert space is isomorphic to one of the form $l_2(\mathbb{A})$. A Hilbert space with finite dimension $n$ is isomorphic to $\mathbb{C}^n$.*

2.2.14. REMARK. We assert that a Hilbert space $\mathscr{H}$ is separable if and only if $\dim \mathscr{H} \leqslant \aleph_0$. In consequence, all separable infinite-dimensional Hilbert spaces have dimension $\aleph_0$, and are therefore isomorphic. In particular, $l_2$ spaces for countably infinite sets, and $L_2$ spaces for Lebesgue measure on measurable subsets of $\mathbb{R}^n$, are all isomorphic.

To prove the above assertion, let $Y$ be an orthonormal basis in $\mathscr{H}$. If $Y$ is countable, $\mathscr{H}$ has a countable everywhere-dense subset, which consists of those finite linear combinations of elements of $Y$ in which each coefficient has rational real and imaginary parts. If $Y$ is uncountable, the open balls with radius $\frac{1}{2}\sqrt{2}$ and centers in $Y$ form an uncountable disjoint family, since

$$\|y_1 - y_2\|^2 = \|y_1\|^2 + \|y_2\|^2 = 2$$

when $y_1$ and $y_2$ are distinct elements of $Y$. An everywhere-dense subset of $\mathscr{H}$ meets each of these balls, and is therefore uncountable. ■

In proving the following result, we describe the *Gram–Schmidt orthogonalization process*, by which a linearly independent sequence of Hilbert space vectors gives rise to an orthonormal sequence. The linearly independent sequence may be finite or (countably) infinite, and the orthonormal sequence has the same number of terms. We recall that the (necessarily closed) subspace generated by a finite set $x_1, \ldots, x_n$ of vectors is denoted by $[x_1, \ldots, x_n]$.

2.2.15. PROPOSITION. *If $(x_1, x_2, x_3, \ldots)$ is a linearly independent sequence of vectors in a Hilbert space $\mathscr{H}$, there is an orthonormal sequence $(y_1, y_2, y_3, \ldots)$ such that $[x_1, \ldots, x_n] = [y_1, \ldots, y_n]$ for each $n = 1, 2, 3, \ldots$.*

*Proof.* We construct $y_1, y_2, y_3, \ldots$ inductively, and start the process by defining $y_1$ to be $\|x_1\|^{-1}x_1$. Now suppose that we have produced an orthonormal set $\{y_1, \ldots, y_{r-1}\}$, with the property that

$$[y_1, \ldots, y_n] = [x_1, \ldots, x_n] \qquad (1 \leqslant n < r).$$

Since

$$x_r \notin [x_1, \ldots, x_{r-1}] = [y_1, \ldots, y_{r-1}],$$

the vector

$$z = x_r - \sum_{j=1}^{r-1} \langle x_r, y_j \rangle y_j$$

is non-zero, and

$$[y_1, \ldots, y_{r-1}, z] = [x_1, \ldots, x_{r-1}, x_r].$$

Moreover, for $k = 1, \ldots, r - 1$,

$$\langle z, y_k \rangle = \langle x_r, y_k \rangle - \sum_{j=1}^{r-1} \langle x_r, y_j \rangle \langle y_j, y_k \rangle = \langle x_r, y_k \rangle - \langle x_r, y_k \rangle = 0.$$

With $y_r = \|z\|^{-1}z$, $\{y_1, \ldots, y_r\}$ is an orthonormal set and $[y_1, \ldots, y_r] = [x_1, \ldots, x_r]$. Hence the inductive process continues (indefinitely, if there are infinitely many $x_j$, but terminating with the construction of $y_n$ if there are just $n$ vectors $x_j$). ■

2.2.16. REMARK. When an orthonormal sequence $\{y_n\}$ is constructed from a linearly independent sequence $\{x_n\}$, as in the proof of the last proposition, we have

$$\langle y_r, x_r\rangle = \langle y_r, z\rangle + \sum_{j=1}^{r-1} \langle y_r, y_j\rangle\langle y_j, x_r\rangle = \langle y_r, z\rangle = \|z\| > 0.$$

It is not difficult to verify that the conditions

$$[y_1, \ldots, y_n] = [x_1, \ldots, x_n], \qquad \langle y_n, x_n\rangle > 0 \qquad (n = 1, 2, 3, \ldots)$$

determine the orthonormal sequence $\{y_n\}$ uniquely. ■

We conclude this section with a brief discussion of certain orthogonal families of functions in $L_2$ spaces, which are encountered in classical analysis and its applications.

The best known examples arise in connection with the theory of Fourier series, for functions in $L_2(-\pi, \pi)$. With $\mathbb{Z}$ the set $\{0, \pm 1, \pm 2, \ldots\}$ of all integers, we can define functions $x_n$ $(n \in \mathbb{Z})$ in $L_2(-\pi, \pi)$ by

$$x_n(s) = \exp(ins) \qquad (-\pi \leqslant s \leqslant \pi).$$

By evaluating the appropriate integrals, we obtain

$$\|x_n\| = \sqrt{2\pi}, \qquad \langle x_m, x_n\rangle = 0 \qquad (m \neq n),$$

so the functions $(2\pi)^{-1/2}x_n$ form an orthonormal set in $L_2(-\pi, \pi)$. We shall note later, as a consequence of Theorem 3.4.14 (see Remark 3.4.15), that linear combinations of these functions are everywhere dense in $L_2(-\pi, \pi)$. Accordingly, the set $\{(2\pi)^{-1/2} x_n : n \in \mathbb{Z}\}$ is an orthonormal basis of $L_2(-\pi, \pi)$. Each function $x$ in $L_2(-\pi, \pi)$ has an expansion

$$x = \sum_{n\in\mathbb{Z}} \langle x, (2\pi)^{-1/2}x_n\rangle(2\pi)^{-1/2}x_n = \sum_{n\in\mathbb{Z}} c_n x_n,$$

convergent in the norm topology on $L_2(-\pi, \pi)$, in which

$$c_n = \frac{1}{2\pi}\langle x, x_n\rangle = \frac{1}{2\pi}\int_{-\pi}^{\pi} x(s)e^{-ins}\,ds.$$

The numbers $c_n$ are called the *Fourier coefficients* of $x$, and the series $\sum c_n x_n$ is called the *Fourier series* of $x$. Given two functions $x$ and $y$ in $L_2(-\pi, \pi)$, with Fourier series $\sum c_n x_n$ and $\sum d_n x_n$, respectively, Parseval's equation assumes the form

$$\int_{-\pi}^{\pi} x(s)\overline{y(s)}\,ds = \frac{1}{2\pi}\sum_{n\in\mathbb{Z}} c_n \bar{d}_n; \tag{6}$$

in particular,

$$\int_{-\pi}^{\pi} |x(s)|^2\,ds = \frac{1}{2\pi}\sum_{n\in\mathbb{Z}} |c_n|^2. \tag{7}$$

The preceding paragraph is concerned with the "complex form" of Fourier series; we now consider briefly the (essentially equivalent) "real form." The functions

$$1,\quad \cos s,\quad \sin s,\quad \cos 2s,\quad \sin 2s,\quad \cos 3s,\quad \sin 3s,\quad \ldots$$

form an orthogonal sequence $(y_0, y_1, y_2, \ldots)$ in $L_2(-\pi, \pi)$, since evaluation of the appropriate integrals shows that

$$\|y_0\| = \sqrt{2\pi}, \qquad \|y_n\| = \sqrt{\pi}\quad (n > 0), \qquad \langle y_m, y_n\rangle = 0\quad (m \neq n).$$

Since

$$y_0 = x_0, \qquad y_{2n-1} = \frac{1}{2}(x_n + x_{-n}), \qquad y_{2n} = \frac{1}{2i}(x_n - x_{-n}),$$

the linear span of the $y_n$ is the same as that of the $x_n$, and is therefore everywhere dense in $L_2(-\pi, \pi)$. It follows that the sequence $\{(2\pi)^{-1/2}y_0, \pi^{-1/2}y_1, \pi^{-1/2}y_2, \ldots\}$ is an orthonormal basis of $L_2(-\pi, \pi)$. The corresponding orthogonal expansion of a function $x$ in $L_2(-\pi, \pi)$ is a series of trigonometrical functions, the "real form" of the Fourier series of $x$. Parseval's equation yields results, analogous (and equivalent) to (6) and (7), in which the integrals representing $\langle x, y\rangle$ and $\|x\|^2$ are expressed in terms of "real" Fourier coefficients.

Certain classical sequences of polynomials form orthogonal sets in appropriate $L_2$ spaces. For example, in $L_2(-1, 1)$, the functions $x_0, x_1, x_2, \ldots$, defined by

$$x_n(s) = s^n \qquad (-1 \leqslant s \leqslant 1),$$

form a linearly independent sequence. By the Gram–Schmidt process, we can construct an orthonormal sequence $(y_0, y_1, y_2, \ldots)$ in $L_2(-1, 1)$, such that

$$[y_0, \ldots, y_n] = [x_0, \ldots, x_n] \qquad (n = 0, 1, 2, \ldots).$$

The linear span of the $y_n$ is the set of all polynomials, and this is everywhere dense in $L_2(-1, 1)$, by the Weierstrass approximation theorem (noted later, in Remark 3.4.15). Accordingly $(y_0, y_1, y_2, \ldots)$ is an orthonormal basis of $L_2(-1, 1)$. It is not difficult to verify that

$$y_n(s) = \sqrt{n + \tfrac{1}{2}}\,P_n(s) \qquad (n = 0, 1, 2, \ldots),$$

where $P_n$ is the *Legendre polynomial*, defined by Rodrigues's formula

$$P_n(s) = \frac{1}{2^n n!} \frac{d^n}{ds^n} \{(s^2 - 1)^n\}. \tag{8}$$

The orthogonality relations

$$\int_{-1}^{1} [P_n(s)]^2 \, ds = \frac{2}{2n+1},$$

$$\int_{-1}^{1} P_n(s) P_m(s) \, ds = 0 \qquad (m \neq n)$$

(which are equivalent to the assertion that the sequence $\{y_n\}$ is orthonormal) are easily deduced from (8), upon repeated integration by parts.

Other classical sequences of polynomials arise in a similar manner. Let $E$ be a Borel subset of $\mathbb{R}$, with positive Lebesgue measure, $w$ a strictly positive Borel measurable function on $E$, such that

$$\int_E s^{2n} w(s) \, ds < \infty \qquad (n = 0, 1, 2, \ldots).$$

The set of (equivalence classes modulo null functions of) Borel measurable functions $x$ on $E$, such that

$$\int_E |x(s)|^2 w(s) \, ds < \infty,$$

is a Hilbert space $\mathscr{H}$, with inner product defined by

$$\langle x, y \rangle = \int_E x(s) \overline{y(s)} w(s) \, ds.$$

The functions $1, s, s^2, s^3, \ldots$ form a linearly independent sequence in $\mathscr{H}$, and by applying the Gram–Schmidt process we obtain an orthonormal sequence $(y_0, y_1, y_2, \ldots)$, in which $y_n$ is a polynomial of degree exactly $n$.

When $E$ is $[-1, 1]$ and $w(s) = (1 - s)^\nu (1 + s)^\mu$, where $\nu, \mu > -1$, we obtain the *Jacobi polynomials* $P_n^{(\nu,\mu)}(s)$. The *Laguerre polynomials* $L_n^\nu(s)$ $(\nu > -1)$ arise when $E = [0, \infty)$ and $w(s) = s^\nu \exp(-s)$. The *Hermite polynomials* correspond to the choice $E = \mathbb{R}$, $w(s) = \exp(-s^2)$. These are the main three classical sequences of polynomials, from which the others can be derived; for example, the Legendre polynomials are a particular case $(\nu = \mu = 0)$ of the Jacobi polynomials.

## 2.3. The weak topology

In this section we prove Riesz's representation theorem (Theorem 2.3.1), which describes the general continuous linear functional on a Hilbert space $\mathscr{H}$.

By means of this result, we establish certain properties of the weak topology on $\mathscr{H}$ (see Section 1.3).

2.3.1. THEOREM. *If $\mathscr{H}$ is a Hilbert space and $y \in \mathscr{H}$, the equation $\varphi_y(x) = \langle x, y \rangle$ $(x \in \mathscr{H})$ defines a continuous linear functional $\varphi_y$ on $\mathscr{H}$, and $\|\varphi_y\| = \|y\|$. Each continuous linear functional on $\mathscr{H}$ arises, in this way, from a unique element $y$ of $\mathscr{H}$.*

*Proof.* For each $y$ in $\mathscr{H}$,

$$|\varphi_y(x)| = |\langle x, y \rangle| \leqslant \|x\| \|y\| \qquad (x \in \mathscr{H}),$$

with equality when $x = y$. Thus $\varphi_y$ is a continuous linear functional on $\mathscr{H}$, and $\|\varphi_y\| = \|y\|$.

If $\varphi$ is a non-zero continuous linear functional on $\mathscr{H}$, the closed subspace $Y = \varphi^{-1}(0)$ is not the whole of $\mathscr{H}$, so $Y^{\perp} \neq \{0\}$. Let $u$ be a unit vector in $Y^{\perp}$, and note that

$$\varphi(\varphi(u)x - \varphi(x)u) = \varphi(u)\varphi(x) - \varphi(x)\varphi(u) = 0$$

for each $x$ in $\mathscr{H}$. It follows that $\varphi(u)x - \varphi(x)u \in Y$, and since $u \in Y^{\perp}$, we have

$$0 = \langle \varphi(u)x - \varphi(x)u, u \rangle = \varphi(u)\langle x, u \rangle - \varphi(x).$$

Hence

$$\varphi(x) = \varphi(u)\langle x, u \rangle = \langle x, y \rangle \qquad (x \in \mathscr{H}),$$

where $y = \overline{\varphi(u)}u$. This shows that $\varphi$ has the form $\varphi_y$ for some $y$ in $\mathscr{H}$ (and the same conclusion is apparent when $\varphi = 0$). If, also, $\varphi = \varphi_z$ with $z$ in $\mathscr{H}$, then

$$\|y - z\| = \|\varphi_{y-z}\| = \|\varphi_y - \varphi_z\| = \|\varphi - \varphi\| = 0,$$

whence $y = z$; so there is only one $y$ in $\mathscr{H}$ for which $\varphi_y = \varphi$. ∎

2.3.2. COROLLARY. *If $\mathscr{H}$ is a Hilbert space, the equation*

$$(Jy)(x) = \langle x, y \rangle \qquad (x, y \in \mathscr{H})$$

*defines a conjugate-linear norm-preserving mapping $J$ from $\mathscr{H}$ onto the Banach dual space $\mathscr{H}^{\#}$.*

*Proof.* Since $Jy$ is the continuous linear functional $\varphi_y$ occurring in Theorem 2.3.1, $J$ is a norm-preserving mapping from $\mathscr{H}$ onto $\mathscr{H}^{\#}$, and it is evident from the conjugate linearity of the inner product in its second variable that $J$ also is conjugate linear. ∎

2.3.3. COROLLARY. *Every Hilbert space is reflexive.*

*Proof.* Suppose that $\mathscr{H}$ is a Hilbert space and $\Phi$ is a continuous linear functional on its Banach dual space $\mathscr{H}^{\#}$. With $J$ defined as in Corollary 2.3.2, the equation

$$\varphi(y) = \overline{\Phi(Jy)} \qquad (y \in \mathscr{H})$$

defines a bounded linear functional $\varphi$ on $\mathscr{H}$. By Theorem 2.3.1, we can choose $z$ in $\mathscr{H}$ so that $\varphi(y) = \langle y, z \rangle$ for each $y$ in $\mathscr{H}$. Every element $\psi$ of $\mathscr{H}^\#$ has the form $Jy$, with $y$ in $\mathscr{H}$, and

$$\Phi(\psi) = \Phi(Jy) = \overline{\varphi(y)} = \overline{\langle y, z \rangle} = \langle z, y \rangle = (Jy)(z) = \psi(z).$$

Since each bounded linear functional $\Phi$ on $\mathscr{H}^\#$ arises in this way from an element $z$ of $\mathscr{H}$, it follows that $\mathscr{H}$ is reflexive. ■

2.3.4. Corollary. *Suppose that $\mathscr{H}$ is a Hilbert space and $x_0 \in \mathscr{H}$. The family of all sets of the form*

$$\{x \in \mathscr{H} : |\langle x - x_0, y_j \rangle| < \varepsilon \ (j = 1, \ldots, n)\},$$

*with $y_1, \ldots, y_n \ (\in \mathscr{H})$ and $\varepsilon \ (> 0)$ preassigned, is a base of neighborhoods of $x_0$ in the weak topology on $\mathscr{H}$. A net $\{x_a\}$ of elements of $\mathscr{H}$ converges weakly to $x_0$ if and only if $\langle x_a, y \rangle \to \langle x_0, y \rangle$ for each $y$ in $\mathscr{H}$. The closed unit ball of $\mathscr{H}$ is weakly compact.*

*Proof.* Since $\mathscr{H}$ is reflexive, its unit ball is weakly compact by Theorem 1.6.7. The remaining assertions in the corollary are simply reinterpretations of the appropriate Banach space definitions, taking into account the information in Theorem 2.3.1 concerning the form of continuous linear functionals on $\mathscr{H}$. ■

2.3.5. Proposition. *If a net $\{x_a\}$ of vectors in a Hilbert space $\mathscr{H}$ converges weakly to an element $x$ of $\mathscr{H}$, and*

$$\|x_a\| \underset{a}{\to} \|x\|,$$

*then $\{x_a\}$ converges to $x$ in the norm topology.*

*Proof.* Since

$$\langle x_a, x \rangle \underset{a}{\to} \langle x, x \rangle = \|x\|^2,$$

we have

$$\|x_a - x\|^2 = \|x_a\|^2 - 2\,\mathrm{Re}\langle x_a, x \rangle + \|x\|^2 \underset{a}{\to} 0. \quad ■$$

## 2.4. Linear operators

We recall from Theorem 1.5.5 that a linear operator $T$, from a normed space $\mathfrak{X}$ into another such space $\mathscr{Y}$, is continuous if and only if it is bounded, in the sense that there is a real number $c$ such that $\|Tx\| \leqslant c\|x\|$ for each $x$ in $\mathfrak{X}$. The set $\mathscr{B}(\mathfrak{X}, \mathscr{Y})$ of all such bounded operators is itself a normed space, when the norm of an element $T$ is defined to be the least such constant $c$; equivalently,

$$\|T\| = \sup\{\|Tx\| : x \in \mathfrak{X},\ \|x\| \leqslant 1\}.$$

By Theorem 1.5.6, $\mathscr{B}(\mathfrak{X}, \mathscr{Y})$ is a Banach space when $\mathscr{Y}$ is a Banach space. In this section we obtain more detailed information concerning bounded linear operators acting on Hilbert spaces.

*General theory.* Suppose that $\mathscr{H}$ and $\mathscr{K}$ are Hilbert spaces. By a *conjugate-bilinear functional* on $\mathscr{H} \times \mathscr{K}$, we mean a complex-valued function $b$ on $\mathscr{H} \times \mathscr{K}$ that is linear in the first variable and conjugate-linear in the second. We say that such a functional $b$ is *bounded* if there is a real number $c$ such that $|b(x, y)| \leqslant c\|x\|\,\|y\|$ for all $x$ in $\mathscr{H}$ and $y$ in $\mathscr{K}$. When this is so, we denote by $\|b\|$ the least possible value of $c$, which is given by

$$\|b\| = \sup\{|b(x, y)| : x \in \mathscr{H}, y \in \mathscr{K}, \|x\| \leqslant 1, \|y\| \leqslant 1\}.$$

When $\mathscr{K} = \mathscr{H}$, we refer to a conjugate-bilinear functional "on $\mathscr{H}$," rather than "on $\mathscr{H} \times \mathscr{H}$."

2.4.1. THEOREM. *If $\mathscr{H}$, $\mathscr{K}$ are Hilbert spaces and $T \in \mathscr{B}(\mathscr{H}, \mathscr{K})$, the equation*

$$b_T(x, y) = \langle Tx, y\rangle \qquad (x \in \mathscr{H},\quad y \in \mathscr{K}) \tag{1}$$

*defines a bounded conjugate-bilinear functional $b_T$ on $\mathscr{H} \times \mathscr{K}$, and $\|b_T\| = \|T\|$. Each bounded conjugate-bilinear functional on $\mathscr{H} \times \mathscr{K}$ arises in this way from a unique element of $\mathscr{B}(\mathscr{H}, \mathscr{K})$.*

*Proof.* Given $T$ in $\mathscr{B}(\mathscr{H}, \mathscr{K})$, it is apparent that $b_T$, as defined in (1), is a conjugate-bilinear functional on $\mathscr{H} \times \mathscr{K}$. The inequalities

$$|b_T(x, y)| = |\langle Tx, y\rangle| \leqslant \|Tx\|\,\|y\| \leqslant \|T\|\,\|x\|\,\|y\|,$$

$$\|Tx\|^2 = \langle Tx, Tx\rangle = b_T(x, Tx) \leqslant \|b_T\|\,\|x\|\,\|Tx\|,$$

show that $b_T$ is bounded and $\|b_T\| = \|T\|$.

If $b$ is a bounded conjugate-bilinear functional on $\mathscr{H} \times \mathscr{K}$ and $x \in \mathscr{H}$, the equation

$$(Ux)(y) = \overline{b(x, y)} \qquad (y \in \mathscr{K})$$

defines a linear functional $Ux$ on $\mathscr{K}$. Since

$$|(Ux)(y)| \leqslant \|b\|\,\|x\|\,\|y\|,$$

$Ux$ is bounded, and $\|Ux\| \leqslant \|b\|\,\|x\|$. From this, together with the linearity of $b$ in its first variable, it is evident that $U$ is a bounded conjugate-linear mapping from $\mathscr{H}$ into the Banach dual space $\mathscr{K}^\sharp$ of $\mathscr{K}$. With $J$ the norm-preserving conjugate-linear mapping from $\mathscr{K}$ onto $\mathscr{K}^\sharp$, as defined in Corollary 2.3.2, $J^{-1}U$ is a bounded linear operator $T$ from $\mathscr{H}$ into $\mathscr{K}$; moreover,

$$\begin{aligned} b_T(x, y) &= \langle Tx, y\rangle = \langle J^{-1}Ux, y\rangle \\ &= \overline{\langle y, J^{-1}Ux\rangle} = \overline{(Ux)(y)} = b(x, y), \end{aligned}$$

so $b = b_T$. If, also, $b = b_S$ for some $S$ in $\mathscr{B}(\mathscr{H}, \mathscr{K})$, we have

$$\|S - T\| = \|b_{S-T}\| = \|b_S - b_T\| = \|b - b\| = 0,$$

and $S = T$. ∎

2.4.2. THEOREM. *Suppose that* $\mathscr{H}$, $\mathscr{K}$, *and* $\mathscr{L}$ *are Hilbert spaces. If* $T \in \mathscr{B}(\mathscr{H}, \mathscr{K})$, *there is a unique element* $T^*$ *of* $\mathscr{B}(\mathscr{K}, \mathscr{H})$ such that

$$\langle T^*x, y\rangle = \langle x, Ty\rangle \qquad (x \in \mathscr{K}, \quad y \in \mathscr{H}). \tag{2}$$

*Moreover,*

(i) $(aS + bT)^* = \bar{a}S^* + \bar{b}T^*$,
(ii) $(RS)^* = S^*R^*$,
(iii) $(T^*)^* = T$,
(iv) $\|T^*T\| = \|T\|^2$,
(v) $\|T^*\| = \|T\|$,

*whenever* $S, T \in \mathscr{B}(\mathscr{H}, \mathscr{K})$, $R \in \mathscr{B}(\mathscr{K}, \mathscr{L})$, *and* $a, b \in \mathbb{C}$.

*Proof.* The equation $b(x, y) = \langle x, Ty\rangle$ defines a conjugate-bilinear functional $b$ on $\mathscr{K} \times \mathscr{H}$. With the notation introduced in Theorem 2.4.1,

$$|b(x, y)| = |\langle Ty, x\rangle| = |b_T(y, x)|,$$

so $b$ is bounded and $\|b\| = \|b_T\| = \|T\|$. By the theorem just cited, there is a unique element $T^*$ of $\mathscr{B}(\mathscr{K}, \mathscr{H})$ for which

$$\langle T^*x, y\rangle = b(x, y) = \langle x, Ty\rangle \qquad (x \in \mathscr{K}, \quad y \in \mathscr{H});$$

and $\|T^*\| = \|b\| = \|T\|$. For each $x$ in $\mathscr{H}$,

$$\|Tx\|^2 = \langle Tx, Tx\rangle = \langle T^*Tx, x\rangle \leqslant \|T^*T\|\,\|x\|^2,$$

so

$$\|T\|^2 \leqslant \|T^*T\| \leqslant \|T^*\|\,\|T\| = \|T\|^2,$$

and (iv) follows.

It remains to prove (i), (ii), and (iii). These follow from the identities

$$\begin{aligned} \langle(\bar{a}S^* + \bar{b}T^*)x, y\rangle &= \bar{a}\langle S^*x, y\rangle + \bar{b}\langle T^*x, y\rangle \\ &= \bar{a}\langle x, Sy\rangle + \bar{b}\langle x, Ty\rangle = \langle x, (aS + bT)y\rangle, \end{aligned}$$

$$\langle S^*R^*z, y\rangle = \langle R^*z, Sy\rangle = \langle z, RSy\rangle,$$

$$\langle Ty, x\rangle = \overline{\langle x, Ty\rangle} = \overline{\langle T^*x, y\rangle} = \langle y, T^*x\rangle$$

($x \in \mathscr{K}$, $y \in \mathscr{H}$, $z \in \mathscr{L}$), together with the fact (proved above) that, for each bounded linear operator $T$, there is *only one* operator $T^*$ satisfying (2). ■

When $T \in \mathscr{B}(\mathscr{H}, \mathscr{K})$, the operator $T^*$ occurring in the preceding theorem is called the (*Hilbert*) *adjoint* of $T$. When dealing with Hilbert space operators, it is understood that the term "adjoint" refers to the Hilbert adjoint $T^*$, rather than the Banach adjoint $T^\sharp$ discussed at the end of Section 1.6, unless there is an explicit indication to the contrary. While $T^*$ maps $\mathscr{K}$ into $\mathscr{H}$, $T^\sharp$ maps the Banach dual space $\mathscr{K}^\sharp$ into $\mathscr{H}^\sharp$. If $J_1 : \mathscr{H} \to \mathscr{H}^\sharp$ and $J_2 : \mathscr{K} \to \mathscr{K}^\sharp$ are the norm-preserving conjugate-linear mappings defined as in Corollary 2.3.2, we have $T^* = J_1^{-1}T^\sharp J_2$, since

$$\langle x, J_1^{-1}T^\sharp J_2 y\rangle = (T^\sharp J_2 y)(x) = (J_2 y)(Tx) = \langle Tx, y\rangle$$

for all $x$ in $\mathscr{H}$ and $y$ in $\mathscr{K}$.

We now specialize to the case in which $\mathscr{K} = \mathscr{H}$. With $T$ in $\mathscr{B}(\mathscr{H})$, the adjoint $T^*$ is again in $\mathscr{B}(\mathscr{H})$. Accordingly, $\mathscr{B}(\mathscr{H})$ is a complex Banach algebra with an "adjoint operation" * satisfying

(i) $(aS + bT)^* = \bar{a}S^* + \bar{b}T^*$,
(ii) $(ST)^* = T^*S^*$,
(iii) $(T^*)^* = T$,
(iv) $\|T^*T\| = \|T\|^2$,

for all $S, T$ in $\mathscr{B}(\mathscr{H})$ and $a, b$ in $\mathbb{C}$. These four basic properties (from which a fifth, $\|T^*\| = \|T\|$, is easily deduced) form part of the basic background material for much of our subsequent work.

When $T \in \mathscr{B}(\mathscr{H})$ and $x, y \in \mathscr{H}$, we have

$$(3) \qquad 4\langle Tx, y\rangle = \langle T(x+y), x+y\rangle - \langle T(x-y), x-y\rangle + i\langle T(x+iy), x+iy\rangle - i\langle T(x-iy), x-iy\rangle.$$

This relation is called the *polarization identity*. It is a particular case of the second relation stated in Proposition 2.1.7 (and is essentially equivalent to it). When $T = I$, it reduces to the expression of inner products in terms of norms, already noted in Proposition 2.1.8.

2.4.3. Proposition. *If $S$ and $T$ are bounded linear operators acting on a Hilbert space $\mathscr{H}$ and $\langle Sx, x\rangle = \langle Tx, x\rangle$ for each $x$ in $\mathscr{H}$, then $S = T$.*

*Proof.* Since $\langle Sx, x\rangle = \langle Tx, x\rangle$ for each vector $x$, it follows by polarization (that is, by means of the polarization identity (3)) that $\langle Sx, y\rangle = \langle Tx, y\rangle$ for all $x$ and $y$ in $\mathscr{H}$. Hence $S$ and $T$ give rise to the same conjugate-bilinear functional on $\mathscr{H}$, and from the uniqueness clause in Theorem 2.4.1, $S = T$. ■

2.4.4. REMARK. For linear operators acting on *real* inner product spaces, the analogue of Proposition 2.4.3 is false. For example, the equation $T(x_1, x_2) = (x_2, -x_1)$ defines a non-zero operator $T$ acting on $\mathbb{R}^2$, and $\langle Tx, x\rangle = 0$ for each $x$ in $\mathbb{R}^2$. ■

2.4.5. PROPOSITION. *If $\mathscr{H}$ and $\mathscr{K}$ are Hilbert spaces and $T \in \mathscr{B}(\mathscr{H}, \mathscr{K})$, then $T$ is an isomorphism from $\mathscr{H}$ onto $\mathscr{K}$ if and only if it is invertible, with $T^{-1} = T^*$.*

*Proof.* The operator $T$ is an isomorphism from $\mathscr{H}$ onto $\mathscr{K}$ if and only if it is both invertible and norm preserving. Accordingly, we may suppose that $T$ has an inverse, and it suffices to show that $T$ preserves norms if and only if $T^*T = I$ (which is equivalent to $T^{-1} = T^*$ when $T$ is invertible). Since

$$\langle T^*Tx, x\rangle - \langle x, x\rangle = \langle Tx, Tx\rangle - \langle x, x\rangle = \|Tx\|^2 - \|x\|^2,$$

for each $x$ in $\mathscr{H}$, the required result follows from Proposition 2.4.3. ■

*Classes of operators.* A bounded linear operator $T$, acting on a Hilbert space $\mathscr{H}$, is said to be *self-adjoint if* $T^* = T$, and *unitary* if $TT^* = T^*T = I$. Both these conditions imply that $T$ is *normal*, by which we mean that $TT^* = T^*T$. We say that $T$ is *positive* if $\langle Tx, x\rangle \geqslant 0$ for each $x$ in $\mathscr{H}$.

For every $T$ in $\mathscr{B}(\mathscr{H})$,

$$\langle T^*Tx, x\rangle = \langle Tx, Tx\rangle \geqslant 0 \qquad (x \in \mathscr{H}),$$

so $T^*T$ is positive; we shall see later (Theorem 4.2.6(iii)) that each positive operator arises in this way.

A conjugate-bilinear functional $b$ on $\mathscr{H}$ is said to be *symmetric* if $b(y, x) = \overline{b(x, y)}$ for all $x$ and $y$ in $\mathscr{H}$, and *positive* if $b(x, x) \geqslant 0$ for each $x$. With $T$ in $\mathscr{B}(\mathscr{H})$, and $b_T$ the conjugate-bilinear functional defined by $b_T(x, y) = \langle Tx, y\rangle$, it is evident that $T$ is positive if and only if $b_T$ is positive; moreover,

$$b_{T^*}(x, y) = \langle T^*x, y\rangle = \langle x, Ty\rangle = \overline{\langle Ty, x\rangle} = \overline{b_T(y, x)},$$

so $b_T = b_{T^*}$ (equivalently, $T$ is self-adjoint) if and only if $b_T$ is symmetric.

2.4.6. PROPOSITION. *Suppose that $T$ is a bounded linear operator acting on a Hilbert space $\mathscr{H}$.*

(i) *$T$ is self-adjoint if and only if $\langle Tx, x\rangle$ is real for each $x$ in $\mathscr{H}$. In particular, positive operators are self-adjoint.*

(ii) *$T$ is unitary if and only if $T$ is a norm-preserving (equivalently, inner product-preserving) mapping from $\mathscr{H}$ onto $\mathscr{H}$.*

(iii) *$T$ is normal if and only if $\|Tx\| = \|T^*x\|$ for each $x$ in $\mathscr{H}$.*

*Proof.* (i) Since

$$\langle Tx, x\rangle - \langle T^*x, x\rangle = \langle Tx, x\rangle - \langle x, Tx\rangle = 2i\,\mathrm{Im}\langle Tx, x\rangle,$$

it follows from Proposition 2.4.3 that $T = T^*$ if and only if $\langle Tx, x\rangle$ is real for each vector $x$.

(ii) An element $T$ of $\mathscr{B}(\mathscr{H})$ is unitary if and only if it is invertible, with inverse $T^*$; so the assertion (ii) is a special case of Proposition 2.4.5.

(iii) Since

$$\langle T^*Tx, x\rangle - \langle TT^*x, x\rangle = \langle Tx, Tx\rangle - \langle T^*x, T^*x\rangle = \|Tx\|^2 - \|T^*x\|^2,$$

if follows from Proposition 2.4.3 that $T^*T = TT^*$ if and only if $\|Tx\| = \|T^*x\|$ for each vector $x$. ∎

From part (ii) of the above proposition, a "unitary operator" acting on a Hilbert space $\mathscr{H}$ is simply an isomorphism from $\mathscr{H}$ onto itself. We shall sometimes describe isomorphisms between *different* Hilbert spaces as unitary operators (or unitary transformations).

2.4.7. Remark. When $\mathscr{H}$ is an infinite-dimensional Hilbert space, a norm-preserving linear operator $T$ acting on $\mathscr{H}$ is not necessarily unitary, since its range may fail to be the whole of $\mathscr{H}$. For an example in which this occurs, consider the operator that acts on the sequence space $l_2$ (see Example 2.1.12), and maps the vector $(x_1, x_2, x_3, \ldots)$ onto $(0, x_1, x_2, \ldots)$. ∎

2.4.8. Lemma. *If $T$ is a bounded normal operator on the Hilbert space $\mathscr{H}$ and*

$$0 < \inf\{\|Tx\| : x \in \mathscr{H},\ \|x\| = 1\} \quad (= a),$$

*then $T$ has a bounded, two-sided inverse, and $\|T^{-1}\| = a^{-1}$.*

*Proof.* By Corollary 1.5.10, $T$ is a bicontinuous linear mapping from $\mathscr{H}$ onto the range $\mathscr{R}(T)$ of $T$, and the inverse mapping $T^{-1}: \mathscr{R}(T) \to \mathscr{H}$ satisfies $\|T^{-1}\| = a^{-1}$; moreover, $\mathscr{R}(T)$ is complete, and is therefore closed in $\mathscr{H}$. It remains to prove that $\mathscr{R}(T) = \mathscr{H}$.

If $\mathscr{R}(T) \neq \mathscr{H}$, there is a unit vector $x$ in $\mathscr{R}(T)^\perp$; by Proposition 2.4.6(iii),

$$\begin{aligned} 0 &= \langle x, TT^*x\rangle = \langle T^*x, T^*x\rangle \\ &= \|T^*x\|^2 = \|Tx\|^2 \geqslant a^2, \end{aligned}$$

a contradiction. Thus $\mathscr{R}(T) = \mathscr{H}$. ∎

The simple properties of the adjoint operation * on $\mathscr{B}(\mathscr{H})$, as set out in (i), ..., (iv) in the discussion preceding Proposition 2.4.3, in some respects resemble those of the process of complex conjugation for elements of the scalar field $\mathbb{C}$. The analogy can usefully be pressed a good deal further. The self-adjoint elements of $\mathscr{B}(\mathscr{H})$ (those for which $T = T^*$) correspond to real numbers (the scalars for which $a = \bar{a}$). Parallel to the expression of a complex number in terms of its real and imaginary parts, each $T$ in $\mathscr{B}(\mathscr{H})$ can be written (uniquely) in the form $H + iK$, with $H$ and $K$ self-adjoint operators acting on $\mathscr{H}$; moreover

$$H = \tfrac{1}{2}(T + T^*), \qquad K = \tfrac{1}{2}i(T^* - T).$$

The operators $H$ and $K$ are sometimes called the "real" and "imaginary" parts of $T$, and denoted by Re $T$ and Im $T$, respectively. It is easily verified that $T$ is normal if and only if $H$ and $K$ commute.

The classes $\mathscr{S}, \mathscr{U}, \mathscr{N}$ and $\mathscr{B}(\mathscr{H})^+$ of all self-adjoint, unitary, normal, and positive operators (respectively) on $\mathscr{H}$ are norm-closed subsets of $\mathscr{B}(\mathscr{H})$. Moreover, $\mathscr{U}$ is a multiplicative group, while $\mathscr{S}$ is a real-linear subspace (that is, $aH + bK \in \mathscr{S}$ whenever $H, K \in \mathscr{S}$ and $a, b \in \mathbb{R}$). From Proposition 2.4.6(i), $\mathscr{B}(\mathscr{H})^+$ is a subset of $\mathscr{S}$; it is apparent that $aH + bK \in \mathscr{B}(\mathscr{H})^+$ whenever $H, K \in \mathscr{B}(\mathscr{H})^+$ and $a, b$ are non-negative real numbers. If both $H \in \mathscr{B}(\mathscr{H})^+$ and $-H \in \mathscr{B}(\mathscr{H})^+$, then $\langle Hx, x\rangle = 0$ for each vector $x$, and $H = 0$ by Proposition 2.4.3; so $\mathscr{B}(\mathscr{H})^+ \cap -\mathscr{B}(\mathscr{H})^+ = \{0\}$. In view of the properties of $\mathscr{B}(\mathscr{H})^+$ just stated, there is a partial order relation $\leqslant$ on $\mathscr{S}$, in which $H \leqslant K$ if and only if $K - H \in \mathscr{B}(\mathscr{H})^+$. As in the case of real numbers, one can add such inequalities between self-adjoint operators and multiply throughout by non-negative scalars. Multiplication by negative real numbers reverses the inequalities. Moreover, if $H, K \in \mathscr{S}$, $T \in \mathscr{B}(\mathscr{H})$, and $H \leqslant K$, then $T^*HT \leqslant T^*KT$; for $K - H \in \mathscr{B}(\mathscr{H})^+$, and therefore

$$\langle T^*(K - H)Tx, x\rangle = \langle (K - H)Tx, Tx\rangle \geqslant 0 \quad (x \in \mathscr{H}),$$

whence $T^*(K - H)T \in \mathscr{B}(\mathscr{H})^+$. For each $H$ in $\mathscr{S}$, the operators $\|H\|I \pm H$ are positive, since

$$\|H\|\langle x, x\rangle \pm \langle Hx, x\rangle \geqslant \|H\|\,\|x\|^2 - \|Hx\|\,\|x\| \geqslant 0 \qquad (x \in \mathscr{H}).$$

It follows that

$$-\|H\|I \leqslant H \leqslant \|H\|I \qquad (H = H^* \in \mathscr{B}(\mathscr{H})),$$

and that each self-adjoint operator $H$ can be expressed as the difference of positive operators $\|H\|I$ and $\|H\|I - H$. Each $T$ in $\mathscr{B}(\mathscr{H})$ has the form $H + iK$, with $H$ and $K$ self-adjoint, and is therefore a linear combination of at most four elements of $\mathscr{B}(\mathscr{H})^+$.

We shall see later, in Theorem 4.1.7, Theorem 6.1.2, and Proposition 4.2.3, that each element of $\mathscr{B}(\mathscr{H})$ is a linear combination of at most four unitary operators, and has a "polar decomposition" analogous to the expression of a complex number in terms of its modulus and argument; moreover, there is an "optimal" way of expressing a self-adjoint operator as a difference of positive operators.

2.4.9. Remark. For the study of bounded linear operators, a Hilbert space is much more convenient than a *real* inner product space. This is due, in part, to the fact that, in contrast with the complex case, a non-self-adjoint operator acting on a real inner product space cannot be expressed as a linear combination (necessarily with real coefficients) of self-adjoint operators. ■

2.4.10. Example. Suppose that $Y$ is an orthonormal basis in a Hilbert space $\mathscr{H}$, $g$ is a bounded complex-valued function on $Y$, and

$$k = \sup\{|g(y)| : y \in Y\}.$$

For each $x$ in $\mathscr{H}$,

$$\sum_{y \in Y} |g(y)\langle x, y\rangle|^2 \leqslant k^2 \sum_{y \in Y} |\langle x, y\rangle|^2 = k^2\|x\|^2;$$

so the equation

$$Tx = \sum_{y \in Y} g(y)\langle x, y\rangle y \qquad (x \in \mathscr{H}) \tag{4}$$

defines a vector $Tx$ in $\mathscr{H}$, and

$$\|Tx\| = \left(\sum_{y \in Y} |g(y)\langle x, y\rangle|^2\right)^{1/2} \leqslant k\|x\|. \tag{5}$$

It is apparent that $Tx$ depends linearly on $x$, so $T$ is a bounded linear operator on $\mathscr{H}$, with $\|T\| \leqslant k$. From (4),

$$Ty = g(y)y \qquad (y \in Y). \tag{6}$$

From this,

$$\|T\| \geqslant \sup\{\|Ty\| : y \in Y\} = \sup\{|g(y)| : y \in Y\} = k,$$

and so

$$\|T\| = \sup\{|g(y)| : y \in Y\}. \tag{7}$$

By the same process, the complex-valued function $\bar{g}$, defined by $\bar{g}(y) = \overline{g(y)}$, gives rise to a bounded linear operator $S$ on $\mathscr{H}$. For all $u$ and $v$ in $\mathscr{H}$,

$$\begin{aligned}
\langle Su, v\rangle &= \langle \sum_{y \in Y} \overline{g(y)}\langle u, y\rangle y, v\rangle \\
&= \sum_{y \in Y} \overline{g(y)}\langle u, y\rangle\langle y, v\rangle \\
&= \langle u, \sum_{y \in Y} g(y)\langle v, y\rangle y\rangle = \langle u, Tv\rangle.
\end{aligned}$$

Hence $S = T^*$ and, in parallel with (4), (5), and (6), we have

$$T^*x = \sum_{y \in Y} \overline{g(y)} \langle x, y \rangle y, \tag{8}$$

$$\|T^*x\| = \left( \sum_{y \in Y} |g(y)\langle x, y \rangle|^2 \right)^{1/2}, \tag{9}$$

$$T^*y = \overline{g(y)}y \qquad (y \in Y). \tag{10}$$

From (5) and (9), $\|Tx\| = \|T^*x\|$ for each vector $x$, so $T$ is normal; alternatively, this can be proved by a simple direct calculation, which shows that

$$TT^*x = T^*Tx = \sum_{y \in Y} |g(y)|^2 \langle x, y \rangle y. \tag{11}$$

Similar calculations show that sums and products of bounded complex-valued functions correspond to sums and products of the associated operators; in particular, all such operators commute.

If $T$ is self-adjoint, it results from (6) and (10) that $g(y) = \overline{g(y)}$ for each $y$ in $Y$; the reverse implication follows from (4) and (8). Thus $T$ is self-adjoint if and only if $g$ is a real-valued function.

From (4)

$$\langle Tx, x \rangle = \sum_{y \in Y} g(y) \langle x, y \rangle \langle y, x \rangle = \sum_{y \in Y} g(y) |\langle x, y \rangle|^2;$$

in particular, $\langle Ty, y \rangle = g(y)$ $(y \in Y)$. Hence $T$ is positive if and only if $g$ takes non-negative real values throughout $Y$.

If $T$ is unitary, it follows from (6) that $|g(y)| = \|Ty\| = \|y\| = 1$ for each $y$ in $Y$. Conversely, if $|g(y)| = 1$ $(y \in Y)$, we deduce from (11) that

$$TT^*x = T^*Tx = \sum_{y \in Y} \langle x, y \rangle y = x \qquad (x \in \mathscr{H}).$$

Hence $T$ is unitary if and only if $|g(y)| = 1$ for each $y$ in $Y$. ■

2.4.11. Example. Suppose that $m$ is a $\sigma$-finite measure defined on a $\sigma$-algebra $\mathscr{S}$ of subsets of a set $S$, $g \in L_\infty$ $(= L_\infty(S, \mathscr{S}, m))$, and $k$ is the essential supremum of $|g|$. For each $x$ in the Hilbert space $\mathscr{H} = L_2(S, \mathscr{S}, m)$, the equation

$$(M_g x)(s) = g(s)x(s) \qquad (s \in S) \tag{12}$$

defines a measurable function $M_g x$ on $S$, and $|(M_g x)(s)| \leqslant k|x(s)|$ almost everywhere. Accordingly, $M_g x \in \mathscr{H}$ $(= L_2)$ and $\|M_g x\| \leqslant k\|x\|$ since

$$\int_S |(M_g x)(s)|^2 \, dm(s) \leqslant k^2 \int_S |x(s)|^2 \, dm(s).$$

It is apparent that $M_g x$ depends linearly on $x$; so $M_g$ is a bounded linear operator on $\mathscr{H}$, and $\|M_g\| \leqslant k$. If $0 < a < k$, the measurable set $\{s \in S: |g(s)| > a\}$ has positive measure, and so has a measurable subset $Y$ such that $0 < m(Y) < \infty$, since $m$ is $\sigma$-finite. The characteristic function $y$ of $Y$ is a non-zero vector in $\mathscr{H}$, $|(M_g y)(s)| \geqslant a|y(s)|$ for each $s$ in $S$; hence $\|M_g y\| \geqslant a\|y\|$, and so $\|M_g\| \geqslant a$. From this, $\|M_g\| = k$; that is,

$$\|M_g\| = \operatorname*{ess\,sup}_{s \in S} |g(s)| = \|g\|_\infty, \tag{13}$$

where $\| \, \|_\infty$ is the usual norm on $L_\infty$.

With $\bar{g}$ defined by $\bar{g}(s) = \overline{g(s)}$ $(s \in S)$, we have $\bar{g} \in L_\infty$ and

$$\begin{aligned}\langle M_{\bar{g}} x, y \rangle &= \int_S (M_{\bar{g}} x)(s) \overline{y(s)} \, dm(s) \\ &= \int_S \overline{g(s)} x(s) \overline{y(s)} \, dm(s) \\ &= \int_S x(s) \overline{(M_g y)(s)} \, dm(s) = \langle x, M_g y \rangle \qquad (x, y \in \mathscr{H});\end{aligned}$$

therefore

$$M_g^* = M_{\bar{g}}. \tag{14}$$

It is apparent that

$$M_{af+bg} = aM_f + bM_g, \qquad M_{fg} = M_f M_g \qquad (f, g \in L_\infty, \quad a, b \in \mathbb{C}). \tag{15}$$

From this, $M_f$ and $M_g$ commute for all $f$ and $g$ in $L_\infty$; in particular, $M_g$ commutes with its adjoint $M_{\bar{g}}$, and is therefore normal.

Since

$$\|M_g - M_g^*\| = \|M_{g - \bar{g}}\| = \operatorname*{ess\,sup}_{s \in S} |g(s) - \overline{g(s)}|,$$

$M_g$ is self-adjoint if and only if $g(s)$ is real for almost all $s$ in $S$. Moreover,

$$\langle M_g x, x \rangle = \int_S g(s) |x(s)|^2 \, dm(s) \qquad (x \in \mathscr{H}),$$

from which it follows that $M_g$ is positive if and only if $g(s) \geqslant 0$ for almost all $s$.

With $u$ in $L_\infty$ defined by $u(s) = 1$ $(s \in S)$, $M_u = I$ and

$$\|I - M_g M_g^*\| = \|I - M_g^* M_g\| = \|M_{u - \bar{g}g}\| = \operatorname*{ess\,sup}_{s \in S} |1 - |g(s)|^2|.$$

Thus $M_g$ is unitary if and only if $|g(s)| = 1$ for almost all $s$ in $S$.

It is well known that a linear operator $T$, acting on a finite-dimensional complex vector space, has at least one eigenvector $x$, with corresponding eigenvalue $\lambda$ (that is, $x \neq 0$ and $Tx = \lambda x$). As indicated in (6), the operator considered in the preceding example has eigenvectors forming an orthonormal basis $Y$. In contrast, the present example permits the construction of self-adjoint and unitary operators that have no eigenvalue. For this purpose, note that the equation $M_g x = cx$ (with $g$ in $L_\infty$, $x$ a non-zero vector in $\mathscr{H}$ $(= L_2)$, and $c$ in $\mathbb{C}$) implies that $g(s) = c$ almost everywhere on the measurable set $\{s \in S : x(s) \neq 0\}$, which has positive measure. Accordingly, $M_g$ has no eigenvalue if $g$ assumes each of its values only on a null set. When $m$ is Lebesgue measure on the $\sigma$-ring $\mathscr{S}$ of Borel subsets of the interval $S = [0, 1]$ and

$$f(s) = s, \qquad g(s) = \exp(is) \qquad (s \in S),$$

$M_f$ is positive, $M_g$ is unitary, and neither has an eigenvalue. ■

## 2.5. The lattice of projections

If $Y$ is a closed subspace of a Hilbert space $\mathscr{H}$, Theorem 2.2.3 asserts that each vector in $\mathscr{H}$ can be expressed uniquely in the form $y + z$, with $y$ in $Y$ and $z$ in $Y^\perp$. From Theorem 1.1.8, the equation

$$E(y + z) = y \qquad (y \in Y, \quad z \in Y^\perp) \tag{1}$$

defines a linear operator $E$ acting on $\mathscr{H}$, the projection onto $Y$, parallel to $Y^\perp$. Moreover, $E^2 = E$,

$$Y = \{Ex : x \in \mathscr{H}\} = \{y \in \mathscr{H} : Ey = y\}, \tag{2}$$

and $Y^\perp = \{z \in \mathscr{H} : Ez = 0\}$. We call $E$ the (*orthogonal*) *projection* from $\mathscr{H}$ onto $Y$. Note that $I - E$ is the orthogonal projection from $\mathscr{H}$ onto $Y^\perp$, because $(Y^\perp)^\perp = Y$, and

$$(I - E)(z + y) = z \qquad (z \in Y^\perp, \quad y \in Y).$$

Since $\langle y, z \rangle = 0$, when $y \in Y$ and $z \in Y^\perp$, we have

$$\|E(y + z)\|^2 = \|y\|^2 \leqslant \|y\|^2 + \|z\|^2 = \|y + z\|^2,$$

$$\langle E(y + z), y + z \rangle = \langle y, y + z \rangle = \|y\|^2 \geqslant 0.$$

It follows that $E$ is bounded, with $\|E\| \leqslant 1$, and is positive (hence, also, self-adjoint). Since $Ey = y$ $(y \in Y)$, $\|E\| = 1$ except in the case in which $Y = \{0\}$ and $E = 0$. Moreover

$$Y = \{x \in \mathscr{H} : \|Ex\| = \|x\|\}. \tag{3}$$

Conversely, suppose that $E \in \mathscr{B}(\mathscr{H})$ and $E^2 = E = E^*$. From Theorem 1.1.8, $E$ is the projection from $\mathscr{H}$ onto the closed subspace $Y$ defined by (2), parallel to the closed subspace $Z = \{z \in \mathscr{H} : Ez = 0\}$. Since

$$Z = \{z \in \mathscr{H} : \langle Ez, x \rangle = 0 \text{ for each } x \text{ in } \mathscr{H}\}$$
$$= \{z \in \mathscr{H} : \langle z, Ex \rangle = 0 \text{ for each } x \text{ in } \mathscr{H}\},$$

it follows from (2) that $Z = Y^{\perp}$. Hence $E$ is the projection onto $Y$, parallel to $Y^{\perp}$; that is, the orthogonal projection from $\mathscr{H}$ onto $Y$.

In the context of Hilbert space theory, it is understood that the term "projection" refers to an orthogonal projection unless there is an explicit statement to the contrary. The following proposition summarizes the results of the preceding discussion.

2.5.1. PROPOSITION. *Relations* (1) *and* (2) *establish a one-to-one correspondence between closed subspaces $Y$ of a Hilbert space $\mathscr{H}$ and projections $E$ acting on $\mathscr{H}$. A projection $E$ is a positive operator, and $\|E\| = 1$ unless $E = 0$. The projections are precisely the self-adjoint idempotents in $\mathscr{B}(\mathscr{H})$.*

The projections acting on a Hilbert space $\mathscr{H}$ inherit from the set of all self-adjoint operators the partial order relation $\leqslant$ described in the discussion preceding Remark 2.4.9.

2.5.2. PROPOSITION. *If $E$ and $F$ are the projections from a Hilbert space $\mathscr{H}$ onto closed subspaces $Y$ and $Z$, respectively, the following conditions are equivalent*:

(i) $Y \subseteq Z$;
(ii) $FE = E$;
(iii) $EF = E$;
(iv) $\|Ex\| \leqslant \|Fx\| \quad (x \in \mathscr{H})$;
(v) $E \leqslant F$.

*Proof.* If $Y \subseteq Z$, then, for each $x$ in $\mathscr{H}$, $Ex \in Y \subseteq Z$, and therefore $FEx = Ex$; so (i) implies (ii). If $FE = E$, then $EF = (FE)^* = E^* = E$, whence (ii) implies (iii). If $EF = E$, then $\|Ex\| = \|EFx\| \leqslant \|Fx\|$ for each $x$ in $\mathscr{H}$, since $\|E\| \leqslant 1$; so (iii) implies (iv). Since

$$\langle Ex, x \rangle = \langle E^2 x, x \rangle = \langle Ex, Ex \rangle = \|Ex\|^2,$$

and similarly $\langle Fx, x \rangle = \|Fx\|^2$, it is apparent that (iv) implies (v). If $E \leqslant F$, then, for each $y$ in $Y$,

$$\|y\|^2 = \langle Ey, y \rangle \leqslant \langle Fy, y \rangle = \|Fy\|^2 \leqslant \|y\|^2;$$

whence $\|Fy\| = \|y\|$, and $y \in Z$ by (3). Hence, (v) implies (i). ∎

When the five equivalent conditions in Proposition 2.5.2 are satisfied, we describe $E$ as a *subprojection* of $F$.

From the equivalence of conditions (i) and (v) in Proposition 2.5.2, it follows that the partial ordering of projections (as self-adjoint operators) corresponds to the partial ordering of closed subspaces by the inclusion relation $\subseteq$. Given any family $\{Y_a\}$ of closed subspaces of a Hilbert space $\mathscr{H}$, there is a greatest closed subspace $\bigwedge Y_a$ that is contained in each $Y_a$ and a smallest closed subspace $\bigvee Y_a$ that contains each $Y_a$. Specifically, $\bigwedge Y_a$ is $\bigcap Y_a$, while $\bigvee Y_a$ is the closed subspace $[\bigcup Y_a]$ generated by $\bigcup Y_a$. From this it follows that each family $\{E_a\}$ of projections acting on $\mathscr{H}$ has a greatest lower bound $\bigwedge E_a$ and a least upper bound $\bigvee E_a$ within the set of projections (ordered as self-adjoint operators). Of course, the projections $\bigwedge E_a$ and $\bigvee E_a$ correspond to the closed subspaces $\bigwedge E_a(\mathscr{H})$ and $\bigvee E_a(\mathscr{H})$, respectively. We write $E \wedge F$ and $E \vee F$ for the lower and upper bounds (often called the *intersection* and *union*) of two projections $E$ and $F$.

Since the mapping $E \to I - E$ reverses the ordering of projections, we have

$$\bigvee(I - E_a) = I - \bigwedge E_a, \qquad \bigwedge(I - E_a) = I - \bigvee E_a \tag{4}$$

for each family $\{E_a\}$ of projections. This gives corresponding relations

$$\bigvee Y_a^{\perp} = (\bigwedge Y_a)^{\perp}, \qquad \bigwedge Y_a^{\perp} = (\bigvee Y_a)^{\perp}$$

(which can easily be verified independently) for each family $\{Y_a\}$ of closed subspaces of $\mathscr{H}$.

Our next few results are concerned with commuting sets of projections.

2.5.3. Proposition. *If $E$ and $F$ are commuting projections acting on a Hilbert space $\mathscr{H}$, corresponding to closed subspaces $Y$ and $Z$, respectively, then*

$$E \vee F = E + F - EF, \qquad E \wedge F = EF, \qquad Y \vee Z = Y + Z.$$

*In particular, the linear subspace $Y + Z$ of $\mathscr{H}$ is closed.*

*Proof.* When $u \in Y \wedge Z$, $Eu = Fu = u$, so $EFu = u$. For each $x$ in $Y^{\perp}$, $Ex = 0$, so $EFx = FEx = 0$; similarly, $EFx = 0$ for each $x$ in $Z^{\perp}$. Since $(Y \wedge Z)^{\perp} = Y^{\perp} \vee Z^{\perp}$, it follows from the linearity and continuity of $EF$ that $EFv = 0$ whenever $v \in (Y \wedge Z)^{\perp}$. We have now shown that

$$EF(u + v) = u \qquad (u \in Y \wedge Z, \quad v \in (Y \wedge Z)^{\perp}),$$

whence $EF$ is the projection from $\mathscr{H}$ onto $Y \wedge Z$. By applying the same result to the commuting projections $I - E$ and $I - F$, we have

$$(I - E) \wedge (I - F) = (I - E)(I - F),$$

and (4) gives

$$E \vee F = I - (I - E) \wedge (I - F) = I - (I - E)(I - F) = E + F - EF.$$

For each $x$ in $Y \vee Z$,

$$x = (E \vee F)x = (E + F - EF)x = y + z,$$

where $y = (E - EF)x \in Y$ and $z = Fx \in Z$. Thus $Y \vee Z \subseteq Y + Z$, and the reverse inclusion is apparent. ■

2.5.4. COROLLARY. *Suppose that $E$ and $F$ are the projections from a Hilbert space $\mathscr{H}$ onto closed subspaces $Y$ and $Z$, respectively. Then $EF = 0$ if and only if $Y$ is orthogonal to $Z$, and when this is so,*

$$E \vee F = E + F, \qquad Y \vee Z = Y + Z.$$

*Proof.* Since $Y = E(\mathscr{H})$, $Z = F(\mathscr{H})$, and $\langle EFu, v\rangle = \langle Fu, Ev\rangle$ for all $u$ and $v$ in $\mathscr{H}$, it is evident that $Y$ is orthogonal to $Z$ if and only if $EF = 0$. When this is so, $FE = (EF)^* = 0\ (= EF)$, and it follows from Proposition 2.5.3 that

$$E \vee F = E + F, \qquad Y \vee Z = Y + Z. \quad ■$$

2.5.5. COROLLARY. *If $E$ and $F$ are the projections from a Hilbert space $\mathscr{H}$ onto closed subspaces $Y$ and $Z$, respectively, and $E \leqslant F$, then $F - E$ is the projection from $\mathscr{H}$ onto $Z \wedge Y^{\perp}$.*

*Proof.* By Proposition 2.5.2, $EF = FE = E$, so the projections $F$ and $I - E$ commute. From Proposition 2.5.3, $F(I - E)\ (= F - E)$ is the projection $F \wedge (I - E)$ from $\mathscr{H}$ onto $Z \wedge Y^{\perp}$. ■

Projections $E$ and $F$, from $\mathscr{H}$ onto closed subspaces $Y$ and $Z$, commute if and only if $Y \wedge (Y \wedge Z)^{\perp}$ and $Z \wedge (Y \wedge Z)^{\perp}$ are orthogonal (loosely, if and only if the spaces $Y$ and $Z$ are "perpendicular"); for these spaces are orthogonal if and only if

$$0 = (E - E \wedge F)(F - E \wedge F) = EF - E \wedge F,$$

and $EF = E \wedge F$ if and only if (see Proposition 2.5.3)

$$EF = E \wedge F = (E \wedge F)^* = FE.$$

2.5.6. PROPOSITION. *If $\{E_a\}$ is an increasing net of projections acting on a Hilbert space $\mathscr{H}$, and if $E = \bigvee E_a$, then $Ex = \lim_a E_a x$ for each $x$ in $\mathscr{H}$.*

*Proof.* Since $\{E_a(\mathscr{H})\}$ is an increasing net of closed subspaces of $\mathscr{H}$, $\bigcup_a E_a(\mathscr{H})$ is a linear subspace of $\mathscr{H}$ and has norm closure $E(\mathscr{H})$. Suppose $x \in \mathscr{H}$ and $\varepsilon > 0$. Since $Ex \in E(\mathscr{H})$, we can choose an element $y$ in one of the subspaces $E_a(\mathscr{H})$ so that $\|Ex - y\| < \varepsilon$. When $b \geqslant a$, we have

$$E_a \leqslant E_b \leqslant E, \qquad y \in E_a(\mathscr{H}) \subseteq E_b(\mathscr{H}) \subseteq E(\mathscr{H}),$$

and thus

$$\begin{aligned}\|Ex - E_b x\| &= \|E(Ex - y) - E_b(Ex - y)\| \\ &\leqslant \|E - E_b\|\,\|Ex - y\| < \varepsilon. \quad ■\end{aligned}$$

2.5.7. Corollary. *If $\{E_a\}$ is a decreasing net of projections acting on a Hilbert space $\mathscr{H}$, and if $E = \bigwedge E_a$, then $Ex = \lim_a E_a x$ for each $x$ in $\mathscr{H}$.*

*Proof.* In view of (4), it suffices to apply Proposition 2.5.6 to the increasing net $\{I - E_a\}$. ■

By an *orthogonal family of projections* we mean a family $(E_a)_{a\in\mathbb{A}}$ of projections such that $E_aE_b = 0$ (equivalently, $E_a(\mathscr{H})$ is orthogonal to $E_b(\mathscr{H})$) whenever $a$ and $b$ are distinct elements of $\mathbb{A}$.

2.5.8. Proposition. *If $(E_a)_{a\in\mathbb{A}}$ is an orthogonal family of projections acting on a Hilbert space $\mathscr{H}$, $E = \bigvee E_a$, and $x \in \mathscr{H}$, then $Ex = \sum E_a x$; the sum converges in the norm topology on $\mathscr{H}$.*

*Proof.* When $\mathbb{A}$ is a finite set, it follows from Corollary 2.5.4, together with a straightforward argument by induction on the number of elements in $\mathbb{A}$, that $E = \sum_{a\in\mathbb{A}} E_a$.

When $\mathbb{A}$ is an infinite set, let $\mathscr{F}$ denote the class of all finite subsets of $\mathbb{A}$; for each $\mathbb{F}$ in $\mathscr{F}$, define $G_{\mathbb{F}} = \sum_{a\in\mathbb{F}} E_a$. By the preceding paragraph, $G_{\mathbb{F}} = \bigvee_{a\in\mathbb{F}} E_a$, so $(G_{\mathbb{F}}, \mathbb{F}\in\mathscr{F}, \supseteq)$ is an increasing net of projections, and

$$\bigvee_{\mathbb{F}\in\mathscr{F}} G_{\mathbb{F}} = \bigvee\left\{\bigvee_{a\in\mathbb{F}} E_a : \mathbb{F}\in\mathscr{F}\right\} = \bigvee_{a\in\mathbb{A}} E_a = E.$$

By Proposition 2.5.6, $Ex$ is the limit, in norm, of the net $(G_{\mathbb{F}}x, F\in\mathscr{F}, \subseteq)$; that is (since $G_{\mathbb{F}}x = \sum_{a\in\mathbb{F}} E_a x$), $\sum_{a\in\mathbb{A}} E_a x$ converges in norm to $Ex$. ■

When $\mathscr{H}$ is a Hilbert space and $x\in\mathscr{H}$, the equation $p_x(T) = \|Tx\|$ defines a semi-norm $p_x$ on $\mathscr{B}(\mathscr{H})$. The family of all such semi-norms separates the points of $\mathscr{B}(\mathscr{H})$, in the sense of Theorem 1.2.6, and so gives rise to a locally convex topology on $\mathscr{B}(\mathscr{H})$, the *strong-operator topology*. In this topology, an element $T_0$ of $\mathscr{B}(\mathscr{H})$ has a base of neighborhoods consisting of all sets of the type

$$V(T_0 : x_1, \ldots, x_m; \varepsilon) = \{T\in\mathscr{B}(\mathscr{H}) : \|(T - T_0)x_j\| < \varepsilon \ (j = 1, \ldots, m)\},$$

where $x_1, \ldots, x_m \in \mathscr{H}$ and $\varepsilon > 0$. In fact, it suffices (and is sometimes convenient) to take $\varepsilon = 1$, since

$$V(T_0 : x_1, \ldots, x_m; \varepsilon) = V(T_0 : \varepsilon^{-1}x_1, \ldots, \varepsilon^{-1}x_m; 1).$$

In a similar way, one can introduce the strong-operator topology on $\mathscr{B}(\mathscr{H}, \mathscr{K})$; the semi-norms and basic neighborhoods are defined as above, but the vectors $Tx$, $(T - T_0)x_j$ lie in $\mathscr{K}$.

The strong-operator topology can be described as the restriction to $\mathscr{B}(\mathscr{H})$ of the point-open topology on mappings from $\mathscr{H}$ into $\mathscr{H}$, with $\mathscr{H}$ in its norm topology. It is apparent that $V(T_0 : x_1, \ldots, x_m; \varepsilon)$ contains the open ball with center $T_0$ and radius $b$, provided that $b\|x_j\| \leqslant \varepsilon$ for each $j = 1, \ldots, m$.

Accordingly, the strong-operator topology is coarser than the norm topology on $\mathscr{B}(\mathscr{H})$; in fact, it is strictly coarser when $\mathscr{H}$ is infinite-dimensional (see Exercise 2.8.32).

In working with the strong-operator topology, it is often possible (and useful) to confine attention to those basic neighborhoods $V(T_0 : x_1, \ldots, x_m; \varepsilon)$ in which $x_1, \ldots, x_m$ are drawn from a suitable preassigned subset $\mathscr{S}$ of $\mathscr{H}$. We mention two cases in which this occurs. First, when $\mathscr{S}$ has (algebraic) linear span $\mathscr{H}$, the sets $V(T_0 : y_1, \ldots, y_n; \delta)$, where $\delta > 0$ and $y_1, \ldots, y_n \in \mathscr{S}$, already form a base of neighborhoods of $T_0$. Second, if $\mathscr{S}$ has closed linear span $\mathscr{H}$, while $\mathscr{B}$ is a bounded subset of $\mathscr{B}(\mathscr{H})$ and $T_0 \in \mathscr{B}$, the sets

$$V(T_0 : y_1, \ldots, y_n; \delta) \cap \mathscr{B} \qquad (\delta > 0, \quad y_1, \ldots, y_n \in \mathscr{S})$$

form a base of neighborhoods of $T_0$ in the (relative) strong-operator topology on $\mathscr{B}$. We shall prove the second of these assertions (the first follows from a similar, but simpler, argument). We may assume that $\|T\| \leqslant M$ for each $T$ in $\mathscr{B}$. Given any positive $\varepsilon$, and $x_1, \ldots, x_m$ in $\mathscr{H}$, each $x_j$ can be approximated within $\varepsilon/4M$ by a finite linear combination of elements of $\mathscr{S}$. Hence we can choose $y_1, \ldots, y_n$ in $\mathscr{S}$ and scalars $a_{jk}$ (some of which may be 0) such that

$$\|x_j - \sum_{k=1}^{n} a_{jk} y_k\| < \frac{\varepsilon}{4M} \qquad (j = 1, \ldots, m).$$

Let $\delta$ be a positive real number such that $2\delta(\sum_{k=1}^{n} |a_{jk}|) < \varepsilon$ for each $j = 1, \ldots, m$. It now suffices to show that

$$V(T_0 : y_1, \ldots, y_n; \delta) \cap \mathscr{B} \subseteq V(T_0 : x_1, \ldots, x_m; \varepsilon);$$

and this results from the fact that, for $j = 1, \ldots, m$,

$$\begin{aligned}
\|(T - T_0)x_j\| &\leqslant \|(T - T_0)(x_j - \sum_{k=1}^{n} a_{jk} y_k)\| + \|\sum_{k=1}^{n} a_{jk}(T - T_0)y_k\| \\
&\leqslant \|T - T_0\| \, \|x_j - \sum_{k=1}^{n} a_{jk} y_k\| + \sum_{k=1}^{n} |a_{jk}| \, \|(T - T_0)y_k\| \\
&< 2M\left(\frac{\varepsilon}{4M}\right) + \delta \sum_{k=1}^{n} |a_{jk}| < \varepsilon,
\end{aligned}$$

when $T$ (as well as $T_0$) lies in $\mathscr{B}$ and $\|(T - T_0)y_k\| < \delta$ $(k = 1, \ldots, n)$.

We can summarize the results of the preceding paragraph as follows: a set of vectors with algebraic linear span $\mathscr{H}$ suffices to determine the strong-operator topology on $\mathscr{B}(\mathscr{H})$; a set of vectors with closed linear span $\mathscr{H}$ suffices to determine the strong-operator topology on bounded subsets of $\mathscr{B}(\mathscr{H})$.

2.5.9. REMARK. From the discussion following Theorem 1.2.18, a net $\{T_j\}$ of elements of $\mathscr{B}(\mathscr{H})$ is strong-operator convergent to $T_0$ ($\in \mathscr{B}(\mathscr{H})$) if and only if, given any $x$ in $\mathscr{H}$ and any positive $\varepsilon$, there is an index $j_0$ such that

$$\|T_j x - T_0 x\| \quad (= p_x(T_j - T_0)) < \varepsilon$$

whenever $j \geqslant j_0$. In other words, $\{T_j\}$ is strong-operator convergent to $T_0$ if and only if $\{T_j x\}$ converges to $T_0 x$ (in the norm topology on $\mathscr{H}$) for each $x$ in $\mathscr{H}$. Similarly, $\{T_j\}$ is a Cauchy net, in the uniform structure associated with the strong-operator topology, if and only if $\{T_j x\}$ is a Cauchy (and hence convergent) net of elements of $\mathscr{H}$ for each $x$ in $\mathscr{H}$. Of course, these characterizations of convergent and Cauchy nets can be verified by direct reference to the basic neighborhoods that determine the strong-operator topology. By considering the appropriate nets of finite subsums, it follows that a sum $\sum T_a$ of elements of $\mathscr{B}(\mathscr{H})$ is strong-operator convergent to $T_0$ if and only if $\sum T_a x$ converges in norm to $T_0 x$ for each $x$ in $\mathscr{H}$. Similar comments apply to $\mathscr{B}(\mathscr{H}, \mathscr{K})$.

The results of Proposition 2.5.6, Corollary 2.5.7, and Proposition 2.5.8 can now be interpreted in terms of strong-operator convergence. An *increasing* net $\{E_a\}$ of projections is strong-operator convergent to $\vee E_a$; a *decreasing* net $\{E_a\}$ of projections is strong-operator convergent to $\wedge E_a$; for an *orthogonal family* $\{E_a\}$ of projections, $\sum E_a$ is strong-operator convergent to $\vee E_a$. ■

2.5.10. REMARK. In observing that the strong-operator topology on $\mathscr{B}(\mathscr{H})$ is locally convex, we have (by implication) noted that the linear space operations

$$(S, T) \to S + T\colon \quad \mathscr{B}(\mathscr{H}) \times \mathscr{B}(\mathscr{H}) \to \mathscr{B}(\mathscr{H}),$$

$$(a, T) \to aT\colon \quad \mathbb{C} \times \mathscr{B}(\mathscr{H}) \to \mathscr{B}(\mathscr{H})$$

are strong-operator continuous. From the form of the basic neighborhoods of $T_0$, and since

$$\|(ST - ST_0)x\| \leqslant \|S\|\,\|(T - T_0)x\|,$$

$$\|(TS - T_0 S)x\| = \|(T - T_0)Sx\|,$$

it follows that the mappings

$$T \to ST, \qquad T \to TS\colon \quad \mathscr{B}(\mathscr{H}) \to \mathscr{B}(\mathscr{H})$$

(of left and right multiplication by a fixed operator $S$) are strong-operator continuous. Since

$$\begin{aligned}\|(ST - S_0 T_0)x\| &= \|S(T - T_0)x + (S - S_0)T_0 x\| \\ &\leqslant \|S\|\,\|(T - T_0)x\| + \|(S - S_0)T_0 x\| \\ &\leqslant k\|(T - T_0)x\| + \|(S - S_0)T_0 x\|\end{aligned}$$

when $S_0, T, T_0 \in \mathscr{B}(\mathscr{H})$ and $S \in (\mathscr{B}(\mathscr{H}))_k = \{A \in \mathscr{B}(\mathscr{H}) : \|A\| \leqslant k\}$, it follows that the mapping

$$(S, T) \to ST : \quad (\mathscr{B}(\mathscr{H}))_k \times \mathscr{B}(\mathscr{H}) \to \mathscr{B}(\mathscr{H})$$

is strong-operator continuous. We can express these results by saying that multiplication is separately continuous in the strong-operator topology, and jointly continuous provided that the first variable is restricted to a bounded set. Similar comments apply to sums and products of operators acting between different Hilbert spaces.

When $\mathscr{H}$ is an infinite-dimensional Hilbert space, neither of the mappings

$$(S, T) \to ST : \quad \mathscr{B}(\mathscr{H}) \times \mathscr{B}(\mathscr{H}) \to \mathscr{B}(\mathscr{H}),$$
$$T \to T^* : \quad \mathscr{B}(\mathscr{H}) \to \mathscr{B}(\mathscr{H})$$

is strong-operator continuous (see Exercises 2.8.32 and 2.8.33). However, the latter mapping is strong-operator continuous on the set $\mathscr{N}$ of normal elements. For this, it suffices (in view of the nature of the basic neighborhoods) to note that, by Proposition 2.4.6(iii),

$$\begin{aligned}
&\|T^*x - T_0^*x\|^2 \\
&\quad = \|T^*x\|^2 + \|T_0^*x\|^2 - 2\,\mathrm{Re}\langle T^*x, T_0^*x\rangle \\
&\quad = \|Tx\|^2 + \|T_0x\|^2 - 2\,\mathrm{Re}\langle T_0^*x, T_0^*x\rangle - 2\,\mathrm{Re}\langle T^*x - T_0^*x, T_0^*x\rangle \\
&\quad = \|Tx\|^2 - \|T_0x\|^2 - 2\,\mathrm{Re}\langle x, (T - T_0)T_0^*x\rangle \\
&\quad \leqslant (\|Tx\| - \|T_0x\|)(\|Tx\| + \|T_0x\|) + 2\|x\|\,\|(T - T_0)T_0^*x\| \\
&\quad \leqslant \|(T - T_0)x\|(\|(T - T_0)x\| + 2\|T_0x\|) + 2\|x\|\,\|(T - T_0)T_0^*x\|,
\end{aligned}$$

when $x \in \mathscr{H}$ and $T, T_0 \in \mathscr{N}$. ■

In connection with the result that follows, we recall the convention of the discussion preceding Proposition 1.2.1, concerning uniform structures and completeness. The completeness of the closed balls, alluded to in its statement, is understood relative to the (linear topological) uniform structure associated with the strong-operator topology.

2.5.11. Proposition. *With $\mathscr{H}$ a Hilbert space, closed balls in $\mathscr{B}(\mathscr{H})$ are complete in the strong-operator topology. If $\{T_n\}$ is a Cauchy sequence in $\mathscr{B}(\mathscr{H})$, relative to this topology, it is strong-operator convergent to an element of $\mathscr{B}(\mathscr{H})$.*

*Proof.* If $\{T_a\}$ is a Cauchy net, in the strong-operator topology on $\mathscr{B}(\mathscr{H})$, and $x \in \mathscr{H}$, then $\{T_a x\}$ converges in norm to some element of $\mathscr{H}$ (Remark 2.5.9), and thus $\{\|T_a x\|\}$ converges. In the case of a Cauchy *sequence* $\{T_n\}$, the convergent sequence $\{\|T_n x\|\}$ is bounded (note, however, that convergent *nets* of real numbers need not be bounded). The principle of uniform boundedness

(Theorem 1.8.9) now shows that a *sequence* in $\mathscr{B}(\mathscr{H})$ is bounded if it is a Cauchy sequence relative to the strong-operator topology. Hence the second assertion in the proposition is a consequence of the first.

To prove the first statement, it suffices to consider only the closed unit ball $(\mathscr{B}(\mathscr{H}))_1$ since every closed ball can be mapped onto $(\mathscr{B}(\mathscr{H}))_1$ by a (uniformly bicontinuous) affine transformation $T \to b(T - T_0)$. If $\{T_a\}$ is a Cauchy net in $(\mathscr{B}(\mathscr{H}))_1$, we can define a mapping $T: \mathscr{H} \to \mathscr{H}$ by $Tx = \lim_a T_a x$. Since $T_a$ is linear and $\|T_a x\| \leqslant \|x\|$ for each $x$ in $\mathscr{H}$, the same is true of $T$; so $T$ lies in $(\mathscr{B}(\mathscr{H}))_1$, and is the strong-operator limit of $\{T_a\}$. ∎

2.5.12. Example. We illustrate some of the results obtained in this section by continuing the discussion of Example 2.4.11. With $\mathscr{H}$ the Hilbert space $L_2$ $(= L_2(S, \mathscr{S}, m))$ associated with a $\sigma$-finite measure, each $g$ in $L_\infty$ gives rise to the bounded linear operator $M_g$ of multiplication by $g$. If $M_g$ is a projection, the equations $M_g^2 = M_g = M_g^*$ imply that $[g(s)]^2 = g(s) = \overline{g(s)}$ for almost all $s$ in $S$. Thus $g(s)$ is 0 or 1 almost everywhere on $S$. By changing the values of $g$ on a null set, which does not alter $M_g$, we may suppose that $g$ is the characteristic function of a measurable set $Y$. Conversely, when $g$ is the characteristic function of a measurable set $Y$, $M_g$ is a projection $P(Y)$, since the above argument reverses.

We assert that

$$P\left(\bigcup_{j=1}^{\infty} Y_j\right) = \bigvee_{j=1}^{\infty} P(Y_j), \qquad P\left(\bigcap_{j=1}^{\infty} Y_j\right) = \bigwedge_{j=1}^{\infty} P(Y_j), \tag{5}$$

whenever $Y_1, Y_2, Y_3, \ldots \in \mathscr{S}$. For this, let $Y = \bigcup_{j=1}^{\infty} Y_j$, and denote by $y, y_1, y_2, \ldots$ the characteristic functions of the sets $Y, Y_1, Y_2, \ldots$, respectively. Since $yy_j = y_j$ for each $j$, we have $P(Y)P(Y_j) = P(Y_j)$; so $P(Y_j) \leqslant P(Y)$, and therefore $\bigvee_{j=1}^{\infty} P(Y_j) \leqslant P(Y)$. From Corollary 2.5.5, $P(Y) - \bigvee_{j=1}^{\infty} P(Y_j)$ is a projection $G$, and each vector $x$ $(\in L_2)$ in the range of $G$ satisfies $P(Y)x = x$, $P(Y_j)x = 0$ $(j = 1, 2, \ldots)$. This implies that $y(s)x(s) = x(s)$ almost everywhere, and $y_j(s)x(s) = 0$ almost everywhere, for each $j$. Accordingly, each of the sets

$$Z_0 = \{s \in S \setminus Y : x(s) \neq 0\}, \qquad Z_j = \{s \in Y_j : x(s) \neq 0\} \qquad (j = 1, 2, \ldots)$$

is null, and therefore so is $\{s \in S\colon x(s) \neq 0\}$ $(= \bigcup_{j=0}^{\infty} Z_j)$. Thus each $x$ in the range of $G$ is a null function (that is, the zero vector in $L_2$); so $G = 0$, and $P(Y) = \bigvee_{j=1}^{\infty} P(Y_j)$. This proves the first of the two equations in (5). Upon replacing $Y_j$ by $S \setminus Y_j$ in that equation and noting that $P(S \setminus Y_j) = I - P(Y_j)$, we obtain

$$\begin{aligned}\bigvee_{j=1}^{\infty} (I - P(Y_j)) &= P\left(\bigcup_{j=1}^{\infty} (S \setminus Y_j)\right)\\ &= P\left(S \setminus \bigcap_{j=1}^{\infty} Y_j\right) = I - P\left(\bigcap_{j=1}^{\infty} Y_j\right).\end{aligned}$$

Hence

$$P\left(\bigcap_{j=1}^{\infty} Y_j\right) = I - \bigvee_{j=1}^{\infty} (I - P(Y_j)) = \bigwedge_{j=1}^{\infty} P(Y_j),$$

and the proof of (5) is complete.

When $\{Y_n\}$ is a pairwise disjoint sequence of measurable sets, $\{P(Y_n)\}$ is an orthogonal family of projections, and in view of Remark 2.5.9, the first relation in (5) becomes

$$P\left(\bigcup_{j=1}^{\infty} Y_j\right) = \sum_{j=1}^{\infty} P(Y_j).$$

When $\{Y_j\}$ is an increasing sequence of measurable sets, the sequence $\{P(Y_j)\}$ of projections is increasing. In this case, from Remark 2.5.9 the first equation in (5) becomes

$$P\left(\bigcup_{j=1}^{\infty} Y_j\right) = \lim_{j\to\infty} P(Y_j), \tag{6}$$

with the convergence in the strong-operator topology. This is a special case of the following result: if a sequence $\{f_j\}$ of functions in $L_\infty$ satisfies $\sup\|f_j\| < \infty$, and converges almost everywhere to a function $f$ (necessarily in $L_\infty$), then $\{M_{f_j}\}$ is strong-operator convergent to $M_f$. Indeed, for each $x$ in $L_2$, we have

$$\lim_{j\to\infty} |f(s) - f_j(s)|^2|x(s)|^2 = 0, \qquad |f(s) - f_j(s)|^2|x(s)|^2 \leqslant 4K^2|x(s)|^2$$

almost everywhere on $S$, where $K = \sup\|f_j\|$; so

$$\lim_{j\to\infty} \|(M_f - M_{f_j})x\|^2 = \lim_{j\to\infty} \int_S |f(s) - f_j(s)|^2|x(s)|^2\, dm(s) = 0,$$

by the dominated convergence theorem. When $f$, $f_j$ are the characteristic functions of $\bigcup_{n=1}^{\infty} Y_n$ and $Y_j$, respectively, we obtain (6). ■

With each bounded linear operator $T$ acting on a Hilbert space $\mathscr{H}$, we associate two closed subspaces of $\mathscr{H}$, the *null space* $\{x \in \mathscr{H} : Tx = 0\}$ and the *range space*, which is the closure $[T(\mathscr{H})]$ of the range $T(\mathscr{H}) = \{Tx : x \in \mathscr{H}\}$ of $T$. The corresponding projections are called the *null projection*, denoted by $N(T)$, and the *range projection*, $R(T)$. When $E$ is a projection, $R(E) = E$ and $N(E) = I - E$.

2.5.13. Proposition. *If $T$ is a bounded linear operator acting on a Hilbert space $\mathscr{H}$, then*

$$R(T) = I - N(T^*), \qquad N(T) = I - R(T^*), \tag{7}$$

$$R(T^*T) = R(T^*), \qquad N(T^*T) = N(T). \tag{8}$$

*Proof.* Since

$$\begin{aligned}\{x\in\mathscr{H}: Tx = 0\} &= \{x\in\mathscr{H}: \langle Tx, y\rangle = 0 \text{ for each } y \text{ in } \mathscr{H}\}\\ &= \{x\in\mathscr{H}: \langle x, T^*y\rangle = 0 \text{ for each } y \text{ in } \mathscr{H}\}\\ &= T^*(\mathscr{H})^\perp = [T^*(\mathscr{H})]^\perp,\end{aligned}$$

it follows that $N(T) = I - R(T^*)$. Upon replacing $T$ by $T^*$ we obtain $N(T^*) = I - R(T)$, which completes the proof of (7).

For each $x$ in $\mathscr{H}$, $\|Tx\|^2 = \langle Tx, Tx\rangle = \langle T^*Tx, x\rangle$, so $Tx = 0$ if (and, obviously, only if) $T^*Tx = 0$; that is, $N(T) = N(T^*T)$. From this, together with the first equation in (7) (applied to both $T^*T$ and $T^*$),

$$R(T^*T) = I - N(T^*T) = I - N(T) = R(T^*). \quad \blacksquare$$

Note that, for a self-adjoint element $K$ of $\mathscr{B}(\mathscr{H})$, (7) gives $R(K) = I - N(K)$.

2.5.14. Proposition. *If $E$ and $F$ are projections acting on a Hilbert space $\mathscr{H}$,*

$$R(E + F) = E \vee F, \qquad R(EF) = E - (E \wedge (I - F)).$$

*Proof.* Since

$$\|Ex\|^2 + \|Fx\|^2 = \langle Ex, x\rangle + \langle Fx, x\rangle = \langle (E + F)x, x\rangle,$$

for each vector $x$, it follows that $(E + F)x = 0$ if and only if $Ex = Fx = 0$. Thus

$$N(E + F) = N(E) \wedge N(F) = (I - E) \wedge (I - F),$$

and (7) and (4) yield

$$R(E + F) = I - (I - E) \wedge (I - F) = E \vee F.$$

The projections $I - E$ and $E \wedge (I - F)$ are mutually orthogonal. If $x\in\mathscr{H}$ and $FEx = 0$, we have

$$Ex = (I - F)Ex \in (E \wedge (I - F))(\mathscr{H}),$$

whence

$$x = (I - E)x + Ex \in (I - E + E \wedge (I - F))(\mathscr{H}).$$

Conversely, each $x$ in the range of $I - E + E \wedge (I - F)$ can be expressed as $y + z$, where $y = (I - E)y$ and $z = Ez = (I - F)z$; and

$$FEx = FEy + FEz = FE(I - E)y + F(I - F)z = 0.$$

The preceding argument shows that $N(FE) = I - E + E \wedge (I - F)$; this, together with (7), gives

$$R(EF) = I - N(FE) = E - E \wedge (I - F). \quad \blacksquare$$

2.5.15. REMARK. If $E$ is a projection acting on a Hilbert space $\mathscr{H}$, the operator $U = I - 2E$ satisfies $U = U^*$, $U^2 = I$, and is therefore both self-adjoint and unitary. Each self-adjoint unitary operator $U$ arises in this way, since it is easily verified that the conditions $U = U^* = U^{-1}$ imply that $\frac{1}{2}(I - U)$ $(= E)$ is a projection. ■

We now give a simple description of the projection from a Hilbert space $\mathscr{H}$ onto a closed subspace $Y$, in terms of an orthonormal basis of $Y$ (of course, $Y$ itself is a Hilbert space).

2.5.16. PROPOSITION. *If $E$ is a projection from a Hilbert space $\mathscr{H}$ onto a closed subspace $Y$, and $(y_a)_{a\in\mathbb{A}}$ is an orthonormal basis of $Y$, then*

$$Ex = \sum_{a\in\mathbb{A}} \langle x, y_a\rangle y_a \qquad (x \in \mathscr{H}).$$

*Proof.* Since $Ex \in Y$, we have

$$\begin{aligned} Ex &= \sum_{a\in\mathbb{A}} \langle Ex, y_a\rangle y_a \\ &= \sum_{a\in\mathbb{A}} \langle x, Ey_a\rangle y_a = \sum_{a\in\mathbb{A}} \langle x, y_a\rangle y_a. \end{aligned}$$ ■

2.5.17. REMARK. If $\{E_a\}$ is an orthogonal family of projections on the Hilbert space $\mathscr{H}$, then $\sum_a E_a = E$ is a projection on $\mathscr{H}$ and $Ex = \sum_a E_a x$ for each $x$ in $\mathscr{H}$ (as noted in Remark 2.5.9). Thus

$$\sum_a \|E_a x\|^2 = \sum_a \langle E_a x, x\rangle = \langle \sum_a E_a x, x\rangle = \langle Ex, x\rangle \leqslant \|x\|^2.$$

(This is really an alternative version of Bessel's inequality – Proposition 2.2.7(i).) If $x$ is in the range of $E$ (in particular, if $E = I$), then

$$\sum_a \|E_a x\|^2 = \langle Ex, x\rangle = \|x\|^2.$$

(This is an alternative version of Parseval's equation – Theorem 2.2.9(ii).) ■

## 2.6. Constructions with Hilbert spaces

In this section we consider subspaces, direct sums, and tensor products of Hilbert spaces, together with related operator-theoretic constructions.

*Subspaces.* Suppose that $E$ is the projection from a Hilbert space $\mathscr{H}$ onto a closed subspace $Y$. By restriction, the inner product on $\mathscr{H}$ gives rise to an inner product on $Y$, relative to which $Y$ itself is a Hilbert space. With $T$ in $\mathscr{B}(\mathscr{H})$, the restriction $ET|Y\,(= ETE|Y)$ is a bounded linear operator $T_Y$ acting on $Y$, the *compression* of $T$ to $Y$. For all $y$ and $z$ in $Y$,

$$\langle T_Y y, z\rangle = \langle ETEy, z\rangle = \langle y, ET^*Ez\rangle,$$

so the adjoint of $T_Y$ is $ET^*E|Y$, the compression of $T^*$ to $Y$.

We say that $Y$ *is invariant under* $T$, or that $T$ *leaves* $Y$ *invariant* (invariant is sometimes replaced by *stable*), if $Ty \in Y$ whenever $y \in Y$. Since

$$Y = \{Ex : x \in \mathscr{H}\} = \{z \in \mathscr{H} : Ez = z\},$$

it is apparent that $T(Y) \subseteq Y$ if and only if $TEx = ETEx$ for each $x$ in $\mathscr{H}$; that is, $T$ leaves $Y$ invariant if and only if $TE = ETE$. Since

$$T^*(I - E) - (I - E)T^*(I - E) = ET^*(I - E) = (TE - ETE)^*,$$

it follows that the orthogonal complement $Y^\perp$ is invariant under $T^*$ if and only if $Y$ is invariant under $T$.

When $Y$ and $Y^\perp$ are both invariant under $T$, we say that $Y$ *reduces* $T$; from the preceding paragraph, this occurs if and only if $Y$ is invariant under both $T$ and $T^*$. In this case $ET = (T^*E)^* = (ET^*E)^* = ETE = TE$. Conversely, if $TE = ET$, then $ET^* = T^*E$, so that $ETE = ET = TE$ and $ET^*E = ET^* = T^*E$. Thus $Y$ reduces $T$ if and only if $T$ and $E$ commute.

*Direct sums.* When $\mathscr{H}_1, \ldots, \mathscr{H}_n$ are Hilbert spaces and $\mathscr{K}$ is the set of all $n$-tuples $\{x_1, \ldots, x_n\}$ with $x_j$ in $\mathscr{H}_j$ $(j = 1, \ldots, n)$, there is a Hilbert space structure on $\mathscr{K}$ in which the algebraic operations, inner product, and norm are defined by

$$a\{x_1, \ldots, x_n\} + b\{y_1, \ldots, y_n\} = \{ax_1 + by_1, \ldots, ax_n + by_n\},$$

$$\langle\{x_1, \ldots, x_n\}, \{y_1, \ldots, y_n\}\rangle = \langle x_1, y_1\rangle + \cdots + \langle x_n, y_n\rangle,$$

$$\|\{x_1, \ldots, x_n\}\| = [\|x_1\|^2 + \cdots + \|x_n\|^2]^{1/2}.$$

The resulting Hilbert space $\mathscr{K}$ is called the (*Hilbert*) *direct sum* of $\mathscr{H}_1, \ldots, \mathscr{H}_n$, and is denoted by $\mathscr{H}_1 \oplus \cdots \oplus \mathscr{H}_n$ or $\sum_1^n \oplus \mathscr{H}_j$.

For each $j = 1, \ldots, n$, the set $\mathscr{H}'_j$, consisting of those $n$-tuples in which all but the $j$th entry are zero, is a closed subspace of $\mathscr{H}_1 \oplus \cdots \oplus \mathscr{H}_n$. The mapping $U_j : \mathscr{H}_j \to \mathscr{H}'_j$, defined by $U_j x = \{0, \ldots, 0, x, 0, \ldots, 0\}$ (with $x$ in the $j$th position) is an isomorphism from $\mathscr{H}_j$ onto $\mathscr{H}'_j$. The subspaces $\mathscr{H}'_1, \ldots, \mathscr{H}'_n$ are pairwise orthogonal, and $\bigvee_{j=1}^n \mathscr{H}'_j = \mathscr{K}$.

Suppose next that $\mathscr{H}_1, \ldots, \mathscr{H}_n$ are mutually orthogonal subspaces of a Hilbert space $\mathscr{H}$, and $\bigvee_{j=1}^n \mathscr{H}_j = \mathscr{H}$. By Proposition 2.5.8, the corresponding pairwise orthogonal projections $E_1, \ldots, E_n$ have sum $I$. The linear mapping $U : \mathscr{H} \to \mathscr{K}$, defined by $Ux = \{E_1 x, \ldots, E_n x\}$, carries $\mathscr{H}_j$ onto $\mathscr{H}'_j$ $(j = 1, \ldots, n)$ and $\mathscr{H}$ onto $\mathscr{K}$ $(= \mathscr{H}_1 \oplus \cdots \oplus \mathscr{H}_n)$, and is unitary since

$$\|Ux\|^2 = \sum_{j=1}^n \|E_j x\|^2 = \|\sum_{j=1}^n E_j x\|^2 = \|x\|^2 \qquad (x \in \mathscr{H}).$$

Its inverse $U^{-1}$ carries an element $\{x_1, \ldots, x_n\}$ of $\mathscr{K}$ onto $x_1 + \cdots + x_n$. In view of this isomorphism, we consider $\mathscr{H}$ as an "internal" direct sum of $\mathscr{H}_1, \ldots, \mathscr{H}_n$, and $\mathscr{K}$ as the "external" direct sum; occasionally, we identify $\mathscr{H}$ with $\mathscr{K}$ and $\mathscr{H}_j$ with $\mathscr{H}'_j$.

If $\mathscr{H}_j$, $\mathscr{K}_j$ are Hilbert spaces and $T_j \in \mathscr{B}(\mathscr{H}_j, \mathscr{K}_j)$ $(j = 1, \ldots, n)$, the equation

$$T\{x_1, \ldots, x_n\} = \{T_1 x_1, \ldots, T_n x_n\} \qquad (x_1 \in \mathscr{H}_1, \ldots, x_n \in \mathscr{H}_n)$$

defines a linear operator $T$ from $\mathscr{H}_1 \oplus \cdots \oplus \mathscr{H}_n$ into $\mathscr{K}_1 \oplus \cdots \oplus \mathscr{K}_n$, the *direct sum* $\sum_1^n \oplus T_j$ of $T_1, \ldots, T_n$. With

$$c = \sup\{\|T_j\| : j = 1, \ldots, n\},$$

we have

$$\begin{aligned}\|T\{x_1, \ldots, x_n\}\| &= [\|T_1 x_1\|^2 + \cdots + \|T_n x_n\|^2]^{1/2} \\ &\leqslant [\|T_1\|^2\|x_1\|^2 + \cdots + \|T_n\|^2\|x_n\|^2]^{1/2} \\ &\leqslant c[\|x_1\|^2 + \cdots + \|x_n\|^2]^{1/2} \\ &= c\|\{x_1, \ldots, x_n\}\|;\end{aligned}$$

so $T$ is bounded, with $\|T\| \leqslant c$. However, for each $j = 1, \ldots, n$ and $x$ in $\mathscr{H}_j$,

$$\begin{aligned}\|T_j x\| &= \|\{0, \ldots, 0, T_j x, 0, \ldots, 0\}\| \\ &= \|T\{0, \ldots, 0, x, 0, \ldots, 0\}\| \\ &\leqslant \|T\| \, \|\{0, \ldots, 0, x, 0, \ldots, 0\}\| = \|T\| \, \|x\|.\end{aligned}$$

Thus $\|T_j\| \leqslant \|T\|$ $(j = 1, \ldots, n)$, whence $c \leqslant \|T\|$, and so $\|T\| = c$.

Since

$$\begin{aligned}&\langle T^*\{y_1, \ldots, y_n\}, \{x_1, \ldots, x_n\}\rangle \\ &= \langle\{y_1, \ldots, y_n\}, T\{x_1, \ldots, x_n\}\rangle \\ &= \langle\{y_1, \ldots, y_n\}, \{T_1 x_1, \ldots, T_n x_n\}\rangle = \sum_{j=1}^{n} \langle y_j, T_j x_j\rangle \\ &= \sum_{j=1}^{n} \langle T_j^* y_j, x_j\rangle = \langle\{T_1^* y_1, \ldots, T_n^* y_n\}, \{x_1, \ldots, x_n\}\rangle,\end{aligned}$$

when $x_j \in \mathscr{H}_j$ and $y_j \in \mathscr{K}_j$ $(j = 1, \ldots, n)$, it follows that

$$T^*\{y_1, \ldots, y_n\} = \{T_1^* y_1, \ldots, T_n^* y_n\}.$$

We have now proved that

$$\Big\| \sum_{j=1}^{n} \oplus T_j \Big\| = \sup\{\|T_j\| : j = 1, \ldots, n\},$$

$$\left(\sum_{j=1}^{n} \oplus T_j\right)^* = \sum_{j=1}^{n} \oplus T_j^*;$$

and it is apparent that

$$\sum_{j=1}^{n} \oplus (aS_j + bT_j) = a\left(\sum_{j=1}^{n} \oplus S_j\right) + b\left(\sum_{j=1}^{n} \oplus T_j\right),$$

$$\left(\sum_{j=1}^{n} \oplus R_j\right)\left(\sum_{j=1}^{n} \oplus S_j\right) = \sum_{j=1}^{n} \oplus R_j S_j,$$

when $S_j, T_j \in \mathscr{B}(\mathscr{H}_j, \mathscr{K}_j)$, $R_j \in \mathscr{B}(\mathscr{K}_j, \mathscr{L}_j)$, and $a, b \in \mathbb{C}$.

So far, we have considered direct sums of *finite* families of Hilbert spaces. With slight modifications, the same ideas apply also to infinite families. Given Hilbert spaces $\mathscr{H}_a$ $(a \in \mathbb{A})$, the direct sum $\sum \oplus \mathscr{H}_a$ consists of all families $\{x_a\}$ such that $x_a \in \mathscr{H}_a$ $(a \in \mathbb{A})$ and $\sum \|x_a\|^2 < \infty$. Given two such families $\{x_a\}$ and $\{y_a\}$, we can apply the inequalities discussed in Example 2.1.12 to the elements $\{\|x_a\|\}$ and $\{\|y_a\|\}$ of $l_2(\mathbb{A})$. We obtain

$$\sum |\langle x_a, y_a \rangle| \leqslant \sum \|x_a\| \, \|y_a\| \leqslant (\sum \|x_a\|^2)^{1/2} (\sum \|y_a\|^2)^{1/2} < \infty,$$

$$\begin{aligned}(\sum \|x_a + y_a\|^2)^{1/2} &\leqslant (\sum (\|x_a\| + \|y_a\|)^2)^{1/2} \\ &\leqslant (\sum \|x_a\|^2)^{1/2} + (\sum \|y_a\|^2)^{1/2} < \infty.\end{aligned}$$

Accordingly, the family $\{x_a + y_a\}$ is in $\sum \oplus \mathscr{H}_a$; it follows easily that $\sum \oplus \mathscr{H}_a$ is a pre-Hilbert space when the algebraic structure, inner product, and norm are defined by

$$\{x_a\} + \{y_a\} = \{x_a + y_a\}, \qquad c\{x_a\} = \{cx_a\},$$

$$\langle \{x_a\}, \{y_a\} \rangle = \sum \langle x_a, y_a \rangle, \qquad \|\{x_a\}\| = [\sum \|x_a\|^2]^{1/2}.$$

The element $\{x_a\}$ of $\sum \oplus \mathscr{H}_a$ is sometimes denoted by $\sum \oplus x_a$.

We assert that $\sum \oplus \mathscr{H}_a$ is complete, relative to the norm just defined, and is therefore a Hilbert space. For this, suppose that $(x^{(n)})$ is a Cauchy sequence in $\sum \oplus \mathscr{H}_a$, so that each $x^{(n)}$ is a family $\{x_a^{(n)}\}$ of the type considered above. Given any positive real number $\varepsilon$, there is a positive integer $n(\varepsilon)$ such that $\|x^{(m)} - x^{(n)}\| \leqslant \varepsilon$ whenever $m, n \geqslant n(\varepsilon)$; that is,

$$\text{(1)} \qquad \sum_{a \in \mathbb{A}} \|x_a^{(m)} - x_a^{(n)}\|^2 \leqslant \varepsilon^2 \qquad (m, n \geqslant n(\varepsilon)).$$

From this, $\|x_a^{(m)} - x_a^{(n)}\| \leqslant \varepsilon$ $(m, n \geqslant n(\varepsilon),\ a \in \mathbb{A})$; so, for each fixed $a$, $\{x_a^{(n)} : n = 1, 2, \ldots\}$ is a Cauchy sequence in $\mathscr{H}_a$, and therefore converges to an element $x_a$ of $\mathscr{H}_a$. With $\mathbb{F}$ a finite subset of $\mathbb{A}$, it follows from (1) that

$$\sum_{a \in \mathbb{F}} \|x_a^{(m)} - x_a^{(n)}\|^2 \leqslant \varepsilon^2 \qquad (m, n \geqslant n(\varepsilon)).$$

When $m \to \infty$, we obtain

$$\sum_{a \in \mathbb{F}} \|x_a - x_a^{(n)}\|^2 \leqslant \varepsilon^2 \qquad (n \geqslant n(\varepsilon)),$$

and since the last inequality is satisfied for every finite subset $\mathbb{F}$ of $\mathbb{A}$, we have

$$\sum_{a\in\mathbb{A}} \|x_a - x_a^{(n)}\|^2 \leqslant \varepsilon^2 \qquad (n \geqslant n(\varepsilon)). \tag{2}$$

This shows that the family $\{x_a - x_a^{(n)}\}$, as well as $\{x_a^{(n)}\}$, is in $\sum\oplus \mathscr{H}_a$ when $n \geqslant n(\varepsilon)$. Accordingly, $\{x_a\}$ $(= \{x_a - x_a^{(n)}\} + \{x_a^{(n)}\})$ is an element $x$ of $\sum\oplus \mathscr{H}_a$, and (2) asserts that $\|x - x^{(n)}\| \leqslant \varepsilon$ whenever $n \geqslant n(\varepsilon)$. Thus $(x^{(n)})$ converges to $x$; so $\sum\oplus \mathscr{H}_a$ is complete, and is therefore a Hilbert space.

With $b$ in $\mathbb{A}$, $\mathscr{H}_b$ is isomorphic to the closed subspace $\mathscr{H}'_b$ of $\sum\oplus \mathscr{H}_a$ consisting of those families $\{x_a\}$ such that $x_a = 0$ whenever $a \neq b$. We obtain a unitary transformation $U_b$, from $\mathscr{H}_b$ onto $\mathscr{H}'_b$, by taking for $U_b x$ the family $\{x_a\}$ in which $x_b$ is $x$ and $x_a = 0$ $(a \neq b)$. The subspaces $\mathscr{H}'_a$ $(a \in \mathbb{A})$ are pairwise orthogonal, and $\vee \mathscr{H}'_a = \sum\oplus \mathscr{H}_a$.

If $\{\mathscr{H}_a\}$ is a family of mutually orthogonal subspaces of a Hilbert space $\mathscr{H}$, and $\vee \mathscr{H}_a = \mathscr{H}$, the corresponding projections form an orthogonal family $\{E_a\}$ with (strong-operator convergent) sum $I$. Just as in the case of finite direct sums, the equation $Ux = \{E_a x\}$ defines an isomorphism $U$ from $\mathscr{H}$ onto $\sum\oplus \mathscr{H}_a$, and we consider $\mathscr{H}$ as an "internal" direct sum of the family $\{\mathscr{H}_a\}$. Moreover, $U^{-1}\{x_a\} = \sum x_a$ when $\{x_a\} \in \sum\oplus \mathscr{H}_a$; the sum converges since $\{x_a\}$ is an orthogonal set in $\mathscr{H}$, and $\sum \|x_a\|^2 < \infty$.

Suppose next that $\mathscr{H}_a$, $\mathscr{K}_a$ are Hilbert spaces, and $T_a \in \mathscr{B}(\mathscr{H}_a, \mathscr{K}_a)$ for each $a$ in $\mathbb{A}$. If

$$\sup\{\|T_a\| : a \in \mathbb{A}\} < \infty,$$

the equation $T\{x_a\} = \{T_a x_a\}$ defines a bounded linear operator $T$ from $\sum\oplus \mathscr{H}_a$ into $\sum\oplus \mathscr{K}_a$. We call $T$ the direct sum $\sum\oplus T_a$ of the family $\{T_a\}$. Just as in the case of finite direct sums, we have

$$\|\sum_{a\in\mathbb{A}} \oplus T_a\| = \sup\{\|T_a\| : a \in \mathbb{A}\},$$

$$\left(\sum_{a\in\mathbb{A}} \oplus T_a\right)^* = \sum_{a\in\mathbb{A}} \oplus T_a^*,$$

$$\sum_{a\in\mathbb{A}} \oplus (cS_a + dT_a) = c\left(\sum_{a\in\mathbb{A}} \oplus S_a\right) + d\left(\sum_{a\in\mathbb{A}} \oplus T_a\right),$$

$$\left(\sum_{a\in\mathbb{A}} \oplus R_a\right)\left(\sum_{a\in\mathbb{A}} \oplus S_a\right) = \sum_{a\in\mathbb{A}} \oplus R_a S_a,$$

when $S_a, T_a \in \mathscr{B}(\mathscr{H}_a, \mathscr{K}_a)$ and $R_a \in \mathscr{B}(\mathscr{K}_a, \mathscr{L}_a)$.

When $\mathscr{H}_a$ is the one-dimensional Hilbert space $\mathbb{C}$, for each $a$ in $\mathbb{A}$, $\sum\oplus \mathscr{H}_a$ reduces to the Hilbert space $l_2(\mathbb{A})$ of Example 2.1.12.

*Tensor products and the Hilbert–Schmidt class.* The material in this subsection will be used in an essential way in later parts of the book (from Chapter 11 onward), but has a relatively minor role until that point. The reader who wishes may bypass it until that stage, and will have only very occasional need to refer back in the meantime.

There are several ways of defining the (Hilbert) tensor product $\mathscr{H}$ of two Hilbert spaces $\mathscr{H}_1$ and $\mathscr{H}_2$, each method having advantages in particular circumstances. Our approach, set out below, emphasizes the "universal" property of the tensor product. The Hilbert space $\mathscr{H}$ is characterized (up to isomorphism) by the existence of a bilinear mapping $p$, from the Cartesian product $\mathscr{H}_1 \times \mathscr{H}_2$ into $\mathscr{H}$, with the following property: each "suitable" bilinear mapping $L$ from $\mathscr{H}_1 \times \mathscr{H}_2$ into a Hilbert space $\mathscr{K}$ has a unique factorization $L = Tp$, with $T$ a bounded linear operator from $\mathscr{H}$ into $\mathscr{K}$.

Before starting the formal development of the theory, we indicate some of the intuitive ideas that underly it. When $x_1 \in \mathscr{H}_1$ and $x_2 \in \mathscr{H}_2$, we shall want to view the element $p(x_1, x_2)$ of $\mathscr{H}$ as a "product," $x_1 \otimes x_2$, of $x_1$ and $x_2$. It turns out that linear combinations of such products form an everywhere-dense subspace of $\mathscr{H}$. The bilinearity of $p$ implies that these products satisfy certain linear relations; for example, as the product notation suggests,

$$(x_1 + y_1) \otimes (x_2 + y_2) - x_1 \otimes x_2 - x_1 \otimes y_2 - y_1 \otimes x_2 - y_1 \otimes y_2 = 0,$$

whenever $x_1, y_1 \in \mathscr{H}_1$ and $x_2, y_2 \in \mathscr{H}_2$. In fact, all the linear relations satisfied by product vectors can be deduced by (possibly repeated) use of the bilinearity of $p$. It turns out that the inner product on $\mathscr{H}$ satisfies (and is determined by) the condition

$$\langle x_1 \otimes x_2, y_1 \otimes y_2 \rangle = \langle x_1, y_1 \rangle \langle x_2, y_2 \rangle;$$

in particular, $\|x_1 \otimes x_2\| = \|x_1\| \, \|x_2\|$. There are various constructions leading to a Hilbert space $\mathscr{H}$ with the required properties. In the method we shall use, the elements of $\mathscr{H}$ are certain complex-valued functions defined on the product $\mathscr{H}_1 \times \mathscr{H}_2$ and conjugate-linear in both variables (and by introducing a concept of "conjugate Hilbert space," these functions are viewed as bilinear functionals). When $v_1 \in \mathscr{H}_1$ and $v_2 \in \mathscr{H}_2$, $v_1 \otimes v_2$ is the function that assigns the value $\langle v_1, x_1 \rangle \langle v_2, x_2 \rangle$ to the element $(x_1, x_2)$ of $\mathscr{H}_1 \times \mathscr{H}_2$.

In the formal development of the theory, we first introduce the class of bilinear mappings used in formulating the universal property mentioned above. The tensor product is then defined, and identified (up to isomorphism) with certain specific Hilbert spaces, such as the completion of the algebraic tensor product and the class of "Hilbert–Schmidt operators" from the conjugate Hilbert space $\bar{\mathscr{H}}_1$ into $\mathscr{H}_2$. We conclude this subsection with a discussion of tensor products of bounded linear operators. It is convenient in the initial stages to consider the tensor product of a finite family of $n$ Hilbert spaces, specializing later to the case in which $n = 2$.

Suppose that $\mathscr{H}_1, \ldots, \mathscr{H}_n$ are Hilbert spaces and $\varphi$ is a mapping from the cartesian product $\mathscr{H}_1 \times \cdots \times \mathscr{H}_n$ into the scalar field $\mathbb{C}$. We describe $\varphi$ as a *bounded multilinear functional* on $\mathscr{H}_1 \times \cdots \times \mathscr{H}_n$ if $\varphi$ is linear in each of its variables (while the other variables remain fixed), and there is a real number $c$ such that

$$|\varphi(x_1, \ldots, x_n)| \leqslant c\|x_1\| \cdots \|x_n\| \qquad (x_1 \in \mathscr{H}_1, \ldots, x_n \in \mathscr{H}_n).$$

When this is so, the least such constant $c$ is denoted by $\|\varphi\|$. Then, $\varphi$ is a continuous mapping from $\mathscr{H}_1 \times \cdots \times \mathscr{H}_n$ into $\mathbb{C}$, relative to the product of the norm topologies on the Hilbert spaces; the estimates required to prove this are much the same as those needed in showing that the mapping

$$(a_1, \ldots, a_n) \to ca_1 \cdots a_n \colon \mathbb{C} \times \cdots \times \mathbb{C} \to \mathbb{C}$$

is continuous, so we omit the details.

In the following proposition, we consider certain sums of positive terms, which may converge or diverge, and a divergent sum is to be interpreted as $+\infty$. In part (ii) of the proposition, inequalities involving $\infty$ are to be understood in the obvious sense, and we adopt the convention that $0 \cdot \infty = 0$. Whether or not the sums considered converge, the manipulations required in the proof are easily justified, in view of the final paragraph of Section 1.2.

2.6.1. PROPOSITION. *Suppose that $\mathscr{H}_1, \ldots, \mathscr{H}_n$ are Hilbert spaces and $\varphi$ is a bounded multilinear functional on $\mathscr{H}_1 \times \cdots \times \mathscr{H}_n$.*

(i) *The sum*

$$\sum_{y_1 \in Y_1} \cdots \sum_{y_n \in Y_n} |\varphi(y_1, \ldots, y_n)|^2 \tag{3}$$

*has the same (finite or infinite) value for all orthonormal bases $Y_1$ of $\mathscr{H}_1, \ldots, Y_n$ of $\mathscr{H}_n$.*

(ii) *If $\mathscr{K}_1, \ldots, \mathscr{K}_n$ are Hilbert spaces, $A_m \in \mathscr{B}(\mathscr{H}_m, \mathscr{K}_m)$ $(m = 1, \ldots, n)$, $\psi$ is a bounded multilinear functional on $\mathscr{K}_1 \times \cdots \times \mathscr{K}_n$, and*

$$\varphi(x_1, \ldots, x_n) = \psi(A_1x_1, \ldots, A_nx_n) \qquad (x_1 \in \mathscr{H}_1, \ldots, x_n \in \mathscr{H}_n),$$

*then*

$$\sum_{y_1 \in Y_1} \cdots \sum_{y_n \in Y_n} |\varphi(y_1, \ldots, y_n)|^2 \leqslant \|A_1\|^2 \cdots \|A_n\|^2 \sum_{z_1 \in Z_1} \cdots \sum_{z_n \in Z_n} |\psi(z_1, \ldots, z_n)|^2,$$

*when $Y_m$ and $Z_m$ are orthonormal bases of $\mathscr{H}_m$ and $\mathscr{K}_m$, respectively $(m = 1, \ldots, n)$.*

*Proof.* In order to prove (i), it is sufficient to show that

$$\sum_{y_1 \in Y_1} \cdots \sum_{y_n \in Y_n} |\varphi(y_1, \ldots, y_n)|^2 \leqslant \sum_{z_1 \in Z_1} \cdots \sum_{z_n \in Z_n} |\varphi(z_1, \ldots, z_n)|^2$$

whenever $Y_m, Z_m$ are orthonormal bases of $\mathscr{H}_m$ $(m = 1, \ldots, n)$, since equality of the two sums then follows by exchanging the roles of $Y_m$ and $Z_m$. The required inequality is a special case of part (ii) of the proposition, with $\psi = \varphi$, $\mathscr{K}_m = \mathscr{H}_m$, and $A_m = I$ $(m = 1, \ldots, n)$.

It now suffices to prove (ii). For this, suppose that $1 \leqslant m \leqslant n$, and choose and fix vectors $y_1$ in $Y_1, \ldots, y_{m-1}$ in $Y_{m-1}$, $z_{m+1}$ in $Z_{m+1}, \ldots, z_n$ in $Z_n$. The mapping

$$z \to \psi(A_1 y_1, \ldots, A_{m-1} y_{m-1}, z, z_{m+1}, \ldots, z_n) : \mathscr{K}_m \to \mathbb{C}$$

is a bounded linear functional on $\mathscr{K}_m$, so there is a vector $w$ in $\mathscr{K}_m$ such that

$$\psi(A_1 y_1, \ldots, A_{m-1} y_{m-1}, z, z_{m+1}, \ldots, z_n) = \langle z, w \rangle \qquad (z \in \mathscr{K}_m).$$

From Parseval's equation

$$\begin{aligned}
&\sum_{y_m \in Y_m} |\psi(A_1 y_1, \ldots, A_{m-1} y_{m-1}, A_m y_m, z_{m+1}, \ldots, z_n)|^2 \\
&\quad = \sum_{y_m \in Y_m} |\langle A_m y_m, w \rangle|^2 = \sum_{y_m \in Y_m} |\langle y_m, A_m^* w \rangle|^2 \\
&\quad = \|A_m^* w\|^2 \leqslant \|A_m\|^2 \|w\|^2 = \|A_m\|^2 \sum_{z_m \in Z_m} |\langle z_m, w \rangle|^2 \\
&\quad = \|A_m\|^2 \sum_{z_m \in Z_m} |\psi(A_1 y_1, \ldots, A_{m-1} y_{m-1}, z_m, z_{m+1}, \ldots, z_n)|^2.
\end{aligned}$$

A further summation now yields

$$\begin{aligned}
&\sum_{y_1 \in Y_1} \cdots \sum_{y_m \in Y_m} \sum_{z_{m+1} \in Z_{m+1}} \cdots \sum_{z_n \in Z_n} |\psi(A_1 y_1, \ldots, A_m y_m, z_{m+1}, \ldots, z_n)|^2 \\
&\quad \leqslant \|A_m\|^2 \sum_{y_1 \in Y_1} \cdots \sum_{y_{m-1} \in Y_{m-1}} \sum_{z_m \in Z_m} \cdots \sum_{z_n \in Z_n} |\psi(A_1 y_1, \ldots, A_{m-1} y_{m-1}, z_m, \ldots, z_n)|^2.
\end{aligned}$$

Thus

$$\begin{aligned}
&\sum_{y_1 \in Y_1} \cdots \sum_{y_n \in Y_n} |\varphi(y_1, \ldots, y_n)|^2 = \sum_{y_1 \in Y_1} \cdots \sum_{y_n \in Y_n} |\psi(A_1 y_1, \ldots, A_n y_n)|^2 \\
&\quad \leqslant \|A_n\|^2 \sum_{y_1 \in Y_1} \cdots \sum_{y_{n-1} \in Y_{n-1}} \sum_{z_n \in Z_n} |\psi(A_1 y_1, \ldots, A_{n-1} y_{n-1}, z_n)|^2 \\
&\quad \leqslant \|A_{n-1}\|^2 \|A_n\|^2 \\
&\qquad \times \sum_{y_1 \in Y_1} \cdots \sum_{y_{n-2} \in Y_{n-2}} \sum_{z_{n-1} \in Z_{n-1}} \sum_{z_n \in Z_n} |\psi(A_1 y_1, \ldots, A_{n-2} y_{n-2}, z_{n-1}, z_n)|^2 \\
&\quad \leqslant \cdots \leqslant \|A_1\|^2 \cdots \|A_n\|^2 \sum_{z_1 \in Z_1} \cdots \sum_{z_n \in Z_n} |\psi(z_1, \ldots, z_n)|^2. \quad \blacksquare
\end{aligned}$$

With $\mathscr{H}_1, \ldots, \mathscr{H}_n$ Hilbert spaces, a mapping $\varphi : \mathscr{H}_1 \times \cdots \times \mathscr{H}_n \to \mathbb{C}$ is described as a *Hilbert–Schmidt functional* on $\mathscr{H}_1 \times \cdots \times \mathscr{H}_n$ if it is a bounded multilinear functional, and the sum (3) is finite for one (and hence each) choice of the orthonormal bases $Y_1$ in $\mathscr{H}_1, \ldots, Y_n$ in $\mathscr{H}_n$.

2.6.2. PROPOSITION. *If $\mathscr{H}_1, \ldots, \mathscr{H}_n$ are Hilbert spaces, the set $\mathscr{H}\mathscr{S}\mathscr{F}$ of all Hilbert–Schmidt functionals on $\mathscr{H}_1 \times \cdots \times \mathscr{H}_n$ is itself a Hilbert space when the linear structure, inner product, and norm are defined by*

$$(a\varphi + b\psi)(x_1, \ldots, x_n) = a\varphi(x_1, \ldots, x_n) + b\psi(x_1, \ldots, x_n),$$

$$\langle \varphi, \psi \rangle = \sum_{y_1 \in Y_1} \cdots \sum_{y_n \in Y_n} \varphi(y_1, \ldots, y_n)\overline{\psi(y_1, \ldots, y_n)}, \tag{4}$$

$$\|\varphi\|_2 = \left[ \sum_{y_1 \in Y_1} \cdots \sum_{y_n \in Y_n} |\varphi(y_1, \ldots, y_n)|^2 \right]^{1/2}, \tag{5}$$

*where $Y_m$ is an orthonormal basis in $\mathscr{H}_m$ $(m = 1, \ldots, n)$. The sum in (4) is absolutely convergent, and the inner product and norm do not depend on the choice of the orthonormal bases $Y_1, \ldots, Y_n$.*

*For each $v(1)$ in $\mathscr{H}_1, \ldots, v(n)$ in $\mathscr{H}_n$, the equation*

$$\varphi_{v(1),\ldots,v(n)}(x_1, \ldots, x_n) = \langle x_1, v(1)\rangle \cdots \langle x_n, v(n)\rangle \qquad (x_1 \in \mathscr{H}_1, \ldots, x_n \in \mathscr{H}_n)$$

*defines an element $\varphi_{v(1),\ldots,v(n)}$ of $\mathscr{H}\mathscr{S}\mathscr{F}$, and*

$$\langle \varphi_{v(1),\ldots,v(n)}, \varphi_{w(1),\ldots,w(n)} \rangle = \langle w(1), v(1)\rangle \cdots \langle w(n), v(n)\rangle,$$

$$\|\varphi_{v(1),\ldots,v(n)}\|_2 = \|v(1)\| \cdots \|v(n)\|.$$

*The set $\{\varphi_{y(1),\ldots,y(n)} : y(1) \in Y_1, \ldots, y(n) \in Y_n\}$ is an orthonormal basis of $\mathscr{H}\mathscr{S}\mathscr{F}$. There is a unitary transformation $U$ from $\mathscr{H}\mathscr{S}\mathscr{F}$ onto $l_2(Y_1 \times \cdots \times Y_n)$, such that $U\varphi$ is the restriction $\varphi | Y_1 \times \cdots \times Y_n$ when $\varphi \in \mathscr{H}\mathscr{S}\mathscr{F}$.*

*Proof.* Having chosen an orthonormal basis $Y_m$ in $\mathscr{H}_m$ $(m = 1, \ldots, n)$, we can associate with *each* bounded multilinear functional $\varphi$ on $\mathscr{H}_1 \times \cdots \times \mathscr{H}_n$ the complex-valued function $U\varphi$ obtained by restricting $\varphi$ to $Y_1 \times \cdots \times Y_n$. Note that $\varphi$ is a Hilbert–Schmidt functional if and only if

$$U\varphi \in l_2(Y_1 \times \cdots \times Y_n).$$

If $U\varphi = 0$, then

$$\varphi(y_1, \ldots, y_n) = 0 \qquad (y_1 \in Y_1, \ldots, y_n \in Y_n).$$

Since $Y_m$ has closed linear span $\mathscr{H}_m$ $(m = 1, \ldots, n)$, it follows from the multilinearity and (joint) continuity of $\varphi$ that $\varphi$ vanishes throughout $\mathscr{H}_1 \times \cdots \times \mathscr{H}_n$.

If $\varphi$ and $\psi$ are Hilbert–Schmidt functionals on $\mathscr{H}_1 \times \cdots \times \mathscr{H}_n$, the same is true of $a\varphi + b\psi$ (as defined in the proposition) for all scalars $a, b$; for $a\varphi + b\psi$ is a bounded multilinear functional, $U\varphi, U\psi \in l_2(Y_1 \times \cdots \times Y_n)$, and therefore

$$U(a\varphi + b\psi) = aU\varphi + bU\varphi \in l_2(Y_1 \times \cdots \times Y_n).$$

The summation occurring in (4) can be written in the form

$$\sum_{y \in Y_1 \times \cdots \times Y_n} (U\varphi)(y)\overline{(U\psi)(y)},$$

and is absolutely convergent with sum $\langle U\varphi, U\psi \rangle$, the inner product in $l_2(Y_1 \times \cdots \times Y_n)$ of $U\varphi$ and $U\psi$.

From the preceding argument, the set $\mathscr{HSF}$ of all Hilbert–Schmidt functionals on $\mathscr{H}_1 \times \cdots \times \mathscr{H}_n$ is a complex vector space, (4) defines an inner product on $\mathscr{HSF}$, the restriction $U|\mathscr{HSF}$ is a one-to-one linear mapping from $\mathscr{HSF}$ into $l_2(Y_1 \times \cdots \times Y_n)$, and $\langle U\varphi, U\psi \rangle = \langle \varphi, \psi \rangle$ when $\varphi, \psi \in \mathscr{HSF}$. Since the inner product on $l_2(Y_1 \times \cdots \times Y_n)$ is definite, so is that on $\mathscr{HSF}$; for if $\varphi \in \mathscr{HSF}$ and $\langle \varphi, \varphi \rangle = 0$, we have $\langle U\varphi, U\varphi \rangle = 0$, whence $U\varphi = 0$ and so $\varphi = 0$. From this, $\mathscr{HSF}$ is a pre-Hilbert space, and it is apparent from (4) that the norm $\| \ \|_2$ in $\mathscr{HSF}$ is given by (5). From Proposition 2.6.1, this norm is independent of the choice of the orthonormal bases $Y_1, \ldots, Y_n$; by polarization, the same is true of the inner product on $\mathscr{HSF}$.

We prove next that $U$ carries $\mathscr{HSF}$ onto the whole of the $l_2$ space. With $f$ in $l_2(Y_1 \times \cdots \times Y_n)$ and $x_m$ in $\mathscr{H}_m$ $(m = 1, \ldots, n)$, the Cauchy–Schwarz inequality and Parseval equation give

$$\begin{aligned}
&\sum_{y_1 \in Y_1} \cdots \sum_{y_n \in Y_n} |f(y_1, \ldots, y_n)\langle x_1, y_1 \rangle \cdots \langle x_n, y_n \rangle| \\
&\quad \leqslant \left[ \sum_{y_1 \in Y_1} \cdots \sum_{y_n \in Y_n} |f(y_1, \ldots, y_n)|^2 \right]^{1/2} \\
&\quad\quad \times \left[ \sum_{y_1 \in Y_1} \cdots \sum_{y_n \in Y_n} |\langle x_1, y_1 \rangle|^2 \cdots |\langle x_n, y_n \rangle|^2 \right]^{1/2} \\
&\quad = \|f\| \left( \sum_{y_1 \in Y_1} |\langle x_1, y_1 \rangle|^2 \right)^{1/2} \cdots \left( \sum_{y_n \in Y_n} |\langle x_n, y_n \rangle|^2 \right)^{1/2} = \|f\| \, \|x_1\| \cdots \|x_n\|.
\end{aligned}$$

From this, the equation

$$\varphi(x_1, \ldots, x_n) = \sum_{y_1 \in Y_1} \cdots \sum_{y_n \in Y_n} f(y_1, \ldots, y_n)\langle x_1, y_1 \rangle \cdots \langle x_n, y_n \rangle$$

defines a bounded multilinear functional $\varphi$ on $\mathscr{H}_1 \times \cdots \times \mathscr{H}_n$, with $\|\varphi\| \leqslant \|f\|$. From orthonormality of the sets $Y_1, \ldots, Y_n$,

$$(U\varphi)(y_1, \ldots, y_n) = \varphi(y_1, \ldots, y_n) = f(y_1, \ldots, y_n) \qquad (y_1 \in Y_1, \ldots, y_n \in Y_n),$$

so $U\varphi = f$. Moreover, $\varphi \in \mathscr{HSF}$ since $U\varphi \in l_2(Y_1 \times \cdots \times Y_n)$, whence $U$ carries $\mathscr{HSF}$ onto the $l_2$ space.

Since $U$ is a norm-preserving linear mapping from $\mathscr{HSF}$ onto $l_2(Y_1 \times \cdots \times Y_n)$, completeness of the $l_2$ space entails completeness of $\mathscr{HSF}$; so $\mathscr{HSF}$ is a Hilbert space, and $U$ is a unitary operator.

When $v(1)\in\mathscr{H}_1,\ldots,v(n)\in\mathscr{H}_n$, $\varphi_{v(1),\ldots,v(n)}$ (as defined in Proposition 2.6.2) is a multilinear functional on $\mathscr{H}_1\times\cdots\times\mathscr{H}_n$, and is bounded since

$$|\varphi_{v(1),\ldots,v(n)}(x_1,\ldots,x_n)|\leqslant\|v(1)\|\cdots\|v(n)\|\,\|x_1\|\cdots\|x_n\|$$

by the Cauchy–Schwarz inequality. Moreover, Parseval's equation gives

$$\begin{aligned}&\sum_{y_1\in Y_1}\cdots\sum_{y_n\in Y_n}|\varphi_{v(1),\ldots,v(n)}(y_1,\ldots,y_n)|^2\\&\quad=\sum_{y_1\in Y_1}\cdots\sum_{y_n\in Y_n}|\langle y_1,v(1)\rangle|^2\cdots|\langle y_n,v(n)\rangle|^2\\&\quad=\left(\sum_{y_1\in Y_1}|\langle y_1,v(1)\rangle|^2\right)\cdots\left(\sum_{y_n\in Y_n}|\langle y_n,v(n)\rangle|^2\right)\\&\quad=\|v(1)\|^2\cdots\|v(n)\|^2.\end{aligned}$$

Hence $\varphi_{v(1),\ldots,v(n)}\in\mathscr{HSF}$ and $\|\varphi_{v(1),\ldots,v(n)}\|_2=\|v(1)\|\cdots\|v(n)\|$. Again, by Parseval's equation and absolute convergence,

$$\begin{aligned}&\langle\varphi_{v(1),\ldots,v(n)},\varphi_{w(1),\ldots,w(n)}\rangle\\&\quad=\sum_{y_1\in Y_1}\cdots\sum_{y_n\in Y_n}\varphi_{v(1),\ldots,v(n)}(y_1,\ldots,y_n)\overline{\varphi_{w(1),\ldots,w(n)}(y_1,\ldots,y_n)}\\&\quad=\sum_{y_1\in Y_1}\cdots\sum_{y_n\in Y_n}\langle y_1,v(1)\rangle\cdots\langle y_n,v(n)\rangle\langle w(1),y_1\rangle\cdots\langle w(n),y_n\rangle\\&\quad=\left(\sum_{y_1\in Y_1}\langle w(1),y_1\rangle\langle y_1,v(1)\rangle\right)\cdots\left(\sum_{y_n\in Y_n}\langle w(n),y_n\rangle\langle y_n,v(n)\rangle\right)\\&\quad=\langle w(1),v(1)\rangle\cdots\langle w(n),v(n)\rangle.\end{aligned}$$

When $y(1)\in Y_1,\ldots,y(n)\in Y_n$, the orthonormality of $Y_1,\ldots,Y_n$ implies that $U\varphi_{y(1),\ldots,y(n)}$ is the function that takes the value 1 at $(y(1),\ldots,y(n))$ and 0 elsewhere on $Y_1\times\cdots\times Y_n$. Thus

$$\{U\varphi_{y(1),\ldots,y(n)}:y(1)\in Y_1,\ldots,y(n)\in Y_n\}$$

is an orthonormal basis of $l_2(Y_1\times\cdots\times Y_n)$, and therefore

$$\{\varphi_{y(1),\ldots,y(n)}:y(1)\in Y_1,\ldots,y(n)\in Y_n\}$$

is such a basis of $\mathscr{HSF}$. ■

In order to simplify the treatment of conjugate-linear mappings, which will be used extensively in this subsection and in Chapter 9, we introduce the notion of the "conjugate" of a Hilbert space $\mathscr{H}$. The algebraic structure and inner product on $\mathscr{H}$ are defined by the mappings

$$\begin{aligned}&(x,y)\to x+y:\quad\mathscr{H}\times\mathscr{H}\to\mathscr{H},\\&(a,x)\to ax:\quad\mathbb{C}\times\mathscr{H}\to\mathscr{H},\\&(x,y)\to\langle x,y\rangle:\quad\mathscr{H}\times\mathscr{H}\to\mathbb{C}.\end{aligned}$$

The *conjugate Hilbert space* $\bar{\mathscr{H}}$ is the same set $\mathscr{H}$, with the algebraic structure and inner product defined by the mappings

$$(x, y) \to x + y: \quad \mathscr{H} \times \mathscr{H} \to \mathscr{H},$$

$$(a, x) \to a \bar{\cdot} x: \quad \mathbb{C} \times \mathscr{H} \to \mathscr{H},$$

$$(x, y) \to \langle x, y \rangle^{-}: \quad \mathscr{H} \times \mathscr{H} \to \mathbb{C},$$

where

$$a \bar{\cdot} x = \bar{a}x \qquad \text{and} \quad \langle x, y \rangle^{-} = \langle y, x \rangle.$$

Of course, the conjugate Hilbert space of $\bar{\mathscr{H}}$ is $\mathscr{H}$.

A subset of a Hilbert space is linearly independent, or orthogonal, or orthonormal, or an orthonormal basis of that space, if and only if it has the same property relative to the conjugate Hilbert space. If $\mathscr{H}_1$ and $\mathscr{H}_2$ are Hilbert spaces and $T$ is a mapping from the set $\mathscr{H}_1$ into the set $\mathscr{H}_2$, linearity of $T: \mathscr{H}_1 \to \mathscr{H}_2$ is equivalent to linearity of $T: \bar{\mathscr{H}}_1 \to \bar{\mathscr{H}}_2$, and corresponds to conjugate-linearity of $T: \mathscr{H}_1 \to \bar{\mathscr{H}}_2$ and of $T: \bar{\mathscr{H}}_1 \to \mathscr{H}_2$. Of course, continuity of $T$ is the same in all four situations (and when $T$ is linear the operators have the same bound), since the norm on $\mathscr{H}_j$ is the same as that on $\bar{\mathscr{H}}_j$.

2.6.3. Definition. Suppose that $\mathscr{H}_1, \ldots, \mathscr{H}_n$ and $\mathscr{K}$ are Hilbert spaces and $L$ is a mapping from $\mathscr{H}_1 \times \cdots \times \mathscr{H}_n$ into $\mathscr{K}$. We describe $L$ as a *bounded multilinear mapping* if it is linear in each of its variables (while the other variables remain fixed), and there is a real number $c$ such that

$$\|L(x_1, \ldots, x_n)\| \leqslant c\|x_1\| \cdots \|x_n\| \qquad (x_1 \in \mathscr{H}_1, \ldots, x_n \in \mathscr{H}_n).$$

In these circumstances, the least such constant $c$ is denoted by $\|L\|$.

By a *weak Hilbert–Schmidt mapping* from $\mathscr{H}_1 \times \cdots \times \mathscr{H}_n$ into $\mathscr{K}$, we mean a bounded multilinear mapping $L$ with the following properties:

(i) for each $u$ in $\mathscr{K}$, the mapping $L_u$ defined by

$$L_u(x_1, \ldots, x_n) = \langle L(x_1, \ldots, x_n), u \rangle$$

is a Hilbert–Schmidt functional on $\mathscr{H}_1 \times \cdots \times \mathscr{H}_n$;

(ii) there is a real number $d$ such that $\|L_u\|_2 \leqslant d\|u\|$ for each $u$ in $\mathscr{K}$.

When these conditions are satisfied, the least possible value of the constant $d$ in (ii) is denoted by $\|L\|_2$. ■

As in the case of multilinear functionals, a bounded multilinear mapping $L: \mathscr{H}_1 \times \cdots \times \mathscr{H}_n \to \mathscr{K}$ is (jointly) continuous relative to the norm topologies on the Hilbert spaces. Condition (ii) is in fact redundant, since it follows from (i), by an application of the closed graph theorem to the mapping $u \to L_u: \bar{\mathscr{K}} \to \mathscr{H}\mathscr{S}\mathscr{F}$ (Exercise 2.8.36). We shall not make use of this implication, and have incorporated (ii) in the definition for convenience.

2.6.4. THEOREM. *Suppose that $\mathscr{H}_1, \ldots, \mathscr{H}_n$ are Hilbert spaces.*

(i) *There is a Hilbert space $\mathscr{H}$ and a weak Hilbert–Schmidt mapping $p: \mathscr{H}_1 \times \cdots \times \mathscr{H}_n \to \mathscr{H}$ with the following property: given any weak Hilbert–Schmidt mapping $L$ from $\mathscr{H}_1 \times \cdots \times \mathscr{H}_n$ into a Hilbert space $\mathscr{K}$, there is a unique bounded linear mapping $T$ from $\mathscr{H}$ into $\mathscr{K}$, such that $L = Tp$; moreover, $\|T\| = \|L\|_2$.*

(ii) *If $\mathscr{H}'$ and $p'$ have the properties attributed in* (i) *to $\mathscr{H}$ and $p$, there is a unitary transformation $U$ from $\mathscr{H}$ onto $\mathscr{H}'$ such that $p' = Up$.*

(iii) *If $v_m, w_m \in \mathscr{H}_m$ and $Y_m$ is an orthonormal basis of $\mathscr{H}_m$ $(m = 1, \ldots, n)$, then*

$$\langle p(v_1, \ldots, v_n), p(w_1, \ldots, w_n) \rangle = \langle v_1, w_1 \rangle \cdots \langle v_n, w_n \rangle,$$

*the set $\{p(y_1, \ldots, y_n): y_1 \in Y_1, \ldots, y_n \in Y_n\}$ is an orthonormal basis of $\mathscr{H}$, and $\|p\|_2 = 1$.*

*Proof.* With $\bar{\mathscr{H}}_m$ the conjugate Hilbert space of $\mathscr{H}_m$, let $\mathscr{H}$ be the set of all Hilbert–Schmidt functionals on $\bar{\mathscr{H}}_1 \times \cdots \times \bar{\mathscr{H}}_n$ with the Hilbert space structure described in Proposition 2.6.2. When $v(1) \in \mathscr{H}_1, \ldots, v(n) \in \mathscr{H}_n$, let $p(v(1), \ldots, v(n))$ be the Hilbert–Schmidt functional $\varphi_{v(1),\ldots,v(n)}$ defined on

$$\bar{\mathscr{H}}_1 \times \cdots \times \bar{\mathscr{H}}_n$$

by

$$\begin{aligned}\varphi_{v(1),\ldots,v(n)}(x_1, \ldots, x_n) &= \langle x_1, v(1) \rangle^- \cdots \langle x_n, v(n) \rangle^- \\ &= \langle v(1), x_1 \rangle \cdots \langle v(n), x_n \rangle.\end{aligned}$$

Since $Y_j$ is an orthonormal basis of $\mathscr{H}_j$ $(j = 1, \ldots, n)$, it follows from Proposition 2.6.2 that the set $\{p(y_1, \ldots, y_n): y_1 \in Y_1, \ldots, y_n \in Y_n\}$ is an orthonormal basis of $\mathscr{H}$, and that

$$\begin{aligned}\langle p(v_1, \ldots, v_n), p(w_1, \ldots, w_n) \rangle &= \langle w_1, v_1 \rangle^- \cdots \langle w_n, v_n \rangle^- \\ &= \langle v_1, w_1 \rangle \cdots \langle v_n, w_n \rangle,\end{aligned}$$

$$\|p(v_1, \ldots, v_n)\|_2 = \|v_1\| \cdots \|v_n\|.$$

From the preceding paragraph, $p: \mathscr{H}_1 \times \cdots \times \mathscr{H}_n \to \mathscr{H}$ is a bounded multilinear mapping: we prove next that it is a weak Hilbert–Schmidt mapping. For this, suppose that $\varphi \in \mathscr{H}$, and consider the bounded multilinear functional $p_\varphi: \mathscr{H}_1 \times \cdots \times \mathscr{H}_n \to \mathbb{C}$ defined by

$$p_\varphi(x_1, \ldots, x_n) = \langle p(x_1, \ldots, x_n), \varphi \rangle.$$

With $y(1)$ in $Y_1, \ldots, y(n)$ in $Y_n$, orthonormality of the bases implies that $\varphi_{y(1),\ldots,y(n)}$ takes the value 1 at $(y(1), \ldots, y(n))$ and 0 elsewhere on

$Y_1 \times \cdots \times Y_n$. Thus

$$\begin{aligned} p_\varphi(y(1),\ldots,y(n)) &= \langle p(y(1),\ldots,y(n)),\varphi\rangle = \langle \varphi_{y(1),\ldots,y(n)},\varphi\rangle \\ &= \sum_{y_1\in Y_1}\cdots\sum_{y_n\in Y_n} \varphi_{y(1),\ldots,y(n)}(y_1,\ldots,y_n)\overline{\varphi(y_1,\ldots,y_n)} \\ &= \overline{\varphi(y(1),\ldots,y(n))}, \end{aligned}$$

$$\sum_{y(1)\in Y_1}\cdots\sum_{y(n)\in Y_n} |p_\varphi(y(1),\ldots,y(n))|^2 = \|\varphi\|_2^2.$$

From this, $p_\varphi$ is a Hilbert–Schmidt functional on $\mathscr{H}_1 \times \cdots \times \mathscr{H}_n$ and $\|p_\varphi\|_2 = \|\varphi\|_2$; so $p\colon \mathscr{H}_1 \times \cdots \times \mathscr{H}_n \to \mathscr{H}$ is a weak Hilbert–Schmidt mapping with $\|p\|_2 = 1$.

Suppose next that $L$ is a weak Hilbert–Schmidt mapping from $\mathscr{H}_1 \times \cdots \times \mathscr{H}_n$ into another Hilbert space $\mathscr{K}$. If $u \in \mathscr{K}$ and $L_u$ is the Hilbert–Schmidt functional occurring in Definition 2.6.3, while $\varphi \in \mathscr{H}$ and $\mathbb{F}$ is a finite subset of $Y_1 \times \cdots \times Y_n$, we have

$$\begin{aligned} &\Big|\Big\langle \sum_{(y_1,\ldots,y_n)\in\mathbb{F}} \varphi(y_1,\ldots,y_n)L(y_1,\ldots,y_n), u\Big\rangle\Big| \\ &\quad\leqslant \sum_{(y_1,\ldots,y_n)\in\mathbb{F}} |\varphi(y_1,\ldots,y_n)|\,|L_u(y_1,\ldots,y_n)| \\ &\quad\leqslant \Big[\sum_{(y_1,\ldots,y_n)\in\mathbb{F}} |\varphi(y_1,\ldots,y_n)|^2\Big]^{1/2}\Big[\sum_{(y_1,\ldots,y_n)\in\mathbb{F}} |L_u(y_1,\ldots,y_n)|^2\Big]^{1/2} \\ &\quad\leqslant \|L_u\|_2\Big[\sum_{(y_1,\ldots,y_n)\in\mathbb{F}} |\varphi(y_1,\ldots,y_n)|^2\Big]^{1/2} \\ &\quad\leqslant \|u\|\,\|L\|_2\Big[\sum_{(y_1,\ldots,y_n)\in\mathbb{F}} |\varphi(y_1,\ldots,y_n)|^2\Big]^{1/2}. \end{aligned}$$

Hence

$$\begin{aligned} &\Big\|\sum_{(y_1,\ldots,y_n)\in\mathbb{F}} \varphi(y_1,\ldots,y_n)L(y_1,\ldots,y_n)\Big\| \\ &\qquad\leqslant \|L\|_2\Big[\sum_{(y_1,\ldots,y_n)\in\mathbb{F}} |\varphi(y_1,\ldots,y_n)|^2\Big]^{1/2}. \end{aligned} \tag{6}$$

Since

$$\sum_{y_1\in Y_1}\cdots\sum_{y_n\in Y_n} |\varphi(y_1,\ldots,y_n)|^2 = \|\varphi\|_2^2 < \infty,$$

it follows from (6) and the Cauchy criterion that the (unordered) sum

$$\sum_{y_1\in Y_1}\cdots\sum_{y_n\in Y_n} \varphi(y_1,\ldots,y_n)L(y_1,\ldots,y_n)$$

converges to an element $T\varphi$ of $\mathscr{K}$, and $\|T\varphi\| \leqslant \|L\|_2 \|\varphi\|_2$. Thus $T$ is a bounded linear operator from $\mathscr{H}$ into $\mathscr{K}$, and $\|T\| \leqslant \|L\|_2$. When $y(1) \in Y_1, \ldots, y(n) \in Y_n$, we have

$$\begin{aligned} Tp(y(1), \ldots, y(n)) &= T\varphi_{y(1),\ldots,y(n)} \\ &= \sum_{y_1 \in Y_1} \cdots \sum_{y_n \in Y_n} \varphi_{y(1),\ldots,y(n)}(y_1, \ldots, y_n) L(y_1, \ldots, y_n) \\ &= L(y(1), \ldots, y(n)). \end{aligned}$$

Since $L$ and $Tp$ are both bounded and multilinear and $Y_m$ has closed linear span $\mathscr{H}_m$ $(m = 1, \ldots, n)$, it follows that $L = Tp$.

The condition $Tp = L$ uniquely determines the bounded linear operator $T$, because the range of $p$ contains the orthonormal basis $p(Y_1 \times \cdots \times Y_n)$ of $\mathscr{H}$. For each $u$ in $\mathscr{K}$, Parseval's equation gives

$$\begin{aligned} \|L_u\|_2^2 &= \sum_{y_1 \in Y_1} \cdots \sum_{y_n \in Y_n} |\langle L(y_1, \ldots, y_n), u \rangle|^2 \\ &= \sum_{y_1 \in Y_1} \cdots \sum_{y_n \in Y_n} |\langle Tp(y_1, \ldots, y_n), u \rangle|^2 \\ &= \sum_{y_1 \in Y_1} \cdots \sum_{y_n \in Y_n} |\langle p(y_1, \ldots, y_n), T^*u \rangle|^2 \\ &= \|T^*u\|^2 \leqslant \|T\|^2 \|u\|^2; \end{aligned}$$

so $\|L\|_2 \leqslant \|T\|$, and thus $\|L\|_2 = \|T\|$.

It remains to prove part (ii) of the theorem. For this, suppose that $\mathscr{H}'$ and $p' : \mathscr{H}_1 \times \cdots \times \mathscr{H}_n \to \mathscr{H}'$ (as well as $\mathscr{H}$ and $p$) have the properties set out in (i). When $\mathscr{K}$ is $\mathscr{H}'$ and $L$ is $p'$, the equation $L = Tp'$ is satisfied when $T$ is the identity operator on $\mathscr{H}'$, and also when $T$ is the projection from $\mathscr{H}'$ onto the closed subspace $[p'(\mathscr{H}_1 \times \cdots \times \mathscr{H}_n)]$ generated by the range $p'(\mathscr{H}_1 \times \cdots \times \mathscr{H}_n)$ of $p'$. From the uniqueness of $T$,

$$[p'(\mathscr{H}_1 \times \cdots \times \mathscr{H}_n)] = \mathscr{H}';$$

moreover,

$$\|p'\|_2 = \|L\|_2 = \|T\| = \|I\| = 1.$$

With the same choice, $\mathscr{K} = \mathscr{H}'$ and $L = p'$, it follows, from the properties of $\mathscr{H}$ and $p$ set out in (i), that there is a bounded linear operator $U$ from $\mathscr{H}$ into $\mathscr{H}'$ such that $p' = Up$ and

$$\|U\| = \|L\|_2 = \|p'\|_2 = 1.$$

The roles of $\mathscr{H}$, $p$ and $\mathscr{H}'$, $p'$ can be reversed in this argument, so there is a bounded linear operator $U'$ from $\mathscr{H}'$ into $\mathscr{H}$ such that $p = U'p'$ and $\|U'\| = 1$. Since

$$U'Up(x_1, \ldots, x_n) = U'p'(x_1, \ldots, x_n) = p(x_1, \ldots, x_n),$$

for all $x_1$ in $\mathscr{H}_1, \ldots, x_n$ in $\mathscr{H}_n$, while

$$[p(\mathscr{H}_1 \times \cdots \times \mathscr{H}_n)] = \mathscr{H},$$

it follows that $U'U$ is the identity operator on $\mathscr{H}$; and similarly, $UU'$ is the identity operator on $\mathscr{H}'$. Finally,

$$\|x\| = \|U'Ux\| \leqslant \|Ux\| \leqslant \|x\| \qquad (x \in \mathscr{H});$$

so $\|Ux\| = \|x\|$, and $U$ is an isomorphism from $\mathscr{H}$ onto $\mathscr{H}'$. ■

By part (ii) of Theorem 2.6.4, the Hilbert space $\mathscr{H}$ appearing in that theorem, together with the multilinear mapping $p: \mathscr{H}_1 \times \cdots \times \mathscr{H}_n \to \mathscr{H}$, is uniquely determined (up to isomorphism) by the "universal" property set out in (i). We describe $\mathscr{H}$ as the *(Hilbert) tensor product* of $\mathscr{H}_1, \ldots, \mathscr{H}_n$, denoted by $\mathscr{H}_1 \otimes \cdots \otimes \mathscr{H}_n$, and refer to $p$ as the canonical (product) mapping from $\mathscr{H}_1 \times \cdots \times \mathscr{H}_n$ into $\mathscr{H}_1 \otimes \cdots \otimes \mathscr{H}_n$. The vector $p(x_1, \ldots, x_n)$ in $\mathscr{H}_1 \otimes \cdots \otimes \mathscr{H}_n$ is usually denoted by $x_1 \otimes \cdots \otimes x_n$. Finite linear combinations of these "simple tensors" form an everywhere-dense subspace of $\mathscr{H}_1 \otimes \cdots \otimes \mathscr{H}_n$; indeed, if $Y_m$ is an orthonormal basis of $\mathscr{H}_m$ ($m = 1, \ldots, n$), then

$$\{y_1 \otimes \cdots \otimes y_n : y_1 \in Y_1, \ldots, y_n \in Y_n\}$$

is an orthonormal basis of $\mathscr{H}_1 \otimes \cdots \otimes \mathscr{H}_n$. Thus

$$\dim(\mathscr{H}_1 \otimes \mathscr{H}_2 \otimes \cdots \otimes \mathscr{H}_n) = \dim \mathscr{H}_1 \dim \mathscr{H}_2 \cdots \dim \mathscr{H}_n.$$

As the notation suggests, the vector $x_1 \otimes \cdots \otimes x_n$ behaves in some respects like a formal product of $x_1, \ldots, x_n$; for example, it results from the multilinearity of $p$, and from Theorem 2.6.4(iii), that

$$\begin{aligned}
(7) \qquad & x_1 \otimes \cdots \otimes x_{m-1} \otimes (ax'_m + bx''_m) \otimes x_{m+1} \otimes \cdots \otimes x_n \\
& = a(x_1 \otimes \cdots \otimes x_{m-1} \otimes x'_m \otimes x_{m+1} \otimes \cdots \otimes x_n) \\
& \quad + b(x_1 \otimes \cdots \otimes x_{m-1} \otimes x''_m \otimes x_{m+1} \otimes \cdots \otimes x_n),
\end{aligned}$$

$$(8) \qquad \langle x_1 \otimes \cdots \otimes x_n, y_1 \otimes \cdots \otimes y_n \rangle = \langle x_1, y_1 \rangle \cdots \langle x_n, y_n \rangle,$$

$$(9) \qquad \|x_1 \otimes \cdots \otimes x_n\| = \|x_1\| \cdots \|x_n\|.$$

In studying tensor products of Hilbert spaces, the properties just listed are usually more important than the detailed constructions employed in the proof of Theorem 2.6.4. Many of the arguments involve two stages; the first stage deals with the linear span $\mathscr{H}_0$ of the simple tensors, and is based on the identities (7)–(9), while the second employs "extension by continuity" from $\mathscr{H}_0$ to its closure $\mathscr{H}_1 \otimes \cdots \otimes \mathscr{H}_n$. Since

$$a(x_1 \otimes x_2 \otimes \cdots \otimes x_n) = (ax_1) \otimes x_2 \otimes \cdots \otimes x_n,$$

$\mathscr{H}_0$ consists of all finite *sums* of simple tensors. In dealing with $\mathscr{H}_0$, it is important to bear in mind that the simple tensors are not linearly independent. Relation (7) can be viewed as the assertion that a certain linear combination of three simple tensors is zero, and repeated application of (7) yields more complicated identities of this type. We shall look at this question in more detail in Proposition 2.6.6. In the meantime, we establish the "associativity" of the tensor product.

2.6.5. PROPOSITION. *If $\mathscr{H}_1, \ldots, \mathscr{H}_{m+n}$ are Hilbert spaces, there is a unique unitary transformation $U$ from $\mathscr{H}_1 \otimes \cdots \otimes \mathscr{H}_{m+n}$ onto*

$$(\mathscr{H}_1 \otimes \cdots \otimes \mathscr{H}_m) \otimes (\mathscr{H}_{m+1} \otimes \cdots \otimes \mathscr{H}_{m+n})$$

*such that*

$$U(x_1 \otimes \cdots \otimes x_{m+n}) = (x_1 \otimes \cdots \otimes x_m) \otimes (x_{m+1} \otimes \cdots \otimes x_{m+n}) \tag{10}$$

*whenever $x_j \in \mathscr{H}_j$ $(j = 1, \ldots, m+n)$.*

*Proof.* Since the set of all simple tensors in $\mathscr{H}_1 \otimes \cdots \otimes \mathscr{H}_{m+n} (= \mathscr{K})$ has linear span everywhere dense in $\mathscr{K}$, there is at most one unitary operator $U$ with the stated property; so it suffices to prove the existence of such an isomorphism. For this, let

$$\mathscr{K}' = (\mathscr{H}_1 \otimes \cdots \otimes \mathscr{H}_m) \otimes (\mathscr{H}_{m+1} \otimes \cdots \otimes \mathscr{H}_{m+n}),$$

and when $x_j \in \mathscr{H}_j$ $(j = 1, \ldots, m+n)$, define

$$p(x_1, \ldots, x_{m+n}) = x_1 \otimes \cdots \otimes x_{m+n} \quad (\in \mathscr{K})$$

$$p'(x_1, \ldots, x_{m+n}) = (x_1 \otimes \cdots \otimes x_m) \otimes (x_{m+1} \otimes \cdots \otimes x_{m+n}) \quad (\in \mathscr{K}').$$

The ranges of $p$ and $p'$ contain orthonormal bases of $\mathscr{K}$ and $\mathscr{K}'$, respectively, and so generate everywhere-dense subspaces $X$ $(\subseteq \mathscr{K})$ and $X'$ $(\subseteq \mathscr{K}')$. If $x_j, y_j \in \mathscr{H}_j$ $(j = 1, \ldots, m+n)$, we have

$$\begin{aligned}
&\langle p(x_1, \ldots, x_{m+n}), p(y_1, \ldots, y_{m+n}) \rangle \\
&= \langle x_1, y_1 \rangle \cdots \langle x_m, y_m \rangle \langle x_{m+1}, y_{m+1} \rangle \cdots \langle x_{m+n}, y_{m+n} \rangle \\
&= \langle x_1 \otimes \cdots \otimes x_m, y_1 \otimes \cdots \otimes y_m \rangle \langle x_{m+1} \otimes \cdots \otimes x_{m+n}, y_{m+1} \otimes \cdots \otimes y_{m+n} \rangle \\
&= \langle p'(x_1, \ldots, x_{m+n}), p'(y_1, \ldots, y_{m+n}) \rangle.
\end{aligned}$$

From this,

$$\begin{aligned}
\Big\| \sum_{k=1}^{q} a_k p(x_1^{(k)}, \ldots, x_{m+n}^{(k)}) \Big\|^2 &= \sum_{k=1}^{q} \sum_{l=1}^{q} a_k \bar{a}_l \langle p(x_1^{(k)}, \ldots, x_{m+n}^{(k)}), p(x_1^{(l)}, \ldots, x_{m+n}^{(l)}) \rangle \\
&= \sum_{k=1}^{q} \sum_{l=1}^{q} a_k \bar{a}_l \langle p'(x_1^{(k)}, \ldots, x_{m+n}^{(k)}), p'(x_1^{(l)}, \ldots, x_{m+n}^{(l)}) \rangle \\
&= \Big\| \sum_{k=1}^{q} a_k p'(x_1^{(k)}, \ldots, x_{m+n}^{(k)}) \Big\|^2,
\end{aligned}$$

whenever $a_k \in \mathbb{C}$ and $x_j^{(k)} \in \mathscr{H}_j$ $(j = 1, \ldots, m+n; k = 1, \ldots, q)$.

The remainder of the argument is of frequently recurring type. The equation

$$U_0\left(\sum_{k=1}^{q} a_k p(x_1^{(k)}, \ldots, x_{m+n}^{(k)})\right) = \sum_{k=1}^{q} a_k p'(x_1^{(k)}, \ldots, x_{m+n}^{(k)})$$

defines a norm-preserving linear mapping $U_0$ from $X$ onto $X'$. The definition is unambiguous since, given two expressions $\sum a_k p(x_1^{(k)}, \ldots, x_{m+n}^{(k)})$ and $\sum b_l p(y_1^{(l)}, \ldots, y_{m+n}^{(l)})$ for a vector $x$ in $X$, it follows (upon replacing $\sum a_k p(x_1^{(k)}, \ldots, x_{m+n}^{(k)})$ by $\sum a_k p(x_1^{(k)}, \ldots, x_{m+n}^{(k)}) - \sum b_l p(y_1^{(l)}, \ldots, y_{m+n}^{(l)})$ in the last chain of equations) that the two corresponding expressions

$$\sum a_k p'(x_1^{(k)}, \ldots, x_{m+n}^{(k)}) \qquad \text{and} \qquad \sum b_l p'(y_1^{(l)}, \ldots, y_{m+n}^{(l)})$$

for $U_0 x$ are equal. By continuity, $U_0$ extends to an isomorphism $U$ from $\mathscr{K}$ onto $\mathscr{K}'$, and

$$\begin{aligned} U(x_1 \otimes \cdots \otimes x_{m+n}) &= U(p(x_1, \ldots, x_{m+n})) \\ &= p'(x_1, \ldots, x_{m+n}) \\ &= (x_1 \otimes \cdots \otimes x_m) \otimes (x_{m+1} \otimes \cdots \otimes x_{m+n}). \quad \blacksquare \end{aligned}$$

By use of the "associativity" established in the preceding proposition, questions concerning the $n$-fold tensor product of Hilbert spaces can usually be reduced to the particular case $n = 2$. Our next few results are directed toward this case. We consider first the question of linear dependence of simple tensors.

2.6.6. PROPOSITION. *Suppose that $\mathscr{H}_1$ and $\mathscr{H}_2$ are Hilbert spaces, $\mathscr{H} = \mathscr{H}_1 \otimes \mathscr{H}_2$, and $\mathscr{H}_0$ is the everywhere-dense subspace of $\mathscr{H}$ generated by the simple tensors.*

(i) *If $x_1, \ldots, x_n \in \mathscr{H}_1, y_1, \ldots, y_n \in \mathscr{H}_2$, then $\sum_{j=1}^{n} x_j \otimes y_j = 0$ if and only if there is an $n \times n$ complex matrix $[c_{jk}]$ such that*

$$\sum_{j=1}^{n} c_{jk} x_j = 0 \qquad (k = 1, \ldots, n),$$

$$\sum_{k=1}^{n} c_{jk} y_k = y_j \qquad (j = 1, \ldots, n).$$

(ii) *If $L$ is a bilinear mapping from $\mathscr{H}_1 \otimes \mathscr{H}_2$ into a complex vector space $\mathscr{K}$, there is a (unique) linear mapping $T$ from $\mathscr{H}_0$ into $\mathscr{K}$ such that $L(x, y) = T(x \otimes y)$ for each $x$ in $\mathscr{H}_1$ and $y$ in $\mathscr{H}_2$.*

*Proof.* (i) If there is a matrix $[c_{jk}]$ with the stated properties, bilinearity of the mapping $(x, y) \to x \otimes y$ implies that

$$\sum_{j=1}^{n} x_j \otimes y_j = \sum_{j=1}^{n} x_j \otimes \left( \sum_{k=1}^{n} c_{jk} y_k \right) = \sum_{j=1}^{n} \sum_{k=1}^{n} c_{jk} x_j \otimes y_k$$

$$= \sum_{k=1}^{n} \left( \sum_{j=1}^{n} c_{jk} x_j \right) \otimes y_k = 0.$$

Conversely, suppose that $\sum_{j=1}^{n} x_j \otimes y_j = 0$. If $v_1, \ldots, v_r$ is an orthonormal basis of the linear subspace of $\mathscr{H}_2$ generated by $y_1, \ldots, y_n$, we can choose an $n \times r$ matrix $A = [a_{jk}]$ and an $r \times n$ matrix $B = [b_{jk}]$ such that

$$y_j = \sum_{l=1}^{r} a_{jl} v_l \qquad (j = 1, \ldots, n),$$

$$v_l = \sum_{k=1}^{n} b_{lk} y_k \qquad (l = 1, \ldots, r).$$

With $[c_{jk}]$ the $n \times n$ matrix $AB$, we have

$$y_j = \sum_{l=1}^{r} a_{jl} \left( \sum_{k=1}^{n} b_{lk} y_k \right) = \sum_{k=1}^{n} c_{jk} y_k \qquad (j = 1, \ldots, n),$$

and

$$0 = \sum_{j=1}^{n} x_j \otimes y_j = \sum_{j=1}^{n} x_j \otimes \left( \sum_{l=1}^{r} a_{jl} v_l \right) = \sum_{l=1}^{r} u_l \otimes v_l,$$

where

$$u_l = \sum_{j=1}^{n} a_{jl} x_j \qquad (l = 1, \ldots, r).$$

For each $m = 1, \ldots, r$,

$$0 = \sum_{l=1}^{r} \langle u_l \otimes v_l, u_m \otimes v_m \rangle = \sum_{l=1}^{r} \langle u_l, u_m \rangle \langle v_l, v_m \rangle = \|u_m\|^2.$$

Thus $u_1 = u_2 = \cdots = u_r = 0$, and

$$\sum_{j=1}^{n} c_{jk} x_j = \sum_{j=1}^{n} \sum_{l=1}^{r} a_{jl} b_{lk} x_j = \sum_{l=1}^{r} b_{lk} u_l = 0 \qquad (k = 1, \ldots, n).$$

(ii) Suppose that $L$ is a bilinear mapping from $\mathscr{H}_1 \times \mathscr{H}_2$ into $\mathscr{K}$. If $x_1, \ldots, x_n \in \mathscr{H}_1, y_1, \ldots, y_n \in \mathscr{H}_2$, and $\sum_{j=1}^{n} x_j \otimes y_j = 0$, we can choose a matrix

$[c_{jk}]$ as in (i). The bilinearity of $L$ then entails

$$\sum_{j=1}^{n} L(x_j, y_j) = \sum_{j=1}^{n} L\left(x_j, \sum_{k=1}^{n} c_{jk} y_k\right)$$
$$= \sum_{k=1}^{n} \sum_{j=1}^{n} c_{jk} L(x_j, y_k) = \sum_{k=1}^{n} L\left(\sum_{j=1}^{n} c_{jk} x_j, y_k\right) = 0.$$

Suppose next that $x_1, \ldots, x_n, u_1, \ldots, u_m \in \mathscr{H}_1, y_1, \ldots, y_n, v_1, \ldots, v_m \in \mathscr{H}_2$, and $\sum_{j=1}^{n} x_j \otimes y_j = \sum_{j=1}^{m} u_j \otimes v_j$. Then

$$\sum_{j=1}^{n} x_j \otimes y_j + \sum_{j=1}^{m} (-u_j) \otimes v_j = 0;$$

the preceding paragraph shows that

$$\sum_{j=1}^{n} L(x_j, y_j) + \sum_{j=1}^{m} L(-u_j, v_j) = 0,$$

and therefore

$$\sum_{j=1}^{n} L(x_j, y_j) = \sum_{j=1}^{m} L(u_j, v_j).$$

From this, it follows that the equation

$$T\left(\sum_{j=1}^{n} x_j \otimes y_j\right) = \sum_{j=1}^{n} L(x_j, y_j)$$

defines (unambiguously) a linear operator $T$ from $\mathscr{H}_0$ into $\mathscr{K}$; and $T(x \otimes y) = L(x, y)$ $(x \in \mathscr{H}_1, y \in \mathscr{H}_2)$. ■

2.6.7. REMARK. The first part of Proposition 2.6.6 asserts, in effect, that the only finite families of simple tensors that have sum zero are those that are "forced" to have zero sum by the bilinearity of the mapping $p: (x, y) \to x \otimes y$. From this, $\mathscr{H}_0$ can be identified with the algebraic tensor product of $\mathscr{H}_1$ and $\mathscr{H}_2$, which was defined, traditionally, as the quotient of the linear space of all formal finite sums of simple tensors by the subspace consisting of those finite sums that must vanish if $p$ is to be bilinear. The second part of the proposition shows that $\mathscr{H}_0$ has the "universal" property that characterizes the algebraic tensor product.

We can identify $\mathscr{H}$ with the completion of its everywhere-dense subspace $\mathscr{H}_0$. Accordingly, the Hilbert tensor product $\mathscr{H}_1 \otimes \mathscr{H}_2$ can be viewed as the completion of the algebraic tensor product $\mathscr{H}_0$, relative to the unique inner product on $\mathscr{H}_0$ that satisfies

$$\langle x_1 \otimes y_1, x_2 \otimes y_2 \rangle = \langle x_1, x_2 \rangle \langle y_1, y_2 \rangle \qquad (x_1, x_2 \in \mathscr{H}_1, \quad y_1, y_2 \in \mathscr{H}_2). \quad ■$$

2.6.8. REMARK. We show that the tensor product of Hilbert spaces $\mathscr{H}$ and $\mathscr{K}$ can be viewed as the $n$-fold direct sum of $\mathscr{H}$ with itself (that is, the direct sum of $n$ copies of $\mathscr{H}$), where $n$ is the (finite or infinite) dimension of $\mathscr{K}$. For this purpose, let $\{y_b : b \in \mathbb{B}\}$ be an orthonormal basis of $\mathscr{K}$, and for each $b$ in $\mathbb{B}$ let $\mathscr{H}_b$ be $\mathscr{H}$. The mapping

$$W_b : x \to x \otimes y_b : \mathscr{H}_b \to \mathscr{H} \otimes \mathscr{K}$$

is a norm-preserving linear operator from $\mathscr{H}_b$ onto a (necessarily closed) subspace $\mathscr{H}'_b$ of $\mathscr{H} \otimes \mathscr{K}$. Since

$$\langle x_1 \otimes y_a, x_2 \otimes y_b \rangle = \langle x_1, x_2 \rangle \langle y_a, y_b \rangle = 0$$

for all $x_1$ and $x_2$ in $\mathscr{H}$, when $a$ and $b$ are distinct elements of $\mathbb{B}$, it follows that the subspaces $\{\mathscr{H}'_b : b \in \mathbb{B}\}$ are pairwise orthogonal. With $\{z_a\}$ an orthonormal basis of $\mathscr{H}$, the closed subspace $\vee \mathscr{H}'_b$ of $\mathscr{H} \otimes \mathscr{K}$ contains the orthonormal basis $\{z_a \otimes y_b\}$, so $\vee \mathscr{H}'_b = \mathscr{H} \otimes \mathscr{K}$. Accordingly, $\mathscr{H} \otimes \mathscr{K}$ is the internal direct sum of its subspaces $\mathscr{H}'_b$ $(b \in \mathbb{B})$, and we have isomorphisms $W$ $(= \sum \oplus W_b$, from $\sum \oplus \mathscr{H}_b$ onto $\sum \oplus \mathscr{H}'_b)$ and $V$ (from $\sum \oplus \mathscr{H}'_b$ onto $\mathscr{H} \otimes \mathscr{K}$), defined by

$$W(\textstyle\sum \oplus x_b) = \sum \oplus x_b \otimes y_b, \qquad V(\sum \oplus u_b) = \sum u_b.$$

Thus $VW$ is an isomorphism $U$, from $\sum \oplus \mathscr{H}_b$ onto $\mathscr{H} \otimes \mathscr{K}$, and

$$U(\textstyle\sum \oplus x_b) = \sum x_b \otimes y_b. \tag{11}$$

From this, each element of $\mathscr{H} \otimes \mathscr{K}$ can be expressed (uniquely, once the orthonormal basis $\{y_b\}$ is specified) in the form $\sum x_b \otimes y_b$, where $x_b \in \mathscr{H} (b \in \mathbb{B})$ and $\sum \|x_b\|^2 < \infty$. When $\mathscr{K}$ is finite dimensional, the elements of $\mathscr{H} \otimes \mathscr{K}$ are *finite* sums of simple tensors; the same is true when $\mathscr{H}$ is finite dimensional, since the roles of $\mathscr{H}$ and $\mathscr{K}$ are interchangeable. ∎

We show next that the tensor product of Hilbert spaces $\mathscr{H}$ and $\mathscr{K}$ can be represented as a certain linear space of operators from the conjugate Hilbert space $\bar{\mathscr{H}}$ into $\mathscr{K}$. For this, note first that the equation

$$b_T(x, y) = \langle Tx, y \rangle \qquad (x \in \mathscr{H}, \quad y \in \mathscr{K})$$

defines a one-to-one linear mapping $T \to b_T$ from $\mathscr{B}(\mathscr{H}, \mathscr{K})$ onto the set of all bounded bilinear functionals on $\mathscr{H} \times \bar{\mathscr{K}}$ (since these are, precisely, the bounded conjugate-bilinear functionals on $\mathscr{H} \times \mathscr{K}$). With $T$ in $\mathscr{B}(\mathscr{H}, \mathscr{K})$, it follows by applying Proposition 2.6.1 to $b_T$ that the (finite or infinite) sum

$$\sum_{x \in X} \sum_{y \in Y} |\langle Tx, y \rangle|^2 \quad \left( = \sum_{y \in Y} \sum_{x \in X} |\langle T^*y, x \rangle|^2 \right) \tag{12}$$

has the same value, for all orthonormal bases $X$ of $\mathscr{H}$ and $Y$ of $\mathscr{K}$. From Parseval's equation, this sum can be written also in the alternative forms

$$\sum_{x \in X} \|Tx\|^2, \qquad \sum_{y \in Y} \|T^*y\|^2. \tag{13}$$

We describe $T$ as a *Hilbert–Schmidt operator* if the value of the sums is finite; equivalently, $T$ is a Hilbert–Schmidt operator if and only if $b_T$ is a Hilbert–Schmidt functional on $\mathscr{H} \times \bar{\mathscr{K}}$.

With $\mathscr{H}\mathscr{S}\mathscr{F}$ the linear space of all Hilbert–Schmidt functionals on $\mathscr{H} \times \bar{\mathscr{K}}$, the Hilbert–Schmidt operators from $\mathscr{H}$ into $\mathscr{K}$ form a linear subspace

$$\mathscr{H}\mathscr{S}\mathscr{O} = \{T \in \mathscr{B}(\mathscr{H}, \mathscr{K}) : b_T \in \mathscr{H}\mathscr{S}\mathscr{F}\}$$

of $\mathscr{B}(\mathscr{H}, \mathscr{K})$. By means of the mapping $T \to b_T$, the Hilbert space structure on $\mathscr{H}\mathscr{S}\mathscr{F}$, as described in Proposition 2.6.2, can be transferred to $\mathscr{H}\mathscr{S}\mathscr{O}$. Accordingly, $\mathscr{H}\mathscr{S}\mathscr{O}$ is a Hilbert space, when the inner product and norm are defined by

$$\langle S, T \rangle = \sum_{x \in X} \sum_{y \in Y} \langle Sx, y \rangle \langle y, Tx \rangle,$$

$$\|T\|_2 = \left[ \sum_{x \in X} \sum_{y \in Y} |\langle Tx, y \rangle|^2 \right]^{1/2},$$

these being independent of the choice of the orthonormal bases $X$ of $\mathscr{H}$ and $Y$ of $\mathscr{K}$. Of course, the mapping $T \to b_T$ is an isomorphism from $\mathscr{H}\mathscr{S}\mathscr{O}$ onto $\mathscr{H}\mathscr{S}\mathscr{F}$. The equality of the four sums appearing in (12) and (13) implies that there are three other, equivalent, expressions for $\|T\|_2$; and similarly, the inner product $\langle S, T \rangle$ can be expressed in the alternative forms

$$\sum_{y \in Y} \sum_{x \in X} \langle T^*y, x \rangle \langle x, S^*y \rangle, \qquad \sum_{x \in X} \langle Sx, Tx \rangle, \qquad \sum_{y \in Y} \langle T^*y, S^*y \rangle.$$

If $\mathscr{H}_0, \mathscr{K}_0$ are Hilbert spaces, $A \in \mathscr{B}(\mathscr{K}, \mathscr{K}_0)$, $B \in \mathscr{B}(\mathscr{H}_0, \mathscr{H})$, and $T$ is a Hilbert–Schmidt operator from $\mathscr{H}$ into $\mathscr{K}$, then $ATB$ is a Hilbert–Schmidt operator from $\mathscr{H}_0$ into $\mathscr{K}_0$, with $\|ATB\|_2 \leqslant \|A\| \, \|T\|_2 \, \|B\|$. For this, let $X_0$ be an orthonormal basis of $\mathscr{H}_0$, and observe that

$$\sum_{x \in X_0} \|TBx\|^2 = \sum_{y \in Y} \|B^*T^*y\|^2,$$

since these sums are the analogues, for $TB$, of the ones in (13). The stated result now follows from the inequalities

$$\begin{aligned}
\sum_{x \in X_0} \|ATBx\|^2 &\leqslant \|A\|^2 \sum_{x \in X_0} \|TBx\|^2 \\
&= \|A\|^2 \sum_{y \in Y} \|B^*T^*y\|^2 \\
&\leqslant \|A\|^2 \|B^*\|^2 \sum_{y \in Y} \|T^*y\|^2 \\
&= \|A\|^2 \|T\|_2^2 \|B\|^2.
\end{aligned}$$

This result can be proved also by means of Proposition 2.6.1.

The identification of $\mathscr{H} \otimes \mathscr{K}$ with the Hilbert space of all Hilbert–Schmidt operators from $\bar{\mathscr{H}}$ into $\mathscr{K}$ is described in the proposition that follows.

2.6.9. PROPOSITION. *If $\mathscr{H}$ and $\mathscr{K}$ are Hilbert spaces, then, for each $x$ in $\mathscr{H}$ and $y$ in $\mathscr{K}$, the equation*

$$T_{x,y}u = \langle u, x\rangle^{-} y = \langle x, u\rangle y \qquad (u \in \bar{\mathscr{H}})$$

*defines a Hilbert–Schmidt operator $T_{x,y}$ from $\bar{\mathscr{H}}$ into $\mathscr{K}$. With $\mathscr{HSO}$ the Hilbert space of all Hilbert–Schmidt operators from $\bar{\mathscr{H}}$ into $\mathscr{K}$, there is a unitary transformation $U$ from $\mathscr{H} \otimes \mathscr{K}$ onto $\mathscr{HSO}$, such that*

$$U(x \otimes y) = T_{x,y} \qquad (x \in \mathscr{H}, \quad y \in \mathscr{K}).$$

*Proof.* As constructed during the proof of Theorem 2.6.4, $\mathscr{H} \otimes \mathscr{K}$ is the Hilbert space $\mathscr{HSF}$ of all Hilbert–Schmidt functionals on $\bar{\mathscr{H}} \times \bar{\mathscr{K}}$. Moreover, when $x \in \mathscr{H}$ and $y \in \mathscr{K}$, $x \otimes y$ $(= p(x, y))$ is the bilinear functional $\varphi_{x,y}$ defined, throughout $\bar{\mathscr{H}} \times \bar{\mathscr{K}}$, by

$$\varphi_{x,y}(u, v) = \langle x, u\rangle\langle y, v\rangle.$$

The discussion preceding Proposition 2.6.9 shows that there is an isomorphism $U$ from $\mathscr{HSF}$ onto $\mathscr{HSO}$ that associates with each Hilbert–Schmidt functional on $\bar{\mathscr{H}} \times \bar{\mathscr{K}}$ the corresponding Hilbert–Schmidt operator from $\bar{\mathscr{H}}$ into $\mathscr{K}$. It is apparent that $T_{x,y}$, as defined in the proposition, is the bounded linear operator from $\bar{\mathscr{H}}$ into $\mathscr{K}$ that corresponds to the bilinear functional $\varphi_{x,y}$. Since $\varphi_{x,y} \in \mathscr{HSF}$, it follows that $T_{x,y} \in \mathscr{HSO}$, and

$$U(x \otimes y) = U\varphi_{x,y} = T_{x,y}. \quad \blacksquare$$

2.6.10. EXAMPLE. With $\mathbb{A}$ and $\mathbb{B}$ arbitrary sets, we can associate with each $x$ in $l_2(\mathbb{A})$ and $y$ in $l_2(\mathbb{B})$ a complex-valued function $p_{x,y}$ defined throughout $\mathbb{A} \times \mathbb{B}$ by

$$p_{x,y}(a, b) = x(a)y(b).$$

We shall show that there is a (unique) unitary transformation $U$ from $l_2(\mathbb{A}) \otimes l_2(\mathbb{B})$ onto $l_2(\mathbb{A} \times \mathbb{B})$ such that

$$U(x \otimes y) = p_{x,y} \qquad (x \in l_2(\mathbb{A}), \quad y \in l_2(\mathbb{B})).$$

For this, note first that $p_{x,y} \in l_2(\mathbb{A} \times \mathbb{B})$ since

$$\sum_{(a,b) \in \mathbb{A} \times \mathbb{B}} |p_{x,y}(a, b)|^2 = \sum_{a \in \mathbb{A}} \sum_{b \in \mathbb{B}} |x(a)y(b)|^2$$

$$= \left(\sum_{a \in \mathbb{A}} |x(a)|^2\right)\left(\sum_{b \in \mathbb{B}} |y(b)|^2\right) < \infty.$$

Moreover,

$$\begin{aligned}\langle p_{x,y}, p_{u,v}\rangle &= \sum_{(a,b)\in\mathbb{A}\times\mathbb{B}} p_{x,y}(a,b)\overline{p_{u,v}(a,b)}\\ &= \sum_{a\in\mathbb{A}}\sum_{b\in\mathbb{B}} x(a)y(b)\overline{u(a)}\,\overline{v(b)}\\ &= \left(\sum_{a\in\mathbb{A}} x(a)\overline{u(a)}\right)\left(\sum_{b\in\mathbb{B}} y(b)\overline{v(b)}\right)\\ &= \langle x,u\rangle\langle y,v\rangle = \langle x\otimes y, u\otimes v\rangle\end{aligned}$$

(the series manipulations being justified by absolute convergence), when $x, u \in l_2(\mathbb{A})$ and $y, v \in l_2(\mathbb{B})$. By expressing norms in terms of inner products, it now follows that, for any finite linear combination of elements $p_{x,y}$,

$$\|\sum_{j=1}^{n} c_j p_{x_j,y_j}\| = \|\sum_{j=1}^{n} c_j x_j \otimes y_j\|. \tag{14}$$

We sketch the remainder of the argument, which follows the same pattern as the second paragraph of the proof of Proposition 2.6.5. The linear span $\mathscr{K}_0$ of $\{p_{x,y}: x\in l_2(\mathbb{A}), y\in l_2(\mathbb{B})\}$ is everywhere dense in $l_2(\mathbb{A}\times\mathbb{B})$ since it contains the usual orthonormal basis (consisting of functions with value 1 at a single point of $\mathbb{A}\times\mathbb{B}$ and 0 elsewhere); and the linear span $\mathscr{H}_0$ of the simple tensors is everywhere dense in $l_2(\mathbb{A})\otimes l_2(\mathbb{B})$. From (14), there is a norm-preserving linear mapping $U_0$ from $\mathscr{K}_0$ onto $\mathscr{H}_0$ such that $U_0 p_{x,y} = x\otimes y$; and this mapping extends by continuity to an isomorphism $U$ from $l_2(\mathbb{A}\times\mathbb{B})$ onto $l_2(\mathbb{A})\otimes l_2(\mathbb{B})$. ■

2.6.11. Example. We now consider the tensor product of the $L_2$ spaces associated with $\sigma$-finite measure spaces $(S,\mathscr{S},m)$ and $(S',\mathscr{S}',m')$. We show that this can be identified with the $L_2$ space of the product measure space $(S\times S', \mathscr{S}\times\mathscr{S}', m\times m')$, in such a way that $x\otimes y$ corresponds to the function $p_{x,y}$ defined throughout $S\times S'$ by

$$p_{x,y}(s,s') = x(s)y(s').$$

For this, note first that $p_{x,y}$ is a complex-valued measurable function when $x\in L_2(S,\mathscr{S},m)$ $(=\mathscr{H})$ and $y\in L_2(S',\mathscr{S}',m')$ $(=\mathscr{H}')$; moreover,

$$p_{x,y}\in L_2(S\times S', \mathscr{S}\times\mathscr{S}', m\times m') \quad (=\mathscr{K}),$$

since

$$\begin{aligned}
\iint_{S\times S'} &|p_{x,y}(s,s')|^2\,dm(s)\,dm'(s') \\
&= \iint_{S\times S'} |x(s)y(s')|^2\,dm(s)\,dm'(s') \\
&= \left(\int_S |x(s)|^2\,dm(s)\right)\left(\int_{S'} |y(s')|^2\,dm'(s')\right) < \infty.
\end{aligned}$$

Also,

$$\begin{aligned}
\langle p_{x,y}, p_{u,v}\rangle &= \iint_{S\times S'} p_{x,y}(s,s')\overline{p_{u,v}(s,s')}\,dm(s)\,dm'(s') \\
&= \iint_{S\times S'} x(s)y(s')\overline{u(s)}\;\overline{v(s')}\,dm(s)\,dm'(s') \\
&= \left(\int_S x(s)\overline{u(s)}\,dm(s)\right)\left(\int_{S'} y(s')\overline{v(s')}\,dm'(s')\right) \\
&= \langle x,u\rangle\langle y,v\rangle = \langle x\otimes y, u\otimes v\rangle,
\end{aligned}$$

whenever $x, u \in \mathscr{H}$ and $y, v \in \mathscr{H}'$. From this, we have

$$\left\|\sum_{j=1}^{n} c_j p_{x_j,y_j}\right\| = \left\|\sum_{j=1}^{n} c_j x_j \otimes y_j\right\|,$$

for every finite linear combination of elements $p_{x,y}$. Accordingly (by the argument already used in the preceding example and in the proof of Proposition 2.6.5), there is a norm-preserving linear mapping $U_0$, from the linear span $\mathscr{K}_0$ of $\{p_{x,y} : x \in \mathscr{H}, y \in \mathscr{H}'\}$ onto the linear span $\mathscr{H}_0$ of the simple tensors in $\mathscr{H} \otimes \mathscr{H}'$, such that $U_0 p_{x,y} = x \otimes y$. Now $\mathscr{H}_0$ is everywhere dense in $\mathscr{H} \otimes \mathscr{H}'$, and $\mathscr{K}_0$ is everywhere dense in $\mathscr{K}$ since it contains the characteristic function of every measurable rectangle of finite measure. Thus $U_0$ extends by continuity to an isomorphism $U$ from $\mathscr{K}$ onto $\mathscr{H} \otimes \mathscr{H}'$. ∎

We now introduce tensor products of bounded linear operators.

2.6.12. Proposition. *If $\mathscr{H}_1, \ldots, \mathscr{H}_n, \mathscr{K}_1, \ldots, \mathscr{K}_n$ are Hilbert spaces and $A_m \in \mathscr{B}(\mathscr{H}_m, \mathscr{K}_m)$ $(m = 1, \ldots, n)$, there is a unique bounded linear operator $A$ from $\mathscr{H}_1 \otimes \cdots \otimes \mathscr{H}_n$ into $\mathscr{K}_1 \otimes \cdots \otimes \mathscr{K}_n$ such that*

$$A(x_1 \otimes \cdots \otimes x_n) = A_1 x_1 \otimes \cdots \otimes A_n x_n \qquad (x_1 \in \mathscr{H}_1, \ldots, x_n \in \mathscr{H}_n).$$

*Proof.* The canonical mapping $p: \mathscr{K}_1 \times \cdots \times \mathscr{K}_n \to \mathscr{K}_1 \otimes \cdots \otimes \mathscr{K}_n$ $(= \mathscr{K})$ is a weak Hilbert–Schmidt mapping, with $\|p\|_2 = 1$. With $u$ in $\mathscr{K}$, and $p_u$

defined by

$$p_u(z_1, \ldots, z_n) = \langle p(z_1, \ldots, z_n), u \rangle,$$

$p_u$ is a Hilbert–Schmidt functional on $\mathscr{K}_1 \times \cdots \times \mathscr{K}_n$, and $\|p_u\|_2 \leqslant \|u\|$. The equation

$$\varphi(x_1, \ldots, x_n) = p(A_1 x_1, \ldots, A_n x_n)$$

defines a bounded multilinear mapping $\varphi\colon \mathscr{H}_1 \times \cdots \times \mathscr{H}_n \to \mathscr{K}$, and

$$\begin{aligned}\varphi_u(x_1, \ldots, x_n) &= \langle \varphi(x_1, \ldots, x_n), u \rangle \\ &= \langle p(A_1 x_1, \ldots, A_n x_n), u \rangle \\ &= p_u(A_1 x_1, \ldots, A_n x_n).\end{aligned}$$

It now follows from Proposition 2.6.1(ii) that $\varphi_u$ is a Hilbert–Schmidt functional on $\mathscr{H}_1 \times \cdots \times \mathscr{H}_n$, with

$$\|\varphi_u\|_2 \leqslant \|A_1\| \cdots \|A_n\| \, \|p_u\|_2 \leqslant \|A_1\| \cdots \|A_n\| \, \|u\|.$$

Accordingly, $\varphi\colon \mathscr{H}_1 \times \cdots \times \mathscr{H}_n \to \mathscr{K}$ is a weak Hilbert–Schmidt mapping, with $\|\varphi\|_2 \leqslant \|A_1\| \cdots \|A_n\|$. By the universal property of the tensor product (see Theorem 2.6.4(i)), there is a unique bounded linear operator $A$, from $\mathscr{H}_1 \otimes \cdots \otimes \mathscr{H}_n$ into $\mathscr{K}$, such that $\varphi = Ap'$, where $p'$ is the canonical mapping from $\mathscr{H}_1 \times \cdots \times \mathscr{H}_n$ into $\mathscr{H}_1 \otimes \cdots \otimes \mathscr{H}_n$. Moreover,

$$\|A\| = \|\varphi\|_2 \leqslant \|A_1\| \cdots \|A_n\|.$$

Also,

$$\begin{aligned}A(x_1 \otimes \cdots \otimes x_n) &= Ap'(x_1, \ldots, x_n) = \varphi(x_1, \ldots, x_n) \\ &= p(A_1 x_1, \ldots, A_n x_n) = A_1 x_1 \otimes \cdots \otimes A_n x_n,\end{aligned}$$

when $x_1 \in \mathscr{H}_1, \ldots, x_n \in \mathscr{H}_n$. ■

The operator $A$ described in Proposition 2.6.12 is called the tensor product of $A_1, \ldots, A_n$ and denoted by $A_1 \otimes \cdots \otimes A_n$. It is apparent that $A_1 \otimes \cdots \otimes A_n$ depends linearly on each $A_m$ and that

$$(A_1 \otimes \cdots \otimes A_n)(B_1 \otimes \cdots \otimes B_n) = A_1 B_1 \otimes \cdots \otimes A_n B_n.$$

Since

$$\begin{aligned}&\langle (A_1 \otimes \cdots \otimes A_n)(x_1 \otimes \cdots \otimes x_n), y_1 \otimes \cdots \otimes y_n \rangle \\ &\quad= \langle A_1 x_1 \otimes \cdots \otimes A_n x_n, y_1 \otimes \cdots \otimes y_n \rangle \\ &\quad= \langle A_1 x_1, y_1 \rangle \cdots \langle A_n x_n, y_n \rangle \\ &\quad= \langle x_1, A_1^* y_1 \rangle \cdots \langle x_n, A_n^* y_n \rangle \\ &\quad= \langle x_1 \otimes \cdots \otimes x_n, A_1^* y_1 \otimes \cdots \otimes A_n^* y_n \rangle \\ &\quad= \langle x_1 \otimes \cdots \otimes x_n, (A_1^* \otimes \cdots \otimes A_n^*)(y_1 \otimes \cdots \otimes y_n) \rangle,\end{aligned}$$

it follows by linearity and continuity that

$$\langle (A_1 \otimes \cdots \otimes A_n)u, v\rangle = \langle u, (A_1^* \otimes \cdots \otimes A_n^*)v\rangle$$

for all vectors $u$ and $v$ in the appropriate tensor product spaces. Thus

$$(A_1 \otimes \cdots \otimes A_n)^* = A_1^* \otimes \cdots \otimes A_n^*. \tag{15}$$

We assert also that

$$\|A_1 \otimes \cdots \otimes A_n\| = \|A_1\| \cdots \|A_n\|. \tag{16}$$

Indeed, given by any unit vectors $x_1$ in $\mathscr{H}_1, \ldots, x_n$ in $\mathscr{H}_n$, we have

$$\begin{aligned} \|A_1 \otimes \cdots \otimes A_n\| &= \|A_1 \otimes \cdots \otimes A_n\| \, \|x_1 \otimes \cdots \otimes x_n\| \\ &\geqslant \|(A_1 \otimes \cdots \otimes A_n)(x_1 \otimes \cdots \otimes x_n)\| \\ &= \|A_1x_1 \otimes \cdots \otimes A_nx_n\| = \|A_1x_1\| \cdots \|A_nx_n\|. \end{aligned}$$

Upon taking the supremum of the right-hand side, as the unit vectors $x_1, \ldots, x_n$ vary, we obtain

$$\|A_1 \otimes \cdots \otimes A_n\| \geqslant \|A_1\| \cdots \|A_n\|;$$

the reverse inequality was noted during the proof of Proposition 2.6.12.

Suppose next that $\mathscr{H}_1, \ldots, \mathscr{H}_{m+n}, \mathscr{K}_1, \ldots, \mathscr{K}_{m+n}$ are Hilbert spaces and that $A_j \in \mathscr{B}(\mathscr{H}_j, \mathscr{K}_j)$ $(j = 1, \ldots, m+n)$. We can construct isomorphisms

$$U: \mathscr{H}_1 \otimes \cdots \otimes \mathscr{H}_{m+n} \to (\mathscr{H}_1 \otimes \cdots \otimes \mathscr{H}_m) \otimes (\mathscr{H}_{m+1} \otimes \cdots \otimes \mathscr{H}_{m+n}),$$

$$V: \mathscr{K}_1 \otimes \cdots \otimes \mathscr{K}_{m+n} \to (\mathscr{K}_1 \otimes \cdots \otimes \mathscr{K}_m) \otimes (\mathscr{K}_{m+1} \otimes \cdots \otimes \mathscr{K}_{m+n}),$$

as in Proposition 2.6.5, and it is at once verified that

$$V(A_1 \otimes \cdots \otimes A_{m+n})U^{-1} = (A_1 \otimes \cdots \otimes A_m) \otimes (A_{m+1} \otimes \cdots \otimes A_{m+n}).$$

This proves the "associativity" of the tensor product of bounded linear operators on Hilbert spaces.

With $\mathscr{H}$ and $\mathscr{K}$ Hilbert spaces, the linear mapping

$$A \to A \otimes I: \mathscr{B}(\mathscr{H}) \to \mathscr{B}(\mathscr{H} \otimes \mathscr{K})$$

preserves operator products, adjoints, and norms; from this last, it is norm continuous. We consider next its continuity properties relative to the strong-operator topology. With $v$ a simple tensor $x \otimes y$ in $\mathscr{H} \otimes \mathscr{K}$,

$$\|(A \otimes I)v - (A_0 \otimes I)v\| = \|(A - A_0)x \otimes y\| = \|(A - A_0)x\| \, \|y\|,$$

for each $A$ and $A_0$ in $\mathscr{B}(\mathscr{H})$. From this, it follows that, if $v_1, \ldots, v_m$ are simple tensors in $\mathscr{H} \otimes \mathscr{K}$ and $\varepsilon > 0$, then the set

$$\{A \in \mathscr{B}(\mathscr{H}): \|(A \otimes I)v_j - (A_0 \otimes I)v_j\| < \varepsilon \ (j = 1, \ldots, m)\}$$

is a strong-operator neighborhood of $A_0$ in $\mathscr{B}(\mathscr{H})$. Since the simple tensors have closed linear span $\mathscr{H} \otimes \mathscr{K}$, they suffice to determine the strong-operator topology on bounded subsets of $\mathscr{H} \otimes \mathscr{K}$; so the preceding sentence implies that the mapping $A \to A \otimes I$ is strong-operator continuous on bounded subsets of $\mathscr{B}(\mathscr{H})$. When one or other of $\mathscr{H}$ and $\mathscr{K}$ is finite dimensional, the simple tensors have algebraic linear span $\mathscr{H} \otimes \mathscr{K}$ (Remark 2.6.8), and therefore suffice to determine the strong-operator topology on (the whole of) $\mathscr{B}(\mathscr{H} \otimes \mathscr{K})$; so, in this case, the mapping $A \to A \otimes I$ is strong-operator continuous on $\mathscr{B}(\mathscr{H})$.

Suppose, finally, that $\{y_b : b \in \mathbb{B}\}$ is an orthonormal basis of $\mathscr{K}$. As noted in Remark 2.6.8, the equation

$$U(\textstyle\sum \oplus x_b) = \sum x_b \otimes y_b$$

defines an isomorphism $U$ from $\sum_{b \in \mathbb{B}} \oplus \mathscr{H}_b$ onto $\mathscr{H} \otimes \mathscr{K}$ when each $\mathscr{H}_b$ is $\mathscr{H}$. With $A$ in $\mathscr{B}(\mathscr{H})$,

$$\begin{aligned}(A \otimes I)U(\textstyle\sum \oplus x_b) &= (A \otimes I)(\sum x_b \otimes y_b)\\ &= \textstyle\sum (A \otimes I)(x_b \otimes y_b)\\ &= \textstyle\sum Ax_b \otimes y_b = U(\sum \oplus Ax_b),\end{aligned}$$

so

$$U^{-1}(A \otimes I)U = \sum_{b \in \mathbb{B}} \oplus A \qquad (A \in \mathscr{B}(\mathscr{H})). \tag{17}$$

*Matrix representations.* We conclude this section with an account of the matrix representations of operators acting on a direct sum $\sum_{b \in \mathbb{B}} \oplus \mathscr{H}_b$, where each $\mathscr{H}_b$ is the same Hilbert space $\mathscr{H}$. Before embarking on this program, we consider the numerical matrices of operators relative to orthonormal bases.

Suppose that $\{y_b : b \in \mathbb{B}\}$ is an orthonormal basis of a Hilbert space $\mathscr{K}$. With $S$ in $\mathscr{B}(\mathscr{K})$, each vector $Sy_b$ has an expansion

$$Sy_b = \sum_{a \in \mathbb{B}} s_{ab} y_a \qquad (b \in \mathbb{B}), \tag{18}$$

in which the coefficients are given by

$$s_{ab} = \langle Sy_b, y_a \rangle. \tag{19}$$

In this way, we associate with each $S$ in $\mathscr{B}(\mathscr{K})$ a complex matrix $[s_{ab}]_{a,b \in \mathbb{B}}$, relative to the orthonormal basis $\{y_b\}$. When the index set $\mathbb{B}$ is finite, *every* complex matrix $[s_{ab}]_{a,b \in \mathbb{B}}$ corresponds, as in (18), to some element $S$ of $\mathscr{B}(\mathscr{K})$. When $\mathbb{B}$ is infinite, however, boundedness of $S$ imposes certain restrictions on its matrix. For example, Parseval's equation gives

$$\sum_{a \in \mathbb{B}} |s_{ab}|^2 = \sum_{a \in \mathbb{B}} |\langle Sy_b, y_a \rangle|^2 = \|Sy_b\|^2 \leqslant \|S\|^2,$$

$$\begin{aligned}\sum_{b \in \mathbb{B}} |s_{ab}|^2 &= \sum_{b \in \mathbb{B}} |\langle Sy_b, y_a \rangle|^2\\ &= \sum_{b \in \mathbb{B}} |\langle y_b, S^* y_a \rangle|^2 = \|S^* y_a\|^2 \leqslant \|S\|^2;\end{aligned}$$

so the "columns" and "rows" of the matrix $[s_{ab}]$ form (bounded sets of) vectors in $l_2(\mathbb{B})$.

The algebraic relations between operators and matrices follow the pattern familiar in the finite-dimensional case. From (19) and (18), the matrix elements $s_{ab}$ depend linearly on $S$, and $[s_{ab}]$ is the zero matrix only when $S = 0$. Since

$$\langle S^* y_b, y_a \rangle = \langle y_b, S y_a \rangle = \overline{\langle S y_a, y_b \rangle} = \bar{s}_{ba},$$

the matrix of $S^*$ has $\bar{s}_{ba}$ in the $(a, b)$ position. If two elements $S$ and $T$ of $\mathscr{B}(\mathscr{H})$ have matrices $[s_{ab}]$ and $[t_{ab}]$, respectively, and $R = ST$, then

$$\langle R y_b, y_a \rangle = \langle ST y_b, y_a \rangle = \langle T y_b, S^* y_a \rangle,$$

and Parseval's equation gives

$$\begin{aligned}\langle R y_b, y_a \rangle &= \sum_{c \in \mathbb{B}} \langle T y_b, y_c \rangle \langle y_c, S^* y_a \rangle \\ &= \sum_{c \in \mathbb{B}} \langle S y_c, y_a \rangle \langle T y_b, y_c \rangle \\ &= \sum_{c \in \mathbb{B}} s_{ac} t_{cb}.\end{aligned}$$

Accordingly, the matrix $[r_{ab}]$ of $R$ $(= ST)$ is given by

$$r_{ab} = \sum_{c \in \mathbb{B}} s_{ac} t_{cb}.$$

The results of the preceding paragraph can be summarized in the assertion that the matrices, corresponding (through a fixed orthonormal basis) to bounded operators on $\mathscr{H}$, form an algebra relative to the usual concepts of sum, product, and scalar multiple of matrices. Moreover, the mapping from bounded operators to the corresponding matrices is an isomorphism.

We now consider operators acting on a direct sum $\sum_{b \in \mathbb{B}} \oplus \mathscr{H}_b$, with each $\mathscr{H}_b$ the same Hilbert space $\mathscr{H}$. For this purpose, we introduce a closed subspace $\mathscr{H}'_a$ of $\sum \oplus \mathscr{H}_b$, and bounded linear operators

$$U_a : \mathscr{H} \to \sum \oplus \mathscr{H}_b, \qquad V_a : \sum \oplus \mathscr{H}_b \to \mathscr{H},$$

for each $a$ in $\mathbb{B}$, as follows. When $x \in \mathscr{H}$ and $u = \{x_b\} \in \sum \oplus \mathscr{H}_b$, $V_a u = x_a$ and $U_a x$ is the family $\{z_b\}$ in which $z_a = x$ and all other $z_b$ are 0; $\mathscr{H}'_a$ is the range of $U_a$, and so consists of all elements $\{z_b\}$ of $\sum \oplus \mathscr{H}_b$ in which $z_b = 0$ when $b \neq a$. Observe that $V_a U_a$ is the identity operator on $\mathscr{H}$ and $U_a V_a$ is the projection $E_a$ from $\sum \oplus \mathscr{H}_b$ onto $\mathscr{H}'_a$. Since the subspaces $\mathscr{H}'_a$ $(a \in \mathbb{B})$ are pairwise orthogonal, and $\vee \mathscr{H}'_a = \sum \oplus \mathscr{H}_b$, it follows that the sum $\sum_{a \in \mathbb{B}} E_a$ is strong-operator convergent to $I$. Note also that $U_a = V_a^*$, since

$$\langle U_a x, \{x_b\} \rangle = \langle x, x_a \rangle = \langle x, V_a \{x_b\} \rangle$$

whenever $x \in \mathscr{H}$ and $\{x_b\} \in \sum \oplus \mathscr{H}_b$.

With each bounded linear operator $T$ acting on $\sum\oplus \mathscr{H}_b$, we associate a matrix $[T_{ab}]_{a,b\in\mathbb{B}}$, with entries $T_{ab}$ in $\mathscr{B}(\mathscr{H})$ defined by

$$T_{ab} = V_a T U_b. \tag{20}$$

If $u = \{x_b\} \in \sum\oplus \mathscr{H}_b$, then $Tu$ is an element $\{y_b\}$ of $\sum\oplus \mathscr{H}_b$, and

$$y_a = V_a Tu = V_a T\left(\sum_b E_b u\right) = \sum_b V_a T U_b V_b u = \sum_b T_{ab} x_b.$$

Thus

$$T(\textstyle\sum\oplus x_b) = \sum\oplus y_b \qquad \text{where} \quad y_a = \displaystyle\sum_{b\in\mathbb{B}} T_{ab} x_b \quad (a\in\mathbb{B}). \tag{21}$$

The usual rules of matrix algebra have natural analogues in this situation. From (20), the matrix elements $T_{ab}$ depend linearly on $T$. Since

$$V_a T^* U_b = U_a^* T^* V_b^* = (V_b T U_a)^* = (T_{ba})^*,$$

the matrix of $T^*$ has $(T_{ba})^*$ in the $(a, b)$ position. If $S$ and $T$ are bounded linear operators acting on $\sum\oplus \mathscr{H}_b$, and $R = ST$, then

$$\begin{aligned} R_{ab} = V_a R U_b = V_a S T U_b &= \sum_{c\in\mathbb{B}} V_a S E_c T U_b \\ &= \sum_{c\in\mathbb{B}} V_a S U_c V_c T U_b = \sum_{c\in\mathbb{B}} S_{ac} T_{cb}, \end{aligned}$$

the sum converging in the strong-operator topology if the index set $\mathbb{B}$ is infinite.

In this way, we establish a one-to-one correspondence between elements of $\mathscr{B}(\sum_{b\in\mathbb{B}} \oplus \mathscr{H}_b)$ and certain matrices $[T_{ab}]_{a,b\in\mathbb{B}}$ with entries $T_{ab}$ in $\mathscr{B}(\mathscr{H})$. When the index set $\mathbb{B}$ is *finite*, *each* such matrix corresponds to some bounded operator $T$ acting on $\sum\oplus \mathscr{H}_b$; indeed, $T$ is defined by (21), and its boundedness follows at once from the relations

$$\begin{aligned} \|\{y_b\}\|^2 = \sum_a \|y_a\|^2 = \sum_a \|\sum_b T_{ab} x_b\|^2 &\leqslant \sum_a \left(\sum_b \|T_{ab}\|\,\|x_b\|\right)^2 \\ &\leqslant \sum_a \left(\sum_b \|T_{ab}\|^2\right)\left(\sum_b \|x_b\|^2\right) = \left(\sum_a \sum_b \|T_{ab}\|^2\right) \|\{x_b\}\|^2. \end{aligned}$$

When the index set $\mathbb{B}$ is *infinite*, it is apparent that some matrices with entries in $\mathscr{B}(\mathscr{H})$ do *not* arise in the above manner from bounded operators. In formal matrix calculations, it is necessary to ensure that no such "unbounded" matrices appear at any stage. While there is no simple general procedure for determining whether or not a given matrix corresponds to a bounded operator, a criterion that is sometimes useful is set out in Proposition 2.6.13 below. In the meantime, we describe certain special types of "bounded" matrices that arise frequently in applications.

In the first place, a matrix $[T_{ab}]$ gives rise to a bounded operator if it has only a finite number of non-zero entries $T_{ab}$. Indeed, the proof used above, for the case in which $\mathbb{B}$ is finite, applies also in the circumstances just described. More generally, the same argument shows that the matrix corresponds to a bounded operator whenever the sum $\sum_a \sum_b \|T_{ab}\|^2$ is finite, since it is easily verified in this case that the series for $y_a$, in (21), is absolutely convergent. However, there are bounded operators on $\sum \oplus \mathscr{H}_b$ (for example, $I$) whose matrices do not satisfy these conditions.

Second, suppose that $\pi$ is a permutation of $\mathbb{B}$, and $\{T_b : b \in \mathbb{B}\}$ is a bounded family of elements of $\mathscr{B}(\mathscr{H})$. Since

$$\sum_{b\in\mathbb{B}} \|T_b x_{\pi(b)}\|^2 \leqslant \sum_{b\in\mathbb{B}} \|T_b\|^2 \|x_{\pi(b)}\|^2 \leqslant \sup_{b\in\mathbb{B}} \|T_b\|^2 \sum_{b\in\mathbb{B}} \|x_b\|^2,$$

the equation $T(\sum \oplus x_b) = \sum \oplus T_b x_{\pi(b)}$ defines a bounded linear operator $T$ on $\sum \oplus \mathscr{H}_b$. The matrix $[T_{ab}]$ of $T$ is given by $T_{ab} = \delta_{\pi(a),b} T_a$, where $\delta_{ab}$ is the Kronecker symbol ($\delta_{ab}$ is 1 when $a = b$, 0 when $a \neq b$). If each $T_b$ is unitary, the same is true of $T$. When $\pi$ is the identity mapping on $\mathbb{B}$, $T$ is the direct sum $\sum \oplus T_b$, and corresponds to the diagonal matrix $[\delta_{ab} T_a]$.

Finally, we consider the matrix representation of certain tensor products of operators. Suppose that $\{y_b : b \in \mathbb{B}\}$ is an orthonormal basis in a Hilbert space $\mathscr{K}$, and $U$ is the isomorphism from $\sum \oplus \mathscr{H}_b$ onto $\mathscr{H} \otimes \mathscr{K}$ (where each $\mathscr{H}_b$ is $\mathscr{H}$), defined by $U(\sum \oplus x_b) = \sum x_b \otimes y_b$. When $A \in \mathscr{B}(\mathscr{H})$, $U^{-1}(A \otimes I)U$ is the direct sum $\sum_{b\in\mathbb{B}} \oplus A$, and has matrix $[\delta_{ab} A]$. This characterizes the operators $T_0$ of the form $A \otimes I$ acting on $\mathscr{H} \otimes \mathscr{K}$ as those for which $U^{-1} T_0 U$ has a diagonal matrix with the same element of $\mathscr{B}(\mathscr{H})$ in each diagonal position. We assert also that an element $T_0$ of $\mathscr{B}(\mathscr{H} \otimes \mathscr{K})$ can be expressed as $I \otimes S$, with $S$ in $\mathscr{B}(\mathscr{K})$, if and only if the matrix of $U^{-1} T_0 U$ has the form $[s_{ab} I]$, with each $s_{ab}$ a scalar. For this, suppose first that $S \in \mathscr{B}(\mathscr{K})$, so that $S$ has a complex matrix $[s_{ab}]$ satisfying (18). With $x$ in $\mathscr{H}$ and $b$ in $\mathbb{B}$, $UU_b x = x \otimes y_b$, so

$$\begin{aligned} V_a(U^{-1}(I \otimes S)U)U_b x &= V_a U^{-1}(I \otimes S)(x \otimes y_b) \\ &= V_a U^{-1}(x \otimes S y_b) = V_a U^{-1}\left(\sum_{c\in\mathbb{B}} s_{cb} x \otimes y_c\right) \\ &= V_a\left(\sum_{c\in\mathbb{B}} \oplus s_{cb} x\right) = s_{ab} x. \end{aligned}$$

Hence $V_a(U^{-1}(I \otimes S)U)U_b = s_{ab} I$, and $U^{-1}(I \otimes S)U$ has matrix $[s_{ab} I]$. Conversely, suppose that $T_0 \in \mathscr{B}(\mathscr{H} \otimes \mathscr{K})$ and the matrix of $U^{-1} T_0 U\ (= T)$ has the form $[s_{ab} I]$. With $x$ a unit vector in $\mathscr{H}$ and $f$ an element of $l_2(\mathbb{B})$, $u = \sum \oplus f(b)x$ is a vector in $\sum \oplus \mathscr{H}_b$ and $Tu = \sum \oplus x_b$, where

$$x_a = \sum_{b\in\mathbb{B}} s_{ab} f(b) x.$$

Hence the sum $\sum_{b\in\mathbb{B}} s_{ab}f(b)$ converges, and

$$\sum_{a\in\mathbb{B}} |\sum_{b\in\mathbb{B}} s_{ab}f(b)|^2 = \sum_{a\in\mathbb{B}} \|x_a\|^2 = \|Tu\|^2$$

$$\leqslant \|T\|^2\|u\|^2 = \|T\|^2 \sum_{b\in\mathbb{B}} |f(b)|^2;$$

so

$$\|\sum_{a\in\mathbb{B}} \left( \sum_{b\in\mathbb{B}} s_{ab}f(b) \right) y_a\| \leqslant \|T\| \, \|\sum_{b\in\mathbb{B}} f(b)y_b\|.$$

Accordingly, the equation

$$S\left( \sum_{b\in\mathbb{B}} f(b)y_b \right) = \sum_{a\in\mathbb{B}} \left( \sum_{b\in\mathbb{B}} s_{ab}f(b) \right) y_a \qquad (f\in l_2(\mathbb{B}))$$

defines a bounded linear operator $S$ on $\mathscr{K}$. Since $S$ has matrix $[s_{ab}]$, $U^{-1}(I\otimes S)U$ has the same matrix $[s_{ab}I]$ as does $U^{-1}T_0U$, so $T_0 = I\otimes S$.

We now establish a criterion for determining whether or not a matrix $[T_{ab}]_{a,b\in\mathbb{B}}$, with entries $T_{ab}$ in $\mathscr{B}(\mathscr{H})$, corresponds to a bounded linear operator acting on $\sum_{b\in\mathbb{B}} \oplus \mathscr{H}_b$, where each $\mathscr{H}_b$ is $\mathscr{H}$. As noted above, each such matrix gives rise to a bounded operator when the index set $\mathbb{B}$ is finite.

2.6.13. Proposition. *Suppose that $\mathscr{H}$ is a Hilbert space and $[T_{ab}]_{a,b\in\mathbb{B}}$ is a matrix, with entries $T_{ab}$ in $\mathscr{B}(\mathscr{H})$. For each finite subset $\mathbb{F}$ of $\mathbb{B}$, let $T(\mathbb{F})$ be the bounded linear operator, corresponding to the matrix $[T_{ab}]_{a,b\in\mathbb{F}}$, that acts on the Hilbert space $\sum_{b\in\mathbb{F}} \oplus \mathscr{H}_b$ (where each $\mathscr{H}_b$ is $\mathscr{H}$).*

(i) $\|T(\mathbb{F}_1)\| \leqslant \|T(\mathbb{F}_2)\|$ *if* $\mathbb{F}_1 \subseteq \mathbb{F}_2$.

(ii) *The matrix $[T_{ab}]_{a,b\in\mathbb{B}}$ corresponds to a bounded linear operator $T$ acting on $\sum_{b\in\mathbb{B}} \oplus \mathscr{H}_b$ if and only if the set $\{\|T(\mathbb{F})\|: \mathbb{F}$ a finite subset of $\mathbb{B}\}$ of real numbers is bounded above. When this is so,*

$$\|T\| = \sup\{\|T(\mathbb{F})\|: \mathbb{F} \text{ a finite subset of } \mathbb{B}\}.$$

*Proof.* Let $\mathscr{F}$ denote the class of all finite subsets of $\mathbb{B}$, and when $\mathbb{F}\in\mathscr{F}$, let $\mathscr{H}(\mathbb{F})$ be the Hilbert space $\sum_{b\in\mathbb{F}} \oplus \mathscr{H}_b$. Note that if $\mathbb{F}\in\mathscr{F}$ and $u$ is an element $\sum_{b\in\mathbb{F}} \oplus x_b$ of $\mathscr{H}(\mathbb{F})$, then

$$\|u\|^2 = \sum_{b\in\mathbb{F}} \|x_b\|^2, \qquad \|T(\mathbb{F})u\|^2 = \sum_{a\in\mathbb{F}} \|\sum_{b\in\mathbb{F}} T_{ab}x_b\|^2.$$

(i) Suppose that $\mathbb{F}_1, \mathbb{F}_2 \in\mathscr{F}$ and $\mathbb{F}_1 \subseteq \mathbb{F}_2$. With $v$ an element $\sum_{b\in\mathbb{F}_1} \oplus x_b$ of $\mathscr{H}(\mathbb{F}_1)$, let $w$ be the element $\sum_{b\in\mathbb{F}_2} \oplus x_b$ of $\mathscr{H}(\mathbb{F}_2)$ obtained by taking $x_b$ to be 0 when $b \in \mathbb{F}_2 \backslash \mathbb{F}_1$. Then

$$\begin{aligned}\|T(\mathbb{F}_1)v\|^2 &= \sum_{a\in\mathbb{F}_1} \|\sum_{b\in\mathbb{F}_1} T_{ab}x_b\|^2 \\ &= \sum_{a\in\mathbb{F}_1} \|\sum_{b\in\mathbb{F}_2} T_{ab}x_b\|^2 \\ &\leqslant \sum_{a\in\mathbb{F}_2} \|\sum_{b\in\mathbb{F}_2} T_{ab}x_b\|^2 \\ &= \|T(\mathbb{F}_2)w\|^2 \leqslant \|T(\mathbb{F}_2)\|^2\|w\|^2 = \|T(\mathbb{F}_2)\|^2\|v\|^2,\end{aligned}$$

and therefore $\|T(\mathbb{F}_1)\| \leqslant \|T(\mathbb{F}_2)\|$.

(ii) Suppose first that the set $\{\|T(\mathbb{F})\| : \mathbb{F} \in \mathscr{F}\}$ has finite supremum $k$. In view of the discussion preceding the proof of (i), it follows that

$$\sum_{a\in\mathbb{F}} \|\sum_{b\in\mathbb{F}} T_{ab}x_b\|^2 \leqslant k^2 \sum_{b\in\mathbb{F}} \|x_b\|^2 \tag{22}$$

whenever $\mathbb{F} \in \mathscr{F}$ and $x_b \in \mathscr{H}_b$ $(b \in \mathbb{F})$. Let $u$ be an element $\sum_{b\in\mathbb{B}} \oplus x_b$ of $\sum_{b\in\mathbb{B}} \oplus \mathscr{H}_b$. Our proof that $[T_{ab}]$ is the matrix of a bounded linear operator is now divided into two stages.

First, we show by the Cauchy criterion that, for each $c$ in $\mathbb{B}$, the sum $\sum_{b\in\mathbb{B}} T_{cb}x_b$ converges to an element $y_c$ of $\mathscr{H}$. For this, suppose $\varepsilon > 0$, and choose $\mathbb{F}_\varepsilon$ in $\mathscr{F}$ so that

$$\sum_{b\in\mathbb{F}} \|x_b\|^2 < \varepsilon^2/k^2 \qquad \text{whenever} \quad \mathbb{F} \in \mathscr{F} \quad \text{and} \quad \mathbb{F} \cap \mathbb{F}_\varepsilon = \varnothing.$$

By enlarging $\mathbb{F}_\varepsilon$ if necessary, we may suppose that $c \in \mathbb{F}_\varepsilon$. When $\mathbb{F} \in \mathscr{F}$ and $\mathbb{F} \cap \mathbb{F}_\varepsilon = \varnothing$, let $\mathbb{F}_0 = \mathbb{F} \cup \{c\}$, and define $x_b' = x_b$ $(b \in \mathbb{F})$, $x_c' = 0$. From (22),

$$\begin{aligned}\|\sum_{b\in\mathbb{F}} T_{cb}x_b\|^2 &= \|\sum_{b\in\mathbb{F}_0} T_{cb}x_b'\|^2 \\ &\leqslant \sum_{a\in\mathbb{F}_0} \|\sum_{b\in\mathbb{F}_0} T_{ab}x_b'\|^2 \\ &\leqslant k^2 \sum_{b\in\mathbb{F}_0} \|x_b'\|^2 = k^2 \sum_{b\in\mathbb{F}} \|x_b\|^2 < \varepsilon^2.\end{aligned}$$

Thus $\|\sum_{b\in\mathbb{F}} T_{cb}x_b\| < \varepsilon$ whenever $\mathbb{F} \in \mathscr{F}$ and $\mathbb{F} \cap \mathbb{F}_\varepsilon = \varnothing$, the Cauchy criterion is satisfied, and $\sum_{b\in\mathbb{B}} T_{cb}x_b$ converges to an element $y_c$ of $\mathscr{H}$.

Second, we prove that

$$\sum_{a\in\mathbb{B}} \|y_a\|^2 \leqslant k^2 \sum_{b\in\mathbb{B}} \|x_b\|^2.$$

For this, suppose that $\mathbb{F}_1, \mathbb{F}_2 \in \mathscr{F}$, let $\mathbb{F} = \mathbb{F}_1 \cup \mathbb{F}_2$, and define $x'_b = x_b$ $(b \in \mathbb{F}_2)$, $x'_b = 0$ $(b \in \mathbb{F} \backslash \mathbb{F}_2)$. From (22),

$$\begin{aligned}\sum_{a\in\mathbb{F}_1} \|\sum_{b\in\mathbb{F}_2} T_{ab}x_b\|^2 &= \sum_{a\in\mathbb{F}_1} \|\sum_{b\in\mathbb{F}} T_{ab}x'_b\|^2 \\ &\leqslant \sum_{a\in\mathbb{F}} \|\sum_{b\in\mathbb{F}} T_{ab}x'_b\|^2 \\ &\leqslant k^2 \sum_{b\in\mathbb{F}} \|x'_b\|^2 \leqslant k^2 \sum_{b\in\mathbb{B}} \|x_b\|^2.\end{aligned}$$

When $\mathbb{F}_2$ increases to $\mathbb{B}$, $\|\sum_{b\in\mathbb{F}_2} T_{ab}x_b\| \to \|y_a\|$; and since $\mathbb{F}_1$ is finite, it results from the preceding inequalities that

$$\sum_{a\in\mathbb{F}_1} \|y_a\|^2 \leqslant k^2 \sum_{b\in\mathbb{B}} \|x_b\|^2.$$

Since the preceding relation has been proved for each $\mathbb{F}_1$ in $\mathscr{F}$,

$$\sum_{a\in\mathbb{B}} \|y_a\|^2 \leqslant k^2 \sum_{b\in\mathbb{B}} \|x_b\|^2.$$

From the two assertions just proved, there is a bounded linear operator $T$, acting on $\sum_{b\in\mathbb{B}} \oplus \mathscr{H}_b$, with

$$\|T\| \leqslant k = \sup\{\|T(\mathbb{F})\| : \mathbb{F} \in \mathscr{F}\}, \tag{23}$$

defined by

$$T(\sum \oplus x_b) = \sum \oplus y_b \qquad \text{where} \quad y_a = \sum_{b\in\mathbb{B}} T_{ab}x_b \quad (a \in \mathbb{B}).$$

Since this is a restatement of (21), $T$ has matrix $[T_{ab}]$.

Conversely, suppose that $[T_{ab}]$ is the matrix of a bounded linear operator $T$. With $\mathbb{F}$ in $\mathscr{F}$ and $u$ an element $\sum_{b\in\mathbb{F}} \oplus x_b$ of $\mathscr{H}(\mathbb{F})$, let $v$ be the vector $\sum_{b\in\mathbb{B}} \oplus x_b$ in $\sum_{b\in\mathbb{B}} \oplus \mathscr{H}_b$, obtained by defining $x_b = 0$ when $b \in \mathbb{B} \backslash \mathbb{F}$. Then

$$\begin{aligned}\|T(\mathbb{F})u\|^2 &= \sum_{a\in\mathbb{F}} \|\sum_{b\in\mathbb{F}} T_{ab}x_b\|^2 \\ &= \sum_{a\in\mathbb{F}} \|\sum_{b\in\mathbb{B}} T_{ab}x_b\|^2 \\ &\leqslant \sum_{a\in\mathbb{B}} \|\sum_{b\in\mathbb{B}} T_{ab}x_b\|^2 \\ &= \|Tv\|^2 \leqslant \|T\|^2\|v\|^2 = \|T\|^2\|u\|^2.\end{aligned}$$

Thus $\|T(\mathbb{F})\| \leqslant \|T\|$, and the set $\{\|T(\mathbb{F})\| : \mathbb{F} \in \mathscr{F}\}$ is bounded above, with supremum at most $\|T\|$. This, with (23), gives

$$\|T\| = \sup\{\|T(\mathbb{F})\| : \mathbb{F} \in \mathscr{F}\}. \quad \blacksquare$$

In the circumstances described in Proposition 2.6.13, we regard the operator $T(\mathbb{F})$ as the "finite diagonal block" of the matrix $[T_{ab}]$ corresponding to the finite subset $\mathbb{F}$ of $\mathbb{B}$. The main result of the proposition is the assertion that $[T_{ab}]$ is the matrix of a bounded linear operator if and only if the set of norms of all the finite diagonal blocks is bounded above.

## 2.7. Unbounded linear operators

In Section 1.5, we discussed linear transformations from one normed space into another. We noted in Theorem 1.5.5 that the continuity of such a transformation is equivalent to its boundedness (on the unit ball), so that we speak, interchangeably, of "continuous" and "bounded" linear transformations. In Section 2.4, we specialized the discussion of bounded linear transformations to Hilbert space. In this section, we take up the study of *discontinuous* (and, necessarily, *unbounded*) linear transformations between Hilbert spaces.

We have only to think of the process of differentiation to be convinced that unbounded linear operators arise in the most natural way and that they are important. Without proceeding carefully, let $\mathscr{D}$ be the linear manifold of all $f$ in $L_2(\mathbb{R})$ (relative to Lebesgue measure) almost everywhere differentiable with derivative $f'$ in $L_2(\mathbb{R})$; and let $D(f)$ be $f'$. Then $D$ is a linear transformation and $D$ is not bounded. (If $f_k(t) = \exp(-k|t|)$, with $k$ a positive integer, then $\|Df_k\|/\|f_k\| = k$.) Although $D$ is defined on a dense submanifold of $L_2(\mathbb{R})$ (as follows from classical approximation results), it is certainly not defined on all of $L_2(\mathbb{R})$. We must expect, then, in dealing with unbounded linear operators, to specify a domain of definition $\mathscr{D}(T)$ for our operator $T$ (and, thereafter, to exercise care not to apply $T$, in a formal way, to each element that suits our convenience).

Not only is the subject of unbounded linear operators natural and important, but the literature devoted to it is vast (almost as a consequence). Not to divert ourselves from the purpose at hand, we restrict the examples and results described in this section to a bare minimum. Unbounded operators will appear again in Section 5.6, when we extend to unbounded self-adjoint operators the spectral theory developed there for bounded self-adjoint operators. A "polar decomposition" for (closed) unbounded operators appears in Section 6.1, and a formulation of Theorem 7.2.1 in terms of unbounded operators appears in Section 7.2; but the essential use of unbounded operator theory occurs, for us, in the presentation in Section 9.2 of modular theory.

For the most part, the naturally arising unbounded operators retain some vestiges of orderly "limit properties" – notably, the possibility of extending them to operators with closed graphs. While this assumption may seem like a negligible replacement for continuity, it can be turned to remarkable advantage, as we shall see. In Section 1.8 we associated a graph $\mathscr{G}(T)$ with a linear transformation $T$, where $\mathscr{G}(T) = \{(x, Tx): x \in \mathscr{D}(T)\}$. The closed graph theorem (1.8.6) tells us that if $T$ is defined on all of $\mathscr{H}$ (mapping into the Hilbert space $\mathscr{K}$) and $\mathscr{G}(T)$ is closed, then $T$ is bounded. (Conversely, if $T$ is bounded and everywhere defined, $\mathscr{G}(T)$ is closed.) This provides us with the possibility of an assumption intermediate between continuity and the totally unrestricted linear operator.

Let $T$ be a linear mapping, with domain $\mathscr{D}(T)$ a linear submanifold (not necessarily closed), of the Hilbert space $\mathscr{H}$ into the Hilbert space $\mathscr{K}$. We say that $T$ is *closed* when $\mathscr{G}(T)$ is closed. The unbounded operators $T$ we consider will usually be *densely defined*, that is, $\mathscr{D}(T)$ is dense in $\mathscr{H}$. Whatever $T$ we consider, it has a graph $\mathscr{G}(T)$, and the closure $\mathscr{G}(T)^-$ of $\mathscr{G}(T)$ will be a linear subspace of $\mathscr{H} \oplus \mathscr{K}$. It may be the case that $\mathscr{G}(T)^-$ is the graph of a linear transformation $\bar{T}$, but it need not be. If it is, $\bar{T}$ "*extends*" $T$ and is closed. We say that $T_0$ *extends* (or *is an extension of*) $T$, and write $T \subseteq T_0$, when $\mathscr{D}(T) \subseteq \mathscr{D}(T_0)$ and $T_0x = Tx$ for each $x$ in $\mathscr{D}(T)$. If $\mathscr{G}(T)^-$ is the graph of a linear transformation $\bar{T}$, clearly $\bar{T}$ is the "smallest" ("minimal") closed extension of $T$. In this case, we say that $T$ is *preclosed* (the term *closable* is also used) and refer to $\bar{T}$ as the *closure* of $T$. If $\mathscr{G}(T)^-$ contains elements $(x, y)$ and $(x, y')$ such that $y \neq y'$ (equivalently, since $\mathscr{G}(T)^-$ is a linear space, if $(0, z) \in \mathscr{G}(T)^-$ with $z$ not 0), then $\mathscr{G}(T)^-$ is not the graph of a (single-valued) mapping and $T$ is not preclosed. This is, of course, the only way in which $T$ can fail to have a closure (for, otherwise, the mapping that sends the first to the second coordinate of $\mathscr{G}(T)^-$ defines $\bar{T}$). Interpreting $\mathscr{G}(T)^-$ as the closure of $\mathscr{G}(T)$ in limit terms, we see that $T$ is preclosed if and only if convergence of the sequence $\{x_n\}$ in $\mathscr{D}(T)$ to 0 and $\{Tx_n\}$ to $z$ implies that $z = 0$.

From the point of view of calculations with an unbounded operator $T$, it is often much easier to study its restriction $T|\mathscr{D}_0$ to a dense linear manifold $\mathscr{D}_0$ in its domain $\mathscr{D}(T)$ than to study $T$ itself. If $T$ is closed and $\mathscr{G}(T|\mathscr{D}_0)^- = \mathscr{G}(T)$, the information obtained in this way is much more applicable to $T$. In this case, we say that $\mathscr{D}_0$ is a *core* for $T$. Each dense linear manifold in $\mathscr{G}(T)$ corresponds to a core for $T$.

2.7.1. Example. With the notation of Example 2.4.10, remove the restriction that $g$ be bounded and let $T$ be defined as in that example for those $x$ in $\mathscr{H}$ such that $\sum_{y \in Y} |g(y)\langle x, y\rangle|^2$ is finite (so that $\mathscr{D}(T)$ consists of such vectors $x$). Of course $\mathscr{D}(T)$ contains the submanifold $\mathscr{D}_0$ of all finite linear combinations of the basis elements in $Y$, from which $\mathscr{D}(T)$ is dense in $\mathscr{H}$. At the same time, with $u$ and $v$ in $\mathscr{D}(T)$, from the Cauchy–Schwarz inequality,

$$\begin{aligned}\sum_{y\in Y} |g(y)\langle au+v,y\rangle|^2 &\leqslant |a|^2 \sum_{y\in Y} |g(y)\langle u,y\rangle|^2 + \sum_{y\in Y} |g(y)\langle v,y\rangle|^2 \\ &\quad + 2|a| \sum_{y\in Y} |g(y)\langle u,y\rangle|\,|g(y)\langle v,y\rangle| \\ &\leqslant |a|^2\|Tu\|^2 + \|Tv\|^2 + 2|a|\,\|Tu\|\,\|Tv\|;\end{aligned}$$

so that $\mathscr{D}(T)$ is a linear manifold. Since $Ty = g(y)y$ for $y$ in $Y$, $T$ is bounded if and only if $g$ is a bounded function. In any event, $T$ is a closed operator. To see this, suppose $\{u_n\}$ is a sequence in $\mathscr{D}(T)$ tending to $u$ in $\mathscr{H}$ and $\{Tu_n\}$ converges to $v$. Then

$$\langle Tu_n, y\rangle = g(y)\langle u_n, y\rangle \to g(y)\langle u, y\rangle.$$

But $\langle Tu_n, y\rangle \to \langle v, y\rangle$, so that $\langle v, y\rangle = g(y)\langle u, y\rangle$; and

$$\sum_{y\in Y} |g(y)\langle u,y\rangle|^2 = \sum_{y\in Y} |\langle v,y\rangle|^2 = \|v\|^2 < \infty.$$

Thus $u \in \mathscr{D}(T)$ and

$$Tu = \sum_{y\in Y} g(y)\langle u,y\rangle y = \sum_{y\in Y} \langle v,y\rangle y = v,$$

so that $\mathscr{G}(T)$ is closed. The submanifold $\mathscr{D}_0$ is easily seen to be a core for $T$. ■

2.7.2. EXAMPLE. With the notation of Example 2.4.11, once again remove the restriction that $g$ be bounded (requiring only that $g$ be measurable and finite almost everywhere) and let $M_g$ be defined as in that example for those $x$ in $\mathscr{H}$ such that $\int_S |(M_g x)(s)|^2\, dm(s)$ is finite (so that $\mathscr{D}(T)$ consists of such $x$). The present example extends the preceding example to the case of non-discrete ($\sigma$-finite) measure spaces (so that the important case in which $Y$ is denumerable is included). Again $\mathscr{D}(T)$ is dense since it contains the submanifold $\mathscr{D}_0$ of measurable functions on $S$ with support in a set of finite measure on which $g$ is essentially bounded. The Cauchy–Schwarz inequality assures us once again that $\mathscr{D}(T)$ is a linear manifold. A more general measure-theoretic argument of the character of that appearing in the preceding example establishes that $M_g$ is closed and $\mathscr{D}_0$ is a core for it. (See the comments following Theorem 5.2.4.) ■

2.7.3. EXAMPLE. If $\mathscr{H}$ is a separable Hilbert space with orthonormal basis $\{y_n\}_{n=1,2,\ldots}$ and $Ty_n = ny_1$, then $T$ extends linearly to the (dense) linear manifold $\mathscr{D}_0$ of finite linear combinations of basis vectors $y_n$. If we denote this extension by $T$ again (so that $\mathscr{D}(T) = \mathscr{D}_0$), then $T$ is densely defined, unbounded, and not preclosed. To see this, it suffices to note that $n^{-1}y_n \to 0$ while $T(n^{-1}y_n) = y_1 \to y_1$. ■

We define the operations of addition and multiplication for unbounded operators so that the domains of the resulting operators consist precisely of those vectors on which the indicated operations can be performed. Thus $\mathscr{D}(A + B) = \mathscr{D}(A) \cap \mathscr{D}(B)$ and $(A + B)x = Ax + Bx$ for $x$ in $\mathscr{D}(A + B)$. Assuming that $\mathscr{D}(B) \subseteq \mathscr{H}$ and $\mathscr{D}(A) \subseteq \mathscr{K}$, where $B$ has its range in $\mathscr{K}$, $AB$ is defined as the linear transformation, with $\{x : x \in \mathscr{D}(B) \text{ and } Bx \in \mathscr{D}(A)\}$ as its domain, assigning $A(Bx)$ to $x$. Of course $\mathscr{D}(aA) = \mathscr{D}(A)$ and $(aA)x = a(Ax)$. More care is needed in defining the adjoint of an unbounded operator.

2.7.4. Definition. If $T$ is a linear transformation with $\mathscr{D}(T)$ dense in the Hilbert space $\mathscr{H}$ and range contained in the Hilbert space $\mathscr{K}$, we define a mapping $T^*$, the adjoint of $T$, as follows. Its domain consists of those vectors $y$ in $\mathscr{K}$ such that, for some vector $z$ in $\mathscr{H}$, $\langle x, z\rangle = \langle Tx, y\rangle$ for all $x$ in $\mathscr{D}(T)$. For such $y$, $T^*y$ is $z$. If $T = T^*$, we say that $T$ is self-adjoint. ■

In connection with this definition, we must note that there is at most one $z$ (for a given $y$) since $x$ can assume values in the dense set $\mathscr{D}(T)$; so $T^*$ is well defined. Note, too, that the existence of $z$ is equivalent to the boundedness of the linear functional $x \to \langle Tx, y\rangle$ on $\mathscr{D}(T)$ (for, given that it is bounded, it has a unique bounded extension from $\mathscr{D}(T)$ to $\mathscr{H}$, and Riesz's representation theorem (2.3.1) provides us with $z$). The formal relation $\langle Tx, y\rangle = \langle x, T^*y\rangle$, familiar from the case of bounded operators, remains valid in the present context only when $x \in \mathscr{D}(T)$ and $y \in \mathscr{D}(T^*)$.

2.7.5. Remark. If $T_0$ is densely defined and $T$ is an extension of $T_0$, then $T_0^*$ is an extension of $T^*$. To see this, suppose that $y \in \mathscr{D}(T^*)$ and $u \in \mathscr{D}(T_0)$. Then

$$\langle T_0 u, y\rangle = \langle Tu, y\rangle = \langle u, T^*y\rangle,$$

so that $y \in \mathscr{D}(T_0^*)$ and $T_0^*y = T^*y$. ■

2.7.6. Remark. If $T$ is densely defined, $T^*$ is a closed linear operator; for, with $u$ and $v$ in $\mathscr{D}(T^*)$ and $x$ in $\mathscr{D}(T)$,

$$\begin{aligned}\langle x, aT^*u + T^*v\rangle &= \langle \bar{a}x, T^*u\rangle + \langle x, T^*v\rangle \\ &= \langle Tx, au\rangle + \langle Tx, v\rangle = \langle Tx, au + v\rangle,\end{aligned}$$

so that $au + v \in \mathscr{D}(T^*)$ and $T^*(au + v) = aT^*u + T^*v$. Thus $\mathscr{D}(T^*)$ is a linear manifold and $T^*$ is a linear operator. If $\{v_n\}$ is a sequence in $\mathscr{D}(T^*)$ converging to $v$ such that $\{T^*v_n\}$ converges to $v'$, then, with $u$ in $\mathscr{D}(T)$, $\langle u, T^*v_n\rangle = \langle Tu, v_n\rangle$; and $\{\langle u, T^*v_n\rangle\}$ converges to $\langle Tu, v\rangle$. Thus $\langle Tu, v\rangle = \langle u, v'\rangle$ for each $u$ in $\mathscr{D}(T)$; so $v \in \mathscr{D}(T^*)$, and $T^*v = v'$. It follows that $T^*$ is closed. ■

2.7.7. REMARK. There are several ways in which we can use the hypothesis that $T$ (that is, $\mathscr{G}(T)$) is closed. The mapping $P$ taking $(u, v)$ to $u$ is a bounded linear transformation of the Hilbert space $\mathscr{G}(T)$ into $\mathscr{H}$. Thus $P$ has a bounded adjoint $P^*$ mapping $\mathscr{H}$ into $\mathscr{G}(T)$. Since $(0, z) \in \mathscr{G}(T)$ only when $z = 0$, $P$ has null space $(0)$. From Proposition 2.5.13, the range of $P^*$ is dense in $\mathscr{G}(T)$. Thus, with $v$ in $\mathscr{H}$, if $P^*(v) = (w, w')$ (in $\mathscr{G}(T)$), then $Tw = w'$ and, for each $u$ in $\mathscr{D}(T)$,

$$\langle v, u \rangle = \langle P^*(v), (u, Tu) \rangle = \langle w, u \rangle + \langle w', Tu \rangle.$$

Hence $\langle Tu, w' \rangle = \langle u, v - w \rangle$ and $w' \in \mathscr{D}(T^*)$. Moreover,

$$T^*w' \ (= T^*Tw) = v - w,$$

so that $(T^*T + I)w = v$. While it is not clear, *a priori*, that $\mathscr{D}(T^*T)$ consists of more than the vector 0, our brief computation, relying on the information that $\mathscr{G}(T)$ is closed, allows us to conclude that $\mathscr{D}(T^*T)$ contains (and, therefore, is) a *core* for $T$ (namely, the first coordinates of the range of $P^*$). We learn, at the same time, that $T^*T + I$ has range $\mathscr{H}$ (for $v$ was an arbitrary element of $\mathscr{H}$). ■

Making use of the preceding remarks, there is no difficulty in proving the main theorem of this section.

2.7.8. THEOREM. *If $T$ is a densely defined transformation from the Hilbert space $\mathscr{H}$ to the Hilbert space $\mathscr{K}$, then*

(i) *if $T$ is preclosed, $(\bar{T})^* = T^*$;*
(ii) *$T$ is preclosed if and only if $\mathscr{D}(T^*)$ is dense in $\mathscr{K}$;*
(iii) *if $T$ is preclosed, $T^{**} = \bar{T}$;*
(iv) *if $T$ is closed, $T^*T + I$ is one-to-one with range $\mathscr{H}$ and positive inverse of bound not exceeding* 1;
(v) *$T^*T$ is self-adjoint when $T$ is closed.*

*Proof.* (i) Since $T \subseteq \bar{T}$; from Remark 2.7.5, $(\bar{T})^* \subseteq T^*$. Suppose $y \in \mathscr{D}(T^*)$. For each $x$ in $\mathscr{D}(\bar{T})$, there is a sequence $\{x_n\}$ of vectors in $\mathscr{D}(T)$ converging to $x$ such that $\{Tx_n\}$ converges to $\bar{T}x$. Thus

$$\langle \bar{T}x, y \rangle = \lim \langle Tx_n, y \rangle = \lim \langle x_n, T^*y \rangle = \langle x, T^*y \rangle,$$

so that $y \in \mathscr{D}((\bar{T})^*)$ and $(\bar{T})^*y = T^*y$. Hence $(\bar{T})^* = T^*$.

(ii) If $T$ is preclosed, from Remark 2.7.7, $\mathscr{G}(\bar{T})$ contains a dense linear manifold (the range of $P^*$) consisting of pairs $(x, \bar{T}x)$ with $\bar{T}x$ in $\mathscr{D}(T^*)$ $(= \mathscr{D}(\bar{T}^*))$. If $y$ is orthogonal to the range of $\bar{T}$, then $0 = \langle \bar{T}x, y \rangle = \langle Tx, y \rangle$ for each $x$ in $\mathscr{D}(T)$; and $y$ is in $\mathscr{D}(T^*)$ ($y$ is annihilated by $T^*$). Thus $\mathscr{D}(T^*)$ contains a dense subset of the range of $\bar{T}$ as well as the orthogonal complement of this range. Since $\mathscr{D}(T^*)$ is a linear manifold, it is dense in $\mathscr{K}$.

Suppose, now, that $\mathscr{D}(T^*)$ is dense in $\mathscr{H}$ and $\{u_n\}$ is a sequence in $\mathscr{D}(T)$ converging to 0 such that $\{Tu_n\}$ converges to $v$. With $y$ in $\mathscr{D}(T^*)$, $\langle Tu_n, y\rangle = \langle u_n, T^*y\rangle$; so $\langle u_n, T^*y\rangle$ converges both to 0 and to $\langle v, y\rangle$. Since $\mathscr{D}(T^*)$ is dense in $\mathscr{H}$, $v = 0$ and $T$ is preclosed.

(iii) If $T$ is preclosed, $\mathscr{D}(T^*)$ is dense, from (ii), and $T^*$ has an adjoint $T^{**}$. If $y \in \mathscr{D}(T^*)$ and $x \in \mathscr{D}(T)$, then $\langle T^*y, x\rangle = \langle y, Tx\rangle$, so that $x \in \mathscr{D}(T^{**})$ and $T^{**}x = Tx$. Thus $T^{**}$ is a closed (from Remark 2.7.6) extension of $T$, and $\bar{T} \subseteq T^{**}$. From Remark 2.7.5, $T^{***} \subseteq \bar{T}^* = T^*$. Since $T^*$ is closed, we have, as well, $T^* \subseteq (T^*)^{**}$. Thus $T^* = T^{***}$.

As noted $\bar{T} \subseteq T^{**}$ (equivalently, $\mathscr{G}(\bar{T}) \subseteq \mathscr{G}(T^{**})$). If $(x, T^{**}x)$, in $\mathscr{G}(T^{**})$, is orthogonal to $\mathscr{G}(\bar{T})$, then $\langle x, u\rangle + \langle T^{**}x, \bar{T}u\rangle = 0$ for each $u$ in $\mathscr{D}(\bar{T})$. This holds, in particular, when $\bar{T}u \in \mathscr{D}(T^*)$ $(= \mathscr{D}(T^{***}))$; and, for such $u$,

$$0 = \langle x, (T^*\bar{T} + I)u\rangle.$$

But, from Remark 2.7.7, $(T^*\bar{T} + I)u$ takes on all values in $\mathscr{H}$. Thus $x = 0$, $\mathscr{G}(T^{**}) = \mathscr{G}(\bar{T})$, and $T^{**} = \bar{T}$.

(iv) We noted in Remark 2.7.7 that the domain of $T^*T$ (and, hence, of $T^*T + I$) is a core for $T$ when $T$ is closed and densely defined. We noted, too, that $T^*T + I$ has $\mathscr{H}$ as its range. If $x \in \mathscr{D}(T^*T + I)$, then

$$\|x\|^2 \leqslant \langle x, x\rangle + \langle Tx, Tx\rangle = \langle (T^*T + I)x, x\rangle \leqslant \|(T^*T + I)x\|\,\|x\|.$$

Thus $T^*T + I$ has (0) as null space, is one-to-one, and has a bounded inverse $H$ of bound not exceeding 1. From this same computation, and since each $z$ in $\mathscr{H}$ has the form $(T^*T + I)x$, it follows that $\langle z, Hz\rangle$ is $\langle (T^*T + I)x, x\rangle$, which is real and non-negative. Thus $H$ is positive.

(v) As noted in (iv), $\mathscr{D}(T^*T)$ is a core for $T$; hence it is dense in $\mathscr{H}$. Since

$$\langle (T^*T + I)x, y\rangle = \langle T^*Tx, y\rangle + \langle x, y\rangle$$

when $x \in \mathscr{D}(T^*T)$, we see that $(T^*T)^*$ and $(T^*T + I)^*$ have the same domain and that $(T^*T)^* + I = (T^*T + I)^*$. With $y$ in $\mathscr{D}(T^*T)$,

$$\langle T^*Tx, y\rangle = \langle x, T^*Ty\rangle,$$

so that $T^*T \subseteq (T^*T)^*$ and $T^*T + I \subseteq (T^*T + I)^*$. It follows that $(T^*T + I)^*$ has $\mathscr{H}$ as its range. If $(T^*T + I)^*y = 0$, then, for each $x$ in $\mathscr{D}(T^*T)$,

$$0 = \langle (T^*T + I)^*y, x\rangle = \langle y, (T^*T + I)x\rangle.$$

Since $T^*T + I$ has range $\mathscr{H}$, $y = 0$. Thus $(T^*T + I)^*$ is one-to-one, extends $T^*T + I$, and has the same range as $T^*T + I$. It follows that

$$T^*T + I = (T^*T + I)^* = (T^*T)^* + I,$$

so that $T^*T = (T^*T)^*$. ■

The statement that $T$ is self-adjoint ($T = T^*$) contains information about the domain of $T$ as well as the formal information that $\langle Tx, y\rangle = \langle x, Ty\rangle$ for all $x$ and $y$ in $\mathscr{D}(T)$. When $\mathscr{D}(T)$ is dense in $\mathscr{H}$ and $\langle Tx, y\rangle = \langle x, Ty\rangle$ for all $x$ and $y$ in $\mathscr{D}(T)$, we say that $T$ is *symmetric*. Equivalently, $T$ is symmetric when $T \subseteq T^*$. Since $T^*$ is closed and $\mathscr{G}(T) \subseteq \mathscr{G}(T^*)$, in this case, $T$ is preclosed if it is symmetric. If $T$ is self-adjoint, $T$ is both symmetric and closed. The operation of differentiation on an appropriate domain provides an example of a closed symmetric operator that is not self-adjoint. In Proposition 2.7.10 we describe conditions that guarantee that a given closed symmetric operator is self-adjoint. If $A \subseteq T$ with $A$ self-adjoint and $T$ symmetric, then $A \subseteq T \subseteq T^*$, so that

$$(T^{**} \subseteq) T^* \subseteq A^* = A \subseteq T \subseteq T^*$$

and $A = T$. It follows that $A$ has no proper symmetric extension. That is, a self-adjoint operator is *maximal symmetric*.

2.7.9. LEMMA. *If $T$ is closed and symmetric, $T \pm iI$ have closed ranges. If $T$ is closed and $0 \leqslant \langle Tz, z\rangle$ for $z$ in $\mathscr{D}(T)$, then $T + I$ has a closed range.*

*Proof.* Suppose $\{x_n\}$ is a sequence in $\mathscr{D}(T)$ such that $\{(T \pm iI)x_n\}$ tends to $y$. Note that, with $z$ in $\mathscr{D}(T)$, $\langle Tz, z\rangle$ is real, so that

$$\|z\|^2 \leqslant (\langle Tz, z\rangle^2 + \langle z, z\rangle^2)^{1/2} = |\langle (T \pm iI)z, z\rangle| \leqslant \|(T \pm iI)z\| \, \|z\|.$$

Thus $\|x_n - x_m\| \leqslant \|(T \pm iI)(x_n - x_m)\|$ and $\{x_n\}$ is convergent. Suppose $x_n \to x$. Since $\{Tx_n\}$ converges to $\mp ix + y$ and $T$ is closed, $x \in \mathscr{D}(T)$ and $Tx = \mp ix + y$. Thus $y = (T \pm iI)x$, and $T \pm iI$ have closed ranges.

Suppose, now, that $T$ is closed and $0 \leqslant \langle Tz, z\rangle$ for each $z$ in $\mathscr{D}(T)$. Then

$$\|z\|^2 \leqslant \langle z, z\rangle + \langle Tz, z\rangle \leqslant \|(T + I)z\| \, \|z\|,$$

for $z$ in $\mathscr{D}(T)$, and, as above, $T + I$ has closed range. ■

2.7.10. PROPOSITION. *If $T$ is a closed symmetric operator on the Hilbert space $\mathscr{H}$, the following assertions are equivalent:*

(i) *$T$ is self-adjoint;*
(ii) *$T^* \pm iI$ have (0) as null space;*
(iii) *$T \pm iI$ have $\mathscr{H}$ as range;*
(iv) *$T \pm iI$ have ranges dense in $\mathscr{H}$.*

*Proof.* (i) → (ii). If $T = T^*$, for each $x$ in $\mathscr{D}(T)$, $\langle Tx, x\rangle = \langle x, Tx\rangle$; and $\langle Tx, x\rangle$ is real. Thus

$$\langle (T^* \pm iI)x, x\rangle = \langle (T \pm iI)x, x\rangle = \langle Tx, x\rangle \pm i\|x\|^2 = 0$$

only if $x = 0$. Hence $T^* \pm iI$ have (0) as null space.

(ii) $\to$ (iii). From Lemma 2.7.9, $T \pm iI$ have closed ranges. Thus, it suffices to show that these ranges are dense in $\mathscr{H}$. If $\langle (T \pm iI)x, y \rangle = 0$ for all $x$ in $\mathscr{D}(T)$, then $\langle Tx, y \rangle = \mp i\langle x, y \rangle$, so that $y \in \mathscr{D}(T^*)$ and $T^*y = \pm iy$. Since $T^* \pm iI$ have (0) as null space, $y = 0$. Hence $T \pm iI$ have dense ranges.

(iii) $\leftrightarrow$ (iv). This follows from the preceding discussion.

(iii) $\to$ (i). Since $T$ is closed and symmetric, $T \subseteq T^*$ and $\mathscr{G}(T)$ is a closed subspace of the closed space $\mathscr{G}(T^*)$. If $(y, T^*y)$ in $\mathscr{G}(T^*)$ is orthogonal to $\mathscr{G}(T)$, then

$$\langle y, x \rangle + \langle T^*y, Tx \rangle = 0$$

for each $x$ in $\mathscr{D}(T)$. Since $T \pm iI$ have range $\mathscr{H}$, there is an $x$ in $\mathscr{D}(T)$ such that ($Tx \in \mathscr{D}(T)$, and) $y = (T + iI)(T - iI)x$ $(=(T^2 + I)x)$. For this $x$,

$$\langle y, y \rangle = \langle y, (T^2 + I)x \rangle = \langle y, x \rangle + \langle T^*y, Tx \rangle = 0.$$

Thus $(y, T^*y) = (0, 0)$, $\mathscr{G}(T) = \mathscr{G}(T^*)$, $T = T^*$, and $T$ is self-adjoint. ■

2.7.11. Remark. If $T$ is self-adjoint, it follows from (iii) of Proposition 2.7.10 and the inequality at the beginning of the proof of Lemma 2.7.9 that $T \pm iI$ have everywhere-defined, bounded inverses with bound not exceeding 1. ■

## 2.8. Exercises

2.8.1. Show that a finite set $\{x_1, \ldots, x_n\}$ of $n$ vectors in a Hilbert space $\mathscr{H}$ is linearly independent if and only if the $n \times n$ matrix that has $\langle x_j, x_k \rangle$ in the $(j, k)$ position is non-singular.

2.8.2. Show that a Hilbert space is uniformly convex (in the sense defined in Exercise 1.9.13).

2.8.3. Show that, if $\mathscr{H}$ is a real Hilbert space, then $\mathscr{H} \times \mathscr{H}$ becomes a (complex) Hilbert space $\mathscr{H}_{\mathbb{C}}$ when its linear structure, inner product, and norm, are defined by

$$(x, y) + (u, v) = (x + u, y + v),$$

$$(a + ib)(x, y) = (ax - by, bx + ay),$$

$$\langle (x, y), (u, v) \rangle = \langle x, u \rangle + \langle y, v \rangle + i\langle y, u \rangle - i\langle x, v \rangle,$$

$$\|(x, y)\|^2 = \|x\|^2 + \|y\|^2,$$

for all $x, y, u, v$ in $\mathscr{H}$ and $a$, $b$ in $\mathbb{R}$.

Prove also that the set $\{(x,0): x \in \mathscr{H}\}$ is a closed real-linear subspace $\mathscr{H}_{\mathbb{R}}$ of $\mathscr{H}_{\mathbb{C}}$, that $\mathscr{H}_{\mathbb{C}} = \{h + ik: h, k \in \mathscr{H}_{\mathbb{R}}\}$, and that the mapping $x \to (x,0)$ is an isometric isomorphism from $\mathscr{H}$ onto (the real Hilbert space) $\mathscr{H}_{\mathbb{R}}$.

2.8.4. Suppose that $\{x_1, x_2, x_3, \ldots\}$ is an orthonormal basis in a Hilbert space $\mathscr{H}$ and that

$$Y = \left\{ y \in \mathscr{H} : \sum_{n=1}^{\infty} \left(1 + \frac{1}{n}\right)^2 |\langle y, x_n \rangle|^2 \leqslant 1 \right\}.$$

Prove that $Y$ is a bounded closed convex set that has no element with greatest norm.

2.8.5. Suppose that $Y$ is a closed convex set in a Hilbert space $\mathscr{H}$, $\mathscr{U}_Y$ is the set of all unitary operators $U$ acting on $\mathscr{H}$ for which $U(Y) = Y$, and

$$Y_0 = \{y \in Y: Uy = y \text{ for each } U \text{ in } \mathscr{U}_Y\}.$$

(i) Prove that $Y_0$ is not empty. [*Hint.* Use Proposition 2.2.1.]

(ii) Show that, if $Y_0$ consists of a single non-zero vector $y_0$, then $Y$ is a subset of the hyperplane

$$\{x \in \mathscr{H}: \mathrm{Re}\langle x - y_0, y_0 \rangle = 0\}.$$

2.8.6. Prove that a bounded sequence of vectors in a Hilbert space has a weakly convergent subsequence.

2.8.7. Suppose that $\mathbb{A}$ is an uncountable set and, for each $a$ in $\mathbb{A}$, $m_a$ is Lebesgue measure on the $\sigma$-algebra $\mathscr{S}_a$ of Borel subsets of the interval $[0,1]$ $(= S_a)$. Show that, if $(S, \mathscr{S}, m)$ is the corresponding infinite-product measure space (see [H: p. 158]), then $L_2(S, \mathscr{S}, m)$ is non-separable.

2.8.8. Suppose that $\mathscr{H}$ is a Hilbert space in which the inner product is denoted by $\langle\ ,\ \rangle$ and that $K \in \mathscr{B}(\mathscr{H})^+$. Show that the equation

$$\langle x, y \rangle_1 = \langle Kx, y \rangle \qquad (x, y \in \mathscr{H})$$

defines an inner product $\langle\ ,\ \rangle_1$ on $\mathscr{H}$. By means of the Cauchy–Schwarz inequality for $\langle\ ,\ \rangle_1$, prove that

$$\|K\| = \min\{a: a \in \mathbb{R},\ K \leqslant aI\}.$$

2.8.9. Let $\mathscr{H}$ be a Hilbert space in which the inner product and norm are denoted by $\langle\ ,\ \rangle$ and $\|\ \|$, respectively. Suppose that $\langle\ ,\ \rangle_1$ is another definite inner product on $\mathscr{H}$ and the corresponding norm $\|\ \|_1$ satisfies $\|x\|_1 \leqslant \|x\|$ for each $x$ in $\mathscr{H}$. Prove that there is a positive self-adjoint operator $K$, acting on $\mathscr{H}$, such that $\|K\| \leqslant 1$, $K$ has null space $\{0\}$, and $\langle x, y\rangle_1 = \langle Kx, y\rangle$ for all $x$, $y$ in $\mathscr{H}$.

2.8.10. Suppose that $\mathscr{H}$ is a Hilbert space in which the inner product and norm are denoted by $\langle\ ,\ \rangle$ and $\|\ \|$, respectively. Let $K$ be a positive element of $\mathscr{B}(\mathscr{H})$, and define an inner product $\langle\ ,\ \rangle_1$ on $\mathscr{H}$ by

$$\langle x, y\rangle_1 = \langle Kx, y\rangle \qquad (x, y \in \mathscr{H}).$$

Let $\|x\|_1 = [\langle x, x\rangle_1]^{1/2}$.

(i) Prove that $\|\ \|_1$ is a norm on $\mathscr{H}$ if and only if $K$ has null space $\{0\}$.

(ii) Show that, if $K$ has null space $\{0\}$, then the norms $\|\ \|$ and $\|\ \|_1$ give rise to the same topology on $\mathscr{H}$ if and only if $K$ has an inverse in $\mathscr{B}(\mathscr{H})$.

(iii) Suppose that $K$ has an inverse in $\mathscr{B}(\mathscr{H})$. If $A^*$ denotes the adjoint of an element $A$ of $\mathscr{B}(\mathscr{H})$ relative to the inner product $\langle\ ,\ \rangle$, find a formula for the adjoint of $A$ relative to the inner product $\langle\ ,\ \rangle_1$.

2.8.11. Suppose that $T$ is a bounded self-adjoint operator acting on a Hilbert space $\mathscr{H}$ and $k$ is a positive real number such that $-kI \leqslant T \leqslant kI$. By using the identity

$$4\,\mathrm{Re}\langle Tx, y\rangle = \langle T(x+y), x+y\rangle - \langle T(x-y), x-y\rangle,$$

show that

$$|\mathrm{Re}\langle Tx, y\rangle| \leqslant \tfrac{1}{2}k\{\|x\|^2 + \|y\|^2\}$$

for all $x$ and $y$ in $\mathscr{H}$. Deduce that $\|T\| \leqslant k$ and that

$$\begin{aligned}\|T\| &= \min\{a : a \in \mathbb{R},\ -aI \leqslant T \leqslant aI\}\\ &= \sup\{|\langle Tx, x\rangle| : x \in \mathscr{H},\ \|x\| = 1\}.\end{aligned}$$

2.8.12. A bounded linear operator $A$, acting on a Hilbert space $\mathscr{H}$, is said to *attain its bound* if $\|Ax\| = \|A\|$ for some unit vector $x$ in $\mathscr{H}$. Give examples of

(a) a bounded self-adjoint operator with an orthonormal basis of eigenvectors,

(b) a bounded self-adjoint operator with no eigenvector,

neither of which attains its bound.

2.8.13. Let $\mathscr{H}$ be a Hilbert space.

(i) Prove that each unit vector $x$ in $\mathscr{H}$ is an extreme point of the unit ball $(\mathscr{H})_1$.

(ii) Prove that each isometric linear operator $V$ from $\mathscr{H}$ into $\mathscr{H}$ is an extreme point of the unit ball $(\mathscr{B}(\mathscr{H}))_1$.

2.8.14. Show that the projection $E$ from a Hilbert space $\mathscr{H}$ onto a closed subspace $\mathscr{K}$ is an extreme point of the set $(\mathscr{B}(\mathscr{H})^+)_1$ of all positive operators in the unit ball of $\mathscr{B}(\mathscr{H})$.

2.8.15. Determine a necessary and sufficient condition for the operator $M_g$, defined in Example 2.4.11, to have a bounded inverse.

2.8.16. Suppose that $T = \sum_{a \in \mathbb{A}} \oplus T_a$, where $T_a \in \mathscr{B}(\mathscr{H}_a)$ for each $a$ in $\mathbb{A}$, and $\sup\{\|T_a\| : a \in \mathbb{A}\} < \infty$. Show that $T$ has a bounded inverse if and only if the following two conditions are satisfied:

(i) each $T_a$ has a bounded inverse,
(ii) $\sup\{\|T_a^{-1}\| : a \in \mathbb{A}\} < \infty$.

2.8.17. Let $\mathscr{P}$ denote the set of all projections from a Hilbert space $\mathscr{H}$ onto its closed subspaces, and suppose that $F \in \mathscr{P}$, $0 \neq F \neq I$. Prove that the mappings

$$E \to E \wedge F, \qquad E \to E \vee F$$

are not continuous, from $\mathscr{P}$ with the norm topology into $\mathscr{P}$ with the strong-operator topology.

2.8.18. Let $\mathscr{S}$ denote the set of all bounded self-adjoint operators acting on a Hilbert space $\mathscr{H}$. If $A, B, C \in \mathscr{S}$, we say that $C$ is a *lower bound* of $\{A, B\}$ if $C \leqslant A$, $C \leqslant B$. We say that $C$ is the *greatest lower bound* of $\{A, B\}$ if it is a lower bound of $\{A, B\}$, and $D \leqslant C$ whenever $D$ is a lower bound of $\{A, B\}$.

(i) Show that, if $A, B \in \mathscr{S}$, then $\{A, B\}$ has a lower bound in $\mathscr{S}$.

(ii) Suppose that $A, B$ are non-zero elements of $\mathscr{B}(\mathscr{H})^+$. Show that there is a vector $x_0$ such that $\langle Ax_0, x_0 \rangle > 0$ and $\langle Bx_0, x_0 \rangle > 0$. Prove that, if $P_0$ is the projection onto the one-dimensional subspace containing $x_0$, $a$ and $b$ are suitable positive real numbers, and $T = aP_0 - b(I - P_0)$, then $T \leqslant A$, $T \leqslant B$, $T \nleqslant 0$. Deduce that 0 is not the greatest lower bound of $\{A, B\}$.

(iii) Suppose that $A, B \in \mathscr{S}$ and $\{A, B\}$ has a greatest lower bound $C$ in $\mathscr{S}$. By applying the result of (ii) to $\{A - C, B - C\}$, show that either $A \leqslant B$ or $B \leqslant A$.

2.8.19. Suppose that $A$ and $B$ are mappings from a Hilbert space $\mathscr{H}$ into itself, and $\langle Ax, y \rangle = \langle x, By \rangle$ for all $x$ and $y$ in $\mathscr{H}$. Prove that $A$ and $B$ are bounded linear operators, and $A = B^*$.

2.8.20. Suppose that $\mathscr{H}$ is a Hilbert space and $A \in \mathscr{B}(\mathscr{H})$. Prove that the following five conditions are equivalent.

(i) $A$ is continuous as a mapping from the unit ball $(\mathscr{H})_1$ (with the weak topology) into $\mathscr{H}$ (with the norm topology).

(ii) If $x, x_1, x_2, \ldots \in \mathscr{H}$ and $\{x_n\}$ is weakly convergent to $x$, then $\{Ax_n\}$ is norm convergent to $Ax$.

(iii) Every bounded sequence $\{x_n\}$ in $\mathscr{H}$ has a subsequence $\{x_{n(k)}\}$ such that $\{Ax_{n(k)}\}$ is norm convergent.

(iv) The set $\{Ax : x \in (\mathscr{H})_1\}$ is relatively compact in the norm topology of $\mathscr{H}$.

(v) The set $\{Ax : x \in (\mathscr{H})_1\}$ is compact in the norm topology of $\mathscr{H}$.

[An element of $\mathscr{B}(\mathscr{H})$ that has any (and, hence, all) of the above properties is described as a *compact* linear operator.]

2.8.21. Prove that the identity operator, acting on an infinite-dimensional Hilbert space, is not compact (in the sense of Exercise 2.8.20).

2.8.22. Suppose that $\mathscr{H}$ is a Hilbert space and

$$\mathscr{I} = \{A \in \mathscr{B}(\mathscr{H}) : A \text{ has finite-dimensional range}\}.$$

(i) Prove that, if $A \in \mathscr{I}$ and $\{y_1, \ldots, y_n\}$ is an orthonormal basis of the range of $A$, there exist vectors $x_1, \ldots, x_n$ in $\mathscr{H}$ such that

$$Ax = \sum_{j=1}^{n} \langle x, x_n \rangle y_n \qquad (x \in \mathscr{H}).$$

(ii) Prove that $\mathscr{I}$ is a two-sided ideal in $\mathscr{B}(\mathscr{H})$ and that every non-zero two-sided ideal in $\mathscr{B}(\mathscr{H})$ contains $\mathscr{I}$.

(iii) Prove that an element $A$ of $\mathscr{B}(\mathscr{H})$ lies in $\mathscr{I}$ if and only if $A$ is continuous as a mapping from $\mathscr{H}$ (with the weak topology) into $\mathscr{H}$ (with the norm topology).

(iv) Prove that the elements of $\mathscr{I}$ are compact linear operators (in the sense of Exercise 2.8.20).

2.8.23. Suppose that $\{A_n\}$ is a sequence of compact linear operators acting on a Hilbert space $\mathscr{H}$, $A \in \mathscr{B}(\mathscr{H})$, and $\|A_n - A\| \to 0$. Prove that $A$ is compact. [*Hint*. Use condition (i) of Exercise 2.8.20 as the defining property of a compact linear operator.]

2.8.24. Suppose that $A$ is a compact linear operator acting on a Hilbert space $\mathscr{H}$ (see Exercise 2.8.20).

(i) Prove that the (closed) range space $[A(\mathscr{H})]$ is separable.

(ii) Suppose that $[A(\mathscr{H})]$ is infinite-dimensional, and let $\{y_1, y_2, y_3, \ldots\}$ be an orthonormal basis of $[A(\mathscr{H})]$. For each positive integer $n$, let $P_n$ be the projection from $\mathscr{H}$ onto the subspace spanned by $y_1, \ldots, y_n$. Prove that $\|A - P_n A\| \to 0$ as $n \to \infty$.

(iii) Deduce that $A$ lies in the norm closure of the ideal $\mathscr{I}$ (see Exercise 2.8.22), and $A^*$ is compact.

2.8.25. Let $\mathscr{K}$ denote the set of all compact linear operators acting on a Hilbert space $\mathscr{H}$. By using the results of the three preceding exercises, show that:

(i) $\mathscr{K}$ is the norm closure of the ideal $\mathscr{I}$ in $\mathscr{B}(\mathscr{H})$;
(ii) $\mathscr{K}$ is a norm closed two-sided ideal in $\mathscr{B}(\mathscr{H})$;
(iii) each non-zero norm closed two-sided ideal in $\mathscr{B}(\mathscr{H})$ contains $\mathscr{K}$.

2.8.26. Suppose that $\{y_1, y_2, y_3, \ldots\}$ is an orthonormal system in a Hilbert space $\mathscr{H}$, and $\{\lambda_1, \lambda_2, \lambda_3, \ldots\}$ is a sequence of real numbers such that $|\lambda_1| \geqslant |\lambda_2| \geqslant |\lambda_3| \geqslant \cdots$. Suppose also that the sequences $\{y_n\}$ and $\{\lambda_n\}$ are *either* both finite and of the same length *or* both infinite, with $\{\lambda_n\}$ converging to 0. Show that the equation

$$Ax = \sum_n \lambda_n \langle x, y_n \rangle y_n \qquad (x \in \mathscr{H})$$

defines a compact self-adjoint operator $A$ on $\mathscr{H}$, with $\|A\| = |\lambda_1|$. [The result of Exercise 2.8.29 below shows that every compact self-adjoint operator has the form just described.]

2.8.27. Suppose that $A$ is the compact self-adjoint operator constructed in Exercise 2.8.26 from an orthonormal system $\{y_1, y_2, y_3, \ldots\}$ in a Hilbert space $\mathscr{H}$ and a real sequence $\{\lambda_1, \lambda_2, \lambda_3, \ldots\}$ that satisfy the conditions set out in that exercise. Extend the orthonormal system to an orthonormal *basis* $\{y_1, y_2, y_3, \ldots\} \cup \{z_a\}$.

(i) Prove that, if $\lambda$ is a non-zero scalar that does not appear in the sequence $\{\lambda_n\}$, the operator $A - \lambda I$ has an inverse in $\mathscr{B}(\mathscr{H})$, and

$$\begin{aligned}(A - \lambda I)^{-1} x &= \sum_n \frac{1}{\lambda_n - \lambda} \langle x, y_n \rangle y_n - \lambda^{-1} \sum_a \langle x, z_a \rangle z_a \\ &= \sum_n \frac{\lambda_n}{\lambda(\lambda_n - \lambda)} \langle x, y_n \rangle y_n - \lambda^{-1} x \qquad (x \in \mathscr{H}).\end{aligned}$$

[*Hint*. Consider the matrix of $A$.]

(ii) Show that, if $\lambda$ is a non-zero scalar that appears in the sequence $\{\lambda_n\}$ and $x \in \mathscr{H}$, the equation

$$(A - \lambda I)z = x$$

has a solution $z$ in $\mathscr{H}$ if and only if $\langle x, y_k \rangle = 0$ for each integer $k$ satisfying $\lambda_k = \lambda$. What is the most general solution $z$, when this condition is satisfied?

2.8.28. Let $A$ be a bounded self-adjoint operator acting on a Hilbert space $\mathscr{H}$.

(i) Show that each eigenvalue of $A$ is real.
(ii) Show that eigenvectors corresponding to distinct eigenvalues of $A$ are orthogonal.

2.8.29. Let $A$ be a compact self-adjoint operator acting on a Hilbert space $\mathscr{H}$.

(i) By using the result of Exercise 2.8.11, show that there exist unit vectors $x_1, x_2, x_3, \ldots$ in $\mathscr{H}$ such that the real sequence $\{\langle Ax_n, x_n \rangle\}$ converges, with limit $\rho$ equal to $\|A\|$ or $-\|A\|$. Show that $\|Ax_n - \rho x_n\| \to 0$ as $n \to \infty$.
(ii) Prove that $Ax = \rho x$ for some unit vector $x$ in $\mathscr{H}$ (so that a non-zero compact self-adjoint operator has a non-zero eigenvalue).
(iii) Show that, if $\lambda$ is a non-zero eigenvalue of $A$, then the null space of $A - \lambda I$ has finite dimension. (We call this finite dimension the *multiplicity* of $\lambda$ as an eigenvalue of $A$.)
(iv) Prove that, if $\varepsilon$ is a positive real number, there are only a finite number of different eigenvalues $\mu$ of $A$ such that $|\mu| > \varepsilon$. Deduce that the distinct non-zero eigenvalues of $A$ *either* form a finite set *or* form a sequence converging to 0.
(v) Let $\{\mu_1, \mu_2, \mu_3, \ldots\}$ be the (finite or infinite) sequence of all distinct non-zero eigenvalues of $A$, arranged so that $|\mu_1| \geqslant |\mu_2| \geqslant |\mu_3| \geqslant \cdots$, and suppose that $\mu_n$ has multiplicity $m(n)$. Let $\{\lambda_1, \lambda_2, \lambda_3, \ldots\}$ be the real sequence consisting of $\mu_1$ ($m(1)$ times), followed by $\mu_2$ ($m(2)$ times), followed by $\mu_3$ ($m(3)$ times), and so on. Let $\{y_1, y_2, y_3, \ldots\}$ be a sequence of unit vectors consisting of an orthonormal basis of the null space of $A - \mu_1 I$, followed by an orthonormal basis of the null space of $A - \mu_2 I$, followed by an orthonormal basis of the null space of $A - \mu_3 I$, and so on. Show that $\{y_n\}$ is an orthonormal system, and $Ay_n = \lambda_n y_n$ for each $n$. Prove that, if $A_0$ is the compact self-adjoint operator defined by

$$A_0 x = \sum_n \lambda_n \langle x, y_n \rangle y_n \qquad (x \in \mathscr{H})$$

(see Exercise 2.8.26), then $A - A_0$ has no non-zero eigenvalue. Deduce that $A = A_0$.
(vi) Show that $A \geqslant 0$ if and only if $\lambda_n \geqslant 0$ for all $n$. Deduce that, in this case, $A$ has a (compact) "positive square root" $A_1$ (that is, $A_1^2 = A$ and $A_1 \geqslant 0$) such that $\|A_1\|^2 = \|A\|$.

2.8.30. Let $A$ be a self-adjoint operator acting on a Hilbert space $\mathscr{H}$, and let $x$ be a unit vector in $\mathscr{H}$ such that $\|Ax\| = \|A\|$.

(i) Show that $x$ is an eigenvector for $A^2$ corresponding to the eigenvalue $\|A\|^2$.

(ii) Show that either $Ax = \|A\|x$ or $Az = -\|A\|z$ for some unit vector $z$. [*Hint*. Consider the vector $\|A\|x - Ax$.]

(iii) Under the added assumption that $A \geqslant 0$, show that $x$ is an eigenvector for $A$ corresponding to the eigenvalue $\|A\|$.

2.8.31. Let $E$ and $F$ be projections acting on a Hilbert space $\mathscr{H}$.

(i) Show that, if $E$ and $F$ commute,

$$(*) \qquad E \vee F \leqslant E + F.$$

(ii) By means of a two-dimensional example, show that (*) need not hold when $E$ and $F$ do not commute.

2.8.32. Suppose that $\mathscr{H}$ is an infinite-dimensional Hilbert space. Let $\{y_1, y_2, y_3, \ldots\}$ be an orthonormal sequence in $\mathscr{H}$. By considering the sequence $\{V_n\}$ in $\mathscr{B}(\mathscr{H})$, where

$$V_n x = \langle x, y_n \rangle y_1 \qquad (x \in \mathscr{H}),$$

prove that the adjoint operation is not strong-operator continuous on (the unit ball of) $\mathscr{B}(\mathscr{H})$. Deduce that the strong-operator topology on $\mathscr{B}(\mathscr{H})$ is strictly coarser than the norm topology.

2.8.33. Let $\mathscr{H}$ be an infinite-dimensional Hilbert space. Given any finite-dimensional subspace $F$ ($\neq \{0\}$) of $\mathscr{H}$, let $A_F = n(I - P_F)$, where $n$ is the dimension of $F$ and $P_F$ is the projection from $\mathscr{H}$ onto $F$. Choose any orthonormal system of $2n$ vectors, $\{x_1, \ldots, x_n, y_1, \ldots, y_n\}$, the first $n$ of which form an orthonormal basis of $F$, and define $V_F$, in $\mathscr{B}(\mathscr{H})$, by

$$V_F x = \frac{1}{n} \sum_{j=1}^{n} \langle x, x_j \rangle y_j.$$

In this way, we obtain nets $\{A_F\}$, $\{V_F\}$, $\{A_F V_F\}$, where the finite-dimensional subspaces $F$ are directed by the inclusion relation. Show that:

(i) $\{A_F\}$ is strong-operator convergent to 0;

(ii) $\{V_F\}$ is bounded, norm convergent to 0, and hence strong-operator convergent to 0;

(iii) $\{A_F V_F\}$ is not strong-operator convergent to 0.

Conclude that multiplication is not (jointly) strong-operator continuous from $\mathscr{B}(\mathscr{H}) \times (\mathscr{B}(\mathscr{H}))_1$ into $\mathscr{B}(\mathscr{H})$.

2.8.34. Show that, if $\mathscr{H}$ is a separable Hilbert space, there is a countable strong-operator dense subset of $\mathscr{B}(\mathscr{H})$.

2.8.35. Show that, if $\mathscr{H}$ is a separable Hilbert space and $\{y_1, y_2, y_3, \ldots\}$ is an orthonormal basis of $\mathscr{H}$, the equation

$$d(S, T) = \sum_{n=1}^{\infty} 2^{-n}\|Sy_n - Ty_n\|$$

defines a translation-invariant metric $d$ on $\mathscr{B}(\mathscr{H})$ (that is,

$$d(S + R, T + R) = d(S, T)$$

for each $R$ in $\mathscr{B}(\mathscr{H})$), and the associated metric topology coincides on bounded subsets of $\mathscr{B}(\mathscr{H})$ with the strong-operator topology.

2.8.36. Suppose that $\mathscr{H}_1, \ldots, \mathscr{H}_n, \mathscr{K}$ are Hilbert spaces and $L: \mathscr{H}_1 \times \cdots \times \mathscr{H}_n \to \mathscr{K}$ is a bounded multilinear mapping. Suppose also that for each $u$ in $\mathscr{K}$, the bounded multilinear functional $L_u$, defined by

$$L_u(x_1, \ldots, x_n) = \langle L(x_1, \ldots, x_n), u \rangle,$$

is a Hilbert–Schmidt functional on $\mathscr{H}_1 \times \cdots \times \mathscr{H}_n$. Prove that the mapping $u \to L_u$ from the conjugate Hilbert space $\bar{\mathscr{K}}$ into the Hilbert space $\mathscr{HSF}$ of Proposition 2.6.2 is linear and has closed graph. Deduce that there is a positive real number $d$ such that

$$\|L_u\|_2 \leqslant d\|u\| \qquad (u \in \mathscr{K}),$$

where $\| \ \|_2$ denotes the usual norm on $\mathscr{HSF}$.

2.8.37. Suppose that $\mathscr{H}$ is a Hilbert space, and let $\mathscr{HSO}$ denote the Hilbert space of all Hilbert–Schmidt operators from $\mathscr{H}$ into $\mathscr{H}$ (see the discussion preceding Proposition 2.6.9).

(i) Prove that $\|T\| \leqslant \|T\|_2$ for all $T$ in $\mathscr{HSO}$, where $\| \ \|$ and $\| \ \|_2$ denote the usual norms on $\mathscr{B}(\mathscr{H})$ and $\mathscr{HSO}$, respectively.

(ii) By identifying $\mathscr{HSO}$ with $\bar{\mathscr{H}} \otimes \mathscr{H}$, prove that the ideal

$$\{A \in \mathscr{B}(\mathscr{H}) : A \text{ has finite-dimensional range}\}$$

in $\mathscr{B}(\mathscr{H})$ is a $\| \ \|_2$-dense subset of $\mathscr{HSO}$.

(iii) Prove that the elements of $\mathscr{HSO}$ are compact linear operators.

2.8.38. Suppose that $\mathscr{H}$ is the $L_2$ space associated with a $\sigma$-finite measure space $(S, \mathscr{S}, m)$. When $y, z \in \mathscr{H}$, define $q_{y,z}$ in $L_2(S \times S, \mathscr{S} \times \mathscr{S}, m \times m)$ by $q_{y,z}(s, t) = z(s)\overline{y(t)}$.

(i) Show that, if $Y$ and $Z$ are orthonormal bases of $\mathscr{H}$, then the set $\{q_{y,z}: y \in Y, z \in Z\}$ is an orthonormal basis of $L_2(S \times S, \mathscr{S} \times \mathscr{S}, m \times m)$.

(ii) Show that, if $k \in L_2(S \times S, \mathscr{S} \times \mathscr{S}, m \times m)$, the equation

$$(T_k x)(s) = \int_S k(s,t)x(t)\,dm(t) \qquad (x \in \mathscr{H})$$

(where the integral exists for almost all $s$ in $S$) defines an element $T_k$ of $\mathscr{B}(\mathscr{H})$. Prove also that $T_k$ is a Hilbert–Schmidt operator acting on $\mathscr{H}$, and that $\|T_k\|_2 = \|k\|$.

(iii) Prove that every Hilbert–Schmidt operator on $\mathscr{H}$ arises, as in (ii), from an element of $L_2(S \times S, \mathscr{S} \times \mathscr{S}, m \times m)$.

2.8.39. Suppose that $(S, \mathscr{S}, m)$ is a $\sigma$-finite measure space, $k \in L_2(S \times S, \mathscr{S} \times \mathscr{S}, m \times m)$, and $T_k$ is the Hilbert–Schmidt operator defined in Exercise 2.8.38(ii). Show that $T_k^* = T_h$, where $h(s,t) = \overline{k(t,s)}$.

2.8.40. Suppose that $\mathscr{H}$ and $\mathscr{K}$ are Hilbert spaces, $A \in \mathscr{B}(\mathscr{H})$, and $B \in \mathscr{B}(\mathscr{K})$. Prove that $A \otimes I$ has an inverse in $\mathscr{B}(\mathscr{H} \otimes \mathscr{K})$ if and only if $A$ has an inverse in $\mathscr{B}(\mathscr{H})$, and that $A \otimes B$ has an inverse in $\mathscr{B}(\mathscr{H} \otimes \mathscr{K})$ if and only if $A$ has an inverse in $\mathscr{B}(\mathscr{H})$ and $B$ has an inverse in $\mathscr{B}(\mathscr{K})$.

2.8.41. Let $\{y_1, y_2, y_3, \ldots\}$ be an orthonormal basis of a Hilbert space $\mathscr{H}$; and let $[a_{j,k}]$, $[b_{j,k}]$ be the matrices, with respect to this basis, of bounded linear operators $A$, $B$ acting on $\mathscr{H}$. Prove that $[a_{j,k}b_{j,k}]$ is the matrix of a bounded linear operator. [*Hint*. Let $P$ be the projection from $\mathscr{H} \otimes \mathscr{H}$ onto the subspace $\mathscr{K}$ spanned by the orthonormal system $\{y_1 \otimes y_1, y_2 \otimes y_2, \ldots\}$, and consider the operator $T$ obtained by restricting $P(A \otimes B)P$ to $\mathscr{K}$.]

2.8.42. Suppose that $A$ is a closed linear operator with domain dense in a Hilbert space $\mathscr{H}$ and with range in $\mathscr{H}$; and let $B$ be in $\mathscr{B}(\mathscr{H})$. Prove that $AB$ is closed. Show also that, if $AB$ is densely defined and bounded, then $AB \in \mathscr{B}(\mathscr{H})$.

2.8.43. Let $\{y_1, y_2, y_3, \ldots\}$ be an orthonormal basis for a Hilbert space $\mathscr{H}$, and let

$$\mathscr{D} = \left\{ x \in \mathscr{H} : \sum_{n=1}^{\infty} n^4 |\langle x, y_n \rangle|^2 < \infty \right\}, \qquad z = \sum_{n=2}^{\infty} n^{-1} y_n.$$

Define $B$ in $\mathscr{B}(\mathscr{H})$ by $Bx = \langle x, z \rangle z$; and define mappings $S$ and $T$ with domain $\mathscr{D}$ by

$$Sx = \sum_{n=2}^{\infty} n^2 \langle x, y_n \rangle y_n, \qquad Tx = Sx + \langle Sx, z \rangle y_1 \qquad (x \in \mathscr{D}).$$

Show that $S$ and $T$ are closed densely defined operators, but that neither $T - S$ nor $BS$ is preclosed.

2.8.44. Suppose that $A$ and $B$ are linear operators with their domains dense in a Hilbert space $\mathscr{H}$ and their ranges in $\mathscr{H}$. Prove that $A^* + B^* \subseteq (A + B)^*$ if $A + B$ is densely defined, and that $B^*A^* \subseteq (AB)^*$ if $AB$ is densely defined.

2.8.45. Suppose that $T$ is a closed linear operator with domain dense in a Hilbert space $\mathscr{H}$ and with range in $\mathscr{H}$. Show that the null space $\{x \in \mathscr{D}(T): Tx = 0\}$ of $T$ is a closed subspace of $\mathscr{H}$.

Let $N(T)$ and $R(T)$ denote the projections whose ranges are, respectively, the null space of $T$ and the closure of the range of $T$. Prove that

$$R(T) = I - N(T^*), \qquad N(T) = I - R(T^*),$$

$$R(T^*T) = R(T^*), \qquad N(T^*T) = N(T).$$

[For the case in which $T \in \mathscr{B}(\mathscr{H})$, these relations have been established in Proposition 2.5.13.]

2.8.46. Let $T$ be a closed operator with domain dense in a separable Hilbert space $\mathscr{H}$ and with range in $\mathscr{H}$. Show that the ranges of $R(T)$ and $R(T^*)$ have the same dimension.

2.8.47. Let $\mathscr{H}$ be a Hilbert space and $E$ be a projection with finite-dimensional range. Let $F$ be a projection such that the dimension of $E(\mathscr{H})$ is less than the dimension of $F(\mathscr{H})$. Show that $(I - E) \wedge F \neq 0$.

2.8.48. Show that, if $T$ is a linear operator with domain dense in a Hilbert space $\mathscr{H}$ and with range in $\mathscr{H}$, and if $\langle Tz, z \rangle$ is real for each $z$ in $\mathscr{D}(T)$, then $T$ is symmetric.

2.8.49. Let $\mathscr{H}$ be the Hilbert space $L_2$, corresponding to Lebesgue measure on the unit interval $[0, 1]$, and let $\mathscr{D}_0$ be the subspace consisting of all complex-valued functions $f$ that have a continuous derivative $f'$ on $[0, 1]$ and satisfy $f(0) = f(1) = 0$. Let $D_0$ be the operator with domain $\mathscr{D}_0$ and with range in $\mathscr{H}$ defined by $D_0 f = f'$. Show that $iD_0$ is a densely defined symmetric operator and that

$$(iD_0)M - M(iD_0) = iI|\mathscr{D}_0,$$

where $M$ is the bounded linear operator defined by

$$(Mf)(s) = sf(s) \qquad (f \in L_2; \quad 0 \leqslant s \leqslant 1).$$

2.8.50. Let $\mathscr{H}$, $\mathscr{D}_0$, and $D_0$ be defined as in Exercise 2.8.49, and let $\mathscr{H}_1 = \{f_1 \in \mathscr{H} : \langle f_1, u \rangle = 0\}$, where $u$ is the unit vector in $\mathscr{H}$ defined by $u(s) = 1$ $(0 \leqslant s \leqslant 1)$. When $f \in \mathscr{H}$, define $Kf$ in $\mathscr{H}$ by

$$(Kf)(s) = \int_0^s f(t)\, dt \qquad (0 \leqslant s \leqslant 1).$$

(i) Prove that $K \in \mathscr{B}(\mathscr{H})$, that $K$ has null space $\{0\}$, and that $K(\mathscr{H}_1) \supseteq \mathscr{D}_0$.

(ii) Show that the equation

$$D_1 K f_1 = f_1 \qquad (f_1 \in \mathscr{H}_1)$$

defines a closed linear operator $D_1$ with domain $\mathscr{D}_1 = K(\mathscr{H}_1)$. Prove also that $D_1$ is the closure of $D_0$.

(iii) Show that the equation

$$D_2(Kf + au) = f \qquad (f \in \mathscr{H}, \quad a \in \mathbb{C})$$

defines a closed linear operator $D_2$, with domain $\mathscr{D}_2 = \{Kf + au : f \in \mathscr{H}, a \in \mathbb{C}\}$, that extends $D_1$.

(iv) Prove that $\langle Kf_1, f \rangle + \langle f_1, Kf + au \rangle = 0$ for all $f_1$ in $\mathscr{H}_1$, $f$ in $\mathscr{H}$, and $a$ in $\mathbb{C}$.

(v) Show that $D_0^* = D_1^* = -D_2$.

(vi) Let $\mathscr{D}_3 = \{Kf_1 + au : f_1 \in \mathscr{H}_1, a \in \mathbb{C}\}$, and let $D_3$ be the restriction $D_2 | \mathscr{D}_3$. Show that $D_3$ is a closed densely defined linear operator, that $D_1 \subseteq D_3 = -D_3^* \subseteq D_2$, and that $iD_3$ is self-adjoint.

# CHAPTER 3

# BANACH ALGEBRAS

In this chapter, we study algebras that have a Banach-space structure relative to which the multiplication is continuous. The operator algebras, which form the principal object of study for us, are a special subclass. Our purpose is to locate those constructs (for example, spectrum and spectral radius), develop the techniques, and prove the results that are natural to this general setting.

## 3.1. Basics

Let $\mathfrak{A}$ be a Banach space (complex or real) and, at the same time, an algebra with identity $I$, in which multiplication is separately continuous (that is, $(A, B) \to AB$ is continuous in $A$ for each fixed $B$ and in $B$ for each fixed $A$). Denote by $L_A$ and $R_B$ the operators on $\mathfrak{A}$ such that $L_A(B) = AB = R_B(A)$. From the continuity assumption, $L_A$ and $R_B$ are in $\mathscr{B}(\mathfrak{A})$. Thus $\|L_A(B)\| \leqslant \|R_B\| \cdot \|A\|$, and $\{\|L_A(B)\| : \|A\| \leqslant 1\}$ is a bounded set of numbers for each $B$ in $\mathfrak{A}$. From the uniform-boundedness principle (see Theorem 1.8.9), $\{\|L_A\| : \|A\| \leqslant 1\}$ is bounded. Similarly $\{\|R_B\| : \|B\| \leqslant 1\}$ is bounded. It follows that the mappings $\mathscr{L} : A \to L_A$ and $\mathscr{R} : B \to R_B$, which are linear isomorphisms (note that $L_A(I) = A$ and $R_B(I) = B$) of $\mathfrak{A}$ into $\mathscr{B}(\mathfrak{A})$, are continuous (bounded). The mapping $\mathscr{L}$ is an algebraic isomorphism, while $\mathscr{R}$ is an algebraic anti-isomorphism. Now,

$$\|A\| = \|L_A(I)\| \leqslant |||L_A||| \cdot \|I\|,$$

where, for the moment, we denote the norm on $\mathscr{B}(\mathfrak{A})$ by $|||\ \ |||$. Thus

$$\|I\|^{-1}\|A\| \leqslant |||L_A||| = |||\mathscr{L}(A)||| \leqslant \|\mathscr{L}\| \cdot \|A\|.$$

Hence $\mathscr{L}(\mathfrak{A})$ is a norm-closed subalgebra of $\mathscr{B}(\mathfrak{A})$ (see Corollary 1.5.10(ii)) and $\mathscr{L}$ is an algebraic isomorphism and homeomorphism of $\mathfrak{A}$ onto $\mathscr{L}(\mathfrak{A})$. As far as the (combined) algebraical and topological properties of $\mathfrak{A}$ and $\mathscr{L}(\mathfrak{A})$ are concerned, they are indistinguishable when identified by the isomorphism $\mathscr{L}$.

The norm on $\mathscr{B}(\mathfrak{A})$ enjoys some special properties: $|||\iota||| = 1$, where $\iota\,(= L_I)$ is the identity mapping on $\mathfrak{A}$; and $|||\varphi\psi||| \leqslant |||\varphi||| \cdot |||\psi|||$. (Note that these properties of the norm on $\mathscr{B}(\mathfrak{A})$ are valid for each normed space $\mathfrak{A}$, from the discussion preceding Theorem 1.5.6.) Thus the norm induced on $\mathscr{L}(\mathfrak{A})$ by that on $\mathscr{B}(\mathfrak{A})$ has these special properties.

If we assume that the norm on $\mathfrak{A}$ satisfies

$$\|AB\| = \|L_A(B)\| \leqslant \|A\| \cdot \|B\|, \tag{1}$$

then, since

$$\begin{aligned}\|AB - A'B'\| &\leqslant \|A\| \cdot \|B - B'\| + \|A - A'\| \cdot \|B'\| \\ &\leqslant \|A\| \cdot \|B - B'\| + \|A - A'\|(\|B\| + 1),\end{aligned}$$

when $\|B - B'\| \leqslant 1$, multiplication is *jointly* continuous on $\mathfrak{A}$. Thus multiplication is jointly continuous on $\mathscr{L}(\mathfrak{A})$, in any event, and since $\mathscr{L}$ is an algebraic isomorphism and homeomorphism, multiplication is jointly continuous on $\mathfrak{A}$—independent of the norm assumption. (The uniform-boundedness principle did the work in getting us *joint* from *separate* continuity of multiplication.) From the joint continuity, it follows at once that the closure of a subalgebra (ideal) is, again, a subalgebra (ideal).

If we assume, now, that the norm on $\mathfrak{A}$ satisfies $\|I\| = 1$ as well as $\|AB\| \leqslant \|A\| \cdot \|B\|$, then from (1),

$$\|L_A(I)\| = \|A\| \leqslant \|\|L_A\|\| \leqslant \|A\|,$$

so that $\mathscr{L}$ is an isometry as well as an algebraic isomorphism. While the natural structural assumption on $\mathfrak{A}$ is that of continuity of multiplication (either joint or separate), the preceding discussion assures us that nothing is lost if we make the convenient normalization assumptions on the norm.

3.1.1. Definition. An algebra $\mathfrak{A}$ (over $\mathbb{R}$ or $\mathbb{C}$) with unit $I$ is said to be a *normed algebra* when $\mathfrak{A}$ is a normed space such that $\|AB\| \leqslant \|A\| \cdot \|B\|$ for all $A$ and $B$ in $\mathfrak{A}$, and $\|I\| = 1$. If $\mathfrak{A}$ is a Banach space relative to this norm, $\mathfrak{A}$ is said to be a *Banach algebra*. ■

3.1.2. Remark. From the discussion preceding Definition 3.1.1, we see that a normed algebra $\mathfrak{A}$ is isometrically, algebraically isomorphic to a subalgebra of $\mathscr{B}(\mathfrak{A})$, and that $\mathscr{B}(\mathfrak{A})$ is a normed algebra. From Theorem 1.5.6, if $\mathfrak{A}$ is a Banach space, then $\mathscr{B}(\mathfrak{A})$ is a Banach space, hence a Banach algebra. If $\mathfrak{A}$ is a normed algebra, completing it to a Banach space $\hat{\mathfrak{A}}$ (see Theorem 1.5.1) allows us to view each $L_A$ as a bounded linear transformation from $\mathfrak{A}$ to $\hat{\mathfrak{A}}$ and extend it (uniquely) in a norm-preserving fashion (see Theorem 1.5.7) to an operator $\hat{L}_A$ on $\hat{\mathfrak{A}}$. The resulting mapping, $\hat{\mathscr{L}}: A \to \hat{L}_A$, is then an isometric algebraic isomorphism of $\mathfrak{A}$ into the Banach algebra $\mathscr{B}(\hat{\mathfrak{A}})$. Moreover, $\hat{\mathscr{L}}$ extends to an isometric linear isomorphism of $\hat{\mathfrak{A}}$ onto the norm closure of $\hat{\mathscr{L}}(\mathfrak{A})$ in $\mathscr{B}(\hat{\mathfrak{A}})$. Thus $\hat{\mathfrak{A}}$ becomes a Banach algebra. We may say, briefly, that the completion of a normed algebra is a Banach algebra. ■

3.1.3. REMARK. The assumption that $\mathfrak{A}$ has an identity is not an essential restriction. If $\mathfrak{A}$ does not have an identity, we can employ the standard process for adjoining an identity to an algebra. We embed $\mathfrak{A}$ in the algebra $\mathfrak{A}_1$ of pairs $(aI, A)$, where $a$ is a scalar and $A \in \mathfrak{A}$. The multiplication $(aI, A) \cdot (bI, B) = (abI, aB + bA + AB)$, addition $(aI, A) + (bI, B) = ((a + b)I, A + B)$, and multiplication by scalars $c(aI, A) = (caI, cA)$ impose the structure of an (associative) algebra on $\mathfrak{A}_1$, and $(I, 0)$ is a unit for it. The algebra $\mathfrak{A}$ appears as (that is, "is isomorphic to") the subalgebra $\{(0, A): A \text{ in } \mathfrak{A}\}$. Defining $\|(aI, A)\|$ to be $|a| + \|A\|$, we map $\mathfrak{A}$ isometrically onto a subalgebra of $\mathfrak{A}_1$ by means of this identification. It is easily checked that $\mathfrak{A}_1$ is a normed algebra; and if $\mathfrak{A}$ is a Banach space, then $\mathfrak{A}_1$ is a Banach algebra (with identity).

Despite the possibility of adjoining an identity to a Banach algebra, it is sometimes artificial and inconvenient to do so (as in Subsection 3.2, *The Banach algebra* $L_1(\mathbb{R})$ *and Fourier analysis*, concerned with certain Banach-algebra generalizations to topological groups of the concept of group algebra). In these cases, one develops the appropriate techniques for dealing with the algebras without identity. For our purposes, this assumption will cause us no difficulty and is a considerable convenience. ■

3.1.4. EXAMPLE. An important class of Banach algebras (Section 3.4 is devoted to a study of their properties) is made up of the algebras of (complex- or real-valued) continuous functions on compact Hausdorff spaces. The algebraic operations are the usual pointwise addition and multiplication of functions. If $X$ is a compact Hausdorff space, we shall denote this algebra of continuous functions (for the most part over $\mathbb{C}$) by $C(X)$. In Example 1.7.2, we studied $C(X)$ as a Banach space with its so-called "supremum norm"

$$\|f\| = \sup\{|f(x)| : x \in X\}.$$

Of course the identity of $C(X)$ is the constant function 1, and $\|1\| = 1$. We note, too, that

$$\begin{aligned}\|f \cdot g\| &= \sup\{|f(x)| \cdot |g(x)| : x \in X\}\\ &\leqslant \sup\{|f(x)| : x \in X\} \cdot \sup\{|g(x)| : x \in X\} = \|f\| \cdot \|g\|.\end{aligned}$$

From this and the fact that $C(X)$, in the given norm, is a Banach space, we see that $C(X)$ is a Banach algebra. While we could have used the fact that each $f$ in $C(X)$ attains its norm at some point of $X$ (that is, $|f(x_0)| = \|f\|$ for some $x_0$ in $X$) in our norm considerations, by avoiding its use, the discussion applies without change with the assumption that $X$ is compact omitted and $C(X)$ denoting the *bounded* continuous functions on $X$. ■

3.1.5. LEMMA. *If $A$ is an element of the Banach algebra $\mathfrak{A}$ and $\|A\| < 1$, then $\sum_{k=0}^{n} A^k$ (where $A^0 = I$) has a limit $B$, as $n$ tends to $\infty$, in $\mathfrak{A}$ and $B(I - A) = (I - A)B = I$.*

*Proof.* Noting that, if $n < m$,

$$\|\sum_{k=0}^{n} A^k - \sum_{k=0}^{m} A^k\| = \|\sum_{k=n+1}^{m} A^k\| \leqslant \sum_{k=n+1}^{m} \|A\|^k,$$

we conclude, from the convergence of the geometric series, $\sum_{k=0}^{\infty} \|A\|^k$ (with $\|A\| < 1$), and the fact that $\mathfrak{A}$ is complete, that $\sum_{k=0}^{n} A^k$ tends to a limit $B$ in $\mathfrak{A}$ as $n \to \infty$. Since

$$\left(\sum_{k=0}^{n} A^k\right)(I - A) = (I - A)\left(\sum_{k=0}^{n} A^k\right) = I - (A^{n+1}),$$

and $\|A^{n+1}\| \leqslant \|A\|^{n+1} \to 0$, we see that $B(I - A) = (I - A)B = I$, by continuity of multiplication in $\mathfrak{A}$. ■

With the notation of Lemma 3.1.5, we write $B = \sum_{n=0}^{\infty} A^n$ and say that $B$ is an inverse (two-sided) to $I - A$ (denoted by $(I - A)^{-1}$).

Numerous consequences result from the small observation of Lemma 3.1.5.

3.1.6. PROPOSITION. *If $\mathfrak{A}$ is a Banach algebra, the set $\mathcal{N}$ of invertible elements is an open subset of $\mathfrak{A}$, and the operation* inv *of inversion on $\mathcal{N}$ is continuous.*

*Proof.* If $A \in \mathcal{N}$, then $L_A$, left multiplication by $A$ on $\mathfrak{A}$, is a continuous mapping of $\mathfrak{A}$ onto $\mathfrak{A}$ with the continuous inverse $L_{A^{-1}}$. Since $B^{-1}A^{-1}$ is an inverse to $AB$, if $A$ and $B$ are in $\mathcal{N}$; $L_A$ maps $\mathcal{N}$ onto $\mathcal{N}$ and $I$ onto $A$. From Lemma 3.1.5, the open ball of radius 1 with center $I$ is a subset of $\mathcal{N}$. Thus, the homeomorphism $L_A$ maps this ball onto an open set in $\mathfrak{A}$ containing $A$ and contained in $\mathcal{N}$. Hence $\mathcal{N}$ is open.

If $\|I - B\| < 1$, writing $A$ for $I - B$, $\sum_{k=0}^{\infty} A^k = (I - A)^{-1} = B^{-1}$; and

$$\|B^{-1} - I^{-1}\| = \|B^{-1} - I\| \leqslant \sum_{k=1}^{\infty} \|A\|^k = \|A\|(1 - \|A\|)^{-1}. \tag{2}$$

In particular, if $\|I - B\|$ $(= \|A\|) < \varepsilon < \frac{1}{2}$, then $\|B^{-1} - I\| < 2\|A\| < 2\varepsilon$. Thus inv is continuous at $I$ on $\mathcal{N}$. For each $A$ in $\mathcal{N}$,

$$\text{inv} = R_{A^{-1}} \circ \text{inv} \circ L_{A^{-1}},$$

and $L_{A^{-1}}$ maps $A$ onto $I$, inv maps $I$ onto $I$, $R_{A^{-1}}$ maps $I$ onto $A^{-1}$, each mapping continuous at the specified element. Thus inv is continuous at $A$ and hence on $\mathcal{N}$. ■

3.1.7. REMARK. The elements that are not invertible in $\mathfrak{A}$ are said to be *singular*. From Proposition 3.1.6, the singular elements form a closed subset of $\mathfrak{A}$. We refer to the invertible elements as *regular* and *non-singular* elements as well.

After showing that inv is continuous at $I$ with the series estimate (2), the proof that inv is continuous on $\mathcal{N}$ is given in a "structural" manner. This structural argument points the way to an estimate for $\|A^{-1} - A_0^{-1}\|$ (in terms of $\|A - A_0\|$), which establishes, again, the continuity of inv. For this, note that

$$\|L_{A_0^{-1}}(A - A_0)\| = \|A_0^{-1}A - I\| \leqslant \|A_0^{-1}\| \cdot \|A - A_0\|,$$

so that, from (2), when $\|A_0^{-1}\| \cdot \|A - A_0\| < 1$,

$$\|A^{-1}A_0 - I\| \leqslant \|A_0^{-1}\| \cdot \|A - A_0\|(1 - \|A_0^{-1}\| \cdot \|A - A_0\|)^{-1};$$

and

$$\text{(3)} \qquad \begin{aligned} \|A^{-1} - A_0^{-1}\| &= \|R_{A_0^{-1}}(A^{-1}A_0 - I)\| \leqslant \|A^{-1}A_0 - I\| \cdot \|A_0^{-1}\| \\ &\leqslant \|A_0^{-1}\|^2\|A - A_0\|(1 - \|A_0^{-1}\| \cdot \|A - A_0\|)^{-1}. \end{aligned}$$

In particular, if $\|A - A_0\| < (2\|A_0^{-1}\|)^{-1}$, then

$$\text{(4)} \qquad \|A^{-1} - A_0^{-1}\| \leqslant 2\|A_0^{-1}\|^2\|A - A_0\|. \quad \blacksquare$$

Making use of the fact that elements near the identity in a Banach algebra are invertible, the possibility of norm-dense, proper ideals and its associated difficulties can be eliminated.

3.1.8. PROPOSITION. *If $\mathcal{I}$ is a proper (left or right) ideal in a Banach algebra $\mathfrak{A}$, then the norm closure $\bar{\mathcal{I}}$ of $\mathcal{I}$ is a proper (left or right) ideal in $\mathfrak{A}$. If $\mathcal{I}$ is a maximal (left, right, or two-sided) ideal in $\mathfrak{A}$, then $\mathcal{I}$ is norm closed. If $\mathcal{I}$ is a proper closed two-sided ideal in $\mathfrak{A}$, then the quotient algebra $\mathfrak{A}/\mathcal{I}$, provided with the quotient Banach space structure, is a Banach algebra.*

*Proof.* If $I \in \bar{\mathcal{I}}$, then there is an $A$ in $\mathcal{I}$ such that $\|I - A\| < 1$. From Lemma 3.1.5, $A\ (= I - (I - A))$ is invertible. Hence $A^{-1}A\ (= AA^{-1})$ is in $\mathcal{I}$. But then $B \cdot I$ (or $I \cdot B$) is in $\mathcal{I}$ for all $B$ in $\mathfrak{A}$, contradicting the assumption that $\mathcal{I}$ is proper. Thus $I \notin \bar{\mathcal{I}}$, and $\bar{\mathcal{I}}$ is a proper ideal. If $\mathcal{I}$ is maximal, then $\mathcal{I} = \bar{\mathcal{I}}$; so that $\mathcal{I}$ is closed.

It follows from Theorem 1.5.3 that $\mathfrak{A}/\mathcal{I}$ is a Banach space in its quotient norm when $\mathcal{I}$ is a closed ideal in $\mathfrak{A}$. If $\mathcal{I}$ is a (proper) closed two-sided ideal in $\mathfrak{A}$, then $\mathfrak{A}/\mathcal{I}$ is a Banach space and (with its natural structure as a quotient algebra) an algebra with identity $I + \mathcal{I}$. We have noted that

$$\inf\{\|I - A\| : A \in \mathcal{I}\} = 1$$

(and, thus, $\|I + \mathcal{I}\| = 1$), since $\mathcal{I}$ is proper. Since

$$\begin{aligned} \|A + \mathcal{I}\| \cdot \|B + \mathcal{I}\| &= \inf\{\|A - C_1\| : C_1 \in \mathcal{I}\} \cdot \inf\{\|B - C_2\| : C_2 \in \mathcal{I}\} \\ &= \inf\{\|A - C_1\| \cdot \|B - C_2\| : C_1, C_2 \in \mathcal{I}\} \\ &\geqslant \inf\{\|AB - (C_1B + AC_2 - C_1C_2)\| : C_1, C_2 \in \mathcal{I}\} \\ &\geqslant \inf\{\|AB - C\| : C \in \mathcal{I}\} = \|(A + \mathcal{I})(B + \mathcal{I})\|, \end{aligned}$$

$\mathfrak{A}/\mathcal{I}$ is a Banach algebra in its quotient norm. $\blacksquare$

## 3.2. The spectrum

Our Banach algebras are intended to provide the general framework for the study of algebras of (linear) operators on a Hilbert space. In the case of a finite-dimensional space (and when they are present in the infinite-dimensional case), the eigenvectors and their associated eigenvalues play an important role in the analysis of the individual operator. The concept of *spectrum*, which we study in this section, is devised as the replacement, in the general setting of Banach algebras, for the set of eigenvalues in the finite-dimensional case. Henceforth our Banach algebras are assumed to be complex.

3.2.1. DEFINITION. If $A$ is an element of the Banach algebra $\mathfrak{A}$, we say that a complex number $\lambda$ is a *spectral value* for $A$ (relative to $\mathfrak{A}$) when $A - \lambda I$ does not have a two-sided inverse in $\mathfrak{A}$. The set of spectral values of $A$ is called the *spectrum* of $A$ and is denoted by $\mathrm{sp}_{\mathfrak{A}}(A)$. ■

When there is no danger of confusion, we write $\mathrm{sp}(A)$ in place of $\mathrm{sp}_{\mathfrak{A}}(A)$.

Before beginning the general study of the spectrum, let us note that it serves the purpose for which it is designed. If $\mathscr{H}$ is a finite-dimensional (Banach) space and $A$ is a linear transformation of $\mathscr{H}$ into itself, then $A - \lambda I$ will fail to have an inverse (in $\mathscr{B}(\mathscr{H})$, the family of all linear transformations of $\mathscr{H}$ into itself) if and only if it annihilates some (unit) vector $x_0$ in $\mathscr{H}$ – that is, if and only if $A$ has some unit eigenvector $x_0$ corresponding to the eigenvalue $\lambda$. At the same time, let us note that while an eigenvalue is always in the spectrum, the reverse need not, in general, be the case.

3.2.2. EXAMPLE. Let $\mathscr{H}$ be the Hilbert space of complex-valued square-integrable functions on $[0, 1]$ relative to Lebesgue measure, and let $A$ be multiplication by the identity transform on $[0, 1]$ (so that $(Af)(x) = xf(x)$). Then $A$ has no eigenvalues; for if $Af = \lambda f$, $f$ must be 0 at all points of $[0, 1]$ other than $\lambda$. Hence $f$ is 0 almost everywhere, and $f$ is the element 0 in $\mathscr{H}$.

Nonetheless, $\mathrm{sp}_{\mathscr{B}(\mathscr{H})}(A) = [0, 1]$. To see this, let $y_n$ be the characteristic function of $[\lambda - 1/2n, \lambda + 1/2n]$, and let $x_n$ be $n^{1/2}y_n$. The obvious modification is made when $\lambda$ is 0 or 1; and, for $\lambda$ in $(0, 1)$, the sequence $\{x_n\}$ has as first element $x_{n_0}$, where $n_0$ is so large that $[\lambda - 1/2n, \lambda + 1/2n] \subseteq [0, 1]$, when $n_0 \leqslant n$. Each $x_n$ is a unit vector, and $\|(A - \lambda I)x_n\| = \sqrt{3}/6n$ (so that $\{x_n\}$ is a sequence of "approximate" eigenvectors for $A$ corresponding to the eigenvalue $\lambda$). If $B$ is a left inverse to $A - \lambda I$, then

$$1 = \|x_n\| = \|B(A - \lambda I)x_n\| \leqslant \|B\| \cdot \|(A - \lambda I)x_n\| = \frac{\|B\|\sqrt{3}}{6n};$$

so that $2n\sqrt{3} \leqslant \|B\|$ for all positive integral $n$. Thus $B$ is not bounded, and

$A - \lambda I$ does not have a two-sided inverse in $\mathscr{B}(\mathscr{H})$. (In essence this is part of the argument of Corollary 1.5.10, which could be applied here.) It follows that $[0,1] \subseteq \mathrm{sp}_{\mathscr{B}(\mathscr{H})}(A)$.

If $\lambda \notin [0,1]$, then $f$, defined as $(x-\lambda)^{-1}$ for $x$ in $[0,1]$, is continuous in $[0,1]$. Multiplication by $f$ on $\mathscr{H}$ is a bounded operator, which is the two-sided inverse (in $\mathscr{B}(\mathscr{H})$) to $A - \lambda I$. Thus $\lambda \notin \mathrm{sp}_{\mathscr{B}(\mathscr{H})}(A)$, and $\mathrm{sp}_{\mathscr{B}(\mathscr{H})}(A) = [0,1]$. ∎

We shall see presently (see Lemma 3.2.13) that for normal operators (see Section 2.4) the situation of Example 3.2.2 holds generally: spectral values correspond to (sequences of) approximate unit eigenvectors.

3.2.3. Theorem. *If $A$ is an element of the Banach algebra $\mathfrak{A}$ then $\mathrm{sp}_{\mathfrak{A}}(A)$ is a non-empty closed subset of the closed disk in $\mathbb{C}$ with center 0 and radius $\|A\|$.*

*Proof.* If $\lambda \notin \mathrm{sp}_{\mathfrak{A}}(A)$, then, by Proposition 3.1.6, $A - \lambda' I$ is invertible for all $\lambda'$ in a small open disk with center $\lambda$. Let $\rho$ be a continuous linear functional on $\mathfrak{A}$. Since

$$\frac{\rho((A-\lambda' I)^{-1}) - \rho((A-\lambda I)^{-1})}{\lambda' - \lambda} = \frac{\rho((\lambda'-\lambda)(A-\lambda' I)^{-1}(A-\lambda I)^{-1})}{\lambda'-\lambda}$$

$$= \rho((A-\lambda' I)^{-1}(A-\lambda I)^{-1})$$

$$\to \rho((A-\lambda I)^{-2})$$

as $\lambda' \to \lambda$, by continuity of inversion (see Proposition 3.1.6) on the set of invertible elements of $\mathfrak{A}$, and the continuity of $\rho$, the function $\lambda \to \rho((A-\lambda I)^{-1})$ is holomorphic on $\mathbb{C} \setminus \mathrm{sp}_{\mathfrak{A}}(A)$. Note, too, that

$$\rho((A-\lambda I)^{-1}) = \lambda^{-1}\rho((\lambda^{-1}A - I)^{-1}) \to 0$$

as $|\lambda| \to \infty$; for $\lambda^{-1}A - I$ is invertible when $\|A\| < |\lambda|$, and

$$(\lambda^{-1}A - I)^{-1} \to -I$$

as $|\lambda| \to \infty$. We see, at the same time, that $\lambda(\lambda^{-1}A - I)\,(= A - \lambda I)$ is invertible when $\|A\| < |\lambda|$; so that $\mathrm{sp}_{\mathfrak{A}}(A)$ is a subset of the closed disk in $\mathbb{C}$ with center 0 and radius $\|A\|$. If $\mathrm{sp}_{\mathfrak{A}}(A)$ were empty, the function $\lambda \to \rho((A-\lambda I)^{-1})$ would be an entire function that vanishes at $\infty$. By Liouville's theorem, this function would vanish everywhere on $\mathbb{C}$. In particular, we would have $\rho(A^{-1}) = 0$, for each continuous linear functional $\rho$ on $\mathfrak{A}$. From the Hahn–Banach theorem (see Corollary 1.2.11), it would follow that $A^{-1} = 0$, a contradiction. Thus $\mathrm{sp}_{\mathfrak{A}}(A)$ is not empty.

We observed, during this argument, that $\mathbb{C} \setminus \mathrm{sp}_{\mathfrak{A}}(A)$ is open, so that $\mathrm{sp}_{\mathfrak{A}}(A)$ is a non-empty closed subset of the disk $\mathbb{C}$ with center 0 and radius $\|A\|$. ∎

Despite the fact that the spectrum of $A$ is not empty, it may consist of just 0. If $\{e_1, e_2\}$ is a basis for two-dimensional Hilbert space and $A$ is the operator on this space that maps $e_1$ to $e_2$ and $e_2$ to 0, then $\mathrm{sp}_{\mathscr{B}(\mathscr{H})}(A)$ consists of just 0. Note, too, from this example, that $A$ may be non-zero and have just 0 in its spectrum.

If each element of $\mathfrak{A}$ other than 0 has an inverse in $\mathfrak{A}$ (so that $\mathfrak{A}$ is a *division algebra*), then, if $\lambda \in \mathrm{sp}_{\mathfrak{A}}(A)$, $A - \lambda I$ must be 0 (being a singular element of $\mathfrak{A}$). Thus, in this case, $\mathfrak{A}$ consists of just scalar multiples of $I$. Since $\mathfrak{A}$ is, then, isomorphic to $\mathbb{C}$, we say, loosely, that $\mathfrak{A}$ *is* $\mathbb{C}$.

3.2.4. COROLLARY. *A (complex) Banach division algebra (or field) is* $\mathbb{C}$.

We noted, in Proposition 3.1.8, that a maximal ideal $\mathscr{M}$ in a Banach algebra $\mathfrak{A}$ is closed. Since $\mathscr{M}$ is maximal, if $\mathfrak{A}$ is commutative, $\mathfrak{A}/\mathscr{M}$ is a field – and a Banach algebra. From the preceding corollary, $\mathfrak{A}/\mathscr{M}$ is $\mathbb{C}$, and the quotient mapping is a continuous multiplicative linear functional on $\mathfrak{A}$ (that is, a homomorphism of $\mathfrak{A}$ onto $\mathbb{C}$). Conversely, if $\rho$ is a homomorphism of $\mathfrak{A}$ onto $\mathbb{C}$, its kernel $\mathscr{M}$ is a *maximal* two-sided ideal in $\mathfrak{A}$ (since $\mathfrak{A}/\mathscr{M}$ is the field $\mathbb{C}$). Hence $\mathscr{M}$ is closed and, from Corollary 1.2.5, $\rho$ is continuous. In general, if $\mathfrak{A}$ is not commutative and $\mathscr{M}$ is a maximal two-sided ideal in $\mathfrak{A}$, we cannot conclude that $\mathfrak{A}/\mathscr{M}$ is a field; so that no multiplicative linear functional need be associated with $\mathscr{M}$.

3.2.5. COROLLARY. *If $\mathfrak{A}$ is a commutative (complex) Banach algebra and $\mathscr{M}$ is a maximal two-sided ideal in $\mathfrak{A}$, then $\mathfrak{A}/\mathscr{M}$ is $\mathbb{C}$ and the quotient mapping from $\mathfrak{A}$ to $\mathfrak{A}/\mathscr{M}$ is a (continuous) multiplicative linear functional on $\mathfrak{A}$. If $\mathfrak{A}$ is an arbitrary (complex) Banach algebra and $\rho$ is a multiplicative linear functional on $\mathfrak{A}$, then $\rho$ is continuous with kernel $\mathscr{M}$ a maximal two-sided ideal in $\mathfrak{A}$ such that $\mathfrak{A}/\mathscr{M}$ is $\mathbb{C}$.*

We saw that $\mathrm{sp}_{\mathfrak{A}}(A)$ is contained in the disk in $\mathbb{C}$ with center 0 and radius $\|A\|$. The radius of the "smallest" disk containing the spectrum will appear in our considerations.

3.2.6. DEFINITION. The *spectral radius* $r_{\mathfrak{A}}(A)$ of an element $A$ of a Banach algebra $\mathfrak{A}$ is

$$\sup\{|\lambda| : \lambda \in \mathrm{sp}_{\mathfrak{A}}(A)\}. \quad \blacksquare$$

3.2.7. REMARK. When no confusion can arise, we write $r(A)$ in place of $r_{\mathfrak{A}}(A)$. As noted in Theorem 3.2.3, $r(A) \leqslant \|A\|$. It is apparent from the definition that $r(A)$ is the radius of the smallest disk in $\mathbb{C}$ with center 0 containing $\mathrm{sp}(A)$. $\blacksquare$

3.2.8. PROPOSITION. *If $A$ and $B$ are elements of a Banach algebra $\mathfrak{A}$, then* $\mathrm{sp}(AB) \cup \{0\} = \mathrm{sp}(BA) \cup \{0\}$, *and* $r(AB) = r(BA)$.

*Proof.* If $\lambda \neq 0$ and $\lambda \in \mathrm{sp}(AB)$, then $AB - \lambda I$ and, hence, $(\lambda^{-1}A)B - I$ are not invertible. On the other hand, if $\lambda \notin \mathrm{sp}(BA)$, then $BA - \lambda I$ and, hence, $B(\lambda^{-1}A) - I$ are invertible. Our task, then, is to show that $I - AB$ is invertible in $\mathfrak{A}$ if and only if $I - BA$ is invertible in $\mathfrak{A}$, for arbitrary elements $A$ and $B$ of $\mathfrak{A}$.

Arguing formally, for the moment,

$$(I - AB)^{-1} = \sum_{n=0}^{\infty} (AB)^n = I + AB + ABAB + \cdots$$

and

$$B(I - AB)^{-1}A = BA + BABA + BABABA + \cdots = (I - BA)^{-1} - I.$$

Thus if $I - AB$ has an inverse, we may hope that $B(I - AB)^{-1}A + I$ is an inverse to $I - BA$. Multiplying, we have

$$\begin{aligned}(I - BA)&[B(I - AB)^{-1}A + I] \\ &= B(I - AB)^{-1}A + I - BAB(I - AB)^{-1}A - BA \\ &= B[(I - AB)^{-1} - AB(I - AB)^{-1}]A + I - BA = I,\end{aligned}$$

and similarly for right multiplication by $I - BA$. ■

3.2.9. Remark. It is apparent that $\mathrm{sp}(A + I) = \{1 + a : a \in \mathrm{sp}(A)\}$. We shall prove the more general result concerning the relation between $\mathrm{sp}(p(A))$ and $\mathrm{sp}(A)$, for an arbitrary polynomial $p$, in the proposition that follows. (We prove the full *spectral mapping theorem* (Theorem 3.3.6) in Section 3.3.) Combining the simple initial observation with the preceding proposition yields the fact that the unit element $I$ of a Banach algebra $\mathfrak{A}$ is not the *commutator* $AB - BA$ of two elements $A$ and $B$ of $\mathfrak{A}$. (If $I = AB - BA$, then $\mathrm{sp}(AB) = 1 + \mathrm{sp}(BA)$, which is not consistent with

$$\mathrm{sp}(AB) \cup \{0\} = \mathrm{sp}(BA) \cup \{0\}.)$$

This fact is familiar in quantum theory where it takes the form

> the commutation relations are not representable in terms of bounded operators.

However, there are unbounded operators whose commutator is $I$ restricted to a dense linear manifold. (See Exercise 2.8.49.) ■

3.2.10. Proposition. *If $A$ is an element of the Banach algebra $\mathfrak{A}$ and $p$ is a polynomial in a single variable, then*

$$\mathrm{sp}(p(A)) = \{p(\lambda) : \lambda \in \mathrm{sp}(A)\} \; (= p(\mathrm{sp}(A))).$$

*If $A$ is invertible, then*

$$\mathrm{sp}(A^{-1}) = \{\lambda^{-1} : \lambda \in \mathrm{sp}(A)\} \ (= (\mathrm{sp}(A))^{-1}).$$

*If $A$ and $B$ are elements of the commutative Banach algebra $\mathfrak{A}$, then*

$$\mathrm{sp}(AB) \subseteq \mathrm{sp}(A)\mathrm{sp}(B), \qquad \mathrm{sp}(A + B) \subseteq \mathrm{sp}(A) + \mathrm{sp}(B),$$

$$r(AB) \leqslant r(A)r(B), \qquad r(A + B) \leqslant r(A) + r(B).$$

*Proof.* If $\lambda \in \mathrm{sp}(A)$, then $A - \lambda I$ does not have a two-sided inverse in $\mathfrak{A}$. Thus one of $(A - \lambda I)\mathfrak{A}$ or $\mathfrak{A}(A - \lambda I)$ is a proper ideal $\mathscr{I}$ in $\mathfrak{A}$. If $p(x) = a_n x^n + \cdots + a_0$, then

$$p(A) - p(\lambda)I = a_n(A^n - \lambda^n I) + \cdots + a_1(A - \lambda I).$$

Noting that

$$\begin{aligned} A^k - \lambda^k I &= (A - \lambda I)(A^{k-1} + \lambda A^{k-2} + \cdots + \lambda^{k-1} I) \\ &= (A^{k-1} + \lambda A^{k-2} + \cdots + \lambda^{k-1} I)(A - \lambda I), \end{aligned}$$

we conclude that $p(A) - p(\lambda)I \in \mathscr{I}$, so that $p(A) - p(\lambda)I$ does not have a two-sided inverse in $\mathfrak{A}$, and $p(\lambda) \in \mathrm{sp}(p(A))$.

If $\gamma \in \mathrm{sp}(p(A))$ and $\lambda_1, \ldots, \lambda_n$ are the $n$ roots of $p(\lambda) - \gamma$, then

$$p(A) - \gamma I = (A - \lambda_1 I) \cdots (A - \lambda_n I),$$

so that at least one of $A - \lambda_1 I, \ldots, A - \lambda_n I$ is not invertible. If $A - \lambda_j I$ is not invertible, then $\lambda_j \in \mathrm{sp}(A)$ and $\gamma = p(\lambda_j) \in p(\mathrm{sp}(A))$. Thus $\mathrm{sp}(p(A)) = p(\mathrm{sp}(A))$.

Suppose $A$ is invertible in $\mathfrak{A}$ (equivalently, $0 \notin \mathrm{sp}_{\mathfrak{A}}(A)$). If $\lambda \neq 0$, then $A^{-1} - \lambda^{-1} I = (\lambda I - A)(\lambda A)^{-1}$, so that $\lambda^{-1} \in \mathrm{sp}(A^{-1})$ if and only if $\lambda \in \mathrm{sp}(A)$. Thus $\mathrm{sp}(A^{-1}) = \mathrm{sp}(A)^{-1}$.

Suppose, now, that $A$ and $B$ are elements of the commutative Banach algebra $\mathfrak{A}$. If $\lambda \in \mathrm{sp}(AB)$, then $AB - \lambda I$ lies in a proper ideal (necessarily, two-sided) $\mathscr{I}$ of $\mathfrak{A}$. Since $\mathfrak{A}$ has an identity, Zorn's lemma, applied to the set of proper ideals in $\mathfrak{A}$ containing $\mathscr{I}$, shows that $\mathscr{I}$ is contained in a maximal ideal $\mathscr{M}$ of $\mathfrak{A}$. From Corollary 3.2.5, $\mathscr{M}$ is the kernel of a multiplicative linear functional $\rho$ on $\mathfrak{A}$. Thus $\lambda = \rho(AB) = \rho(A)\rho(B)$. Since $A - \rho(A)I$ and $B - \rho(B)I$ are in the kernel $\mathscr{M}$ of $\rho$, $\rho(A) \in \mathrm{sp}(A)$ and $\rho(B) \in \mathrm{sp}(B)$. Thus $\lambda \in \mathrm{sp}(A)\mathrm{sp}(B)$ and $\mathrm{sp}(AB) \subseteq \mathrm{sp}(A)\mathrm{sp}(B)$.

Again, if $\lambda \in \mathrm{sp}(A + B)$, there is a multiplicative linear functional $\rho$ on our commutative $\mathfrak{A}$ such that $\rho(A + B) = \lambda$. As $\rho(A) \in \mathrm{sp}(A)$, $\rho(B) \in \mathrm{sp}(B)$, and $\lambda = \rho(A) + \rho(B)$; $\lambda \in \mathrm{sp}(A) + \mathrm{sp}(B)$, and $\mathrm{sp}(A + B) \subseteq \mathrm{sp}(A) + \mathrm{sp}(B)$. The inequalities for the spectral radius are immediate consequences of the corresponding relations for the spectra. ∎

3.2.11. Remark. We make special note of the fact established at the end of the proof of Proposition 3.2.10. If $A$ is an element of a commutative Banach algebra $\mathfrak{A}$ and $\lambda \in \mathrm{sp}_{\mathfrak{A}}(A)$, then there is a multiplicative linear functional $\rho$ on $\mathfrak{A}$ such that $\rho(A) = \lambda$. Conversely, if $\rho$ is a (non-zero) multiplicative linear functional on $\mathfrak{A}$ (not necessarily commutative), then $\rho(A) \in \mathrm{sp}_{\mathfrak{A}}(A)$ for each $A$ in $\mathfrak{A}$. For this last assertion, note that $A - \rho(A)I$ is in the kernel of $\rho$, a proper two-sided ideal in $\mathfrak{A}$. ■

The examples that follow illustrate the concepts of spectrum and spectral radius in the Banach algebra $\mathscr{B}(\mathscr{H})$ of bounded operators on the Hilbert space $\mathscr{H}$.

3.2.12. Example. Let $\{e_n\}$ be an orthonormal basis for a separable Hilbert space $\mathscr{H}$. Recalling Example 2.4.10, we have a bounded operator $A$ on $\mathscr{H}$ such that $Ae_n = \lambda_n e_n$, where $\{\lambda_n\}$ is an arbitrary bounded (denumerable) subset of $\mathbb{C}$. We saw that $\|A\| = \sup\{|\lambda_n|\}$, that $A$ is normal, in general, and self-adjoint exactly when all $\lambda_n$ are real, unitary when all $\lambda_n$ have modulus 1, and positive when all $\lambda_n$ are real and non-negative. Since each $\lambda_n$ is an eigenvalue (with eigenvector $e_n$), $\{\lambda_n\} \subseteq \mathrm{sp}_{\mathscr{B}(\mathscr{H})}(A)$. From Theorem 3.2.3, $\mathrm{sp}(A)$ is closed, so that $\{\lambda_n\}^-$, the closure of $\{\lambda_n\}$, is contained in $\mathrm{sp}(A)$. If $\lambda$ is not in this closure, then $\inf\{|\lambda - \lambda_n|\} > 0$, and $\{(\lambda_n - \lambda)^{-1}\}$ is a bounded subset of $\mathbb{C}$. Thus there is a bounded operator $B$ on $\mathscr{H}$ such that $Be_n = (\lambda_n - \lambda)^{-1}e_n$. Since $(A - \lambda I)e_n = (\lambda_n - \lambda)e_n$, we have $B(A - \lambda I)e_n = e_n$ and $(A - \lambda I)Be_n = e_n$ for all $n$. Thus $B$ is a two-sided inverse in $\mathscr{B}(\mathscr{H})$ to $A$ and $\lambda \notin \mathrm{sp}_{\mathscr{B}(\mathscr{H})}(A)$. Hence $\{\lambda_n\}^- = \mathrm{sp}_{\mathscr{B}(\mathscr{H})}(A)$.

If $\{\lambda_n\}$ is an enumeration of the rationals in $[0, 1]$, then $\mathrm{sp}(A) = [0, 1]$. In Example 3.2.2 we considered an operator with spectrum $[0, 1]$ but no eigenvectors. Although the present example and Example 3.2.2 exhibit self-adjoint operators with the same spectrum, and, in a sense still to be made precise, both of these operators have spectra without "multiplicity"; these operators are quite different structurally. One has an orthonormal basis of eigenvectors, while the other has not a single eigenvector. In the finite-dimensional case, self-adjoint operators having the same spectrum, each without multiplicity, have identical structure (are "unitarily equivalent"). ■

With the aid of an extension of the "approximate eigenvector" technique encountered in Example 3.2.2, we shall be able to extend to the spectra of self-adjoint, positive, and unitary operators, the information we have about the eigenvalues for the corresponding operators with an orthonormal basis of eigenvectors.

3.2.13. Lemma. *If $\mathscr{H}$ is a Hilbert space and $A$ is a normal operator in $\mathscr{B}(\mathscr{H})$, then $\lambda \in \mathrm{sp}(A)$ if and only if there is a sequence $\{x_n\}$ of unit vectors in $\mathscr{H}$ such that $\|(A - \lambda I)x_n\| \to 0$ as $n \to \infty$.*

*Proof.* Since $A$ is normal, $A - \lambda I$ is a normal operator in $\mathscr{B}(\mathscr{H})$. From Lemma 2.4.8, $A - \lambda I$ fails to have a bounded two-sided inverse (that is,

$\lambda \in \mathrm{sp}(A)$) if and only if

$$\inf\{\|(A - \lambda I)x\| : \|x\| = 1, x \in \mathscr{H}\} = 0.$$

Thus $\lambda \in \mathrm{sp}(A)$ if and only if there is a sequence $\{x_n\}$ of unit vectors in $\mathscr{H}$ such that $\|(A - \lambda I)x_n\| \to 0$ as $n \to \infty$. ∎

3.2.14. THEOREM. *If $\mathscr{H}$ is a Hilbert space and $T \in \mathscr{B}(\mathscr{H})$, then*

(i) $\mathrm{sp}(T)$ *consists of real numbers if $T$ is self-adjoint;*
(ii) $\mathrm{sp}(T)$ *consists of non-negative real numbers if $T$ is a positive operator;*
(iii) $\mathrm{sp}(T) \subseteq \{0, 1\}$ *if $T$ is a projection;*
(iv) $\mathrm{sp}(T)$ *consists of complex numbers of modulus* 1 *if $T$ is a unitary operator;*
(v) $\mathrm{sp}(T^*)$ *consists of the complex conjugates of numbers in* $\mathrm{sp}(T)$.

*Proof.* Suppose $T$ is self-adjoint and $\lambda \in \mathrm{sp}(T)$. From Lemma 3.2.13, there is a sequence of unit vectors $x_n$ such that $\|(T - \lambda I)x_n\| \to 0$ as $n \to \infty$. Then $\langle (T - \lambda I)x_n, x_n \rangle \to 0$ as $n \to \infty$. Since $\langle \lambda x_n, x_n \rangle = \lambda$, $\langle Tx_n, x_n \rangle$ tends to $\lambda$. But $\langle Tx_n, x_n \rangle$ is real. Thus $\lambda$ is real, and (i) follows.

In the same way, if $T \geqslant 0$, then $\langle Tx_n, x_n \rangle \geqslant 0$ and $\lambda \geqslant 0$. Thus (ii) is established. If $T$ is a projection, then $T^2 = T$, so that

$$T(T - \lambda I)x_n = (1 - \lambda)Tx_n \to 0.$$

Thus $(1 - \lambda)\lambda x_n \to 0$. But $\|x_n\| = 1$. Hence $(1 - \lambda)\lambda = 0$, and $\lambda$ is either 0 or 1, so that (iii) is established.

If $T$ is unitary, then $1 = \langle x_n, x_n \rangle = \langle Tx_n, Tx_n \rangle$. Since $\langle \lambda x_n, \lambda x_n \rangle = |\lambda|^2$, and $\langle Tx_n, Tx_n \rangle - \langle \lambda x_n, \lambda x_n \rangle \to 0$ as $n \to \infty$, $1 = |\lambda|^2$ and (iv) follows.

From the properties of the adjoint operation on $\mathscr{B}(\mathscr{H})$ (see Theorem 2.4.2), $B^*$ is a bounded inverse to $T^* - \bar{\lambda} I$ if and only if $B$ is a bounded inverse to $T - \lambda I$, and (v) follows. ∎

With the added assumption that $T$ is normal, the converses to (i)–(iv) of Theorem 3.2.14 are valid. It is more convenient to establish these after the spectral theory of normal operators has been developed (see Theorem 4.4.5).

3.2.15. PROPOSITION. *If $\mathscr{H}$ is a Hilbert space and $A$ is a self-adjoint operator in $\mathscr{B}(\mathscr{H})$, then at least one of $\|A\|$ or $-\|A\|$ is in* $\mathrm{sp}(A)$.

*Proof.* By working with $\|A\|^{-1}A$ in place of $A$, we may assume that $\|A\| = 1$. In this case, there is a sequence $\{x_n\}$ of unit vectors such that $\|Ax_n\| \to 1$ as $n \to \infty$. Thus

$$\|(I - A^2)x_n\|^2 = \|x_n\|^2 + \|A^2 x_n\|^2 - 2\,\mathrm{Re}\langle A^2 x_n, x_n \rangle \leqslant 2 - 2\|Ax_n\|^2 \to 0$$

as $n \to \infty$. From Lemma 3.2.13, $1 \in \mathrm{sp}(A^2)$; and by Proposition 3.2.10, $1 \in (\mathrm{sp}(A))^2$. Thus 1 or $-1$ is in $\mathrm{sp}(A)$. ∎

It follows from the preceding proposition that $r(A) = \|A\|$ when $A$ is self-adjoint. More generally, $r(T) = \|T\|$ when $T$ is normal. These facts will follow directly from a general formula for the spectral radius that will be developed in Section 3.3 (see Theorem 3.3.3 and Proposition 4.1.1(i)).

3.2.16. EXAMPLE. There is no difficulty extending the construction used in Example 3.2.2 to identify the spectrum of more general "multiplication operators" (see Example 2.4.11). If $(X, \mu)$ is a $\sigma$-finite measure space and $f$ is an essentially bounded measurable function on $X$, then $M_f(g) = fg$ defines a bounded operator $M_f$ on $L_2(X, \mu)$. The *essential range* $\mathrm{sp}(f)$ of $f$ is the set of complex numbers $\lambda$ such that $\mu(f^{-1}(\mathcal{O})) > 0$ for each open subset $\mathcal{O}$ of $\mathbb{C}$ containing $\lambda$.

Suppose $\lambda \in \mathrm{sp}(f)$. For each positive integer $n$, let $y_n$ be the characteristic function of a measurable subset of $f^{-1}(\mathcal{O}_n)$ of finite positive $\mu$-measure $a_n$, where $\mathcal{O}_n$ is the open disk in $\mathbb{C}$ with center $\lambda$ and radius $n^{-1}$. Then $\{x_n\}$ is a sequence of unit vectors, where $x_n = a_n^{-1/2} y_n$, and

$$\|(M_f - \lambda I)x_n\|^2 = \int_{f^{-1}(\mathcal{O}_n)} |f(p) - \lambda|^2 a_n^{-1} y_n(p)\, d\mu(p) \leqslant \frac{1}{n^2}.$$

Thus $\|(M_f - \lambda I)x_n\| \to 0$ as $n \to \infty$, and $\lambda \in \mathrm{sp}(M_f)$.

Conversely, if $\lambda \notin \mathrm{sp}(f)$ there is a disk $\mathcal{O}_n$ of radius $n^{-1}$ with center $\lambda$ such that $\mu(f^{-1}(\mathcal{O}_n)) = 0$. Then $1/(f - \lambda)$ is a measurable function $g$ with an essential bound $n$, and $M_g$ is a two-sided bounded inverse to $M_f - \lambda I$. Thus $\lambda \notin \mathrm{sp}(M_f)$. It follows that $\mathrm{sp}(M_f) = \mathrm{sp}(f)$ (and that $r(M_f) = \|f\|_\infty = \|M_f\|$). ■

3.2.17. EXAMPLE. Let $\mathscr{H}$ be a separable Hilbert space and $\{e_n : n = 0, \pm 1, \pm 2, \ldots\}$ an orthonormal basis for $\mathscr{H}$. The transformation $U$ on $\mathscr{H}$ such that $Ue_n = e_{n+1}$ is a unitary operator. From Theorem 3.2.14, $\mathrm{sp}(U)$ is a subset of $\mathbb{C}_1$, the complex numbers of modulus 1. If $\lambda \in \mathbb{C}_1$, then, in a formal sense, $\sum_{n=-\infty}^{\infty} \lambda^{-n} e_n$ is an "eigenvector" for $U$ corresponding to the "eigenvalue" $\lambda$. Although this sum is not a "genuine" element of $\mathscr{H}$, its partial sums, multiplied by suitable normalizing factors, provide us with a sequence of approximating eigenvectors. Specifically, let $x_n$ be $(2n+1)^{-1/2} \sum_{k=-n}^{n} \lambda^{-k} e_k$. Then $\|x_n\| = 1$ and

$$\begin{aligned}
\|(U - \lambda I)x_n\| &= (2n+1)^{-1/2} \Big\| \sum_{k=-n}^{n} \lambda^{-k} e_{k+1} - \sum_{k=-n}^{n} \lambda^{-(k-1)} e_k \Big\| \\
&= (2n+1)^{-1/2} \|\lambda^{-n} e_{n+1} - \lambda^{n+1} e_{-n}\| \\
&= 2^{1/2}(2n+1)^{-1/2} \to 0
\end{aligned}$$

as $n \to \infty$. Thus $\lambda \in \mathrm{sp}(U)$, and $\mathrm{sp}(U) = \mathbb{C}_1$.

Let $\mathscr{H}'$ be $L_2(\mathbb{C}_1)$ relative to Lebesgue measure on $\mathbb{C}_1$ normalized so that the total measure on $\mathbb{C}_1$ is 1. Then $\{z^n : n = 0, \pm 1, \ldots\}$ (where $z$ denotes the identity transform on $\mathbb{C}_1$) is an orthonormal basis for $\mathscr{H}'$ [the Weierstrass approximation theorem (see Remark 3.4.15) is used to show that this system generates $\mathscr{H}'$].

There is a unitary transformation $V$ of $\mathscr{H}'$ onto $\mathscr{H}$ such that $V(z^n) = e_n$. The multiplication operator $M_z$ on $\mathscr{H}'$ has spectrum $\mathbb{C}_1$, from Example 3.2.16. Note that $VM_zV^{-1} = U$. From this "unitary equivalence" of the "two-sided shift" operator with "multiplication by $z$," we can deduce the spectral properties of one from those of the other. ■

3.2.18. Example. With $\mathscr{H}$ a separable Hilbert space and $\{e_n : n = 0, 1, 2, \ldots\}$ an orthonormal basis for $\mathscr{H}$, let $W$ be the bounded operator on $\mathscr{H}$ such that $We_n = e_{n+1}$. In this case, $W$, the "one-sided shift" operator, is not a unitary operator on $\mathscr{H}$, since $e_0$ is not in its range (although $W$ is a unitary transformation of $\mathscr{H}$ onto the range of $W$). Note that if $|\lambda| = 1$, then $a_n \sum_{k=0}^{n} \lambda^{-k} e_k$ $(= x_n)$ is a unit vector in $\mathscr{H}$, where $a_n = (n+1)^{-1/2}$, and

$$\begin{aligned}\|(W - \lambda I)x_n\| &= a_n \Big\| \sum_{k=0}^{n} \lambda^{-k} e_{k+1} - \sum_{k=0}^{n} \lambda^{-(k-1)} e_k \Big\| \\ &= a_n \|\lambda^{-n} e_{n+1} - \lambda e_0\| = a_n \sqrt{2} \to 0\end{aligned}$$

as $n \to \infty$, since $a_n \to 0$. Thus $\lambda \in \mathrm{sp}(W)$. If $|\lambda| < 1$, then $W^*x = \lambda x$, where $x = \sum_{k=0}^{\infty} \lambda^k e_k$, for $\langle W^*x, e_k \rangle = \langle x, We_k \rangle = \lambda^{k+1}$. Thus $\lambda \in \mathrm{sp}(W^*)$ and $\bar{\lambda} \in \mathrm{sp}(W)$ (see Theorem 3.2.14(v)). Since $\|W\| = 1$, $\mathrm{sp}(W)$ is contained in the closed disk of radius 1 with center 0 in $\mathbb{C}$ (see Theorem 3.2.3). Thus $\mathrm{sp}(W)$ is this closed disk. ■

3.2.19. Example. Returning to $\mathscr{H}$ and $U$ of Example 3.2.17, we let $\mathfrak{A}$ be the Banach subalgebra of $\mathscr{B}(\mathscr{H})$ consisting of the norm closure of the algebra of polynomials of a single variable in $U$ (and $I$). If $U$ has an inverse in $\mathfrak{A}$, then that inverse must be $U^*$. Each polynomial in $U$, and hence each element of $\mathfrak{A}$, however, maps the closed subspace generated by $\{e_n : n = 1, 2, \ldots\}$ into itself, whereas $U^*$ does not map this space into itself ($U^*e_1 = e_0$). Thus $U^* \notin \mathfrak{A}$ and $U$ has no inverse in $\mathfrak{A}$. Stated in terms of spectrum, we have $0 \in \mathrm{sp}_{\mathfrak{A}}(U)$ but $0 \notin \mathrm{sp}_{\mathscr{B}(\mathscr{H})}(U)$.

Of course, as we pass from a Banach algebra to a Banach subalgebra, an element of the subalgebra may "lose its inverse." Thus, in theory, the spectrum may "grow" on passage to a subalgebra. The present example indicates that this increase of spectrum can occur in practice. We shall note (Proposition 4.1.5) that no change occurs in the spectrum when passing from a $C^*$-algebra to a $C^*$-subalgebra. This fact plays a crucial role in the application of spectral theory to $C^*$-algebras. ■

3.2.20. PROPOSITION. *The non-zero multiplicative linear functionals on a Banach algebra $\mathfrak{A}$ form a weak* compact subset of the unit ball of $\mathfrak{A}^{\#}$.*

*Proof.* From Remark 3.2.11, if $\rho$ is a non-zero multiplicative linear functional on $\mathfrak{A}$, then $\rho(A) \in \mathrm{sp}_{\mathfrak{A}}(A)$, for each $A$ in $\mathfrak{A}$. Thus, from Theorem 3.2.3, $|\rho(A)| \leqslant \|A\|$, and $\rho$ lies in the unit ball of $\mathfrak{A}^{\#}$.

The set of elements $\rho$ in $\mathfrak{A}^{\#}$ such that $\rho(AB) - \rho(A)\rho(B) = 0$ is weak* closed. The intersection of these sets (as $A$ and $B$ range through $\mathfrak{A}$) is the weak* closed set of multiplicative linear functionals in $\mathfrak{A}^{\#}$. The further condition, $\rho(I) = 1$, singles out the weak* closed subset consisting of non-zero multiplicative linear functionals on $\mathfrak{A}$ – a subset of the unit ball of $\mathfrak{A}^{\#}$. From Theorem 1.6.5(i), the unit ball of $\mathfrak{A}^{\#}$ is weak* compact, as is this closed subset. ∎

*The Banach algebra $L_1(\mathbb{R})$ and Fourier analysis.* In this subsection, we study the maximal ideals of the special Banach algebra $L_1(\mathbb{R})$ provided with *convolution* multiplication. By letting $L_1(\mathbb{R})$ act on $L_2(\mathbb{R})$ as a convolution–multiplication algebra, we define an algebra $\mathscr{A}_1(\mathbb{R})$ of operators acting on the Hilbert space $L_2(\mathbb{R})$. We adjoin $I$ to $\mathscr{A}_1(\mathbb{R})$ and take the (operator) norm closure to obtain another algebra $\mathfrak{A}_0(\mathbb{R})$ of operators on $L_2(\mathbb{R})$. The algebra $\mathfrak{A}_0(\mathbb{R})$ is an example of a class of operator algebras, abelian $C^*$-algebras, whose general properties will be studied intensively in Chapter 4. For the present, we identify the maximal ideals of $\mathfrak{A}_0(\mathbb{R})$ and use this information to develop some of the basic theory of Fourier transforms. The Banach algebra $L_1(\mathbb{R})$ and its ideal structure is the general framework for this theory. We shall have occasion to use Fourier transforms in Sections 9.2, 13.2, and 13.3. The results we obtain here on the ideal theory of $L_1(\mathbb{R})$ will play an important role in the analysis of the (continuous) homomorphisms of the additive group $\mathbb{R}$ into the group of unitary operators on a Hilbert space (see Stone's theorem (5.6.36)).

The various algebras we define may be viewed as generalizations of the complex group algebra of a finite group to the case of the group $\mathbb{R}$. The methods we describe apply to more general (locally compact) topological groups and can be extended without great difficulty to such abelian groups.

3.2.21. DEFINITION. With $f$ and $g$ measurable functions on $\mathbb{R}$, the *convolution* of $f$ and $g$ is the function $f * g$ whose domain consists of those real numbers $s$ for which the integral $\int_{\mathbb{R}} f(t)g(s-t)\,dt$ converges and whose value at $s$ is this integral. ∎

Since Lebesgue measure on $\mathbb{R}$ is invariant under the transformations $t \to -t$ and $t \to t+s$, for each real $s$, we have, for each $h$ in $L_1(\mathbb{R})$,

$$\int_{\mathbb{R}} h(s-t)\,dt = \int_{\mathbb{R}} h(s+t)\,dt = \int_{\mathbb{R}} h(t)\,dt.$$

3.2.22. PROPOSITION. (i) *If $f$ and $g$ are measurable functions on $\mathbb{R}$, then $f*g = g*f$.*

(ii) *If $f \in L_1(\mathbb{R})$ and $g \in L_p(\mathbb{R})$ (where $1 \leqslant p$), then $f*g \in L_p(\mathbb{R})$ and*

$$\|f*g\|_p \leqslant \|f\|_1 \cdot \|g\|_p. \tag{1}$$

(iii) *If $f \in L_1(\mathbb{R})$, $g, h \in L_p(\mathbb{R})$, and $a \in \mathbb{C}$, then both $f*(a \cdot g + h)$ and $a \cdot f*g + f*h$ are in $L_p(\mathbb{R})$ and*

$$f*(a \cdot g + h) = a \cdot f*g + f*h.$$

(iv) *If $f, g \in L_1(\mathbb{R})$ and $h \in L_p(\mathbb{R})$, then both $(f*g)*h$ and $f*(g*h)$ are in $L_p(\mathbb{R})$ and*

$$(f*g)*h = f*(g*h).$$

(v) *Provided with the mappings $(f,g) \to f*g$ and $f \to \|f\|_1$, $L_1(\mathbb{R})$ is a commutative Banach algebra.*

*Proof.* (i)

$$\begin{aligned}(f*g)(s) &= \int_{\mathbb{R}} f(t)g(s-t)\,dt = \int_{\mathbb{R}} f(s+t)g(-t)\,dt \\ &= \int_{\mathbb{R}} g(t)f(s-t)\,dt = (g*f)(s).\end{aligned}$$

(ii) If $h_s(t) = g(s-t)$ and $\mu(S) = \int_S |f(t)|\,dt$ for each measurable subset $S$ of $\mathbb{R}$, then $h_s \in L_p(\mathbb{R})$ and $\mu$ is a finite measure on $\mathbb{R}$. Applying the Hölder inequality [R: p. 62, Theorem 3.5] to $h_s$ and the constant function 1 relative to $\mu$, we have

$$\int |g(s-t)| \cdot 1\,d\mu(t) \leqslant \left(\int |h_s(t)|^p |f(t)|\,dt\right)^{1/p} \|f\|_1^{(p-1)/p}$$

and

$$\left(\int |g(s-t)| \cdot |f(t)|\,dt\right)^p \leqslant \|f\|_1^{p-1} \int |g(s-t)|^p |f(t)|\,dt.$$

Now $(s,t) \to |g(s-t)|^p f(t)$ is in $L_1(\mathbb{R} \times \mathbb{R})$ since

$$\begin{aligned}\int\left(\int |g(s-t)|^p |f(t)|\,ds\right)dt &= \int |f(t)| \left(\int |g(s-t)|^p\,ds\right)dt \\ &= \|g\|_p^p \int |f(t)|\,dt = \|f\|_1 \cdot \|g\|_p^p.\end{aligned}$$

Thus, using Fubini's theorem,

$$\int |(f*g)(s)|^p\,ds = \int \left| \int f(t)g(s-t)\,dt \right|^p ds$$

$$\leqslant \|f\|_1^{p-1} \int \left( \int |g(s-t)|^p |f(t)|\,dt \right) ds$$

$$= \|f\|_1^{p-1} \int \left( \int |g(s-t)|^p |f(t)|\,ds \right) dt = \|f\|_1^p \cdot \|g\|_p^p,$$

from which (1) follows.

(iii) From (ii), $f*(a\cdot g + h)$ and $a\cdot f*g + f*h$ are in $L_p(\mathbb{R})$. In particular

$$\int f(t)[a\cdot g(s-t) + h(s-t)]\,dt$$

and

$$a\int f(t)g(s-t)\,dt + \int f(t)h(s-t)\,dt$$

converge for almost all $s$, and (iii) follows.

(iv) From (ii), $(f*g)*h$ and $f*(g*h)$ are in $L_p(\mathbb{R})$. Now

$$(f*(g*h))(s) = \int f(t)\left(\int g(r)h(s-t-r)\,dr\right)dt$$

and

$$((f*g)*h)(s) = \int\left(\int f(t)g(r-t)\,dt\right)h(s-r)\,dr.$$

Since $|f|$ and $|g|$ are in $L_1(\mathbb{R})$, and $|h| \in L_p(\mathbb{R})$; when $f$, $g$, and $h$ are replaced by their respective absolute values, the last two integrals converge for almost all $s$. Fubini's theorem applies, and, for almost every $s$,

$$((f*g)*h)(s) = \int f(t)\left(\int g(r-t)h(s-r)\,dr\right)dt.$$

Since

$$\int g(r-t)h(s-r)\,dr = \int g(r)h(s-t-r)\,dr,$$

$(f*g)*h = f*(g*h)$.

(v) From (i)–(iv), the $L_1$-norm and convolution multiplication provide $L_1(\mathbb{R})$ with the structure of a commutative Banach algebra (without unit). ■

If we define $L_f(g)$ to be $f*g$, where $f \in L_1(\mathbb{R})$ and $g \in L_p(\mathbb{R})$, (ii) and (iii) of Proposition 3.2.22 tell us that $L_f$ is a bounded linear operator on $L_p(\mathbb{R})$ and $\|L_f\| \leqslant \|f\|_1$. From (iv) of that proposition (and right-distributivity of convolution multiplication – proved as in (iii)), we have that the mapping $f \to L_f$ is a homomorphism of the algebra $L_1(\mathbb{R})$ into $\mathscr{B}(L_p(\mathbb{R}))$. In particular, if $p = 2$, the image $\mathscr{A}_1(\mathbb{R})$ of $L_1(\mathbb{R})$ under this homomorphism is an algebra of operators on the Hilbert space $L_2(\mathbb{R})$. We denote by $\mathfrak{A}_1(\mathbb{R})$ the norm closure of $\mathscr{A}_1(\mathbb{R})$.

3.2.23. PROPOSITION. *For each $f$ in $L_1(\mathbb{R})$, $\frac{1}{2} \leqslant \|L_f - I\|$, so that $I \notin \mathfrak{A}_1(\mathbb{R})$. The linear space $\mathfrak{A}_0(\mathbb{R})$ generated by $I$ and $\mathfrak{A}_1(\mathbb{R})$ is a norm-closed commutative algebra of operators on $L_2(\mathbb{R})$ and $\mathfrak{A}_1(\mathbb{R})$ is a (proper) maximal ideal in $\mathfrak{A}_0(\mathbb{R})$.*

*Proof.* Let $v_n(t)$ be $(n/2)^{1/2}$ for $t$ in $[-n^{-1}, n^{-1}]$ and 0 for other values of $t$. If $f$ in $L_1(\mathbb{R})$ is such that $\|L_f - I\| < \frac{1}{2}$, then

$$\frac{1}{4} > \|f * v_n - v_n\|_2^2 = \int \left| \int f(s-t)v_n(t)\,dt - v_n(s) \right|^2 ds$$

$$\geqslant \int_{-1/n}^{1/n} \left| \int_{-1/n}^{1/n} \left(\frac{n}{2}\right)^{1/2} f(s-t)\,dt - \left(\frac{n}{2}\right)^{1/2} \right|^2 ds$$

$$= \frac{n}{2} \int_{-1/n}^{1/n} \left| \int_{-1/n}^{1/n} f(s-t)\,dt - 1 \right|^2 ds,$$

for all positive integers $n$. Since $f \in L_1(\mathbb{R})$, we can choose $n$ so large that $\int_{-2/n}^{2/n} |f(t)|\,dt \leqslant \frac{1}{2}$. With $s$ in $[-n^{-1}, n^{-1}]$,

$$\left| \int_{-1/n}^{1/n} f(s-t)\,dt \right| \leqslant \int_{-2/n}^{2/n} |f(t)|\,dt \leqslant \frac{1}{2},$$

so that $\frac{1}{4} \leqslant |\int_{-1/n}^{1/n} f(s-t)\,dt - 1|^2$ and

$$\frac{1}{4} \leqslant \frac{n}{2} \int_{-1/n}^{1/n} \left| \int_{-1/n}^{1/n} f(s-t)\,dt - 1 \right|^2 ds < \frac{1}{4}.$$

It follows, from this contradiction, that $\frac{1}{2} \leqslant \|L_f - I\|$ for each $f$ in $L_1(\mathbb{R})$, so that $I \notin \mathfrak{A}_1(\mathbb{R})$.

From Corollary 1.5.4, $\mathfrak{A}_0(\mathbb{R})$ is norm closed. It is a commutative algebra and $\mathfrak{A}_1(\mathbb{R})$ is a (proper) maximal ideal in it. ■

We denote by $\rho_\infty$ the multiplicative linear functional on $\mathfrak{A}_0(\mathbb{R})$ that assigns 1 to $I$ and 0 to each element of $\mathfrak{A}_1(\mathbb{R})$. There are other non-zero multiplicative linear functionals on (equivalently, by Corollary 3.2.5, proper maximal ideals in) $\mathfrak{A}_0(\mathbb{R})$. Our present goal is to describe each of these and establish a homeomorphism between the set of these functionals with the weak* topology (see the discussion preceding Proposition 1.3.5) and $\mathbb{R}$.

Although $L_1(\mathbb{R})$ has no unit, the sequence $\{u_n\}$, appearing in the lemma that follows, functions as a unit for $L_1(\mathbb{R})$ in many circumstances. The sequence $\{u_n\}$ is referred to as an *approximate identity* for $L_1(\mathbb{R})$. We shall encounter "approximate identities" again when we study the ideal theory of $C^*$-algebras (notably Lemma 4.2.11 and Proposition 4.2.12).

3.2.24. LEMMA. *If $\{u_n\}$ is a sequence of positive functions $u_n$ in $L_1(\mathbb{R})$ such that $\|u_n\|_1 = 1$ and $u_n(t) = 0$ when $t \notin [-n^{-1}, n^{-1}]$, then $\|f * u_n - f\|_p \to 0$ for each $f$ in $L_p(\mathbb{R})$.*

*Proof.* Assume, first, that $f$ is continuous on $\mathbb{R}$ and vanishes outside the finite interval $[a, b]$. Choose $\varepsilon$ positive. There is a positive integer $m$ such that $|f(s) - f(t)| < \varepsilon$ when $|t - s| < m^{-1}$. If $m < n$,

$$\begin{aligned}|(f * u_n)(s) - f(s)| &= \left| \int f(t)u_n(s-t)\,dt - \int f(s)u_n(t)\,dt \right| \\ &\leqslant \int |f(t) - f(s)|u_n(s-t)\,dt \leqslant \varepsilon.\end{aligned}$$

In addition $f * u_n - f$ vanishes outside of $[a - n^{-1}, b + n^{-1}]$. Hence

$$\|f * u_n - f\|_p \leqslant \varepsilon(b - a + 1)^{1/p},$$

and $\|f * u_n - f\|_p \to 0$ as $n \to \infty$. For an arbitrary $f$ in $L_p(\mathbb{R})$, we can choose $\{f_m\}$ so that $\|f - f_m\|_p \to 0$, $\|f_m\|_p \leqslant \|f\|_p$, where each $f_m$ is continuous and vanishes outside a finite interval. Then

$$\begin{aligned}\|f * u_n - f\|_p &\leqslant \|f * u_n - f_m * u_n\|_p + \|f_m * u_n - f_m\|_p + \|f_m - f\|_p \\ &\leqslant \|u_n\|_1 \cdot \|f - f_m\|_p + \|f_m * u_n - f_m\|_p + \|f_m - f\|_p \\ &\leqslant 2\|f - f_m\|_p + \|f_m * u_n - f_m\|_p.\end{aligned}$$

Choosing $m$ large enough, for all $n$, we have

$$\|f * u_n - f\|_p \leqslant 2\varepsilon/3 + \|f_m * u_n - f_m\|_p.$$

From what we have proved, with $n$ large enough, $\|f_m * u_n - f_m\|_p \leqslant \varepsilon/3$; and $\|f * u_n - f\|_p \leqslant \varepsilon$. ■

If $f \in L_1(\mathbb{R})$ and $\{u_n\}$ is as described in Lemma 3.2.24, then $\|L_f u_n - f\|_1 \to 0$. Thus $L_f u_n = 0$ for all $n$ if and only if $f = 0$. Since $u_n$ can be chosen in $L_p(\mathbb{R})$ for all $n$ and $p$, the homomorphism $f \to L_f : L_1(\mathbb{R}) \to \mathscr{B}(L_p(\mathbb{R}))$ is an isomorphism. From (1), this isomorphism does not increase norm.

We denote by $f^*$ the function whose value at $s$ is $\overline{f(-s)}$. If $g$ and $h$ are in $L_2(\mathbb{R})$, then

$$L_{f^*} = L_f^* \tag{2}$$

since

$$\langle L_{f^*}g, h\rangle = \langle f^* * g, h\rangle = \langle g * f^*, h\rangle = \int\left(\int g(t)f^*(s-t)\,dt\right)\overline{h(s)}\,ds$$

$$= \int g(t)\left(\int \overline{f(t-s)h(s)}\,ds\right)dt = \langle g, h * f\rangle = \langle g, L_f h\rangle.$$

Thus $\mathscr{A}_1(\mathbb{R})$, $\mathfrak{A}_1(\mathbb{R})$, and $\mathfrak{A}_0(\mathbb{R})$ contain $A^*$ when they contain $A$. The algebra $\mathfrak{A}_0(\mathbb{R})$ is an *abelian $C^*$-algebra* (with unit). Such algebras will be studied in Section 4.4.

We shall prove in Theorem 3.2.26 that the non-zero multiplicative linear functionals on $L_1(\mathbb{R})$ are in (natural) one-to-one correspondence with $\mathbb{R}$ – more precisely, with $\hat{\mathbb{R}}$, the *dual group* of $\mathbb{R}$. The elements of $\hat{\mathbb{R}}$ are the continuous homomorphisms (*characters*) of the (additive) group $\mathbb{R}$ into the (circle) group $\mathbb{T}_1$ of complex numbers of modulus 1 (and the product in $\hat{\mathbb{R}}$ is pointwise multiplication of characters). We describe the characters of $\mathbb{R}$ in the lemma that follows.

3.2.25. Lemma. *For each real number $r$, the equation*

$$\xi_r(t) = e^{itr}$$

*defines a character of $\mathbb{R}$. The mapping $r \to \xi_r$ is an isomorphism of (the group) $\mathbb{R}$ onto $\hat{\mathbb{R}}$.*

*Proof.* As defined, $\xi_r$ is clearly a character of $\mathbb{R}$. It remains to show that each character $\xi$ of $\mathbb{R}$ has the form $\xi_r$ for some real number $r$. Note, for this, that the continuity of $\xi$ at 0 implies that $\xi$ maps some interval $[-a, a]$, with $a$ positive, into the neighborhood

$$V = \{e^{is} : -\tfrac{1}{2}\pi < s < \tfrac{1}{2}\pi\}$$

of 1 in $\mathbb{T}_1$. The equation $\arg(\exp is) = s$ defines a homeomorphism arg from $V$ onto the interval $(-\tfrac{1}{2}\pi, \tfrac{1}{2}\pi)$. The continuous mapping

$$\eta = \arg \circ \xi : \quad [-a, a] \to (-\tfrac{1}{2}\pi, \tfrac{1}{2}\pi)$$

is additive, in the sense that $\eta(s + t) = \eta(s) + \eta(t)$ when $s$, $t$, and $s + t$ all lie in $[-a, a]$. From this it follows easily that

$$\eta(ma/n) = (m/n)\eta(a) = (ma/n)r_0$$

(where $r_0 = a^{-1}\eta(a)$), when $m$ and $n$ are integers and $|m| < n$. By continuity of $\eta$, we have that $\eta(t) = tr_0$ (and therefore $\xi(t) = \exp itr_0$) when $-a \leqslant t \leqslant a$. Accordingly, the two homomorphisms $\xi$ and $t \to \exp itr_0$ coincide on the interval $[-a, a]$ (which generates $\mathbb{R}$ as an additive group), and so coincide throughout $\mathbb{R}$. ∎

The function $\delta_r$ that takes the value 1 at $r$ and 0 at other points of $\mathbb{R}$ corresponds, of course, to the element 0 in $L_1(\mathbb{R})$. If we treat $\mathbb{R}$ as a discrete space (for the purpose of heuristics) and replace integration by discrete summation, then $\delta_r * f$ becomes $f_r$, where $f_r(t) = f(t - r)$. (That is, convolution by $\delta_r$ is translation by $-r$.) Of course $f_r \in L_p(\mathbb{R})$ if $f \in L_p(\mathbb{R})$ and $\|f_r\|_p = \|f\|_p$ so that $f \to f_r$ is an isometry of $L_p(\mathbb{R})$ onto itself. If $\rho$ is a multiplicative linear functional on $L_1(\mathbb{R})$, then $\rho(f_r)/\rho(f) = \rho(\delta_r)$ $(= \xi(r))$, provided $\rho(f) \neq 0$. Again, in a purely formal sense, $\delta_r * \delta_s = \delta_{r+s}$, so that $\xi$ is a homomorphism on $\mathbb{R}$. While $\delta_r$ is not available to us in a rigorous presentation, $f_r$ is. To reconstruct $\rho(f)$ from $\xi$, we replace $f$ by $\sum_r f(r)\,\delta_r$ or, more appropriately, by $\int f(r)\,\delta_r\,dr$. Then $\rho(f)$ is

$$\int f(r)\rho(\delta_r)\,dr = \int f(r)\xi(r)\,dr = \int f(r)\exp irq\,dr,$$

where $\xi(r) = \exp irq$ (as in Lemma 3.2.25). The accurate version of the preceding discussion appears in the theorem that follows.

3.2.26. THEOREM. *The non-zero linear functional $\rho$ on $L_1(\mathbb{R})$ is multiplicative if and only if there is a unique real number $r$ such that*

$$\rho(f) = \int f(t)e^{itr}\,dt = \hat{f}(r). \tag{3}$$

*Proof.* We show, first, that $\rho$, as defined in (3), is multiplicative. (It is non-zero, for if $\rho(f) = 0$ for each $f$ in $L_1(\mathbb{R})$, then the function $t \to \exp itr$ is 0 almost everywhere, which is absurd.) If $f, g \in L_1(\mathbb{R})$,

$$\begin{aligned} \widehat{f * g}(r) &= \int (f * g)(t)e^{itr}\,dt = \int\left(\int f(s)g(t-s)\,ds\right)e^{itr}\,dt \\ &= \int f(s)\left(\int g(t-s)e^{itr}\,dt\right)ds = \int f(s)\left(\int g(t)e^{i(t+s)r}\,dt\right)ds \\ &= \int f(s)e^{isr}\,ds \cdot \int g(t)e^{itr}\,dt = \hat{f}(r)\cdot\hat{g}(r). \end{aligned} \tag{4}$$

Assume, now, that $\rho$ is multiplicative. If we adjoin a unit to $L_1(\mathbb{R})$, as described in Remark 3.1.3, and extend $\rho$ linearly to the Banach algebra so obtained, after assigning 1 to the unit element, the extended functional is multiplicative. Proposition 3.2.20 applies, and the extended functional lies in the unit ball. In particular, we conclude that $\rho$ is a bounded linear functional (with norm not exceeding 1) on $L_1(\mathbb{R})$.

Since $\rho \neq 0$, there is a function $f$ in $L_1(\mathbb{R})$ such that $\rho(f) \neq 0$. Define $\xi(r)$ to be $\rho(f_r)/\rho(f)$, and note that, with $h$ and $g$ in $L_1(\mathbb{R})$,

$$(5) \qquad (h_r * g)(s) = \int h(t-r)g(s-t)\,dt = \int h(t)g(s-t-r)\,dt$$
$$= \int h(t)g_r(s-t)\,dt = (h * g_r)(s).$$

It follows that

$$(6) \qquad \rho(h_r)\rho(g) = \rho(h)\rho(g_r),$$

so that $\xi$ does not depend on the choice of $f$ (provided $\rho(f) \neq 0$). If $\rho(f_r) = 0$ for some $r$, then

$$0 = \rho(f_r)\rho(g) = \rho(f)\rho(g_r);$$

and $\rho(g_r) = 0$ for each $g$ in $L_1(\mathbb{R})$. In particular,

$$0 = \rho[(f_{-r})_r] = \rho(f),$$

contradicting the choice of $f$. Thus $\rho(f_r) \neq 0$ for all $r$, and

$$\xi(r+s) = \rho(f_{r+s})\rho(f)^{-1} = \rho[(f_r)_s]\rho(f_r)^{-1}\rho(f_r)\rho(f)^{-1} = \xi(s)\cdot\xi(r).$$

Since $\|f_r\|_1 = \|f\|_1$ and $\rho$ is bounded, $\{|\xi(r)| : r \in \mathbb{R}\}$ is a bounded set of positive numbers ($\xi(r) \neq 0$ as $\rho(f_r) \neq 0$). But $\xi(nr) = \xi(r)^n$ for each integer $n$, so that $|\xi(r)| = 1$. Hence $\xi$ is a homomorphism of $\mathbb{R}$ into $\mathbb{T}_1$.

The continuity of $\xi$ on $\mathbb{R}$ will follow once we establish the continuity of $\xi$ at 0; for $|\xi(r) - \xi(s)| = |\xi(r-s) - 1|$. The set of continuous functions vanishing outside some finite interval in $\mathbb{R}$ is dense in $L_1(\mathbb{R})$, so that there is such an $f$ with $\rho(f)$ not zero. Suppose $f$ vanishes outside of $[-n+1, n-1]$. If $|r| < 1$,

$$\|f_r - f\|_1 = \int_{-n}^{n} |f(t-r) - f(t)|\,dt;$$

and this integral is small for small $r$ by (uniform) continuity of $f$ on $[-n, n]$. Now,

$$|\xi(r) - 1| = |\rho(f_r - f)| \cdot |\rho(f)|^{-1} \leqslant \|f_r - f\|_1 \cdot |\rho(f)|^{-1},$$

which proves the continuity of $\xi$ at 0, hence on $\mathbb{R}$. By Lemma 3.2.25, there is an $r$ in $\mathbb{R}$ such that $\xi(t) = \exp itr$ for each $t$ in $\mathbb{R}$. To prove (3), note that, from Theorem 1.7.8, there is a function $g$ in $L_\infty(\mathbb{R})$ such that $\rho(f) = \int f(t)g(t)\,dt$ for each $f$ in $L_1(\mathbb{R})$. Hence

$$\int f(t)g(t)e^{isr}\,dt = \xi(s)\rho(f) = \rho(f_s) = \int f(t-s)g(t)\,dt = \int f(t)g(t+s)\,dt$$

for all $f$ in $L_1(\mathbb{R})$, and $g_{-s} = (\exp isr) \cdot g$ almost everywhere. If $h(t) = [\exp(-itr)]g(t)$, it follows that $h$ is a bounded measurable function that is (essentially) invariant under all translations. Thus there is a constant $c$ to which $h$ is equal almost everywhere. Hence $g(t) = c \exp itr$ for almost all $t$, and

$$c^{-1}\rho(f) = c^{-1}\int f(t)c\exp itr\,dt = \hat{f}(r). \tag{7}$$

Using (4), (7), and the assumption that $\rho$ is a multiplicative, we have

$$c^{-1}\rho(f)^2 = (c^{-1}\rho)(f*f) = \widehat{f*f}(r) = \hat{f}(r)^2 = c^{-2}\rho(f)^2.$$

Since $\rho(f) \neq 0$, $c = 1$ and (3) follows from (7).

If $\rho(f) = \hat{f}(r) = \hat{f}(r')$ for each $f$ in $L_1(\mathbb{R})$, then $\exp itr = \exp itr'$ for almost all $t$. Choosing $t$ small enough, we conclude that $r = r'$. ■

In the theorem that follows, we use the relation between $L_1(\mathbb{R})$ and $\mathfrak{A}_0(\mathbb{R})$ to identify the space $\mathscr{M}_0(\mathbb{R})$ of non-zero multiplicative linear functionals on $\mathfrak{A}_0(\mathbb{R})$ different from $\rho_\infty$. We show that $\mathscr{M}_0(\mathbb{R})$ with its weak* topology is homeomorphic to $\mathbb{R}$.

3.2.27. THEOREM. *There is a homeomorphism $\Lambda$ of $\mathscr{M}_0(\mathbb{R})$ onto $\mathbb{R}$ such that, if $\rho_0 \in \mathscr{M}_0(\mathbb{R})$ and $r = \Lambda(\rho_0)$,*

$$\rho_0(L_f) = \hat{f}(r) \tag{8}$$

*for each $f$ in $L_1(\mathbb{R})$.*

*Proof.* The restriction of $\rho_0$ to $\mathscr{A}_1(\mathbb{R})$ gives rise to a non-zero multiplicative linear functional $\rho$ on $L_1(\mathbb{R})$ though the isomorphism $f \to L_f$ of $L_1(\mathbb{R})$ with $\mathscr{A}_1(\mathbb{R})$ described in the comments following Lemma 3.2.24. From Theorem 3.2.26, there is a unique real number $r$ such that

$$\rho_0(L_f) = \rho(f) = \hat{f}(r) \tag{9}$$

for each $f$ in $L_1(\mathbb{R})$. We define $\Lambda(\rho_0)$ to be $r$. Since distinct functionals in $\mathscr{M}_0(\mathbb{R})$ give rise to distinct functionals on $L_1(\mathbb{R})$ by the process just described, (3) of Theorem 3.2.26 implies that $\Lambda$ is a one-to-one mapping into $\mathbb{R}$.

Given $r$ in $\mathbb{R}$, (9) defines a non-zero multiplicative linear functional $\rho_1$ on $\mathscr{A}_1(\mathbb{R})$. Although $|\rho_1(L_f)| \leqslant \|f\|_1$, it is not evident that

$$|\rho_1(L_f)| \leqslant \|L_f\|. \tag{10}$$

We prove (10), from which it will follow that $\rho_1$ has a (unique) bounded extension to $\mathfrak{A}_1(\mathbb{R})$ and, thence, to a functional $\rho_0$ in $\mathscr{M}_0(\mathbb{R})$ (by taking the extension to be 1 at $I$). This will show that $\Lambda$ maps $\mathscr{M}_0(\mathbb{R})$ onto $\mathbb{R}$.

To prove (10), we note, first, that $\mathscr{M}_0(\mathbb{R})$ is not empty. With $f$ a non-zero element of $L_1(\mathbb{R})$, from (2) and the comments following Proposition 3.2.22,

$$L_{f^**f} = L_{f^*}L_f = L_f^*L_f;$$

and $L_f^* L_f$ is a non-zero positive operator $A$ in $\mathscr{A}_1(\mathbb{R})$. From Proposition 3.2.15, $\|A\|$ is in the spectrum of $A$ relative to $\mathscr{B}(L_2(\mathbb{R}))$, and, hence, relative to the smaller algebra $\mathfrak{A}_0(\mathbb{R})$. From Remark 3.2.11, there is a multiplicative linear functional $\tau$ on $\mathfrak{A}_0(\mathbb{R})$ assigning $\|A\|$ to $A$. Since $0 \neq \tau \neq \rho_\infty$, $\tau \in \mathscr{M}_0(\mathbb{R})$. From the foregoing, $\tau$ corresponds to some real number $t_0$ and

$$\tau(L_f) = \int f(s) \exp ist_0 \, ds$$

for each $f$ in $L_1(\mathbb{R})$. If $\xi(t) = \exp itr'$ and $M_\xi(f) = \xi \cdot f$, then $M_\xi$ is an isometry with inverse $M_{\bar{\xi}}$ when viewed as a mapping on either $L_1(\mathbb{R})$ or $L_2(\mathbb{R})$. Since

$$\begin{aligned}(M_\xi L_f M_{\bar{\xi}}(g))(s) &= e^{isr'}(f * M_{\bar{\xi}} g)(s) = e^{isr'} \int f(t+s)(M_{\bar{\xi}} g)(-t)\, dt \\ &= e^{isr'} \int f(t+s) e^{itr'} g(-t)\, dt \\ &= \int e^{i(t+s)r'} f(t+s) g(-t)\, dt = (L_{M_\xi f}(g))(s);\end{aligned}$$

we have,

$$L_{M_\xi f} = M_\xi L_f M_{\bar{\xi}} \tag{11}$$

for each $f$ in $L_1(\mathbb{R})$. Thus $\|L_{M_\xi f}\| = \|L_f\|$. If $\|L_f\| \leqslant 1$, then $\|L_{M_\xi f}\| \leqslant 1$ and

$$\left|\int f(s) e^{is(t_0 + r')}\, ds\right| = \left|\int (M_\xi f)(s) e^{ist_0}\, ds\right| = |\tau(L_{M_\xi f})| \leqslant 1. \tag{12}$$

If we replace $r'$ by $r - t_0$ in (12), we have $|\rho_1(L_f)| \leqslant 1$, and (10) follows. Thus $\Lambda$ is a one-to-one mapping of $\mathscr{M}_0(\mathbb{R})$ onto $\mathbb{R}$.

We prove next that $\Lambda$ is continuous, where $\mathscr{M}_0(\mathbb{R})$ is provided with the weak* topology. Suppose $\rho$ and $\tau$ correspond to $r$ and $t$, respectively. Let $f(s)$ be $\exp(-isr)$ for $s$ in $[0, 1]$ and 0 for other $s$. Then

$$\rho(L_f) - \tau(L_f) = \int_0^1 (1 - e^{is(t-r)})\, ds = 1 + \frac{i}{u}(e^{iu} - 1);$$

where $u = t - r$. If $\pi \leqslant |u|$, then

$$\frac{1}{3} \leqslant 1 - \frac{|e^{iu} - 1|}{|u|} \leqslant |\rho(L_f) - \tau(L_f)|.$$

Since $\mathrm{Re}[\rho(L_f) - \tau(L_f)] = 1 - u^{-1} \sin u$, which has an alternating series expansion, $u^2/3! - u^4/5! + \cdots$, whose terms are monotone decreasing to 0 when $|u| \leqslant \pi$,

$$\frac{u^2}{12} \leqslant \frac{u^2}{3!} - \frac{u^4}{5!} \leqslant |\mathrm{Re}[\rho(L_f) - \tau(L_f)]| \leqslant |\rho(L_f) - \tau(L_f)|$$

for such $u$. Thus $|t - r|$ is small when $|\rho(L_f) - \tau(L_f)|$ is small.

The space $\mathscr{M}(\mathbb{R})$ of non-zero multiplicative linear functionals on $\mathfrak{A}_0(\mathbb{R})$ is weak* compact, from Proposition 3.2.20, so that it is $\mathscr{M}_0(\mathbb{R})$ "compactified" by adjoining the point $\rho_\infty$. Let $\Lambda'$ be the one-to-one mapping of $\mathscr{M}(\mathbb{R})$ onto the one-point compactification $\{\mathbb{R}, \infty\}$ of $\mathbb{R}$ that restricts to $\Lambda$ on $\mathscr{M}_0(\mathbb{R})$ (and assigns $\infty$ to $\rho_\infty$). If we prove that $\Lambda'$ is continuous at $\rho_\infty$, then, since $\Lambda$ is continuous on $\mathscr{M}_0(\mathbb{R})$, $\Lambda'$ is a continuous one-to-one mapping between the compact spaces $\mathscr{M}(\mathbb{R})$ and $\{\mathbb{R}, \infty\}$ and, hence, $\Lambda'$ is a homeomorphism. It will follow that $\Lambda$ is a homeomorphism of $\mathscr{M}_0(\mathbb{R})$ onto $\mathbb{R}$. To prove the continuity of $\Lambda'$ at $\rho_\infty$, choose an integer $n$ greater than 1 and let $f$ be the characteristic function of $[0, n^{-1}]$. If $\tau \in \mathscr{M}_0(\mathbb{R})$, $t = \Lambda(\tau)$, and $|\tau(L_f) - \rho_\infty(L_f)| < n^{-3}$, then

$$\tau(L_f) = \int_0^{1/n} e^{ist}\, ds = t^{-1} \sin n^{-1}t - it^{-1}[\cos n^{-1}t - 1].$$

From the earlier alternating series computation (with $n^{-1}t$ for $u$),

$$|1 - nt^{-1} \sin n^{-1}t| \leqslant t^2/6n^2,$$

when $|t| \leqslant n\pi$, so that

$$(1/n)(1 - t^2/6n^2) \leqslant t^{-1} \sin n^{-1}t \leqslant |\tau(L_f)| \leqslant n^{-3}.$$

Thus $4n^2 \leqslant 6(n^2 - 1) \leqslant t^2$, and $2n \leqslant |t|$. We have shown that $|\Lambda(\tau)|$ is large if $|\tau(L_f)|$ is small, for $\tau$ in $\mathscr{M}_0(\mathbb{R})$. Hence $\Lambda'$ is continuous at $\rho_\infty$ on $\mathscr{M}(\mathbb{R})$. ■

The function $\hat{f}$ corresponding to $f$ in $L_1(\mathbb{R})$ (defined by (3)) is called the *Fourier transform* of $f$. It will be convenient to use this transform with other normalizations – that is, multiplied by constants (such as $(2\pi)^{-1/2}$), appropriate to the circumstances in which it is used. For the present, we call attention to some of the special information we have obtained about Fourier transforms.

3.2.28. Corollary. *The Fourier transform $f \to \hat{f}$ defined on $L_1(\mathbb{R})$ has the following properties*:

(i) *It is linear.*

(ii) (*Uniqueness.*) *It is one-to-one.*

(iii) (*Riemann–Lebesgue lemma.*) *It maps $L_1(\mathbb{R})$ into the space of continuous functions on $\mathbb{R}$ vanishing at $\infty$.*

(iv) (*Convolution formula.*) $\widehat{f * g}(r) = \hat{f}(r)\hat{g}(r)$.

*Proof.* (i) This assertion amounts to the linearity of $\rho$ in (3).

(ii) From (i), it suffices to prove that $f$ is the element 0 in $L_1(\mathbb{R})$ if $\hat{f}(r) = 0$ for all real $r$. Suppose $f \neq 0$. In the third paragraph of the proof of Theorem

3.2.27, we show that there is a $\tau$ in $\mathscr{M}_0(\mathbb{R})$ such that $\tau(L_f) \neq 0$. But $\tau(L_f) = \hat{f}(t)$ for some real $t$.

(iii) Again, from Theorem 3.2.27, $\Lambda$ is a homeomorphism, so that

$$r \xrightarrow{\Lambda^{-1}} \rho_0 \to \rho_0(L_f) = \hat{f}(r)$$

is continuous, where continuity of the second mapping is a consequence of the definition of the weak* topology on $\mathscr{M}_0(\mathbb{R})$. Since $\Lambda'^{-1}$ is continuous on $\{\mathbb{R}, \infty\}$, $\Lambda'^{-1}(\infty) = \rho_\infty$, and $\rho_\infty(L_f) = 0$; $f$ vanishes at $\infty$.

(iv) This assertion is a restatement of (4). ■

One important aspect of the Fourier transform that has not appeared in our discussion up to this point concerns "inversion," the process by which a function $f$ may be "reconstructed" from $\hat{f}$. Some of the known results in this area are delicate. We shall need only an easily available sampling (for application in Section 13.2), which we prove in Theorem 3.2.30. In preparation for the proof of that theorem, we recall that the operation $f \to f_r$ of "translating" functions is an isometry of $L_p(\mathbb{R})$ onto itself. Another fact about this operation is proved in the lemma that follows.

3.2.29. LEMMA. *If $f \in L_p(\mathbb{R})$, then*

$$\|f_t - f_{t_0}\|_p \to 0 \qquad (t \to t_0). \tag{13}$$

*Proof.* Since

$$\|f_t - f_{t_0}\|_p = \|(f_t - f_{t_0})_{-t_0}\|_p = \|f_{t-t_0} - f\|_p,$$

(13) will follow from

$$\|f_t - f\|_p \to 0 \qquad (t \to 0). \tag{14}$$

If we have proved (14) for each function $g$ in a dense subset $\mathscr{F}$ of $L_p(\mathbb{R})$, then, given a positive $\varepsilon$ and choosing $g$ in $\mathscr{F}$ such that $\|g - f\|_p < \varepsilon/3$, we have

$$\|f_t - f\|_p \leqslant \|f_t - g_t\|_p + \|g_t - g\|_p + \|g - f\|_p \leqslant \tfrac{2}{3}\varepsilon + \|g_t - g\|_p.$$

By assumption (on $\mathscr{F}$), there is a positive $\delta$ such that $\|g_t - g\|_p < \varepsilon/3$ when $|t| < \delta$, and, in this case, $\|f_t - f\|_p < \varepsilon$.

For $\mathscr{F}$, we choose the set of continuous functions on $\mathbb{R}$, each of which vanishes outside some finite interval. With $f$ in $\mathscr{F}$, (14) follows from uniform continuity of $f$ and the fact that $f_t - f$ vanishes outside a fixed finite interval for all small $t$. ■

It will be useful for the proof that follows to recall that (from an integration by parts)

$$\int_0^\infty \frac{1 - \cos t}{t^2}\, dt = \int_0^\infty \frac{\sin t}{t}\, dt,$$

and that

$$\int_0^\infty \frac{\sin t}{t}\,dt = \int_0^\infty \left(\int_0^\infty e^{-st}\sin t\,ds\right)dt = \int_0^\infty \left(\int_0^\infty e^{-st}\sin t\,dt\right)ds = \frac{\pi}{2}.$$

We are now in a position to prove our *inversion theorem.*

3.2.30. THEOREM. *If* $f \in L_1(\mathbb{R})$ *and*

$$\hat{f}(p) = (2\pi)^{-1/2}\int_{\mathbb{R}} f(s)e^{isp}\,ds,$$

*then*

$$f(s) = (2\pi)^{-1/2}\lim_{a\to\infty}\int_{-a}^{a}\left(1 - \frac{|p|}{a}\right)e^{-isp}\hat{f}(p)\,dp \tag{15}$$

*for each s at which f is continuous if* $f \in L_\infty(\mathbb{R})$. *If* $\hat{f} \in L_1(\mathbb{R})$, *then, for almost all s,*

$$f(s) = (2\pi)^{-1/2}\int_{\mathbb{R}} e^{-isp}\hat{f}(p)\,dp. \tag{16}$$

*Proof.* Let $h_a(p)$ be $(2\pi)^{-1/2}(1 - |p|/a)$ when $|p| \leqslant a$ and 0 when $a < |p|$. Then, since $h_a(p) = h_a(-p)$,

$$\hat{h}_a(t) = (2\pi)^{-1/2}\int_{\mathbb{R}} h_a(p)e^{itp}\,dp = 2(2\pi)^{-1/2}\int_0^\infty h_a(p)\cos tp\,dp = \frac{1 - \cos at}{\pi a t^2}$$

when $t \neq 0$; and $\hat{h}_a(0) = a(2\pi)^{-1}$. We have

$$\begin{aligned}
(2\pi)^{-1/2}\int_{-a}^{a}\left(1 - \frac{|p|}{a}\right)e^{-isp}\hat{f}(p)\,dp &= \int_{\mathbb{R}} h_a(p)e^{-isp}(2\pi)^{-1/2}\left(\int_{\mathbb{R}} f(t)e^{itp}\,dt\right)dp \\
&= (2\pi)^{-1/2}\int_{\mathbb{R}} h_a(p)\left(\int_{\mathbb{R}} f(t)e^{i(t-s)p}\,dt\right)dp \\
&= (2\pi)^{-1/2}\int_{\mathbb{R}} f(t)\left(\int_{\mathbb{R}} h_a(p)e^{i(t-s)p}\,dp\right)dt \\
&= \int_{\mathbb{R}} f(t)\hat{h}_a(t-s)\,dt \\
&= \int_{\mathbb{R}} f(t+s)\hat{h}_a(t)\,dt \ (= g_a(s)).
\end{aligned}$$

Now

$$(17)\quad g_a(s) = \int_{\mathbb{R}} f(t+s)\frac{1-\cos at}{\pi a t^2}\,dt = \pi^{-1}\int_{\mathbb{R}} f\left(\frac{t}{a}+s\right)\frac{1-\cos t}{t^2}\,dt$$

and, if $f \in L_\infty(\mathbb{R})$ and is continuous at $s$, then

$$\left|f\left(\frac{t}{a}+s\right)\right|\frac{1-\cos t}{t^2} \leqslant \|f\|_\infty \frac{1-\cos t}{t^2} \quad (\in L_1(\mathbb{R}))$$

for almost all $t$. Moreover, for all $t$,

$$f\left(\frac{t}{a}+s\right)\frac{1-\cos t}{t^2} \to f(s)\frac{1-\cos t}{t^2} \qquad (a \to \infty).$$

Thus, from the dominated convergence theorem and (17),

$$g_a(s) \to f(s)\cdot \pi^{-1}\int_{\mathbb{R}} \frac{1-\cos t}{t^2}\,dt = f(s) \qquad (a \to \infty).$$

Noting that

$$\int_{\mathbb{R}} \hat{h}_a(t)\,dt = 2\int_0^\infty \hat{h}_a(t)\,dt = \frac{2}{\pi}\int_0^\infty \frac{1-\cos at}{at^2}\,dt = 1,$$

we have

$$(18)\quad |g_a(s) - f(s)| \leqslant \int_{\mathbb{R}} |f(t+s) - f(s)|\hat{h}_a(t)\,dt = \int_{\mathbb{R}} |(f_{-t} - f)(s)|\hat{h}_a(t)\,dt.$$

Integrating both sides of (18) with respect to $s$ and using Fubini's theorem, we have

$$(19)\qquad \|g_a - f\|_1 \leqslant \int_{\mathbb{R}} \|f_{-t} - f\|_1 \hat{h}_a(t)\,dt.$$

Given a positive $\varepsilon$, there is a positive $r$ such that $\|f_{-t} - f\|_1 < \varepsilon/2$ when $|t| \leqslant r$, from Lemma 3.2.29. Thus, from (19),

$$\begin{aligned}
\|g_a - f\|_1 &\leqslant \int_{-r}^{r} \|f_{-t} - f\|_1 \hat{h}_a(t)\,dt + \int_{\mathbb{R}\setminus[-r,r]} \|f_{-t} - f\|_1 \hat{h}_a(t)\,dt \\
&\leqslant \frac{\varepsilon}{2}\int_{-r}^{r} \hat{h}_a(t)\,dt + 2\|f\|_1 \int_{\mathbb{R}\setminus[-r,r]} \frac{1-\cos at}{\pi a t^2}\,dt \\
&\leqslant \frac{\varepsilon}{2}\int_{\mathbb{R}} \hat{h}_a(t)\,dt + \frac{4\|f\|_1}{\pi a}\int_{\mathbb{R}\setminus[-r,r]} \frac{dt}{t^2}.
\end{aligned}$$

If $a$ is large, $\|g_a - f\|_1 < \varepsilon$, so that, as $a \to \infty$, $\{g_a\}$ converges in $L_1(\mathbb{R})$ to $f$. For some sequence of integers $\{n(j)\}$, monotone increasing to $\infty$, $\{g_{n(j)}\}$ converges to $f$ almost everywhere. If we assume, now, that $\hat{f} \in L_1(\mathbb{R})$, then $p \to h_{n(j)}(p)e^{-isp}\hat{f}(p)$ is a sequence of functions with absolute values dominated by $|\hat{f}|$ and tending to $(2\pi)^{-1/2}e^{-isp}\hat{f}(p)$ for each $p$ and each $s$. Hence $\{g_{n(j)}(s)\}$ tends to $(2\pi)^{-1/2}\int_{\mathbb{R}} e^{-isp}\hat{f}(p)\,dp$ for each $s$, and (16) is valid for almost every $s$. ■

We turn now to the $L_2$ theory of Fourier transforms. The theorem that follows, Plancherel's theorem, will be used in Section 13.2.

3.2.31. THEOREM. *There is a unitary operator $T$ on $L_2(\mathbb{R})$ such that*

$$(Tf)(p) = (2\pi)^{-1/2}\int_{\mathbb{R}} e^{isp}f(s)\,ds \qquad (p \in \mathbb{R})$$

*when $f$ is continuous with support in a finite interval. If $Rg = \overline{T(\bar{g})}$ for all $g$ in $L_2(\mathbb{R})$, then $R = T^*$.*

*Proof.* If $f$ is a continuous complex-valued function on $\mathbb{R}$ with support in a finite interval, then so is the function $g$ defined by

$$g(s) = \int_{\mathbb{R}} \overline{f(t)}f(s+t)\,dt.$$

Thus $g \in L_1(\mathbb{R})$, and, by Fubini's theorem,

$$\begin{aligned}\hat{g}(p) &= (2\pi)^{-1/2}\int_{\mathbb{R}} e^{isp}\left\{\int_{\mathbb{R}} \overline{f(t)}f(s+t)\,dt\right\}ds \\ &= \int_{\mathbb{R}}\left\{(2\pi)^{-1/2}\int_{\mathbb{R}} e^{i(s+t)p}f(s+t)\,ds\right\}e^{-itp}\overline{f(t)}\,dt \\ &= \hat{f}(p)\int_{\mathbb{R}} e^{-itp}\overline{f(t)}\,dt = (2\pi)^{1/2}|\hat{f}(p)|^2.\end{aligned}$$

Since $g$ is continuous at 0, it follows from the inversion theorem (3.2.30(15)) that

$$\begin{aligned}\int_{\mathbb{R}} |f(t)|^2\,dt = g(0) &= \lim_{a\to\infty}(2\pi)^{-1/2}\int_{-a}^{a}\left[1 - \frac{|p|}{a}\right]\hat{g}(p)\,dp \\ &= \lim_{n\to\infty}\int_{-n}^{n}\left[1 - \frac{|p|}{n}\right]|\hat{f}(p)|^2\,dp \\ &= \lim_{n\to\infty}\int_{\mathbb{R}} k_n(p)\,dp,\end{aligned}$$

where

$$k_n(p) = \begin{cases} [1 - |p|/n]|\hat{f}(p)|^2 & (|p| \leqslant n) \\ 0 & (|p| > n). \end{cases}$$

Since $\{k_n\}$ is an increasing sequence of $L_1$ functions with pointwise limit $|\hat{f}(p)|^2$, it follows from the monotone convergence theorem that $\hat{f} \in L_2(\mathbb{R})$, and

$$\int_{\mathbb{R}} |f(t)|^2 \, dt = \int_{\mathbb{R}} |\hat{f}(p)|^2 \, dp.$$

The set $\mathscr{K}(\mathbb{R})$ of continuous functions with support in a finite interval forms a dense linear subspace of $L_2(\mathbb{R})$. It follows from what we have proved thus far that there is an isometric linear operator $T$ acting on $L_2(\mathbb{R})$, such that $Tf = \hat{f}$ when $f \in \mathscr{K}(\mathbb{R})$.

The mapping $f \to \bar{f}$ is a conjugate-linear isometry of $L_2(\mathbb{R})$ onto $L_2(\mathbb{R})$, so that $R$, as defined in the statement of this theorem, is an isometric linear mapping on $L_2(\mathbb{R})$. To verify that $R = T^*$, it will suffice to prove that $\langle Tf, g \rangle = \langle f, Rg \rangle$ for $f$, $g$ in $\mathscr{K}(\mathbb{R})$ (since $T$ and $R$ are bounded and $\mathscr{K}(\mathbb{R})$ is dense in $L_2(\mathbb{R})$). For such $f$ and $g$, we have, with the aid of Fubini's theorem,

$$\begin{aligned} \langle Tf, g \rangle &= \int_{-\infty}^{\infty} (Tf)(t)\bar{g}(t) \, dt = \int_{-\infty}^{\infty} \left( (2\pi)^{-1/2} \int_{-\infty}^{\infty} f(s) e^{ist} \, ds \right) \bar{g}(t) \, dt \\ &= \int_{-\infty}^{\infty} f(s) \left( (2\pi)^{-1/2} \int_{-\infty}^{\infty} \bar{g}(t) e^{ist} \, dt \right) ds = \langle f, \overline{T(\bar{g})} \rangle \\ &= \langle f, Rg \rangle. \end{aligned}$$

Now the null space of $R$ is the orthogonal complement of the range of $T$, from Proposition 2.5.13. Thus, since $R$ is an isometry, $T$ maps $L_2(\mathbb{R})$ onto $L_2(\mathbb{R})$. Hence $T$ and $R$ are unitary operators. ∎

### 3.3. The holomorphic function calculus

We develop, in very brief form, the theory of holomorphic functions of a complex variable with values in a Banach space, together with the associated holomorphic function calculus of a Banach algebra element. One of our main aims is to derive the *spectral radius formula* (Theorem 3.3.3). Although no other serious use of the Banach-algebra-valued holomorphic function calculus will be made, it is a powerful technique in the subject; and its results follow easily from classical complex function theory and the basic principles of functional analysis.

*Holomorphic functions.* We need some concept of line integral of a Banach-space-valued function for this program. It will suffice for our purposes to use a "Riemann sum" limit, $\int_C f(z)\,dz$, of a continuous Banach-space-valued function $f$ on a "smooth" curve $C\,(= t \to z(t),\, a \leqslant t \leqslant b$, where $z$ is a continuously differentiable complex-valued function on $[a,b])$. The integral $\int_C f(z)\,dz\; (= \int_a^b f(z(t))z'(t)\,dt)$ is the norm limit of Riemann sums of the form

$$\text{(1)} \qquad \sum_{j=1}^{n} f(z(t_j'))[z(t_j) - z(t_{j-1})],$$

where, as customary,

$$a = t_0 < t_1 < \cdots < t_n = b, \qquad t_{j-1} \leqslant t_j' \leqslant t_j,$$

the limit taken as $\max\{|t_j - t_{j-1}| : j = 1, \ldots, n\}$ tends to 0. This theory is a straightforward extension of the classical theory (as presented, for example, in [R, Chapter 10]). By considering the Riemann sums (1), it follows that

$$\text{(2)} \qquad A\left(\int_C f(z)\,dz\right) = \int_C A(f(z))\,dz,$$

where $A$ is a bounded linear transformation from the range space of $f$ to some other Banach space. Our applications of (2) occur most often in the case where $A$ is a continuous linear functional. From (1) we also have

$$\text{(3)} \qquad \left\|\int_C f(z)\,dz\right\| \leqslant \int_a^b \|f(z(t))\| \cdot |z'(t)|\,dt = \int_C \|f(z)\| \cdot |dz|.$$

With $\mathfrak{A}$ a Banach space and $\mathcal{O}$ an open subset of $\mathbb{C}$, a mapping $f$ from $\mathcal{O}$ into $\mathfrak{A}$ is said to be *holomorphic* (on $\mathcal{O}$) when it is differentiable at each point $z_0$ of $\mathcal{O}$, in the sense that the limit of the usual difference quotient exists in the norm topology on $\mathfrak{A}$. In this case, we denote the limit by $f'(z_0)$. If $f$ is holomorphic on $\mathcal{O}$ with values in $\mathfrak{A}$ and $\rho$ is a bounded linear functional on $\mathfrak{A}$, then $\rho \circ f$ is a (classical) *complex-valued* holomorphic function on $\mathcal{O}$. This fact, coupled with the Hahn–Banach theorem and the other basic principles of linear functional analysis, provides one of the main methods for proving results in the Banach-space-valued holomorphic function theory. (An illustration of this technique occurs in the proof of Theorem 3.2.3, where we proved in the course of the argument that $z \to (zI - T)^{-1}$ is holomorphic on the set on which it is defined.) Applying this method to a function $f$ holomorphic in a region $\mathcal{O}$ containing a curve $C$ for which the classical Cauchy theorem and formula are valid, we have, as a consequence of the classical results,

$$0 = \int_C \rho(f(z))\,dz = \rho\left(\int_C f(z)\,dz\right)$$

and (for each $z_0$ "interior" to $C$)

$$0 = \rho(f(z_0)) - \frac{1}{2\pi i}\int_C \frac{\rho(f(z))}{z - z_0}\,dz = \rho\left(f(z_0) - \frac{1}{2\pi i}\int_C \frac{f(z)}{z - z_0}\,dz\right),$$

for every bounded linear functional $\rho$ on $\mathfrak{A}$. Thus

$$\int_C f(z)\,dz = 0 \tag{4}$$

and

$$f(z_0) = \frac{1}{2\pi i}\int_C \frac{f(z)}{z - z_0}\,dz \tag{5}$$

for each $z_0$ "interior" to $C$. For the purposes of application in this subject, the curves and regions needed present none of the possible topological intricacies appearing in the more general versions of the classical results: piecewise "smooth" curves and regions with "uncomplicated" boundaries suffice. Without the detailed discussion of such matters (appropriate to classical complex function theory), we speak of *smooth closed curves*.

The method of applying bounded functionals combined with an application of the Hahn–Banach theorem yields the fact that if a power series $\sum_{n=0}^{\infty} T_n(z - z_0)^n$, with coefficients $T_n$ in $\mathfrak{A}$, converges to 0 in norm throughout some open disk containing $z_0$, then each $T_n$ is 0. Repetition of the classical argument establishes the following theorems.

3.3.1. THEOREM. *Each function $f$ holomorphic on an open set $\mathcal{O}$ in $\mathbb{C}$ and taking values in a Banach space $\mathfrak{A}$ can be represented as a power series $\sum_{n=0}^{\infty} T_n(z - z_0)^n$, with coefficients in $\mathfrak{A}$, throughout the largest open disk with center $z_0$ contained in $\mathcal{O}$, for each $z_0$ in $\mathcal{O}$.*

3.3.2. THEOREM. *If $|z - z_0| < (\overline{\lim}\|T_n\|^{1/n})^{-1}$, then $\sum_{n=0}^{\infty}\|T_n\| \cdot |z - z_0|^n$ converges. If $(\overline{\lim}\|T_n\|^{1/n})^{-1} < |z - z_0|$, then $\sum_{n=0}^{\infty} T_n(z - z_0)^n$ does not converge in norm.*

By analogy with the classical situation, we refer to $(\overline{\lim}\|T_n\|^{1/n})^{-1}$ as the *radius of convergence* of the power series $\sum_{n=0}^{\infty} T_n(z - z_0)^n$. We use these considerations to establish the spectral radius formula in the theorem that follows.

3.3.3. THEOREM (Spectral radius formula). *An element $A$ of a Banach algebra $\mathfrak{A}$ has spectral radius $r(A)$ given by the formula*

$$r(A) = \lim_{n\to\infty} \|A^n\|^{1/n}. \tag{6}$$

*In particular, $\{\|A^n\|^{1/n}\}$ has a limit.*

*Proof.* The existence of the limit in (6) is established by proving that

$$\overline{\lim}\|A^n\|^{1/n} \leqslant r(A) \leqslant \underline{\lim}\,\|A^n\|^{1/n} \tag{7}$$

(so that $\underline{\lim}\,\|A^n\|^{1/n} = \overline{\lim}\,\|A^n\|^{1/n}$), which proves (6) as well. If $a \in \mathrm{sp}(A)$, then $a^n \in \mathrm{sp}(A^n)$ (see Proposition 3.2.10), so that (from Theorem 3.2.3) $|a^n| \leqslant \|A^n\|$. Hence $|a| \leqslant \underline{\lim}\,\|A^n\|^{1/n}$ and $r(A) \leqslant \underline{\lim}\,\|A^n\|^{1/n}$.

The argument used in the proof of Theorem 3.2.3 shows that $z \to (I - zA)^{-1}$ $(= f(z))$ is holomorphic where defined (with derivative $A(1 - zA)^{-2}$) and that the set of such $z$ is open and contains all (small) $z$ such that $\|zA\| < 1$. Thus, for small $z$, $f$ is defined and is represented by the power series $\sum_{n=0}^{\infty} A^n z^n$. From Theorem 3.3.1 (and the comment preceding it on the uniqueness of series representation), this series represents $f$ on the largest open disk with center 0 on which $f$ is defined. On the other hand, Theorem 3.3.2 informs us that this series fails to converge for $z$ of modulus exceeding $(\overline{\lim}\,\|A^n\|^{1/n})^{-1}$. Thus, if $0 \leqslant a' < \overline{\lim}\,\|A^n\|^{1/n}$, there is an $a$ such that $a' < |a|$ for which $I - a^{-1}A$ and, hence, $A - aI$ fail to have inverses in $\mathfrak{A}$. Therefore $a \in \mathrm{sp}(A)$ and $a' < r(A)$. Since $a'$ is an arbitrary non-negative number less than $\overline{\lim}\,\|A^n\|^{1/n}$, the inequality (7) follows. ■

In case $\lim \|A^n\|^{1/n} = 0$, the inverse of $\overline{\lim}\,\|A^n\|^{1/n}$ is interpreted, as is customary, as $\infty$. When this occurs, $r(A) = 0$ and $\mathrm{sp}(A)$ consists of 0 alone. We discussed an instance of this in the comment following Theorem 3.2.3. An element $A$ in $\mathfrak{A}$ for which $r(A) = 0$ is said to be a *generalized nilpotent* in $\mathfrak{A}$. If $A^n = 0$ for some positive integer $n$, we say that $A$ is *nilpotent* (so that a nilpotent element in $\mathfrak{A}$ is, in particular, a generalized nilpotent).

One notes from Theorem 3.3.3 that $r_{\mathscr{B}}(A) = r_{\mathfrak{A}}(A)$ when $A$ lies in the Banach subalgebra $\mathscr{B}$ of $\mathfrak{A}$. If $A$ and $B$ are commuting elements of $\mathfrak{A}$, applying Proposition 3.2.10 to the *commutative* Banach subalgebra $\mathscr{B}$ of $\mathfrak{A}$ that $A$ and $B$ generate, together with this observation, we have the following result.

3.3.4. COROLLARY. *If $A$ and $B$ are commuting elements in the Banach algebra $\mathfrak{A}$, then $r(AB) \leqslant r(A)r(B)$ and $r(A + B) \leqslant r(A) + r(B)$.*

*The holomorphic function calculus.* Turning now to the case where $\mathfrak{A}$ is a Banach algebra, we note that

$$A^n = \frac{1}{2\pi i}\int_C z^n(zI - A)^{-1}\,dz, \tag{8}$$

where $n$ is a positive integer, $A \in \mathfrak{A}$, and $C$ is a smooth closed curve whose interior contains $\mathrm{sp}(A)$. To see this, observe that $z \to (zI - A)^{-1}$ is holomorphic on $\mathbb{C}\backslash\mathrm{sp}(A)$ (as proved in Theorem 3.2.3). Employing the Cauchy theorem (see (4)) in the case of the $\mathfrak{A}$-valued function $z \to z^n(zI - A)^{-1}$, we may replace $C$ by the (circular) perimeter of a disk with center 0 and large radius. Assuming $C$ is this circle and $z$ is on $C$,

$$z^{n-1}(I - Az^{-1})^{-1} = z^{n-1}\sum_{k=0}^{\infty} A^k z^{-k} = \sum_{k=0}^{\infty} A^k z^{n-k-1}, \tag{9}$$

where convergence is in the norm topology (and uniform on $C$). It follows from (3) and this convergence that term-by-term integration of (9) is justified. Now

$$\int_C A^k z^{n-k-1}\,dz = \left(\int_C z^{n-k-1}\,dz\right) A^k = 0 \cdot A^k = 0$$

unless $k = n$, in which case the integral is $2\pi i A^n$. This proves (8). It follows that

$$f(A) = \frac{1}{2\pi i}\int_C f(z)(zI - A)^{-1}\,dz \tag{10}$$

for each polynomial $f$, when $C$ is as in (8).

With the foregoing in mind, we take (10) as the definition of $f(A)$ for holomorphic functions $f$. More precisely, when $f$ is holomorphic in an open set containing $\mathrm{sp}(A)$, we can choose a smaller open set $\mathcal{O}$ containing $\mathrm{sp}(A)$ whose boundary consists of a finite number of closed piecewise linear curves $C_1, \ldots, C_n$. If $C$ denotes the collection of these curves oriented in the customary way in complex function theory, then (10) defines $f(A)$. To find the smaller open set with boundary as described, an argument involving a square grid in the plane (with squares of diameter less than the distance from $\mathrm{sp}(A)$ to the boundary of the initial open set) will suffice. Since the integral in (10) converges in norm, from our discussion of line integrals, it represents an element $f(A)$ in $\mathfrak{A}$. From (4) (Cauchy's theorem) $f(A)$ is independent of the curve $C$ (consisting of a finite number of smooth closed curves constituting the boundary of an open set in which $f$ is holomorphic).

Let $\mathscr{H}(A)$ be the set of functions holomorphic in some open set containing $\mathrm{sp}(A)$ (the open set may vary with the function). The following two results constitute a "calculus" of such functions—the *holomorphic function calculus.*

3.3.5. THEOREM. *The mapping $f \to f(A)$ is a homomorphism from $\mathscr{H}(A)$ into $\mathfrak{A}$ for each $A$ in the Banach algebra $\mathfrak{A}$. If $f$ is represented by the power series $\sum_{n=0}^{\infty} a_n z^n$ throughout an open set containing* $\mathrm{sp}(A)$, *then*

$$f(A) = \sum_{n=0}^{\infty} a_n A^n. \tag{11}$$

*Proof.* Since

$$(af + g)(A) = \frac{1}{2\pi i}\int_C (af + g)(z)(zI - A)^{-1}\,dz = af(A) + g(A),$$

the mapping $f \to f(A)$ is linear. The proof that $f(A) \cdot g(A) = (f \cdot g)(A)$ requires more effort. Let $\mathcal{O}$ be an open set, containing $\mathrm{sp}(A)$, on which both $f$ and $g$ are holomorphic. We can choose open sets $\mathcal{O}_1$ and $\mathcal{O}_2$ such that

$$\mathrm{sp}(A) \subseteq \mathcal{O}_1, \qquad \mathcal{O}_1 \cup C_1 \subseteq \mathcal{O}_2, \qquad \mathcal{O}_2 \cup C_2 \subseteq \mathcal{O},$$

where $C_1$ and $C_2$, the boundaries of $\mathcal{O}_1$ and $\mathcal{O}_2$, consist of a finite number of smooth closed curves. Then

$$\begin{aligned}
f(A) \cdot g(A) &= \left(\frac{1}{2\pi i}\int_{C_1} f(z)(zI - A)^{-1}\,dz\right)\left(\frac{1}{2\pi i}\int_{C_2} g(w)(wI - A)^{-1}\,dw\right) \\
&= \left(\frac{1}{2\pi i}\right)^2 \int_{C_1}\int_{C_2} f(z)g(w)(zI - A)^{-1}(wI - A)^{-1}\,dz\,dw \\
&= \left(\frac{1}{2\pi i}\right)^2 \int_{C_1}\int_{C_2} f(z)g(w)\frac{[(zI - A)^{-1} - (wI - A)^{-1}]}{w - z}\,dz\,dw \\
&= \frac{1}{2\pi i}\int_{C_1} f(z)(zI - A)^{-1}\left(\frac{1}{2\pi i}\int_{C_2}\frac{g(w)}{w - z}\,dw\right)dz \\
&\quad - \left(\frac{1}{2\pi i}\right)^2 \int_{C_2} g(w)(wI - A)^{-1}\left(\int_{C_1}\frac{f(z)}{w - z}\,dz\right)dw \\
&= \frac{1}{2\pi i}\int_{C_1} f(z)g(z)(zI - A)^{-1}\,dz - \left(\frac{1}{2\pi i}\right)^2 \int_{C_2} g(w)(wI - A)^{-1}(0)\,dw \\
&= (f \cdot g)(A).
\end{aligned}$$

To prove the last assertion of the theorem, we may assume that $f$ is defined on the disk of convergence of the series. Let $C$ be a circle with center at 0 containing $\mathrm{sp}(A)$ in its interior and contained in an open set on which $f$ is holomorphic and represented by $\sum_{n=0}^{\infty} a_n z^n$. Then this series converges uniformly on $C$, so that, from (8),

$$\begin{aligned}
f(A) &= \frac{1}{2\pi i}\int_C f(z)(zI - A)^{-1}\,dz \\
&= \sum_{n=0}^{\infty} a_n\left(\frac{1}{2\pi i}\int_C z^n(zI - A)^{-1}\,dz\right) \\
&= \sum_{n=0}^{\infty} a_n A^n. \quad \blacksquare
\end{aligned}$$

In our next result, we identify the spectrum of $f(A)$. The special case where $f$ is a polynomial has been treated in Proposition 3.2.10.

3.3.6. THEOREM (Spectral mapping theorem). *If $A$ is an element of a Banach algebra $\mathfrak{A}$ and $f$ is holomorphic on an open neighborhood of* $\mathrm{sp}(A)$, *then*

$$\mathrm{sp}(f(A)) = \{f(a) : a \in \mathrm{sp}(A)\} \quad (= f(\mathrm{sp}(A))). \tag{12}$$

*Proof.* Suppose $a \in \mathrm{sp}(A)$, so that either $\mathfrak{A}(A - aI)$ or $(A - aI)\mathfrak{A}$ is a proper (left or right) ideal in $\mathfrak{A}$. Say, $\mathfrak{A}(A - aI)$ is a proper (left) ideal in $\mathfrak{A}$.

Then

$$2\pi i[f(A) - f(a)I] = \int_C f(z)[(zI - A)^{-1} - (z - a)^{-1}I]\,dz$$

$$= \left(\int_C [(zI - A)(z - a)]^{-1} f(z)\,dz\right)(A - aI) \in \mathfrak{A}(A - aI)$$

and $f(a) \in \mathrm{sp}(f(A))$. Thus

$$f(\mathrm{sp}(A)) \subseteq \mathrm{sp}(f(A)). \tag{13}$$

If $b \notin f(\mathrm{sp}(A))$, then $(f - b)^{-1}$ $(= g)$ is holomorphic on an open neighborhood of $\mathrm{sp}(A)$. From Theorem 3.3.5, $g(A)$ is a two-sided inverse to $f(A) - bI$ in $\mathfrak{A}$ (since $g \cdot (f - b)$ is 1 on an open neighborhood of $\mathrm{sp}(A)$). Hence $b \notin \mathrm{sp}(f(A))$, and

$$\mathrm{sp}(f(A)) \subseteq f(\mathrm{sp}(A)). \tag{14}$$

Combining (13) and (14), we have (12). ■

An interesting and simple corollary of the holomorphic function calculus and the spectral mapping theorem asserts that if the Banach algebra $\mathfrak{A}$ has an element $A$ whose spectrum is not connected, then $\mathfrak{A}$ has an idempotent $E$ different from 0 and $I$. To see this, suppose that $\mathrm{sp}(A) = S_1 \cup S_2$, where $S_1$ and $S_2$ are disjoint closed sets. Since $\mathrm{sp}(A)$ is compact (see Theorem 3.2.3), both $S_1$ and $S_2$ are compact. It follows that there are disjoint open sets $\mathcal{O}_1$ and $\mathcal{O}_2$ such that $S_1 \subseteq \mathcal{O}_1$ and $S_2 \subseteq \mathcal{O}_2$. The function $f$ taking the values 1 on $\mathcal{O}_1$ and 0 on $\mathcal{O}_2$ is holomorphic on $\mathcal{O}_1 \cup \mathcal{O}_2$, an open neighborhood of $\mathrm{sp}(A)$. Then $f(A)$ is an idempotent $E$ since $E = f(A) = (f^2)(A) = f(A) \cdot f(A) = E^2$, and $\mathrm{sp}(E) = \{0, 1\}$. Thus $E$ is neither 0 nor $I$.

3.3.7. Corollary. *If $A$ is in the Banach algebra $\mathfrak{A}$ and* $\mathrm{sp}(A)$ *is not connected, then $\mathfrak{A}$ contains an idempotent different from* 0 *and $I$.*

The composite-function result that follows is an important addition to the function calculus.

3.3.8. Theorem. *If $A$ is an element of the Banach algebra $\mathfrak{A}$, $g \in \mathscr{H}(A)$, and $f \in \mathscr{H}(g(A))$, then $f \circ g \in \mathscr{H}(A)$ and $(f \circ g)(A) = f(g(A))$.*

*Proof.* By assumption, $f$ is holomorphic on an open set $\mathcal{O}_1$ containing $\mathrm{sp}(g(A))$ and $g$ is holomorphic on an open set $\mathcal{O}_2$ containing $\mathrm{sp}(A)$. From Theorem 3.3.6,

$$g(\mathrm{sp}(A)) = \mathrm{sp}(g(A)) \subseteq \mathcal{O}_1,$$

so that $\mathrm{sp}(A) \subseteq g^{-1}(\mathcal{O}_1)$. By continuity of $g$, $g^{-1}(\mathcal{O}_1) \cap \mathcal{O}_2$ is an open set $\mathcal{O}$ (containing $\mathrm{sp}(A)$) on which $f \circ g$ is holomorphic. Thus $f \circ g \in \mathscr{H}(A)$.

Choose open sets $\mathscr{U}$ and $\mathscr{U}_1$ with boundaries $C$ and $C_1$ consisting of a finite number of smooth closed curves such that

$$\mathrm{sp}(A) \subseteq \mathscr{U} \subseteq \mathscr{U} \cup C \subseteq \mathscr{O}$$

and

$$\mathrm{sp}(g(A)) \subseteq g(\mathscr{U} \cup C) \subseteq \mathscr{U}_1 \subseteq \mathscr{U}_1 \cup C_1 \subseteq \mathscr{O}_1.$$

By continuity of $g$ on $\mathscr{O}$, there is an open set $\mathscr{U}'$ such that $\mathscr{U} \cup C \subseteq \mathscr{U}' \subseteq \mathscr{O}$ and $g(\mathscr{U}') \subseteq \mathscr{U}_1$. Then, for each $w$ on $C_1$, $h_w$ is holomorphic on $\mathscr{U}'$, where $h_w(z) = [w - g(z)]^{-1}$. From Theorem 3.3.5, $h_w(A) = [wI - g(A)]^{-1}$ for each $w$ on $C_1$. It follows now that

$$\begin{aligned} f(g(A)) &= \frac{1}{2\pi i}\int_{C_1} f(w)[wI - g(A)]^{-1}\,dw \\ &= \frac{1}{2\pi i}\int_{C_1} f(w)h_w(A)\,dw \\ &= \frac{1}{2\pi i}\int_{C_1} f(w)\left(\frac{1}{2\pi i}\int_C h_w(z)(zI - A)^{-1}\,dz\right)dw \\ &= \frac{1}{2\pi i}\int_C (zI - A)^{-1}\left(\frac{1}{2\pi i}\int_{C_1} f(w)[w - g(z)]^{-1}\,dw\right)dz \\ &= \frac{1}{2\pi i}\int_C (f\circ g)(z)(zI - A)^{-1}\,dz = (f\circ g)(A). \quad \blacksquare \end{aligned}$$

We conclude this section with a result that allows us to treat convergence in the holomorphic function calculus.

3.3.9. Proposition. *If $\mathscr{O}$ is an open set containing* $\mathrm{sp}(A)$, *where $A$ is an element of the Banach algebra $\mathfrak{A}$, and $\{f_n\}$ is a sequence of functions holomorphic on $\mathscr{O}$ and converging uniformly to $f$ on compact subsets of $\mathscr{O}$, then $f \in \mathscr{H}(A)$ and* $\|f_n(A) - f(A)\| \to 0$ as $n \to \infty$.

*Proof.* Choose an open set $\mathscr{U}$ with boundary $C$ consisting of a finite number of smooth closed curves such that

$$\mathrm{sp}(A) \subseteq \mathscr{U} \subseteq \mathscr{U} \cup C \subseteq \mathscr{O}.$$

Since $\{f_n\}$ converges uniformly to $f$ on compact subsets of $\mathscr{O}$, $f$ is holomorphic on $\mathscr{O}$. Thus $f \in \mathscr{H}(A)$ and $\{f_n\}$ converges to $f$ uniformly on $C$. It follows that

$$\begin{aligned} 2\pi\|f_n(A) - f(A)\| &= \left\|\int_C [f_n(z) - f(z)](zI - A)^{-1}\,dz\right\| \\ &\leqslant k|C| \cdot \|f_n - f\|_C \to 0, \end{aligned}$$

as $n \to \infty$, where $k = \sup\{\|(zI - A)^{-1}\| : z \in C\}$, $|C|$ denotes the length of $C$, and $\|f_n - f\|_C = \{\sup|f_n(z) - f(z)| : z \in C\}$. ■

## 3.4. The Banach algebra $C(X)$

In Example 3.1.4 we introduced the algebra $C(X)$ of continuous complex-valued functions on the compact Hausdorff space $X$ together with its (supremum) norm and established that it is a Banach algebra. From the point of view of $C^*$-algebras, $C(X)$ is, by far, the most important example of a commutative Banach algebra. We shall see in Section 4.4 that $C(X)$ is *the* example of a commutative $C^*$-algebra. In its function algebra form it provides the basis for the spectral theory and "function calculus" of a self-adjoint (or normal) operator.

Our purpose in this section is to study $C(X)$ both with respect to its Banach-algebra structure and with respect to its order structure. We begin by identifying the closed ideals (and, hence, the maximal ideals in $C(X)$).

3.4.1. THEOREM. *If $\mathscr{I}$ is a closed ideal in $C(X)$, there is a closed subset $S$ of $X$ such that $\mathscr{I}$ is the set of all functions vanishing on $S$. If $S$ is a closed subset of $X$, the set of all functions vanishing on $S$ is a closed ideal in $C(X)$. The maximal ideals in $C(X)$ are those closed ideals for which the corresponding closed subset of $X$ (on which all the functions of the ideal vanish) consists of a single point.*

*Proof.* Since the set of points at which a continuous function vanishes is closed and the set of points $S$ at which all the functions of $\mathscr{I}$ vanish is an intersection of such sets, $S$ is a closed subset of $X$. We use the assumption that $\mathscr{I}$ is closed to show that a function $f$ that vanishes on $S$ is in $\mathscr{I}$. Note that this has the implication that $\mathscr{I} = C(X)$ if $S$ is null.

Suppose, then, that $f \in C(X)$ and $f$ vanishes on $S$. Given any positive $\varepsilon$, let $F_\varepsilon$ be the set of all points $p$ at which $|f(p)| \geqslant \varepsilon$, so that $F_\varepsilon$ is compact, and does not meet $S$. We shall construct an element $g_\varepsilon$ of $\mathscr{I}$ such that $0 \leqslant g_\varepsilon(p) \leqslant 1$ for all $p$ in $X$, while $g_\varepsilon$ is 1 throughout $F_\varepsilon$. Once this is done, we have $fg_\varepsilon \in \mathscr{I}$ (since $g_\varepsilon \in \mathscr{I}$); moreover, $\|f - fg_\varepsilon\| \leqslant \varepsilon$, because $|1 - g_\varepsilon(p)|$ never exceeds 1, and is zero at all points where $|f(p)| \geqslant \varepsilon$. Since $\mathscr{I}$ is closed, we can conclude that $f \in \mathscr{I}$.

We now construct $g_\varepsilon$ with the properties set out above. By definition of $S$, and since $F_\varepsilon$ does not meet $S$, for each point $p$ of $F_\varepsilon$, there is an $f_p$ in $\mathscr{I}$ such that $f_p(p) \neq 0$. Thus $\bar{f}_p \cdot f_p$ is (strictly) positive on some open neighborhood of $p$. A finite number of such neighborhoods cover the compact set $F_\varepsilon$. If $p_1, \ldots, p_n$ are the corresponding points, then $|f_{p_1}|^2 + \cdots + |f_{p_n}|^2$ is a function $h_\varepsilon$, in $\mathscr{I}$, which is non-negative on $X$ and positive throughout $F_\varepsilon$. From the compactness of $F_\varepsilon$, $h_\varepsilon$ has infimum $c$ ($> 0$) on $F_\varepsilon$. The equation $k_\varepsilon(p) = \max(h_\varepsilon(p), c)$ defines an element $k_\varepsilon$ of $C(X)$, and

$$k_\varepsilon(p) > 0, \qquad k_\varepsilon(p) \geqslant h_\varepsilon(p) \geqslant 0$$

for all $p$ in $X$, while $k_\varepsilon$ coincides with $h_\varepsilon$ on $F_\varepsilon$. Since $k_\varepsilon^{-1} \in C(X)$ and $h_\varepsilon \in \mathscr{I}$, it now suffices to take $h_\varepsilon k_\varepsilon^{-1}$ for $g_\varepsilon$.

The set of functions vanishing on an arbitrary subset of $X$ will be a closed ideal $\mathscr{I}$ in $C(X)$, but the set of points at which all the functions of $\mathscr{I}$ vanish will be a closed subset containing that set, its closure. To establish this Galois-like correspondence between closed subsets of $X$ and closed ideals in $C(X)$, note that if $S$ is a closed subset of $X$ and $\mathscr{I}$ is the closed ideal of functions in $C(X)$ vanishing on $S$, the Tietze extension theorem tells us that, given a point $p$ not in $S$, there is a continuous function $f$ on $X$ that is 0 on $S$ and 1 at $p$. Thus $f \in \mathscr{I}$ and $p$ is not in the closed set corresponding to $\mathscr{I}$. Hence the closed set corresponding to $\mathscr{I}$ is $S$.

Of course, now, the maximal ideals in $C(X)$ are those whose corresponding closed set consists of a point. ■

3.4.2. COROLLARY. *Each non-zero multiplicative linear functional $\rho$ on $C(X)$ corresponds to a point $p_0$ in $X$; and $\rho(f) = f(p_0)$ for each $f$ in $C(X)$.*

*Proof.* The kernel $\mathscr{M}$ of $\rho$ is a proper ideal (since $\rho \neq 0$) and is a maximal ideal of $C(X)$. From Theorem 3.4.1, $\mathscr{M}$ is the set of functions in $C(X)$ vanishing at some point $p_0$ of $X$. Since $\rho(1^2) = \rho(1)^2 = \rho(1) \neq 0$, $\rho(1) = 1$; and $f - \rho(f)1 \in \mathscr{M}$ for each $f$ in $C(X)$. Thus $f(p_0) = \rho(f)$ for each $f$ in $C(X)$. ■

We say that the functional $\rho$ in Corollary 3.4.2 is *evaluation* at $p_0$. The closed subset corresponding to an ideal (or just the common zeros of any set of functions) is referred to as the *kernel* of that ideal (or that set of functions). The (closed) ideal of functions vanishing on a set of points in $X$ is referred to as the *hull* of that set.

Corollary 3.4.2 (or Theorem 3.4.1) provides us with a means of recapturing the topological space $X$ from the algebraic structure of $C(X)$.

3.4.3. THEOREM. *A mapping $\varphi$ of $C(X)$ onto $C(Y)$, with $X$ and $Y$ compact Hausdorff spaces, is an algebraic isomorphism if and only if there is a homeomorphism $\eta$ of $Y$ onto $X$ such that $\varphi(f) = f \circ \eta$ for each $f$ in $C(X)$.*

*Proof.* If $\eta$ is a homeomorphism of $Y$ onto $X$ and $\varphi(f) = f \circ \eta$, then for each $f$ in $C(X)$, $f \circ \eta \in C(Y)$, and $\varphi$ is an algebraic isomorphism of $C(X)$ onto $C(Y)$.

Suppose, now, that $\varphi$ is an algebraic isomorphism of $C(X)$ onto $C(Y)$. If $\mathscr{M}$ is a maximal ideal in $C(Y)$, then $\varphi^{-1}(\mathscr{M})$ is a maximal ideal in $C(X)$. If $p$ is the point in $Y$ corresponding to $\mathscr{M}$, denote by $\eta(p)$ the point in $X$ corresponding to $\varphi^{-1}(\mathscr{M})$. Since each maximal ideal $\mathscr{M}_0$ in $C(X)$ has the form $\varphi^{-1}(\varphi(\mathscr{M}_0))$, with $\varphi(\mathscr{M}_0)$ a maximal ideal in $C(Y)$, $\eta$ is a one-to-one mapping of $Y$ onto $X$. If $f \in C(X)$, $f - f(\eta(p))1$ vanishes at $\eta(p)$ in $X$ for each $p$ in $Y$. Thus, by definition of $\eta$, $\varphi(f - f(\eta(p))1)$ vanishes at $p$. Hence $\varphi(f)(p) = f \circ \eta(p)$ for all $p$ in $Y$, and $\varphi(f) = f \circ \eta$. Since $X$ is a compact Hausdorff space, it is completely regular.

Hence sets of the form $f^{-1}(\mathcal{O})$, with $\mathcal{O}$ an open subset of $\mathbb{C}$ and $f$ in $C(X)$, form a subbase for the open sets in $X$. Now $\eta^{-1}(f^{-1}(\mathcal{O})) = \varphi(f)^{-1}(\mathcal{O})$; and $\varphi(f)^{-1}(\mathcal{O})$ is an open set in $Y$, since $\varphi(f)$ is a continuous function on $Y$. As the inverse images of subbasic open sets in $X$, under $\eta$, are open in $Y$, $\eta$ is continuous. Symmetrically, $\eta^{-1}$ is continuous; and $\eta$ is a homeomorphism of $Y$ onto $X$ such that $\varphi(f) = f \circ \eta$ for each $f$ in $C(X)$. ■

3.4.4. Remark. It follows from Theorem 3.4.3 that $\varphi$ maps real functions in $C(X)$ onto real functions and positive functions onto positive functions. It follows, too, that $\varphi$ is an isometry. Thus the assumption that $\varphi$ is an algebraic isomorphism entails very strict response from $\varphi$ in terms of other structure that $C(X)$ possesses (in particular, the norm and order structure on $C(X)$). These consequences of the assumption that $\varphi$ is an algebraic isomorphism become more apparent when we note that $\varphi$ "preserves" spectrum, that is $\mathrm{sp}_{C(X)}(f) = \mathrm{sp}_{C(Y)}(\varphi(f))$, and that the spectrum of $f$ relative to $C(X)$ is the range of the function $f$ (see Example 3.2.16, in this connection, where the spectrum of $M_f$ is the essential range of $f$). That $\varphi$ preserves spectrum is a consequence of the fact that spectrum is defined in terms of inverses and $\varphi$ preserves inverses. Of course, $f - \lambda 1$ fails to have an inverse in $C(X)$ if and only if it vanishes at some point of $X$, that is, if and only if $\lambda$ is in the range of $f$. The property of being real or positive for $f$ and the norm of $f$ are all determined by the range of $f$. We see, at the same time, that $\varphi$ preserves the operation of complex conjugation on $C(X)$ (that is, $\varphi(\bar{f}) = \overline{\varphi(f)}$)—for $\varphi$ maps real functions onto real functions. Note the formal similarity between the operation of complex conjugation of functions on $C(X)$ and the adjoint operation on $\mathcal{B}(\mathcal{H})$ (see Theorem 2.4.2). Both are conjugate-linear, involutory, (anti-)automorphisms on their respective algebras. With Theorem 4.4.3, this similarity becomes more than "formal." ■

When Theorem 4.4.3 has been established, the view of a $C^*$-algebra as a non-commutative generalization of $C(X)$ (that is, as a "non-commutative function algebra") will be quite plausible. Up to this point, we have been studying the Banach-algebra structure of $C(X)$. In application to non-commutative $C^*$-algebras, extending the order properties of $C(X)$, rather than its Banach-algebra structure, proves to be the more fruitful procedure.

The order structure of $C(X)$ is a natural partial ordering of its real-linear subspace $C(X, \mathbb{R})$, the continuous real-valued functions on $X$. It is introduced by means of the "cone" $\mathcal{P}$ of positive functions, which has the properties (of a *cone*):

(i) if $f$ and $-f$ are in $\mathcal{P}$, then $f = 0$;
(ii) if $a$ is a positive scalar and $f \in \mathcal{P}$, then $af \in \mathcal{P}$;
(iii) $f + g \in \mathcal{P}$ if $f$ and $g$ are in $\mathcal{P}$.

A real vector space $\mathscr{V}$ with such a cone is said to be a *partially ordered vector space*. Defining (as is usual in $C(X, \mathbb{R})$), $f \leqslant g$ when $g - f \in \mathscr{P}$, induces a partial ordering on $\mathscr{V}$. An element $I$ in $\mathscr{V}$ (the constant function 1 in $C(X, \mathbb{R})$) is said to be an *order unit* when, given any $f$ in $\mathscr{V}$, we have $-aI \leqslant f \leqslant aI$ for a suitable positive scalar $a$ (depending on $f$—in the case of $C(X, \mathbb{R})$, we may choose $a$ to be $\|f\|$).

3.4.5. DEFINITION. If $\mathscr{V}$ is a partially ordered vector space with order unit $I$, a linear functional $\rho$ on $\mathscr{V}$ is said to be *positive* when $\rho(f) \geqslant 0$ if $f \geqslant 0$. If, in addition, $\rho(I) = 1$, $\rho$ is said to be a *state* of $\mathscr{V}$. If $\rho$ is an extreme point of the (convex) family $\mathscr{S}(\mathscr{V})$ of states of $\mathscr{V}$, we say that $\rho$ is a *pure state* of $\mathscr{V}$. ■

Note that the set of positive linear functionals on $\mathscr{V}$ is a cone relative to which the dual space of $\mathscr{V}$ becomes a partially ordered vector space (generally without an order unit). A simple modification of the condition for a state to be pure proves useful to us.

3.4.6. LEMMA. *If $\mathscr{V}$ is a partially ordered vector space with order unit $I$, a state $\rho$ of $\mathscr{V}$ is pure if and only if each positive functional $\tau$ on $\mathscr{V}$ such that $\tau \leqslant \rho$ is a scalar multiple of $\rho$.*

*Proof.* If the stated condition holds for $\rho$ and $\rho = a\rho_1 + (1 - a)\rho_2$, with $0 < a < 1$ and $\rho_1$ and $\rho_2$ states of $\mathscr{V}$, then $0 \leqslant a\rho_1 \leqslant \rho$, so that $a\rho_1 = b\rho$. Since $\rho_1(I) = \rho(I) = 1$, $a = b$ and $\rho_1 = \rho$. Similarly $\rho_2 = \rho$, and $\rho$ is pure.

On the other hand, if $\rho$ is pure and $0 \leqslant \tau \leqslant \rho$, then $0 \leqslant \tau(I) \leqslant \rho(I) = 1$. If $\tau(I) = 0$, then, for each $f$ in $\mathscr{V}$, $0 = \tau(-aI) \leqslant \tau(f) \leqslant \tau(aI) = 0$ for some scalar $a$, and $\tau(f) = 0$; so $\tau = 0$ $(= 0 \cdot \rho)$. If $\tau(I) = 1$ $(= \rho(I))$, a similar argument shows that the positive functional $\rho - \tau$ is 0 (and $\tau = 1 \cdot \rho$). Finally, if $0 < \tau(I) < 1$, we have $\rho = (1 - b)\rho_1 + b\rho_2$, where $b = \tau(I)$ and $\rho_1, \rho_2$ are the states defined by $\rho_1 = (1 - b)^{-1}(\rho - \tau)$, $\rho_2 = b^{-1}\tau$. Since $\rho$ is pure, $\rho_2 = \rho$, and $\tau = b\rho$. In each case, $\tau$ is a multiple of $\rho$. ■

By expressing elements of $C(X)$ in terms of their real and imaginary parts, it is apparent that each (real-)linear functional on $C(X, \mathbb{R})$ extends uniquely to a linear functional on $C(X)$. A linear functional on $C(X)$ is said to be positive (or a state, or a pure state) if its restriction to $C(X, \mathbb{R})$ is positive (or a state, or a pure state). Since the states of $C(X, \mathbb{R})$ form a convex set with the pure states as extreme points, the same is true of $C(X)$.

The significance of states and pure states for $C^*$-algebras will appear in Sections 4.3 and 4.5. For the present, we establish that the pure states of $C(X)$ are precisely the (non-zero) multiplicative linear functionals on $C(X)$.

3.4.7. THEOREM. *A non-zero functional $\rho$ on $C(X)$ is a pure state of $C(X)$ if and only if it is multiplicative.*

*Proof.* If $\rho$ is multiplicative, from Corollary 3.4.2, there is a point $p_0$ in $X$ such that $\rho(f) = f(p_0)$ for all $f$ in $C(X)$. Since each continuous function vanishing at $p_0$ is a linear combination of positive continuous functions vanishing at $p_0$; if $0 < \tau \leqslant \rho$ for some linear functional $\tau$ on $C(X)$, then $\tau(f) = 0$ when $f(p_0) = 0$. Thus $\tau$ and $\rho$ are linear functionals on $C(X)$ with the same (maximal linear) null space and $\tau$ is a scalar multiple of $\rho$. From Lemma 3.4.6, $\rho$ is a pure state of $C(X)$.

Suppose $\rho$ is a pure state of $C(X)$. If $0 \leqslant f \leqslant 1$ and $\tau(g) = \rho(fg)$, then $\tau$ is a linear functional on $C(X)$ such that $0 \leqslant \tau \leqslant \rho$. Thus $\tau = a\rho$. If $\rho(h) = 0$, then $\rho(fh) = \tau(h) = a\rho(h) = 0$. Since each function $g$ in $C(X)$ is a linear combination of functions between 0 and 1, $\rho(gh) = 0$ for all $g$ in $C(X)$. Thus the null space of $\rho$ is an ideal—clearly maximal since the null space of $\rho$ is a maximal linear subspace. As $\rho(I) = 1$, $\rho$ is multiplicative. ∎

The linear order structure in $C(X)$ is strong enough to characterize it. We say that a mapping $\varphi$ between two partially ordered vector spaces (or two $C(X)$ spaces) that is a linear isomorphism of one onto the other is a *linear order isomorphism* when $\varphi(f) \geqslant 0$ if and only if $f \geqslant 0$.

3.4.8. COROLLARY. *A linear order isomorphism $\varphi$ of $C(X)$ onto $C(Y)$ such that $\varphi(1) = 1$ is an algebraic isomorphism.*

*Proof.* If $\rho_0$ is a pure state of $C(Y)$ corresponding to the point $p_0$ of $Y$, by Theorem 3.4.7, then $\rho_0 \circ \varphi$ is a pure state of $C(X)$ corresponding to a point $\eta(p_0)$ of $X$. Now, $\eta$ is a one-to-one mapping of $Y$ onto $X$, and $f(\eta(p_0)) = \rho_0(\varphi(f)) = \varphi(f)(p_0)$ for all $f$ in $C(X)$ and $p_0$ in $Y$. Thus $\varphi(f) = f \circ \eta$. As in Theorem 3.4.3, $\eta$ is a homeomorphism of $Y$ onto $X$; and $\varphi$ is an algebraic isomorphism. ∎

3.4.9. REMARK. The partial ordering on $C(X)$ induces a lattice structure on the set of real-valued functions. If $f$ and $g$ are two such functions, we define $(f \vee g)(p)$ to be $\max\{f(p), g(p)\}$ and $(f \wedge g)(p)$ to be $\min\{f(p), g(p)\}$ for each $p$ in $X$. Then

$$f \vee g = \tfrac{1}{2}(f + g) + \tfrac{1}{2}|f - g|, \qquad f \wedge g = \tfrac{1}{2}(f + g) - \tfrac{1}{2}|f - g|;$$

so that $f \vee g$ and $f \wedge g$ are in $C(X)$. Clearly, $f \vee g$ is the smallest function greater than both $f$ and $g$, and $f \wedge g$ is the largest function less than both $f$ and $g$. Moreover, $f = f^+ - f^-$, where $f^+ = f \vee 0$ and $f^- = -(f \wedge 0)$; so that $f$ is the difference of two positive functions in $C(X)$ (with disjoint supports).

The lattice structure assures us that $C(X)$ has the *Riesz decomposition property*: if $f \leqslant g_1 + g_2$, where $f, g_1$, and $g_2$ are positive functions in $C(X)$, then $f = f_1 + f_2$, where $0 \leqslant f_1 \leqslant g_1$ and $0 \leqslant f_2 \leqslant g_2$. To see this, let $f_1$ be $f \wedge g_1$ and $f_2$ be $f - f_1$. Then $0 \leqslant f_1 \leqslant g_1$, $f = f_1 + f_2$, and $f_2$ is positive. It remains to note that $f_2 \leqslant g_2$. If

$$g_2(p) < f_2(p) = f(p) - f_1(p) = f(p) - \min\{f(p), g_1(p)\}$$

for some $p$ in $X$, then, since $0 \leqslant g_2(p)$, $\min\{f(p), g_1(p)\} = g_1(p)$; and $g_1(p) + g_2(p) < f(p)$—contradicting the hypothesis, $f \leqslant g_1 + g_2$.

We shall note that the Riesz decomposition property causes the dual of $C(X)$ to have a lattice structure. To establish this, we must single out those functionals on $C(X)$ that preserve conjugation (that is, functionals $\rho$ such that $\rho(\bar{f}) = \overline{\rho(f)}$, for each $f$ in $C(X)$). We say that such a functional is *hermitian* (compare the introductory remarks in Section 4.3). Note that if $\rho$ is a linear functional on $C(X)$, then $\rho^*(f) = \overline{\rho(\bar{f})}$ defines a linear functional $\rho^*$ on $C(X)$. If $\rho_1 = \frac{1}{2}(\rho + \rho^*)$ and $\rho_2 = -(i/2)(\rho - \rho^*)$, then $\rho_1$ and $\rho_2$ are hermitian and $\rho = \rho_1 + i\rho_2$, so that each functional on $C(X)$ is a linear combination of two hermitian functionals. If $\rho$ is continuous, then so are $\rho^*$, $\rho_1$, and $\rho_2$. In general, little is lost by restricting attention to hermitian functionals. In this same line, let us note that each state $\rho$ of $C(X)$ is continuous (and has norm 1). To see this, we observe that $-\|f\|1 \leqslant f \leqslant \|f\|1$ for each real function $f$ in $C(X)$; so that $-\|f\| \leqslant \rho(f) \leqslant \|f\|$. For arbitrary $g$ in $C(X)$, we see that $(h, k) \to \rho(\bar{k} \cdot h)$ defines an inner product on $C(X)$; so that, by the Cauchy–Schwarz inequality (see Proposition 2.1.1(i)), when $\|g\| \leqslant 1$,

$$|\rho(1 \cdot g)|^2 \leqslant \rho(|g|^2) \cdot \rho(1^2) \leqslant 1. \quad \blacksquare$$

3.4.10. PROPOSITION. *The hermitian linear functionals in the (norm-)dual of $C(X)$ form a lattice.*

*Proof.* With $\rho_1$ and $\rho_2$ hermitian linear functionals on $C(X)$ and $f$ positive in $C(X)$, define

$$(1) \qquad (\rho_1 \vee \rho_2)(f) = \sup\{\rho_1(f_1) + \rho_2(f_2) : f_1, f_2 \in C(X), \; 0 \leqslant f_1, 0 \leqslant f_2, f = f_1 + f_2\}.$$

Note that, with $f_1, f_2$ as in (1),

$$|\rho_1(f_1) + \rho_2(f_2)| \leqslant \|\rho_1\| \cdot \|f_1\| + \|\rho_2\| \cdot \|f_2\| \leqslant (\|\rho_1\| + \|\rho_2\|)\|f\|,$$

so that $(\rho_1 \vee \rho_2)(f)$ is finite. The decomposition $f = f + 0 = 0 + f$ establishes that $\rho_1 \leqslant \rho_1 \vee \rho_2$ and $\rho_2 \leqslant \rho_1 \vee \rho_2$. If $\rho_1 \leqslant \tau$ and $\rho_2 \leqslant \tau$ for some hermitian linear functional $\tau$ on $C(X)$, then

$$\rho_1(f_1) + \rho_2(f_2) \leqslant \tau(f_1) + \tau(f_2) = \tau(f),$$

so that $\rho_1 \vee \rho_2 \leqslant \tau$. Once we show that $\rho_1 \vee \rho_2$ is linear on the cone of positive elements, and, by routine technique, that its extension to $C(X)$ is linear, we shall have that that extension is a (bounded) hermitian linear functional on $C(X)$ that is a least upper bound for $\rho_1$ and $\rho_2$.

We assume, until further notice, that all functions appearing in the argument are positive functions in $C(X)$. Choose $g_1, g_2$ and $h_1, h_2$ with $g$ equal to $g_1 + g_2$ and $h$ equal to $h_1 + h_2$ such that $\rho_1(g_1) + \rho_2(g_2)$ approximates $(\rho_1 \vee \rho_2)(g)$ and $\rho_1(h_1) + \rho_2(h_2)$ approximates $(\rho_1 \vee \rho_2)(h)$. Then

$$g_1 + h_1 + g_2 + h_2 = g + h,$$

and

$$\begin{aligned}\rho_1(g_1 + h_1) + \rho_2(g_2 + h_2) &= \rho_1(g_1) + \rho_2(g_2) + \rho_1(h_1) + \rho_2(h_2)\\ &\leqslant (\rho_1 \vee \rho_2)(g + h).\end{aligned}$$

Thus

$$(\rho_1 \vee \rho_2)(g) + (\rho_1 \vee \rho_2)(h) \leqslant (\rho_1 \vee \rho_2)(g + h). \tag{2}$$

Suppose $g+h=f_1+f_2$ and $\rho_1(f_1)+\rho_2(f_2)$ approximates $(\rho_1 \vee \rho_2)(g+h)$. The Riesz decomposition property permits us to express $f_1$ as $f_{11} + f_{12}$ and $f_2$ as $f_{21} + f_{22}$ with $f_{11} + f_{21}$ equal to $g$ and $f_{12} + f_{22}$ equal to $h$. The see this, note that $f_1 \leqslant g + h$ so that $f_1 = f_{11} + f_{12}$, where $f_{11} \leqslant g$ and $f_{12} \leqslant h$. Thus $f_2 = g + h - f_1 = g - f_{11} + h - f_{12}$. Letting $f_{21}$ be $g - f_{11}$ and $f_{22}$ be $h - f_{12}$, we have the desired decomposition. Hence

$$\begin{aligned}\rho_1(f_1) + \rho_2(f_2) &= \rho_1(f_{11}) + \rho_1(f_{12}) + \rho_2(f_{21}) + \rho_2(f_{22})\\ &\leqslant (\rho_1 \vee \rho_2)(g) + (\rho_1 \vee \rho_2)(h).\end{aligned}$$

Thus

$$(\rho_1 \vee \rho_2)(g + h) \leqslant (\rho_1 \vee \rho_2)(g) + (\rho_1 \vee \rho_2)(h). \tag{3}$$

Combining (2) and (3), we have the additivity of $\rho_1 \vee \rho_2$ on the cone of positive elements.

If $0 < a, f_1 + f_2 = f$, and $\rho_1(f_1) + \rho_2(f_2)$ approximates $(\rho_1 \vee \rho_2)(f)$, then $af_1 + af_2 = af$, so that

$$a(\rho_1(f_1) + \rho_2(f_2)) \leqslant (\rho_1 \vee \rho_2)(af).$$

Thus

$$a(\rho_1 \vee \rho_2)(f) \leqslant (\rho_1 \vee \rho_2)(af).$$

But then

$$a^{-1}(\rho_1 \vee \rho_2)(af) \leqslant (\rho_1 \vee \rho_2)(a^{-1}af) = (\rho_1 \vee \rho_2)(f).$$

Thus $a(\rho_1 \vee \rho_2)(f) = (\rho_1 \vee \rho_2)(af)$.

Suppose, now, that $f$ is an arbitrary real-valued function in $C(X)$. We can express $f$ as $f_1 - f_2$, with $f_1$ and $f_2$ positive functions in $C(X)$ (see Remark 3.4.9). Define $(\rho_1 \vee \rho_2)(f)$ to be $(\rho_1 \vee \rho_2)(f_1) - (\rho_1 \vee \rho_2)(f_2)$. If $f = g_1 - g_2$

with $g_1$ and $g_2$ positive functions in $C(X)$, then $f_1 + g_2 = g_1 + f_2$, so that

$$\begin{aligned}(\rho_1 \vee \rho_2)(f_1 + g_2) &= (\rho_1 \vee \rho_2)(f_1) + (\rho_1 \vee \rho_2)(g_2) = (\rho_1 \vee \rho_2)(g_1 + f_2) \\ &= (\rho_1 \vee \rho_2)(g_1) + (\rho_1 \vee \rho_2)(f_2),\end{aligned}$$

and

$$(\rho_1 \vee \rho_2)(f_1) - (\rho_1 \vee \rho_2)(f_2) = (\rho_1 \vee \rho_2)(g_1) - (\rho_1 \vee \rho_2)(g_2).$$

Thus $\rho_1 \vee \rho_2$ as extended to the space of real-valued functions in $C(X)$ is well defined (single valued). The additivity and positive homogeneity of $\rho_1 \vee \rho_2$ on the positive cone establish the additivity and homogeneity of $\rho_1 \vee \rho_2$ on the space of real-valued functions in $C(X)$. For arbitrary $g$ in $C(X)$, let $(\rho_1 \vee \rho_2)(g)$ be $(\rho_1 \vee \rho_2)(g_1) + i(\rho_1 \vee \rho_2)(g_2)$, where $g = g_1 + ig_2$ with $g_1$ and $g_2$ real-valued functions in $C(X)$.

There is no difficulty in seeing that $-((-\rho_1) \vee (-\rho_2))$ is a greatest lower bound for $\rho_1$ and $\rho_2$ in the space of hermitian linear functionals on $C(X)$.

The decomposition of an arbitrary function in $C(X)$ as a linear combination of positive functions combined with the observation (noted above)

$$|(\rho_1 \vee \rho_2)(f)| \leqslant (\|\rho_1\| + \|\rho_2\|)\|f\|$$

for positive functions shows us that $\rho_1 \vee \rho_2$, as extended to $C(X)$, is bounded. ■

3.4.11. PROPOSITION. *If $\rho$ is a (bounded) hermitian linear functional on $C(X)$, $\rho^+ = \rho \vee 0$, and $\rho^- = -(\rho \wedge 0)$, then*

$$\rho = \rho^+ - \rho^-, \qquad \rho^+ \wedge \rho^- = 0,$$

*and*

$$\|\rho\| = \|\rho^+\| + \|\rho^-\| = \rho^+(1) + \rho^-(1).$$

*Proof.* With $f$ a positive function in $C(X)$,

$$\begin{aligned}\rho^+(f) &= \sup\{\rho(f_1) : 0 \leqslant f_1 \leqslant f, f_1 \in C(X)\}, \\ -\rho^-(f) &= (\rho \wedge 0)(f) = -((-\rho) \vee 0)(f) \\ &= -\sup\{-\rho(g) : 0 \leqslant g \leqslant f, g \in C(X)\} \\ &= \inf\{\rho(g) : 0 \leqslant g \leqslant f, g \in C(X)\} \qquad (4) \\ &= \inf\{\rho(f - f_1) : 0 \leqslant f_1 \leqslant f, f_1 \in C(X)\} \\ &= \rho(f) - \sup\{\rho(f_1) : 0 \leqslant f_1 \leqslant f, f_1 \in C(X)\} \\ &= \rho(f) - \rho^+(f).\end{aligned}$$

Thus $\rho = \rho^+ - \rho^-$.

Again with $f$ a positive function in $C(X)$,

$$\begin{aligned}(\rho^{+} \wedge \rho^{-})(f) &= \inf\{\rho^{+}(f_1) + \rho^{-}(f - f_1) : 0 \leqslant f_1 \leqslant f, f_1 \in C(X)\}\\ &= \inf\{\rho(f_1) + \rho^{-}(f) : 0 \leqslant f_1 \leqslant f, f_1 \in C(X)\}\\ &= \rho^{-}(f) + \inf\{\rho(f_1) : 0 \leqslant f_1 \leqslant f, f_1 \in C(X)\} = 0\end{aligned}$$

(see (4) above).

If $\tau$ is a positive linear functional on $C(X)$ and $\tau(1) = 0$, then $\tau = 0$ (since 1 is an order unit), so that $\|\tau\| = \tau(1) = 0$. If $\tau(1) \neq 0$, then $\tau(1)^{-1}\tau$ is a state of $C(X)$, and, as noted in Remark 3.4.9, $\|\tau(1)^{-1}\tau\| = 1$, so that $\|\tau\| = \tau(1)$. Thus

$$\begin{aligned}\|\rho\| \leqslant \|\rho^{+}\| + \|\rho^{-}\| &= \rho^{+}(1) + \rho^{-}(1)\\ &= \sup\{\rho(f) : 0 \leqslant f \leqslant 1, f \in C(X)\}\\ &\quad - \inf\{\rho(g) : 0 \leqslant g \leqslant 1, g \in C(X)\}.\end{aligned}$$

For suitable $f$ and $g$ in $C(X)$, satisfying $0 \leqslant f \leqslant 1$ and $0 \leqslant g \leqslant 1$, $\rho(f) - \rho(g)$ $(= \rho(f - g))$ approximates $\rho^{+}(1) + \rho^{-}(1)$, and

$$\rho(f) - \rho(g) \leqslant \|\rho\| \cdot \|f - g\| \leqslant \|\rho\|.$$

Hence $\rho^{+}(1) + \rho^{-}(1) \leqslant \|\rho\|$, and this, with the previous inequalities, gives

$$\|\rho\| = \|\rho^{+}\| + \|\rho^{-}\| = \rho^{+}(1) + \rho^{-}(1). \quad ■$$

3.4.12. Remark. The decomposition of a bounded hermitian linear functional $\rho$ on $C(X)$ as $\rho^{+} - \rho^{-}$ proved in Proposition 3.4.11 is a special case (the commutative case) of the decomposition of positive linear functionals on a $C^*$-algebra established in Theorem 4.3.6. It is proved there that the decomposition with the norm property $\|\rho\| = \|\rho^{+}\| + \|\rho^{-}\|$ is unique. Of course, the special case of $C(X)$ is covered by this result. Nonetheless, it is useful to see an argument for the uniqueness in this simpler case. With the assurance that a precise proof is given for the general case, we can enjoy the luxury of arguing loosely (and possibly, therefore, more lucidly).

Suppose, then, that $\rho = \rho^{+} - \rho^{-}$ and $\|\rho\| = \|\rho^{+}\| + \|\rho^{-}\|$. Choose $f_0$ in $C(X)$ so that $\|f_0\| = 1$ and $\|\rho\| = \rho(f_0) = \rho^{+}(f_0) - \rho^{-}(f_0)$. (This can be done only approximately, strictly speaking.) From this, $\rho^{+}(f_0) = \|\rho^{+}\|$ and $\rho^{-}(f_0) = -\|\rho^{-}\|$. Also $\rho^{+}(f_0^{+}) = \|\rho^{+}\|$ and $\rho^{-}(f_0^{+}) = 0$, where $f_0^{+} = f_0 \vee 0$. If $f$ is a positive function in $C(X)$ and $0 \leqslant g \leqslant f$, with $g$ in $C(X)$, then $\rho(g) = \rho^{+}(g) - \rho^{-}(g) \leqslant \rho^{+}(g) \leqslant \rho^{+}(f)$. Since $\rho^{+}(1) = \|\rho^{+}\|$, $\rho^{+}(1 - f_0^{+}) = 0$. As $1 - f_0^{+} \geqslant 0$, it follows from the Cauchy–Schwarz inequality applied to $\rho^{+}$ (see the end of Remark 3.4.9) that

$$\rho^{+}((1 - f_0^{+})f) \; (= \rho^{+}((1 - f_0^{+})^{1/2}(1 - f_0^{+})^{1/2}f)) = 0.$$

Thus $\rho^{+}(f) = \rho^{+}(ff_0^{+})$. Similarly $\rho^{-}(f_0^{+}f) = 0$, since $\rho^{-}(f_0^{+}) = 0$. Thus

$\rho(f_0^+ f) = \rho^+(ff_0^+) = \rho^+(f)$. Now $0 \leqslant f_0^+ f \leqslant f$, so that

$$\rho^+(f) = \sup\{\rho(g) : 0 \leqslant g \leqslant f, f \in C(X)\},$$

which identifies $\rho^+$ with the construction in Proposition 3.4.11. Thus the decomposition $\rho = \rho^+ - \rho^-$, where $\|\rho\| = \|\rho^+\| + \|\rho^-\|$, and $\rho^+$, $\rho^-$ are positive, is unique. If we replace the unwarranted assumption that $\|\rho\|$ is attained at some $f_0$ by an approximation and argue along precisely the same line, using estimates, a valid proof of the uniqueness results. ■

3.4.13. Remark. If $\tau$ is a positive linear functional on $C(X)$, the set $\mathcal{I}$ of functions $f$ in $C(X)$ such that $\tau(|f|^2) = 0$ forms a closed ideal in $C(X)$ (use Cauchy–Schwarz for $\tau$ to prove this). From Theorem 3.4.1, $\mathcal{I}$ is the set of functions in $C(X)$ vanishing on a closed subset of $X$. We refer to this closed subset as the *support* of $\tau$ and denote it by $s(\tau)$.

The measure-theoretic approach views $\tau$ as "integration" relative to an associated positive measure $\mu$ on $C(X)$ (through the Riesz representation theorem). In that context, $s(\tau)$ is the support of the measure $\mu$. If $\rho$ is a bounded hermitian linear functional on $C(X)$, then $\rho$ corresponds to a "signed" measure $\mu$, $\rho^+$ to $\mu^+$, and $\rho^-$ to $\mu^-$, where $\mu = \mu^+ - \mu^-$. The decomposition of $\mu$ as $\mu^+ - \mu^-$ is the *Hahn–Jordan decomposition* of $\mu$. ■

With the aid of the information we have accumulated concerning functionals (in particular, states) on $C(X)$, we can prove a crucial approximation theorem.

3.4.14. Theorem. *If $\mathcal{A}$ is a norm closed subalgebra of $C(X)$, $1 \in \mathcal{A}$, $\bar{f} \in \mathcal{A}$ if $f \in \mathcal{A}$, and, for each pair of distinct points $p$ and $p'$ in $X$, there is a function $f$ in $\mathcal{A}$ such that $f(p) \neq f(p')$, then $\mathcal{A} = C(X)$.*

*Proof.* Suppose $\mathcal{A} \neq C(X)$. From the Hahn–Banach theorem (see Corollary 1.6.3), there is a non-zero bounded linear functional $\rho$ on $C(X)$ that annihilates $\mathcal{A}$. We may assume that $\rho$ is hermitian, since $\rho^*$ (see Remark 3.4.9) also annihilates $\mathcal{A}$. The set of all hermitian functionals with norm at most 2 and annihilating $\mathcal{A}$ is a weak* closed bounded convex subset of the (norm) dual of $C(X)$. Hence this set is weak* compact and is the closed convex hull of its extreme points (from Corollary 1.6.6). Let $\eta$ be such an extreme point. Then $\eta = \eta^+ - \eta^-$, where $\eta^+$ and $\eta^-$ are positive linear functionals on $C(X)$ such that $2 = \|\eta\| = \|\eta^+\| + \|\eta^-\|$ (from Proposition 3.4.11). Since $\eta$ is 0 on $\mathcal{A}$, and $1 \in \mathcal{A}$,

$$0 = \eta(1) = \eta^+(1) - \eta^-(1) = \|\eta^+\| - \|\eta^-\|.$$

Thus

$$1 = \|\eta^+\| = \eta^+(1) = \|\eta^-\| = \eta^-(1),$$

so that $\eta^+$ and $\eta^-$ are states of $\mathcal{A}$. We shall prove that $\eta^+$ and $\eta^-$ are pure states of $C(X)$ (based on the extremal property of $\eta$). If this has been established, then, from Theorem 3.4.7 and Corollary 3.4.2, there are points $p_+$ and $p_-$ in $X$ such that $\eta^+(f) = f(p_+)$ and $\eta^-(f) = f(p_-)$ for all $f$ in $C(X)$. Since $\eta \neq 0$; $\eta^+ \neq \eta^-$ and $p_+ \neq p_-$. But

$$0 = \eta(g) = \eta^+(g) - \eta^-(g) = g(p_+) - g(p_-)$$

for each $g$ in $\mathscr{A}$, contradicting the assumption that for some $g$ in $\mathscr{A}$, $g(p_+) \neq g(p_-)$.

It remains to show that $\eta^+$ and $\eta^-$ are pure states of $C(X)$. Using the Cauchy–Schwarz inequality as in Remark 3.4.9, we see that the set $\mathscr{I}$ of functions $f$ in $C(X)$ such that $\eta^+(|f|^2) = 0$ is a norm closed ideal in $C(X)$ on which $\eta^+$ vanishes (see Remark 3.4.13, where $\mathscr{I}$ is used to define the support of $\eta^+$). Suppose we have proved that $\eta^+$ and $\eta^-$ are multiplicative on $\mathscr{A}$. Then $\mathscr{I} \cap \mathscr{A}$ is a maximal ideal in $\mathscr{A}$. From Zorn's lemma, $\mathscr{I}$ has an extension to a maximal ideal $\mathscr{M}$ in $C(X)$, and $\mathscr{M} \cap \mathscr{A} = \mathscr{I} \cap \mathscr{A}$. Thus there is a multiplicative linear functional extending the restriction of $\eta^+$ to $\mathscr{A}$. If there were two distinct such extensions, we would, again, have two distinct points of $X$ at which all the functions of $\mathscr{A}$ take the same values; so there is one and only one multiplicative extension of $\eta^+$ to $C(X)$ from its restriction to $\mathscr{A}$. The set of all states of $C(X)$ that agree with $\eta^+$ on $\mathscr{A}$ is convex, weak* compact, and, hence, the closed convex hull of its extreme points. Let $\tau$ be one of these extreme points. If $\tau = \frac{1}{2}(\rho_1 + \rho_2)$, with $\rho_1$ and $\rho_2$ states of $C(X)$, then $\frac{1}{2}\rho_1$ and $\frac{1}{2}\rho_2$ are majorized by $\eta^+$ on $\mathscr{A}$ (since their sum is $\tau$, hence, $\eta^+$ on $\mathscr{A}$). Thus $\rho_1$ and $\rho_2$ annihilate the same positive functions in $\mathscr{A}$ as $\eta^+$ does; from which (again by using the Cauchy–Schwarz inequality), $\rho_1$ and $\rho_2$ annihilate $\mathscr{I} \cap \mathscr{A}$. As $\rho_1(1) = \rho_2(1) = \eta^+(1)$ and $\mathscr{I} \cap \mathscr{A}$ is the null space of $\eta^+$ (under the assumption currently in force, that $\eta^+$ is multiplicative on $\mathscr{A}$), it follows that $\rho_1$, $\rho_2$, and $\eta^+$ coincide on $\mathscr{A}$. Thus $\tau = \rho_1 = \rho_2$ (from the extremal property of $\tau$). Hence $\tau$ is a pure state of $C(X)$ that agrees with $\eta^+$ on $\mathscr{A}$. From Theorem 3.4.7, $\tau$ is multiplicative on $C(X)$, so that it is the unique multiplicative linear functional on $C(X)$ that agrees with $\eta^+$ on $\mathscr{A}$. It follows that the set of states of $C(X)$ that agree with $\eta^+$ on $\mathscr{A}$ has a single extreme point and hence, consists of this extreme point $\eta^+$. Thus $\eta^+$ is multiplicative on $C(X)$.

The foregoing is predicated on our showing that $\eta^+$ is multiplicative on $\mathscr{A}$, which we proceed to do. With $0 \leqslant f \leqslant 1$ and $f$ in $\mathscr{A}$, if $\eta^+(f)$ $(= \eta^+((f^{1/2})^2)) = 0$, then $f^{1/2} \in \mathscr{I}$; so that $f \in \mathscr{I}$ and for each $g$ in $C(X)$, $fg \in \mathscr{I}$ and $\eta^+(fg) = 0 = \eta^+(f)\eta^+(g)$. Similarly, if $\eta^+(f) = 1$, then $1 - f \in \mathscr{I}$ and $\eta^+(fg) = \eta^+(g) = \eta^+(f)\eta^+(g)$. We assume $0 \leqslant f \leqslant 1$, $f \in \mathscr{A}$, and $0 \neq \eta^+(f) = a \neq 1$. For $g$ in $C(X)$, let $\eta_1^+(g)$ be $a^{-1}\eta^+(fg)$, $\eta_1^-(g)$ be $a^{-1}\eta^-(fg)$, $\eta_1$ be $\eta_1^+ - \eta_1^-$, $\eta_2^+(g)$ be $(1-a)^{-1}\eta^+((1-f)g)$, $\eta_2^-(g)$ be $(1-a)^{-1}\eta^-((1-f)g)$, and $\eta_2$ be $\eta_2^+ - \eta_2^-$. Then $\eta_1^+, \eta_1^-, \eta_2^+$, and $\eta_2^-$ are states, $\eta_1$ and $\eta_2$ are 0 on $\mathscr{A}$, and $a\eta_1 + (1-a)\eta_2 = \eta$. Since $\eta_1$ and $\eta_2$ have bound not exceeding 2 and $\eta$ is extreme in the set of hermitian linear functionals of norm not exceeding 2 that are 0 on $\mathscr{A}$, we have that $\eta = \eta_1 = \eta_2$. Thus $\eta = \eta_1^+ - \eta_1^-$ and $2 = \|\eta\| = \|\eta_1^+\| + \|\eta_1^-\|$. From the uniqueness of the decomposition of $\eta$ as $\eta^+ - \eta^-$ (see Remark 3.4.12), $\eta_1^+ = \eta^+$, and $a^{-1}\eta^+(fg) = \eta^+(g)$ for all $g$ in $C(X)$. Thus $\eta^+(fg) = a\eta^+(g) = \eta^+(f)\eta^+(g)$, when $f \in \mathscr{A}$, $0 \leqslant f \leqslant 1$, and $g \in C(X)$. It follows at once that $\eta^+$ and $\eta^-$ are multiplicative on $\mathscr{A}$. ∎

3.4.15. Remark. The preceding theorem is known as the Stone–Weierstrass theorem. It was stated by M. H. Stone [22: Theorem 10] in the form given and proved (as it usually is) by making use of the special case discovered by K. Weierstrass where $X$ is a closed bounded interval in $\mathbb{R}$ and $\mathscr{A}$ is the norm closure of the set of polynomials on $X$. This special case (the Weierstrass approximation theorem) can be reformulated as the assertion that each continuous function on a closed bounded interval is the uniform (norm) limit of polynomials on that interval. The Stone generalization allows us to conclude, for example, that the same is true for continuous functions and polynomials in $n$ variables on a closed bounded cube in $\mathbb{R}^n$. Another immediate application includes the fact that (complex) linear combinations of the functions $z \to z^n$ ($n = 0, \pm 1, \pm 2, \ldots$) on the circle $C_1$ with center 0 and radius 1 in $\mathbb{C}$ are norm dense in $C(C_1)$ (since they form an algebra separating points, containing the constants and closed under complex conjugation). As uniform approximation implies $L_2$ approximation for finite measure spaces, and the continuous functions are dense in $L_2$ (in the $L_2$ norm), it follows that $\mathscr{A}$ is ($L_2$-)dense in $L_2(C_1)$. With $f$ in $L_2(C_1)$, let $(Uf)(s)$ be $f(\exp is)$ for $s$ in $(-\pi, \pi)$. Then $U$ is a unitary transformation of $L_2(C_1)$ onto $L_2(-\pi, \pi)$ carrying the function $z \to z^n$ onto $s \to \exp ins$. Thus (complex) linear combinations of the functions $\{\exp ins : n = 0, \pm 1, \pm 2, \ldots\}$ (and, consequently, of 1, $\cos s$, $\sin s$, $\cos 2s$, $\sin 2s, \ldots$) are ($L_2$-)dense in $L_2(-\pi, \pi)$.

An application of the Stone–Weierstrass theorem allows us to conclude that $C(X)$ is norm separable (that is, has a countable (norm) dense subset – equivalently, is countably generated as a Banach algebra) when $X$ is a compact metric space. Note for this last that $X$ contains a countable dense subset $\{p_n\}$ of points (obtained by choosing one point in each set of a finite covering of $X$ by balls of radius $1/n$ for each positive integer $n$). If $d$ is the metric on $X$ and $f_n(p) = d(p_n, p)$, then the set $\{f_n\}$ separates points of $X$ (for if $\{p_{n'}\}$ tends to $p$ and $f_n(p) = f_n(q)$ for all $n$, then $\{p_{n'}\}$ tends to $q$, and $p = q$). Thus $\{f_n\}$ generates a norm-dense (complex) subalgebra of $C(X)$, and the finite complex–rational linear combinations of finite products of functions in $\{f_n\}$ is a (countable) norm-dense subset of this algebra, hence of $C(X)$.

Concerning the hypotheses of the Stone–Weierstrass theorem, the example of the subalgebra of functions in $C(X)$ vanishing at some point $p$ in $X$ underlines the necessity of assuming that the constants are in the subalgebra. This assumption appears, first, at the stage where we decompose $\eta$ and conclude that $\eta^+$ and $\eta^-$ are distinct and *non-zero*.

The necessity of assuming something that restricts the class of closed subalgebras containing the constants even further (in our case, the assumption is that $\mathscr{A}$ is stable under complex conjugation) is illustrated by taking $X$ to be the closed disk $D_1$ of radius 1 with center 0 in $\mathbb{C}$ and $\mathscr{A}$ to be the norm closed subalgebra of $C(D_1)$ generated by $z$, the identity transform on $D_1$, and 1. Then $\mathscr{A}$ is the algebra of functions continuous on $D_1$ and holomorphic on the open disk. In particular $\mathscr{A}$ does not contain $\bar{z}$; although $1 \in \mathscr{A}$ and $z$ alone "separates" points of $D_1$. The assumption that $\mathscr{A}$ is stable under complex conjugation appears, at the outset, when we locate a *hermitian* linear functional annihilating $\mathscr{A}$. ∎

The result that follows is crucial for our study of the spectral theory of self-adjoint (and normal) operators on a Hilbert space (see Theorem 5.2.1). It describes the topological property of $X$ associated with a special (and extraordinary) order property of $C(X)$. When $C(X)$ is a boundedly complete lattice (see the statement of the theorem following), $X$ will be "extremely disconnected" – the closure of each open set is open (as well as closed). Such an $X$ (with more than one point) is certainly disconnected – even "totally disconnected" (each pair of points can be separated by sets that are both open and closed: *clopen sets*). What is not apparent, though, is that an extremely disconnected space is "more than" totally disconnected.

The canonical example of a totally disconnected (infinite) compact Hausdorff space $X$ is the countable product of two-point spaces $\{0, 1\}$ (with the product topology). Points of $X$ can be represented as sequences of zeros and ones. Convergence is coordinatewise. If $(a_1, a_2, \ldots)$ is a point of $X$ and

$$\eta(a_1, a_2, \ldots) = a_1 2^{-1} + a_2 2^{-2} + \cdots,$$

then $\eta$ is a continuous mapping of $X$ onto $[0, 1]$. With $r$ an irrational number in $[0, 1]$, let $\mathcal{O}$ be $\eta^{-1}((r, 1])$. If

$$r = a_1 2^{-1} + a_2 2^{-2} + \cdots,$$

then $(b_1, b_2, \ldots) \in \mathcal{O}$ if and only if

$$a_1 = b_1, \ldots, a_k = b_k, \qquad a_{k+1} < b_{k+1}$$

for some positive $k$. Since $r$ is irrational, $a_k$ is 0 and 1 for an infinite number of $k$. Thus, with $k$ large and $a_{k+1}$ equal to 0, $(a_1, a_2, \ldots, a_k, 1, 0, 0, \ldots)$ is near $(a_1, a_2, \ldots)$ and lies in $\mathcal{O}$. Hence $(a_1, a_2, \ldots) \in \mathcal{O}^-$, the closure of $\mathcal{O}$. If $j$ is large and $a_{j+1} = 1$, then $(a_1, a_2, \ldots, a_j, 0, 0, \ldots)$ is near $(a_1, a_2, \ldots)$ and is not in $\mathcal{O}^-$ (because no point of $\mathcal{O}$ agrees with it on the first $j + 1$ coordinates). Thus $(a_1, a_2, \ldots)$ is not interior to $\mathcal{O}^-$, and $\mathcal{O}^-$ is not open.

If no examples of (infinite) extremely disconnected compact Hausdorff spaces are apparent, it is because these spaces, and their associated algebras of continuous functions are primarily a (useful) mathematical artifice rather than a naturally occurring geometrical construct. With the aid of Theorem 5.2.1 and Example 5.1.6, we shall have examples of such spaces. For the moment, we content ourselves with showing that $X$ is extremely disconnected if $C(X)$ is a boundedly complete lattice.

3.4.16. THEOREM. *If each set of functions in $C(X)$ that has an upper bound in $C(X)$ has a least upper bound in $C(X)$ (so that $C(X)$ is a boundedly complete lattice), then each open set in $X$ has an open closure (so that $X$ is extremely disconnected).*

*Proof.* Let $\mathcal{O}$ be an open subset of $X$, $\mathcal{O}^-$ its closure, $\mathcal{F}$ the family of functions $f$ in $C(X)$ such that $0 \leqslant f \leqslant 1$ and $f(p') = 0$ if $p' \notin \mathcal{O}$, and $f_0$ the least upper bound of $\mathcal{F}$ in $C(X)$. Since 1 is an upper bound for $\mathcal{F}$, $f_0 \leqslant 1$. If $p \in \mathcal{O}$ there is an $f$ in $\mathcal{F}$ such that $f(p) = 1$, so that $f_0(p) = 1$ for each $p$ in $\mathcal{O}$ (hence, for each $p$ in $\mathcal{O}^-$). If $p' \notin \mathcal{O}^-$, there is a $g$ in $C(X)$ such that $0 \leqslant g \leqslant 1$, $g(p') = 0$, and $g(p) = 1$ for $p$ in $\mathcal{O}^-$. Thus $g$ is an upper bound for $\mathcal{F}$, and $f_0 \leqslant g$. It follows that $f_0$ is 1 on $\mathcal{O}^-$ and 0 on $X \backslash \mathcal{O}^-$. As $f_0$ is continuous, $\mathcal{O}^-$ is open. ■

## 3.5. Exercises

3.5.1. In Remark 3.1.3 the algebra $\mathfrak{A}_1$ obtained from a normed algebra $\mathfrak{A}$ is defined. Complete the indicated check that $\mathfrak{A}_1$ is a normed algebra, and that $\mathfrak{A}_1$ is a Banach algebra when $\mathfrak{A}$ is.

3.5.2. Let $\mathfrak{A}$ be a Banach algebra over $\mathbb{R}$, and let $\mathfrak{A}_{\mathbb{C}}$ be the Banach space "complexification" of $\mathfrak{A}$ described in Exercise 1.9.6. Define the product $(A, B)(C, D)$ to be $(AC - BD, AD + BC)$. Show that:

(i) $\|(A, B)(C, D)\| \leqslant 2\|(A, B)\| \, \|(C, D)\|$;

(ii) $\mathfrak{A}_{\mathbb{C}}$ has another norm $|||\ \ |||$ relative to which it is a Banach algebra, the identity mapping from $(\mathfrak{A}_{\mathbb{C}}, \|\ \ \|)$ onto $(\mathfrak{A}_{\mathbb{C}}, |||\ \ |||)$ is a homeomorphism, and the mapping $A \to (A, 0)$ is an isometric (real) algebraic isomorphism.

3.5.3. Let $\{\mathfrak{A}_a : a \in \mathbb{A}\}$ be a family of Banach algebras, and let $\mathfrak{A}$ be the set of functions $f$ on $\mathbb{A}$ such that $f(a) \in \mathfrak{A}_a$ and

$$\sup\{\|f(a)\| : a \in \mathbb{A}\} (= \|f\|) < \infty.$$

(i) Show that $\mathfrak{A}$, provided with the operations of pointwise addition, multiplication, multiplication by scalars, and with the norm described, is a Banach algebra. (It is usually denoted by "$\sum_{a \in \mathbb{A}} \oplus \mathfrak{A}_a$." Compare it with the $l_\infty$ spaces of Example 1.7.1.)

(ii) With $\mathbb{N}$ in place of $\mathbb{A}$, show that $\mathfrak{A}_0$, the set of functions $f$ in $\mathfrak{A}$ such that $\lim_{n \to \infty} \|f(n)\| = 0$ is a (norm-)closed proper two-sided ideal in $\mathfrak{A}$.

(iii) Determine whether or not $\mathfrak{A}_0$ is a maximal (two-sided) ideal in $\mathfrak{A}$.

3.5.4. With the notation of Exercise 1.9.19:

(i) Show that $l_\infty$ and $c$, provided with pointwise multiplication, are (commutative) Banach algebras.

(ii) Show that $c_0$ is a (norm-)closed proper ideal in $l_\infty$.

(iii) Describe a multiplicative linear functional on $c$ with $c_0$ as its null space; and conclude that $c_0$ is a maximal ideal in $c$.

(iv) Is $c_0$ a maximal ideal in $l_\infty$? Explain. Reconsider Exercise 3.5.3(iii). [*Hint*. Consider "dimension" in combination with Corollary 3.2.5.]

3.5.5. Let $\beta(\mathbb{N})$ be the space of non-zero multiplicative linear functionals on $l_\infty(\mathbb{N}, \mathbb{C})$ $(= l_\infty)$ topologized with the weak* topology. Define $\hat{f}(\rho)$ to be $\rho(f)$ for $\rho$ in $\beta(\mathbb{N})$ and $f$ in $l_\infty$.

(i) Note that $\beta(\mathbb{N})$ is a compact Hausdorff space; and show that the mapping, $f \to \hat{f}$, is an algebraic isometric isomorphism of $l_\infty$ into $C(\beta(\mathbb{N}))$.

(ii) Show that $\hat{\bar{f}} = \bar{\hat{f}}$. [*Hint*. Consider the spectrum of a real-valued $f$ in $l_\infty$ relative to $l_\infty$.]

(iii) Show that $\hat{1} = 1$, where 1 denotes the constant function "one" in both $l_\infty$ and $C(\beta(\mathbb{N}))$, and that $\hat{l}_\infty$ is (norm) closed in $C(\beta(\mathbb{N}))$. Conclude that $\hat{l}_\infty = C(\beta(\mathbb{N}))$. (The space $\beta(\mathbb{N})$ is called "the $\beta$-compactification of $\mathbb{N}$.")

3.5.6. With the notation of Exercise 3.5.5:

(i) Show that $\beta(\mathbb{N})$ is extremely disconnected.

(ii) Let $n^\#(f)$ be $f(n)$ for $n$ in $\mathbb{N}$ and $f$ in $l_\infty$; denote by $\mathbb{N}$, again, the subset $\{n^\# : n \in \mathbb{N}\}$ of $\beta(\mathbb{N})$. Show that $\mathbb{N}$ is dense in $\beta(\mathbb{N})$.

(iii) Show that the one-point subset $\{n^\#\}$ of $\beta(\mathbb{N})$ is open in $\beta(\mathbb{N})$. [*Hint*. Consider $\hat{\chi}_n$, where $\chi_n(m)$ is 0 unless $n = m$, and $\chi_n(n) = 1$.]

(iv) Show that $\rho(f) = 0$ when $\rho \in \beta(\mathbb{N}) \backslash \mathbb{N}$ and $f \in c_0$.

3.5.7. Let $G$ be a countable (discrete) group and $\{G_n\}$ an ascending sequence of finite subgroups with union $G$. (Such a group is said to be *locally finite*.) Let $\rho$ be an element of $(l_\infty(G, \mathbb{C})^\#)_1$ such that $\rho(\{1, 1, \ldots\}) = 1$; and define $\rho_n$ to be $|G_n|^{-1} \sum_{g \in G_n} T_g^\#(\rho)$, where $|G_n|$ is the number of elements in $G_n$, and the bounded linear operator $T_g$, acting on $l_\infty(G, \mathbb{C})$, is given by $[T_g(f)](g_0) = f(g^{-1}g_0)$.

(i) Show that $\rho_n \in (l_\infty(G, \mathbb{C})^\#)_1$ and that $\rho_n(\{1, 1, \ldots\}) = 1$.

(ii) Show that the weak* closure of $\{\rho_n : n \in \mathbb{N}\}$ contains an element $\rho_0$ such that $T_g^\# \rho_0 = \rho_0$ for every $g$ in $G$, and such that $\rho_0(\{1, 1, \ldots\}) = 1$. (The functional $\rho_0$ is called an *invariant mean* on $G$.) [*Hint*. With $f$ in $l_\infty(G, \mathbb{C})$, define a function $\varphi$ on $\mathbb{N}$ by $\varphi(n) = \rho_n(f)$. Fix $x_0$ in $\beta(\mathbb{N}) \backslash \mathbb{N}$; and let $\rho_0(f)$ be $\hat{\varphi}(x_0)$.]

3.5.8. Let $A$ be an element of the Banach algebra $\mathfrak{A}$; and let $\mathcal{O}$ be an open subset of $\mathbb{C}$ containing $\mathrm{sp}_{\mathfrak{A}} A$. Show that there is a positive $\delta$ such that $\mathrm{sp}_{\mathfrak{A}} B \subseteq \mathcal{O}$ if $B \in \mathfrak{A}$ and $\|A - B\| < \delta$.

3.5.9. Let $\{A_n\}$ be a sequence of invertible elements in a Banach algebra $\mathfrak{A}$ and suppose $\{A_n\}$ has a limit $A$ that commutes with each $A_n$.

(i) Show, by example, that $A$ need not be invertible.
(ii) Show that $A$ is invertible if $\{r(A_n^{-1})\}$ is bounded. [*Hint*. Use Corollary 3.3.4.]

3.5.10. Given an element $A$ of a Banach algebra $\mathfrak{A}$ and a positive number $\varepsilon$, show that:

(i) if $A$ is singular (that is $0 \in \mathrm{sp}\, A$), there is a positive number $\delta_0$ such that if $B$ in $\mathfrak{A}$ commutes with $A$ and $\|A - B\| < \delta_0$, then there is a $\lambda$ in $\mathrm{sp}\, B$ with $|\lambda| < \varepsilon$;
(ii) if $\lambda \in \mathrm{sp}\, A$ there is a positive number $\delta_\lambda$ such that if $B$ in $\mathfrak{A}$ commutes with $A$ and $\|A - B\| < \delta_\lambda$, then, for some $\lambda'$ in $\mathrm{sp}\, B$, $|\lambda - \lambda'| < \varepsilon$;
(iii) there is a positive $\delta$ such that if $\|A - B\| < \delta$ and $AB = BA$, with $B$ in $\mathfrak{A}$, then $\mathrm{sp}\, A \subseteq S_\varepsilon(B)$, where $S_\varepsilon(B) = \{\lambda : |\lambda - \lambda'| < \varepsilon$ for some $\lambda'$ in $\mathrm{sp}\, B\}$.

3.5.11. Nilpotent and generalized nilpotent elements are defined following the proof of Theorem 3.3.3 (the spectral radius formula). Use Proposition 3.2.10 to show, without the aid of the spectral radius formula, that:

(i) each nilpotent element in a Banach algebra has spectrum consisting of 0 alone;
(ii) each generalized nilpotent element in a Banach algebra has spectrum consisting of 0 alone.

3.5.12. Let $\{y_1, y_2, \ldots\}$ be an orthonormal basis for the (complex, separable) Hilbert space $\mathscr{H}$. Let $A$ be the linear transformation of $\mathscr{H}$ into $\mathscr{H}$ that maps $y_n$ onto $a_n y_{\pi(n)}$, where $\{a_n\}$ is a bounded sequence of complex numbers and $\pi$ is a one-to-one mapping of $\{1, 2, \ldots\}$ into itself.

(i) Determine $\|A\|$.
(ii) Describe $A^k$, where $k$ is a positive integer.
(iii) Determine $\|A^k\|$.
(iv) Determine $\|A\|$ and $\|A^k\|$ under the additional assumptions that $|a_n| \leqslant |a_m|$ and $\pi(m) \leqslant \pi(n)$ when $m \leqslant n$.
(v) Show that $A$ is nilpotent when $a_n = 0$ for all $n$ larger than some $k$ and $j < \pi(j)$ for $j$ in $\{1, \ldots, k\}$.
(vi) Use Exercises 2.8.22(iv) and 2.8.23 to show that $A$ is compact when $\{a_n\}$ tends to 0.

(vii) Show that $A$ is a compact generalized nilpotent operator on $\mathscr{H}$ with infinite-dimensional range when $a_n = 1/n$ and $\pi(m) = m + 1$.

3.5.13. As in Exercise 3.5.12, let $\{y_n\}$ be an orthonormal basis for the Hilbert space $\mathscr{H}$ and let $Ay_n$ be $a_n y_{n+1}$, where $a_n = 2^{-k}$ and $k$ is the largest positive integer such that $2^k$ divides $n$.

(i) Show that

$$a_1 a_2 \cdots a_{2^n - 1} = \prod_{j=1}^{n-1} 2^{-j2^{n-j-1}} (= b_n);$$

and deduce that $b_n \leqslant \|A^{2^n - 1}\|$. [*Hint*. Observe that $2^j$ has $2^{n-j-1}$ odd multiples among the numbers $1, 2, \ldots, 2^n - 1$.]

(ii) Show that $A$ is *not* a generalized nilpotent operator.

(iii) Let $A_k y_n$ be $a_n y_{n+1}$ except when $n$ is an odd multiple of $2^k$, in which case let $A_k y_n$ be 0. Show that each $A_k$ is nilpotent and that $\|A - A_k\| \to 0$.

(iv) Deduce that the norm limit of nilpotent operators need not be generalized nilpotent. What relation does this conclusion have to the "spectral (semi-)continuity" considerations of Exercises 3.5.8 and 3.5.10?

3.5.14. With the notation of Exercise 3.5.12, show that:

(i) $\{a_n\}$ tends to 0 if $A$ is compact (compare (vi) of Exercise 3.5.12);

(ii) $A$ is a compact generalized nilpotent operator on $\mathscr{H}$ when $j < \pi(j)$ for all $j$ and $\{a_n\}$ tends to 0 (compare (vii) of Exercise 3.5.12).

3.5.15. Does each compact operator have a (non-zero) eigenvector? (Example? Proof? See Exercise 2.8.29(ii).)

3.5.16. Can a non-zero self-adjoint operator be a generalized nilpotent operator? (Example? Proof?)

3.5.17. Let $A$ be a compact operator acting on a Hilbert space $\mathscr{H}$, $\lambda$ be a non-zero complex number, and $\{x_n\}$ be a sequence of vectors in $\mathscr{H}$ such that $\{(A - \lambda I)x_n\}$ converges in norm to a vector $y$ in $\mathscr{H}$.

(i) Show that if $\{\|x_n\|\}$ is unbounded, $(A - \lambda I)x = 0$ for some unit vector $x$ in $\mathscr{H}$.

(ii) Show that if $\{\|x_n\|\}$ is bounded, $y$ is in the range of $A - \lambda I$. [*Hint*. Use Exercise 1.9.17(ii).]

(iii) Show that the range of $A - \lambda I$ is closed if the null space of $A - \lambda I$ is (0).

(iv) Show that if $\lambda \in \operatorname{sp} A$ either $\lambda$ is an eigenvalue of $A$ or $\bar{\lambda}$ is an eigenvalue of $A^*$.

3.5.18. Let $\mathscr{H}$ be a Hilbert space and $A$ be an operator on $\mathscr{H}$ such that $Ax = x$ for some unit vector $x$ in $\mathscr{H}$.

(i) Show that $A^*y = y$ for some unit vector $y$ in $\mathscr{H}$ when the range of $A$ is finite dimensional.

(ii) Show that if $A$ is compact, there is a sequence $\{A_n\}$ of operators on $\mathscr{H}$ converging to $A$ in norm, each $A_n$ having finite-dimensional range, and such that $A_n x = x$ for all $n$. [*Hint*. Use Exercise 2.8.24(ii).]

(iii) Show that if $A$ is compact, $A^*y = y$ for some unit vector $y$ in $\mathscr{H}$.

(iv) Show that if $A$ is compact and $0 \neq \lambda \in \operatorname{sp} A$, then $\lambda$ is an eigenvalue of $A$.

(v) Suppose that $Ax_n = \lambda_n x_n$ $(n = 1, 2, \ldots)$, where $\{\lambda_n\}$ is a sequence of distinct non-zero eigenvalues of $A$ and each $x_n$ is a unit vector. Show that $\{x_n\}$ is linearly independent and that $\langle Ay_n, y_n \rangle = \lambda_n$, where $\{y_n\}$ is the orthonormal sequence obtained from $\{x_n\}$ by the Gram–Schmidt orthogonalization process.

(vi) Deduce from (iv) that $\operatorname{sp} A$ is *either* a finite set *or* can be arranged as a sequence of complex numbers tending to 0 and that each non-zero element in $\operatorname{sp} A$ is an eigenvalue of $A$ of finite multiplicity when $A$ is compact.

3.5.19. Let $A$ be the operator defined in Exercise 3.5.12. Assume that $j < \pi(j)$ for all $j$ and $a_n \to 0$.

(i) Show that $A$ has no non-zero eigenvalues.

(ii) Show that $A$ has no eigenvalues if $a_n \neq 0$ for all $n$.

(iii) Deduce again (see Exercise 3.5.14(ii)), with the aid of Exercise 3.5.18, that $A$ is a (compact) generalized nilpotent operator.

3.5.20. Let $\mathscr{H}$ be $L_2(0, 1)$ relative to Lebesgue measure, and let $(Kf)(s)$ be $\int_0^s f(t)\,dt$ for each $f$ in $\mathscr{H}$.

(i) Show that $K$ is a Hilbert–Schmidt operator and a compact operator on $\mathscr{H}$.

(ii) Show that $K$ has no eigenvalues.

(iii) Deduce that $K$ is a compact generalized nilpotent operator.

3.5.21. With the notation of Example 3.2.17, $\sum_{n=-\infty}^{\infty} \lambda^{-n} e_n$ is an "eigenvector" for $U$ in a formal sense even when $|\lambda| \neq 1$. Pursue the argument of that example to see how it fails to establish that *each* $\lambda$ in $\mathbb{C}$ is in $\operatorname{sp}(U)$.

3.5.22. Let $X$ be a compact Hausdorff space and $\mathscr{A}$ be a subalgebra of $C(X)$. Suppose $\|\ \|'$ and $\|\ \|''$ are norms on $\mathscr{A}$ such that $\mathscr{A}$ is complete relative to $\|\ \|'$ and $\mathscr{A}$ has a completion $\mathscr{A}_0$ in $C(X)$ relative to $\|\ \|''$. Let $\|f\|$ be $\sup\{|f(x)| : x \in X\}$ for $f$ in $C(X)$.

(i) Show that $\|f\| \leqslant \|f\|'$ and $\|g\| \leqslant \|g\|''$ for $f$ in $\mathscr{A}$ and $g$ in $\mathscr{A}_0$.

(ii) Show that $\mathscr{A}$ is complete relative to the norm that assigns $\|f\|' + \|f\|''$ to $f$ in $\mathscr{A}$.

(iii) Show that $\| \|'$ and $\| \|''$ are equivalent norms on $\mathscr{A}$ (that is, the identity mapping is a homeomorphism from $\mathscr{A}$ relative to $\| \|'$ to $\mathscr{A}$ relative to $\| \|''$) when $\mathscr{A}$ is complete relative to $\| \|''$.

(iv) Without the assumption that $\mathscr{A}$ is complete relative to $\| \|''$, show that the identity mapping from $(\mathscr{A}, \| \|')$ to $(\mathscr{A}, \| \|'')$ is continuous.

3.5.23. Let $\mathscr{B}$ be a commutative Banach algebra, and let $\mathscr{R}$ be the intersection of the maximal ideals of $\mathscr{B}$. (The ideal $\mathscr{R}$ is called the *radical* of $\mathscr{B}$, and $\mathscr{B}$ is said to be semi-simple when $\mathscr{R} = (0)$.) Let $\mathfrak{A}$ be an arbitrary Banach algebra and $\varphi$ be a homomorphism of $\mathfrak{A}$ into $\mathscr{B}$ mapping the identity of $\mathfrak{A}$ onto that of $\mathscr{B}$. Let $\mathscr{K}$ be the kernel of $\varphi$.

(i) Show that $\varphi$ maps $\mathscr{K}^-$, the norm closure of $\mathscr{K}$, into $\mathscr{R}$. Deduce that $\mathscr{K}$ is closed when $\mathscr{R} = (0)$.

(ii) Show that if $\mathscr{B}$ is semi-simple, then $\mathscr{B}$ is isomorphic with a subalgebra $\mathscr{A}_1$ of $C(X)$, where $X$ is the compact Hausdorff space of non-zero multiplicative linear functionals on $\mathscr{B}$ with the weak* topology. [*Hint*. Let $\hat{A}(\rho)$ be $\rho(A)$, when $A \in \mathscr{B}$ and $\rho \in X$.]

(iii) Show that if $\mathscr{B}$ is semi-simple, then $\varphi$ is continuous. [*Hint*. Use the mapping of (ii) and this mapping composed with the quotient mapping $\mathfrak{A} \to \mathfrak{A}/\mathscr{K}$ together with Exercise 3.5.22(iv).]

3.5.24. Establish the following inclusions for a pair of commuting elements $A$ and $B$ of an arbitrary Banach algebra $\mathfrak{A}$:

$$\mathrm{sp}_{\mathfrak{A}}(AB) \subseteq \mathrm{sp}_{\mathfrak{A}}(A)\,\mathrm{sp}_{\mathfrak{A}}(B), \qquad \mathrm{sp}_{\mathfrak{A}}(A + B) \subseteq \mathrm{sp}_{\mathfrak{A}}(A) + \mathrm{sp}_{\mathfrak{A}}(B).$$

[*Hint*. Apply Proposition 3.2.10 to a maximal abelian subalgebra of $\mathfrak{A}$.]

3.5.25. Let $A$ be a normal operator acting on a Hilbert space $\mathscr{H}$.

(i) Show that $A + aI$ is a normal operator for each scalar $a$ and that $A^{-1}$ is normal when $A$ is invertible.

(ii) Is the sum of two normal operators necessarily normal?

(iii) Use Theorem 2.4.2(iv) to show that $r_{\mathscr{B}(\mathscr{H})}(A) = \|A\|$.

3.5.26. Let $A_n$ be a bounded operator on a Hilbert space $\mathscr{H}_n$ for $n$ in $\{1, 2, \ldots\}$, and suppose that $\{\|A_n\|\}$ is bounded. Let $A$ be $\sum \oplus A_n$ and $\mathscr{H}$ be $\sum \oplus \mathscr{H}_n$. Show that:

(i) (*) $$[\cup_{n=1}^{\infty} \mathrm{sp}_{\mathscr{B}(\mathscr{H}_n)}(A_n)]^- \subseteq \mathrm{sp}_{\mathscr{B}(\mathscr{H})}(A);$$

(ii) the inclusion (*) becomes equality when each $A_n$ is normal.

3.5.27. Let $\mathscr{H}_n$ be an $n$-dimensional Hilbert space, $\{e_1, \ldots, e_n\}$ be an orthonormal basis for $\mathscr{H}_n$, $N_n$ be the linear operator on $\mathscr{H}_n$ determined by $N_n e_1 = 0$ and $N_n e_j = e_{j-1}$ if $1 < j \leqslant n$, and $A_n$ be $I - N_n$.

(i) Show that $\|A_n\| \leqslant 2$ and that $\mathrm{sp}_{\mathscr{B}(\mathscr{H}_n)}(A_n) = \{1\}$ for each $n$ in $\{1, 2, \ldots\}$.

(ii) Show that $A_n$ is invertible in $\mathscr{B}(\mathscr{H}_n)$ with inverse $I + N_n + N_n^2 + \cdots + N_n^{n-1}$ $(= B_n)$ and that $\|B_n\| \geqslant \sqrt{n}$.

(iii) Show that $0 \in \mathrm{sp}_{\mathscr{B}(\mathscr{H})}(A)$, where $A = \sum \oplus A_n$ and $\mathscr{H} = \sum \oplus \mathscr{H}_n$, and deduce that the inclusion (*) of Exercise 3.5.26 is strict in the present case.

3.5.28. Suppose the singular element $A$ in the Banach algebra $\mathfrak{A}$ is a norm limit of the sequence $\{A_n\}$ of regular elements in $\mathfrak{A}$ and $\mathfrak{A}_0$ is a Banach subalgebra of $\mathfrak{A}$. Show that:

(i) $\{\|A_n^{-1}\|\}$ tends to $\infty$;

(ii) if $B_n = (A_n^{-1} - I)/\|A_n^{-1} - I\|$ (when $A_n^{-1} \neq I$), then $\{AB_n\}$ and $\{B_n A\}$ tend to 0; (When a sequence such as $\{B_n\}$, not tending to 0, with $\{AB_n\}$ and $\{B_n A\}$ tending to 0, exists, we say that $A$ is a *(two-sided) topological divisor of zero*.) [*Hint*. Note that $A(A_n^{-1} - I) = (A - A_n)(A_n^{-1} - I) - A_n + I$.]

(iii) if $B$ is a topological divisor of 0 in $\mathfrak{A}$, then $B$ has no inverse in $\mathfrak{A}$;

(iv) if $B \in \mathfrak{A}_0$ and $\lambda$ is a boundary point of $\mathrm{sp}_{\mathfrak{A}_0}(B)$, then $\lambda \in \mathrm{sp}_{\mathfrak{A}}(B)$;

(v) if $B \in \mathfrak{A}_0$ and either one of $\mathrm{sp}_{\mathfrak{A}_0}(B)$ or $\mathrm{sp}_{\mathfrak{A}}(B)$ is real, then $\mathrm{sp}_{\mathfrak{A}_0}(B) = \mathrm{sp}_{\mathfrak{A}}(B)$.

3.5.29. Let $A$ be an element of a Banach algebra $\mathfrak{A}$, and let $f$ be in $\mathscr{H}(A)$. Suppose $f(z) = 0$ for each $z$ in $\mathrm{sp}(A)$.

(i) Show that $f(A) = 0$ if $\mathrm{sp}(A)$ is infinite and connected.

(ii) Find an example where $\mathrm{sp}(A)$ is connected but not infinite and $f(A) \neq 0$.

(iii) Find an example where $\mathrm{sp}(A)$ is infinite but not connected and $f(A) \neq 0$.

3.5.30. Let $\mathscr{H}$ be a Hilbert space and $\lambda$ be an isolated point of $\mathrm{sp}\, A$ with $\mathrm{sp}\, A \backslash \{\lambda\}$ non-empty and $A$ in $\mathscr{B}(\mathscr{H})$. Let $D$ be an open disk with center $\lambda$ and $\mathscr{O}$ be an open set disjoint from $D$ containing $\mathrm{sp}\, A \backslash \{\lambda\}$. Does the idempotent $g(A)$, obtained from the function $g$ that takes the value 1 on $D$ and 0 on $\mathscr{O}$, necessarily project onto a subspace of eigenvectors for $A$ corresponding to $\lambda$?

3.5.31. Let $\mathfrak{A}_1$ and $\mathfrak{A}_2$ be Banach algebras with units $I_1$ and $I_2$, respectively. Let $\varphi$ be a homomorphism or anti-homomorphism (that is, $\varphi(AB) = \varphi(B)\varphi(A)$) of $\mathfrak{A}_1$ into $\mathfrak{A}_2$ such that $\varphi(I_1) = I_2$.

(i) Show that $\mathrm{sp}_{\mathfrak{A}_2}(\varphi(A)) \subseteq \mathrm{sp}_{\mathfrak{A}_1}(A)$ for each $A$ in $\mathfrak{A}_1$.

(ii) Show, by example, that the inclusion of spectra described in (i) can be proper.

3.5.32. Let $\mathcal{M}$ be a closed left ideal in a Banach algebra $\mathfrak{A}$, and let $\mathcal{N}$ be a closed right ideal in $\mathfrak{A}$. Let $\mathcal{X}$ and $\mathcal{Y}$ be the quotient Banach spaces $\mathfrak{A}/\mathcal{M}$ and $\mathfrak{A}/\mathcal{N}$, respectively.

(i) Show that $\varphi(A)$ and $\psi(A)$ are bounded linear operators on $\mathcal{X}$ and $\mathcal{Y}$, respectively, for each $A$ in $\mathfrak{A}$, where

$$\varphi(A)(B + \mathcal{M}) = AB + \mathcal{M}, \qquad \psi(A)(B + \mathcal{N}) = BA + \mathcal{N}.$$

(ii) Show that $\varphi$ is a bounded homomorphism and $\psi$ is a bounded anti-homomorphism of $\mathfrak{A}$ into $\mathcal{B}(\mathcal{X})$ and $\mathfrak{A}$ into $\mathcal{B}(\mathcal{Y})$, respectively, and that $\varphi(I)$ and $\psi(I)$ are the respective identity operators on $\mathcal{X}$ and $\mathcal{Y}$.

3.5.33. With $\mathbb{Z}$ the additive group of integers and $f$, $g$ complex-valued functions on $\mathbb{Z}$, let $f * g$ be the function whose domain consists of those integers $s$ for which $\sum_{t \in \mathbb{Z}} f(t)g(s-t)$ converges and whose value at $s$ is the sum. (We call $f * g$ the *convolution* of $f$ and $g$.) Show that:

(i) $f * g = g * f$;

(ii) if $f \in l_1(\mathbb{Z})$ and $g \in l_p(\mathbb{Z})$ (where $1 \leqslant p$), then $f * g \in l_p(\mathbb{Z})$ and

$$\|f * g\|_p \leqslant \|f\|_1 \cdot \|g\|_p;$$

(iii) if $f \in l_1(\mathbb{Z})$, $g$ and $h$ are in $l_p(\mathbb{Z})$, and $a \in \mathbb{C}$, then $f * (a \cdot g + h)$ and $a \cdot f * g + f * h$ are in $l_p(\mathbb{Z})$ and

$$f * (a \cdot g + h) = a \cdot f * g + f * h;$$

(iv) if $f, g \in l_1(\mathbb{Z})$ and $h \in l_p(\mathbb{Z})$, then $(f * g) * h$ and $f * (g * h)$ are in $l_p(\mathbb{Z})$ and

$$(f * g) * h = f * (g * h).$$

(v) Conclude that $l_1(\mathbb{Z})$, provided with the mappings $(f, g) \to f * g$ and $f \to \|f\|_1$, is a commutative Banach algebra with unit $e$, where $e(0) = 1$ and $e(t) = 0$ if $t \neq 0$.

3.5.34. (i) Show that for each $z$ in $\mathbb{T}_1$ the equation

$$\xi_z(t) = z^t \qquad (t \in \mathbb{Z})$$

defines a homomorphism $\xi_z$ of the additive group $\mathbb{Z}$ into $\mathbb{T}_1$. (We call such a homomorphism a *character* of $\mathbb{Z}$.)

(ii) Show that the set $\hat{\mathbb{Z}}$ of characters of $\mathbb{Z}$ provided with the multiplication $(\xi \cdot \xi')(t) = \xi(t) \cdot \xi'(t)$ for $t$ in $\mathbb{Z}$ is a group. (We call $\hat{\mathbb{Z}}$ the *dual group* of $\mathbb{Z}$.)

(iii) Show that the mapping $z \to \xi_z$ is an isomorphism of $\mathbb{T}_1$ onto $\hat{\mathbb{Z}}$.

3.5.35. Using the notation of Exercises 3.5.33 and 3.5.34, let $\xi$ ($=\xi_z$) be a character of $\mathbb{Z}$. Show that:

(i) the formula

$$(*) \qquad \rho_z(f) = \sum_{t\in\mathbb{Z}} f(t)\xi_z(t) \; (=\hat{f}(z)) \qquad (f\in l_1(\mathbb{Z}))$$

defines a non-zero multiplicative linear functional on $l_1(\mathbb{Z})$;

(ii) each such functional on $l_1(\mathbb{Z})$ corresponds to a (unique) point $z$ in $\mathbb{T}_1$ by means of $(*)$;

(iii) the function $\hat{f}$ defined in $(*)$ is continuous on $\mathbb{T}_1$, and the mapping $z\to\rho_z$ of $\mathbb{T}_1$ onto the set $\mathscr{M}(\mathbb{Z})$ of non-zero multiplicative linear functionals on $l_1(\mathbb{Z})$ is a homeomorphism of $\mathbb{T}_1$ onto $\mathscr{M}(\mathbb{Z})$ with its weak* topology.

3.5.36. Let $\eta(s)$ be $\exp is$ for $s$ in $[-\pi,\pi)$, and let $m(S)$ be the Lebesgue measure of $\eta^{-1}(S)$ divided by $2\pi$ when $S$ is a subset of $\mathbb{T}_1$ such that $\eta^{-1}(S)$ is Lebesgue measurable.

(i) With the notation of Exercise 3.5.35, show that the mapping $f\to\hat{f}$ of $l_1(\mathbb{Z})$ into $C(\mathbb{T}_1)$ is linear and satisfies

$$\sum_{t\in\mathbb{Z}} |f(t)|^2 = \int_{\mathbb{T}_1} |\hat{f}(z)|^2 \, dm(z).$$

(ii) Show that the mapping $f\to\hat{f}$ of (i) extends to a unitary transformation of $l_2(\mathbb{Z})$ onto $L_2(\mathbb{T}_1, m)$.

[Compare the results of (i) and (ii) with the Plancherel theorem (3.2.31).]

3.5.37. With the measure $m$ on $\mathbb{T}_1$ introduced in Exercise 3.5.36, define the convolution of $m$-measurable functions $f$ and $g$ on $\mathbb{T}_1$ as the function $f*g$ whose domain consists of those $w$ in $\mathbb{T}_1$ for which the integral $\int_{\mathbb{T}_1} f(z)g(z^{-1}w)\,dm(z)$ converges and whose value at $w$ is this integral. Show that (i)–(v) of Proposition 3.2.22 hold for this convolution (where $\mathbb{T}_1$ replaces $\mathbb{R}$ in those assertions).

3.5.38. With $t$ in $\mathbb{Z}$ and $z$ in $\mathbb{T}_1$, let $\xi_t(z)$ be $z^t$.

(i) Show that $\xi_t$ is a continuous homomorphism of $\mathbb{T}_1$ into $\mathbb{T}_1$ (a *character* of $\mathbb{T}_1$) and that each such homomorphism has the form $\xi_t$ for some $t$ in $\mathbb{Z}$. [*Hint.* Recall the form of the characters of $\mathbb{R}$.]

(ii) Show that the set $\hat{\mathbb{T}}_1$ of characters of $\mathbb{T}_1$ provided with pointwise multiplication is a group and that the mapping $t\to\xi_t$ is an isomorphism of $\mathbb{Z}$ onto $\hat{\mathbb{T}}_1$.

3.5.39. Using the notation of Exercises 3.5.37 and 3.5.38, let $\rho_t$ be given by the formula

$$(*) \qquad \rho_t(f) = \int_{\mathbb{T}_1} f(z)\xi_t(z)\,dm(z) \ (= \hat{f}(t)) \qquad (f \in L_1(\mathbb{T}_1, m)).$$

(i) Show that $\rho_t$ is a non-zero multiplicative linear functional on $L_1(\mathbb{T}_1, m)$ for each $t$ in $\mathbb{Z}$.

(ii) Show that each non-zero multiplicative linear functional on $L_1(\mathbb{T}_1, m)$ is $\rho_t$ for some $t$ in $\mathbb{Z}$. [*Hint*. Make use of the fact that $\{\xi_r : r \in \mathbb{Z}\}$ spans a dense linear suspace of $L_1(\mathbb{T}_1, m)$ and $\xi_r * \xi_s = 0$ when $r \neq s$ to give a shorter argument than the one patterned on the proof of Theorem 3.2.26.]

(iii) Show that the set $\mathscr{M}(\mathbb{T}_1)$ of non-zero multiplicative linear functionals on $L_1(\mathbb{T}_1, m)$ is *not* weak* compact. Deduce that $L_1(\mathbb{T}_1, m)$ does not have an identity element. [*Hint*. Compute $\rho_r(\xi_{-t})$ for $r$ and $t$ in $\mathbb{Z}$.]

(iv) Let $\mathscr{A}_0(\mathbb{T}_1)$ be the Banach algebra obtained from $L_1(\mathbb{T}_1, m)$ by adjoining an identity $I$ (as in Remark 3.1.3). Let $\rho_\infty$ be the (non-zero) multiplicative linear functional on $\mathscr{A}_0(\mathbb{T}_1)$ that assigns 1 to $I$ and 0 to each $f$ in $L_1(\mathbb{T}_1, m)$. Denote by $\rho_t'$ the (unique) non-zero multiplicative linear functional on $\mathscr{A}_0(\mathbb{T}_1)$ extending $\rho_t$. Show that $\{\rho_\infty, \rho_t' : t \in \mathbb{Z}\}$ $(= \mathscr{M}_0(\mathbb{T}_1))$ is the set of all non-zero multiplicative linear functionals on $\mathscr{A}_0(\mathbb{T}_1)$ and that the mapping $\Lambda'$ of $\mathscr{M}_0(\mathbb{T}_1)$ onto the one-point compactification $\{\mathbb{Z}, \infty\}$ of $\mathbb{Z}$ that assigns $\infty$ to $\rho_\infty$ and $t$ to $\rho_t'$ is a homeomorphism of $\mathscr{M}_0(\mathbb{T}_1)$ with its weak* topology onto $\{\mathbb{Z}, \infty\}$.

(v) Show that

$$\lim_{|t| \to \infty} |\hat{f}(t)| = 0 \qquad (f \in L_1(\mathbb{T}_1, m)).$$

3.5.40. Let $Sf$ be $\hat{f}$ (as defined in Exercise 3.5.39) for $f$ in $L_2(\mathbb{T}_1, m)$ $(\subseteq L_1(\mathbb{T}_1, m))$, and let $T$ be the unitary transformation of $l_2(\mathbb{Z})$ onto $L_2(\mathbb{T}_1, m)$ described in Exercise 3.5.36(ii).

(i) Show that $S$ is a unitary transformation of $L_2(\mathbb{T}_1, m)$ onto $l_2(\mathbb{Z})$.

(ii) Let $U$ be the (self-adjoint) unitary operator on $L_2(\mathbb{T}_1, m)$ that maps $\xi_t$ onto $\xi_{-t}$ for each $t$ in $\mathbb{Z}$. Show that

$$TSU(f) = f \qquad (f \in L_2(\mathbb{T}_1, m))$$

and

$$SUT(g) = g \qquad (g \in l_2(\mathbb{Z})).$$

(iii) Deduce that

$$T^* = T^{-1} = SU$$

and

$$S^* = S^{-1} = UT.$$

3.5.41. Let $L_1(\mathbb{R})$ denote the Banach algebra formed by providing $L_1(\mathbb{R})$ with convolution multiplication. With $f$ in $L_1(\mathbb{R})$, denote by $f_r$ the "translate" of $f$ by the real number $r$ ($f_r(t) = f(t-r)$). Let $\mathscr{I}$ be a norm-closed linear subspace of $L_1(\mathbb{R})$.

(i) Show that if $\mathscr{I}$ is an ideal in $L_1(\mathbb{R})$, then $\mathscr{I}$ is invariant under translations (that is, $f_r \in \mathscr{I}$ when $f \in \mathscr{I}$). [*Hint*. Use the approximate identity of Lemma 3.2.24 and (5) of Section 3.2.]

(ii) Show that if $\mathscr{I}$ is invariant under translations, it is a (norm-closed) ideal in $L_1(\mathbb{R})$. [*Hint*. Use the Hahn–Banach theorem to support the "view" of $g * f$ as $\int g(t) f_t \, dt$ and recall the identification of the dual of $L_1$.]

(iii) Show that if the span of the translates of a function $f$ in $L_1(\mathbb{R})$ is norm dense in $L_1(\mathbb{R})$, $\hat{f}$ vanishes nowhere on $\mathbb{R}$. (With its converse, this is one of the main theorems in a body of work known as "Wiener's Tauberian theorems.")

3.5.42. Let $L_1(\mathbb{T}_1, m)$ be the Banach algebra described in Exercise 3.5.37. With $f$ in $L_1(\mathbb{T}_1, m)$, denote by $f_w$ the "translate" of $f$ by $w$ ($f_w(z) = f(z\bar{w})$). Let $\mathscr{I}$ be a norm-closed linear subspace of $L_1(\mathbb{T}_1, m)$.

(i) Show that the sequence $\{v_n\}$ satisfies

$$\|f * v_n - f\|_p \to 0$$

as $n \to \infty$, where $v_n = 2\pi(u_n \circ \eta^{-1})$, $\{u_n\}$ is the "approximate identity" described in Lemma 3.2.24 and $\eta$ is defined by $\eta(s) = \exp is$.

(ii) Show that if $\mathscr{I}$ is an ideal in $L_1(\mathbb{T}_1, m)$, then $\mathscr{I}$ is invariant under translation.

(iii) Show that if $\mathscr{I}$ is invariant under translation, it is a (norm-closed) ideal in $L_1(\mathbb{T}_1, m)$.

(iv) Show that if the span of the translates of a function $f$ in $L_1(\mathbb{T}_1, m)$ is norm dense in $L_1(\mathbb{T}_1, m)$, then $\hat{f}(t) \neq 0$ for each $t$ in $\mathbb{Z}$.

3.5.43. Let $\mathfrak{A}$ be a Banach algebra and $A$ be an element of $\mathfrak{A}$ such that $\mathrm{sp}_{\mathfrak{A}}(A) \subseteq \mathbb{C} \backslash \mathbb{R}^-$, where $\mathbb{R}^- = \{z : z \in \mathbb{C},\ z = -|z|\}$.

(i) Show that there is an element $A_0$ in $\mathfrak{A}$ such that $(A_0)^2 = A$. (The element $A_0$ is said to be a *square root* of $A$ in $\mathfrak{A}$.)

(ii) Deduce that if $B \in \mathfrak{A}$ and $\|I - B\| < 1$, then $B$ has a square root in $\mathfrak{A}$.

3.5.44. Let $A$ be an element of the Banach algebra $\mathfrak{A}$.

(i) Show that if $\mathrm{sp}_{\mathfrak{A}}(A) \subseteq \{\lambda : \lambda \in \mathbb{R}, 0 < \lambda\}$, then $A$ has a square root in $\mathfrak{A}$ with positive spectrum.

(ii) If the hypothesis of (i) is weakened to

$$\mathrm{sp}_{\mathfrak{A}}(A) \subseteq \{\lambda : \lambda \in \mathbb{R},\ 0 \leqslant \lambda\} \ (= \mathbb{R}^+),$$

does $A$ still have a positive square root in $\mathfrak{A}$? [*Hint*. Consider

$$\begin{bmatrix} 0 & 1 \\ 0 & 0 \end{bmatrix}$$

in the Banach algebra of complex $2 \times 2$ matrices.]

(iii) Show that if

$$\mathrm{sp}_{\mathfrak{A}}(A) \subseteq \mathbb{R}^+$$

and $\|B\| = r_{\mathfrak{A}}(B)$ for each $B$ in the Banach subalgebra $\mathfrak{A}_0$ of $\mathfrak{A}$ generated by $A$ and $I$, then $A$ has a square root in $\mathfrak{A}_0$ with spectrum (relative to $\mathfrak{A}_0$) in $\mathbb{R}^+$.

[*Hint*. Study the square roots of $A + n^{-1}I$ as $n \to \infty$ and use Remark 3.2.11 applied to $\mathfrak{A}_0$.]

3.5.45. Let $\rho$ be a hermitian functional on $C(X)$, where $X$ is a compact Hausdorff space. Suppose $\|\rho\| = \rho(1)$.

(i) Use Proposition 3.4.11 to show that $\rho$ is a positive linear functional on $C(X)$.

(ii) Show that $\rho$ is a positive linear functional on $C(X)$ without using Proposition 3.4.11.

3.5.46. Let $X$ be a compact Hausdorff space and $\mathscr{A}$ be a closed subalgebra of $C(X)$ containing the constant functions and containing $\bar{f}$ when $f \in \mathscr{A}$.

(i) Show that if $f$ is a real-valued function in $\mathscr{A}$, $\lambda \in \mathrm{sp}_{\mathscr{A}}(f)$, and $\lambda$ is not real, then there is a real-valued function $g$ in $\mathscr{A}$ such that $i \in \mathrm{sp}_{\mathscr{A}}(g)$.

(ii) With $g$ as in (i), show that $\sum_{n=0}^{\infty}(-ig)^n/n!\,(=\exp(-ig))$ is an element of $\mathscr{A}$ of norm 1.

(iii) Use a multiplicative linear functional on $\mathscr{A}$ to show that $e \in \mathrm{sp}_{\mathscr{A}}[\exp(-ig)]$.

(iv) Conclude that $\mathrm{sp}_{\mathscr{A}}(f) \subseteq \mathbb{R}$ for each real-valued function $f$ in $\mathscr{A}$.

(v) Conclude that $\mathrm{sp}_{\mathscr{A}}(f) \subseteq \mathbb{R}$, again, with the aid of Exercise 3.5.28(v).

3.5.47. Let $C(X)$ and $\mathscr{A}$ be as in Exercise 3.5.46 and let the partial ordering of the (real) algebra $\mathscr{A}_r$ of real-valued functions in $\mathscr{A}$ be that induced by the partial ordering of $C(X, \mathbb{R})$.

(i) Show that each (non-zero) multiplicative linear functional $\rho$ on $\mathscr{A}$ has an extension $\rho'$ to $C(X)$ such that $\rho'$ is a state of $C(X)$.

(ii) Show that $\rho$ is a pure state of $\mathscr{A}$.

(iii) Show that the set $\mathscr{E}$ of state extensions of $\rho$ to $C(X)$ is convex and weak* compact.

(iv) Show that each extreme point of $\mathscr{E}$ is a pure state of $C(X)$.

(v) Conclude that $\rho$ has an extension to $C(X)$ that is a (non-zero) multiplicative linear functional on $C(X)$.

3.5.48. Let $C(X)$ and $\mathscr{A}$ be as in Exercise 3.5.46.

(i) Show that $\mathrm{sp}_{\mathscr{A}}(f) = \mathrm{sp}_{C(X)}(f)$ for each $f$ in $\mathscr{A}$.
(ii) Use Exercise 3.5.44 to show that if $f$ is a positive function in $\mathscr{A}$, then the (unique) positive square root of $f$ in $C(X)$ lies in $\mathscr{A}$.
(iii) Conclude that $|f| \in \mathscr{A}$ when $f \in \mathscr{A}$.

3.5.49. Let $X$ be a compact Hausdorff space, and let $\mathscr{L}$ be a subset of $C(X, \mathbb{R})$ such that, with $f$ and $g$ in $\mathscr{L}$, $\mathscr{L}$ contains $f \vee g$ and $f \wedge g$ (in this case, $\mathscr{L}$ is said to be a *sublattice* of $C(X, \mathbb{R})$). Suppose that for each pair $r$, $s$ of real numbers and each pair $p$, $q$ of distinct elements of $X$ there is an $f$ in $\mathscr{L}$ such that $f(p) = r$ and $f(q) = s$. Show that $\mathscr{L}$ is norm dense in $C(X, \mathbb{R})$.

3.5.50. Combine the results of Exercises 3.5.48 and 3.5.49 to give another proof of the Stone–Weierstrass theorem (3.4.14).

# CHAPTER 4

# ELEMENTARY $C^*$-ALGEBRA THEORY

In this chapter we study a special class of Banach algebras, termed $C^*$-algebras, the ones that have an involution with properties parallel to those of the adjoint operation on Hilbert space operators. With $X$ a compact Hausdorff space and $\mathscr{H}$ a Hilbert space, $C(X)$ and $\mathscr{B}(\mathscr{H})$ are examples of $C^*$-algebras, and so is each norm-closed subalgebra of $\mathscr{B}(\mathscr{H})$ that contains the adjoint of each of its members. Two basic representation theorems (4.4.3 and 4.5.6) assert that, up to isomorphism, these are the only examples; every $C^*$-algebra can be viewed as a normed-closed self-adjoint subalgebra of $\mathscr{B}(\mathscr{H})$, for an appropriate choice of $\mathscr{H}$, and every abelian $C^*$-algebra is isomorphic to one of the form $C(X)$.

Earlier sections of the chapter are devoted to studying the spectral theoretic properties of certain special elements in $C^*$-algebras and the order structure in such algebras and in their Banach dual spaces. These are basic tools, both for the proofs of the representation theorems just cited, and also for all the subsequent theory.

## 4.1. Basics

By an *involution* on a complex Banach algebra $\mathfrak{A}$, we mean a mapping $A \to A^*$, from $\mathfrak{A}$ into $\mathfrak{A}$, such that

(i) $(aS + bT)^* = \bar{a}S^* + \bar{b}T^*$,
(ii) $(ST)^* = T^*S^*$,
(iii) $(T^*)^* = T$,

whenever $S, T \in \mathfrak{A}$ and $a, b \in \mathbb{C}$ and $\bar{a}, \bar{b}$ denote the conjugate complex numbers. A *$C^*$-algebra* is a complex Banach algebra (with a unit element $I$) with an involution that satisfies the additional condition

(iv) $\|T^*T\| = \|T\|^2 \qquad (T \in \mathfrak{A})$.

This last condition ensures that the involution in a $C^*$-algebra preserves norm (and is therefore continuous); for

$$\|T\|^2 = \|T^*T\| \leqslant \|T^*\|\,\|T\|,$$

whence $\|T\| \leqslant \|T^*\|$, and we obtain the reverse inequality upon replacing $T$ by $T^*$.

We have already encountered several examples of $C^*$-algebras. If $\mathscr{H}$ is a Hilbert space, $\mathscr{B}(\mathscr{H})$ is a $C^*$-algebra, with the adjoint operation as its involution; indeed, the defining conditions (i)–(iv) above are abstracted from the properties of adjoints of Hilbert space operators, as set out in Theorem 2.4.2 and the discussion that follows it. Several Banach algebras of complex-valued functions are $C^*$-algebras, with an involution that assigns to an element $f$ the conjugate complex function $\bar{f}$, defined by $\bar{f}(x) = \overline{f(x)}$. In this way, the Banach algebra $C(X)$, of all continuous functions on a compact Hausdorff space $X$, becomes a $C^*$-algebra. The same applies to the Banach algebra $l_\infty(X)$ of all bounded functions on an arbitrary set $X$ (with pointwise algebraic operations and supremum norm), and to the Banach algebra $L_\infty(S, \mathscr{S}, m)$ of all essentially bounded measurable functions (with pointwise algebraic operations and essential supremum norm), associated with a measure space $(S, \mathscr{S}, m)$.

We now introduce some terminology, concerning elements of a Banach algebra $\mathfrak{A}$ with involution, and note certain immediate consequences of conditions (i)–(iv) above. Motivated by the example of the algebra $\mathscr{B}(\mathscr{H})$, we refer to $A^*$ as the *adjoint* of $A$ $(\in \mathfrak{A})$, and describe $A$ as *self-adjoint* if $A = A^*$, *normal* if $A$ commutes with $A^*$, *unitary* if $A^*A = AA^* = I$. With $S = I^*$ and $T = I$, it follows from (ii) and (iii) that $I^* = I$; so the unit element $I$ is both self-adjoint and unitary. The set of all self-adjoint elements of $\mathfrak{A}$ is a real vector space, while the unitary elements form a multiplicative group, the unitary group of $\mathfrak{A}$. Each $A$ in $\mathfrak{A}$ can be expressed (uniquely) in the form $H + iK$, where $H$ $(= \frac{1}{2}(A + A^*))$ and $K$ $(= \frac{1}{2}i(A^* - A))$ are self-adjoint elements of $\mathfrak{A}$, the "real" and "imaginary" parts of $A$; moreover, $A$ is normal if and only if $H$ and $K$ commute. From (ii), $A$ is invertible if and only if $A^*$ is invertible, and then $(A^{-1})^* = (A^*)^{-1}$. By applying this result, with $aI - A$ and its adjoint $\bar{a}I - A^*$ in place of $A$ and $A^*$, it follows that the spectra of $A$ and $A^*$ satisfy

$$\mathrm{sp}(A^*) = \{\bar{a} : a \in \mathrm{sp}(A)\}.$$

Accordingly, these elements have the same spectral radius, $r(A^*) = r(A)$.

If $\mathfrak{A}$ and $\mathscr{B}$ are Banach algebras with involutions, a mapping $\varphi$ from $\mathfrak{A}$ into $\mathscr{B}$ is described as a * *homomorphism* if it is a homomorphism (that is, it is linear, multiplicative, and carries the unit of $\mathfrak{A}$ onto that of $\mathscr{B}$) with the additional property that $\varphi(A^*) = \varphi(A)^*$ for each $A$ in $\mathfrak{A}$. If, further, $\varphi$ is one-to-one, it is described as a * *isomorphism*. Although we impose no continuity condition in these definitions, we shall see later (Theorem 4.1.8) that * homomorphisms do not increase norm and * isomorphisms are norm preserving, when $\mathfrak{A}$ and $\mathscr{B}$ are $C^*$-algebras.

If $\mathfrak{A}$ is a Banach algebra with involution, a subset $\mathscr{F}$ of $\mathfrak{A}$ is said to be *self-adjoint* if it contains the adjoint of each of its members. A self-adjoint subalgebra of $\mathfrak{A}$ is termed a * *subalgebra*. If the involution is continuous (in particular, if $\mathfrak{A}$ is a $C^*$-algebra), the closure of a * subalgebra is again a * subalgebra. It is clear that a closed * subalgebra $\mathscr{B}$ of $\mathfrak{A}$ that contains the unit of $\mathfrak{A}$ is itself a Banach algebra with involution; if, further, $\mathfrak{A}$ is a $C^*$-algebra, then so is $\mathscr{B}$. In this last case, we describe $\mathscr{B}$ as a *$C^*$-subalgebra* of $\mathfrak{A}$.

The proposition that follows extends, to appropriate elements of a $C^*$-algebra, the information concerning Hilbert space operators contained in Theorem 3.2.14, together with Proposition 3.2.15 and the comments following it.

4.1.1. PROPOSITION. *Suppose that $A$ is an element of a $C^*$-algebra $\mathfrak{A}$.*

(i) *If $A$ is normal, $r(A) = \|A\|$.*

(ii) *If $A$ is self-adjoint,* $\mathrm{sp}(A)$ *is a compact subset of the real line $\mathbb{R}$, and contains at least one of the two real numbers $\pm \|A\|$.*

(iii) *If $A$ is unitary, $\|A\| = 1$ and* $\mathrm{sp}(A)$ *is a compact subset of the unit circle* $\{a \in \mathbb{C} : |a| = 1\}$.

*Proof.* (i) With $H$ self-adjoint in $\mathfrak{A}$ and $n$ a positive integer, $\|H^{2n}\| = \|(H^n)^*H^n\| = \|H^n\|^2$. By induction on $m$, $\|H^q\| = \|H\|^q$ when $q$ has the form $2^m$ $(m = 1, 2, \ldots)$; so, by Theorem 3.3.3,

$$r(H) = \lim_{q \to \infty} \|H^q\|^{1/q} = \|H\|.$$

With $A$ normal and $H$ the self-adjoint element $A^*A$, it follows from the preceding argument, together with Corollary 3.3.4 and the $C^*$ property of the norm, that

$$\begin{aligned}\|A\|^2 = \|A^*A\| &= r(A^*A)\\ &\leqslant r(A^*)r(A) = r(A)^2 \leqslant \|A\|^2;\end{aligned}$$

so $r(A) = \|A\|$.

(ii) With $A$ self-adjoint in $\mathfrak{A}$, $\mathrm{sp}(A)$ is compact (Theorem 3.2.3) and so contains a scalar with absolute value $r(A)$; and $r(A) = \|A\|$, from part (i) of the present proposition. Consequently, if suffices to prove that $\mathrm{sp}(A) \subseteq \mathbb{R}$. For this, suppose that $c \in \mathrm{sp}(A)$, where $c = a + ib$. For each integer $n$, let $B_n = A - aI + inbI$, and observe that

$$i(n+1)b = a + ib - a + inb \in \mathrm{sp}(B_n).$$

Accordingly

$$\begin{aligned}(n^2 + 2n + 1)b^2 &= |i(n+1)b|^2 \leqslant [r(B_n)]^2 \leqslant \|B_n\|^2\\ &= \|B_n^* B_n\| = \|(A - aI - inbI)(A - aI + inbI)\|\\ &= \|(A - aI)^2 + n^2b^2 I\| \leqslant \|A - aI\|^2 + n^2 b^2.\end{aligned}$$

Thus $(2n+1)b^2 \leqslant \|A - aI\|^2$ $(n = 1, 2, \ldots)$; so $b = 0$, and $c = a \in \mathbb{R}$.

(iii) With $A$ unitary in $\mathfrak{A}$, it results from the $C^*$ property of the norm that

$$\|A\|^2 = \|A^*A\| = \|I\| = 1,$$

so $\|A\| = 1$. With $a$ in $\mathrm{sp}(A)$, we have

$$a^{-1} \in \mathrm{sp}(A^{-1}) = \mathrm{sp}(A^*);$$

hence

$$|a| \leqslant \|A\| = 1, \qquad |a|^{-1} \leqslant \|A^*\| = 1,$$

and thus $|a| = 1$. ■

4.1.2. Corollary. *If $A$ is a normal element of a $C^*$-algebra $\mathfrak{A}$, and $A^k = 0$ for some positive integer $k$, then $A = 0$.*

*Proof.* Since $A^n = 0$ when $n \geqslant k$, it results from Proposition 4.1.1(i) that $\|A\| = r(A) = \lim \|A^n\|^{1/n} = 0$. ■

Suppose that $A$ is a self-adjoint element of a $C^*$-algebra $\mathfrak{A}$, and denote by $C(\mathrm{sp}(A))$ the $C^*$-algebra of all continuous complex-valued functions on the spectrum $\mathrm{sp}(A)$. We now introduce the (continuous) *function calculus* for $A$, a mapping that associates with each $f$ in $C(\mathrm{sp}(A))$ an element $f(A)$ of $\mathfrak{A}$. The existence and properties of this mapping are the subject of Theorems 4.1.3, 4.1.6, and 4.1.8(ii) and Propositions 4.1.4 and 4.2.3(i) below. At a later stage (Theorem 4.4.5) we shall construct a similar function calculus for a normal element of a $C^*$-algebra. For self-adjoint elements, the two methods lead to the same function calculus; Remark 4.4.6, Theorem 4.4.8, and Example 4.4.9 provide information that, even in the self-adjoint case, is not contained in the other results just cited.

4.1.3. Theorem. *If $A$ is a self-adjoint element of a $C^*$-algebra $\mathfrak{A}$, there is a unique continuous mapping $f \to f(A)$: $C(\mathrm{sp}(A)) \to \mathfrak{A}$ such that*

(i) *$f(A)$ has its elementary meaning when $f$ is a polynomial.*

*Moreover, when $f, g \in C(\mathrm{sp}(A))$ and $a, b \in \mathbb{C}$,*

(ii) $\|f(A)\| = \|f\|$;
(iii) $(af + bg)(A) = af(A) + bg(A)$;
(iv) $(fg)(A) = f(A)g(A)$;
(v) $\bar{f}(A) = [f(A)]^*$, *where $\bar{f}$ denotes the conjugate complex function; in particular, $f(A)$ is self-adjoint if and only if $f$ takes real values throughout* $\mathrm{sp}(A)$;
(vi) *$f(A)$ is normal;*
(vii) *$f(A)B = Bf(A)$ whenever $B \in \mathfrak{A}$ and $AB = BA$.*

*Proof.* By Proposition 4.1.1(ii), $\mathrm{sp}(A)$ is a compact subset of the real line. The Weierstrass approximation theorem shows that the set $P$ of all polynomials with complex coefficients, considered as a subset of $C(\mathrm{sp}(A))$, is everywhere dense. If $p$ is such a polynomial, say

$$p(t) = a_0 + a_1 t + a_2 t^2 + \cdots + a_n t^n,$$

then

$$p(A) = a_0 I + a_1 A + a_2 A^2 + \cdots + a_n A^n,$$

$$[p(A)]^* = \bar{a}_0 I + \bar{a}_1 A + \bar{a}_2 A^2 + \cdots + \bar{a}_n A^n.$$

Hence $p(A)$ and $[p(A)]^*$ commute; that is, $p(A)$ is normal. From the spectral mapping theorem for polynomials (see Proposition 3.2.10), together with Proposition 4.1.1(i),

$$\begin{aligned}\|p(A)\| = r(p(A)) &= \max\{|s| : s \in \mathrm{sp}(p(A))\} \\ &= \max\{|p(t)| : t \in \mathrm{sp}(A)\} = \|p\|\end{aligned}$$

(the norm of $p$ as an element of $C(\mathrm{sp}(A))$). If two distinct polynomials $p$ and $q$ are identically equal on $\mathrm{sp}(A)$, we can replace $p$ by $p - q$ in the above argument, and deduce that $p(A) = q(A)$; of course, this question arises only when $\mathrm{sp}(A)$ is finite.

The preceding argument shows that the linear mapping $p \to p(A) : P \to \mathfrak{A}$ is well defined and continuous (in fact, isometric). Since $\mathfrak{A}$ is complete and $P$ is everywhere dense in $C(\mathrm{sp}(A))$, there is a unique extension to a continuous mapping $f \to f(A) \colon C(\mathrm{sp}(A)) \to \mathfrak{A}$.

We have now proved the existence of a unique continuous mapping $f \to f(A)$ satisfying condition (i) in the theorem. In view of the above argument, each of the remaining properties (ii)–(vii) is easily verified when $f$ and $g$ are polynomials, and by continuity remains valid for all $f$ and $g$ in $C(\mathrm{sp}(A))$. ■

Clauses (i)–(v) of Theorem 4.1.3 amount to the assertion that the function calculus $f \to f(A) \colon C(\mathrm{sp}(A)) \to \mathfrak{A}$ is an isometric * isomorphism that carries the identity mapping on $\mathrm{sp}(A)$ to the self-adjoint element $A$ of $\mathfrak{A}$. Since $C(\mathrm{sp}(A))$ is a complete metric space, the same is true of its image $\{f(A) : f \in C(\mathrm{sp}(A))\}$ in $\mathfrak{A}$, so this set is an abelian $C^*$-subalgebra $\mathfrak{A}(A)$ of $\mathfrak{A}$, containing $I$ and $A$. Since polynomials form an everywhere-dense subset of $C(\mathrm{sp}(A))$, each element of $\mathfrak{A}(A)$ is the limit of a sequence of polynomials in $A$. A closed subalgebra $\mathscr{B}$ of $\mathfrak{A}$ that contains $I$ and $A$ necessarily contains all polynomials in $A$ and therefore contains $\mathfrak{A}(A)$. We have now proved the following result.

4.1.4. Proposition. *If $A$ is a self-adjoint element of a $C^*$-algebra $\mathfrak{A}$, the set $\{f(A) : f \in C(\mathrm{sp}(A))\}$ is an abelian $C^*$-subalgebra $\mathfrak{A}(A)$ of $\mathfrak{A}$, and is the smallest closed subalgebra of $\mathfrak{A}$ that contains $I$ and $A$. Each element of $\mathfrak{A}(A)$ is the limit of a sequence of polynomials in $A$.*

Suppose that $\mathfrak{A}$ is a complex Banach algebra, $\mathscr{B}$ is a closed subalgebra that contains the unit $I$ of $\mathfrak{A}$, and $B \in \mathscr{B}$. If $a \in \mathrm{sp}_{\mathfrak{A}}(B)$, then $aI - B$ has no inverse in $\mathfrak{A}$; accordingly, it has no inverse in $\mathscr{B}$, so $a \in \mathrm{sp}_{\mathscr{B}}(B)$. Hence $\mathrm{sp}_{\mathfrak{A}}(B) \subseteq \mathrm{sp}_{\mathscr{B}}(B)$, and Example 3.2.19 shows that strict inclusion can occur. By use of the function calculus described in Theorem 4.1.3, we now show that the two spectra coincide when $\mathfrak{A}$ and $\mathscr{B}$ are $C^*$-algebras.

4.1.5. PROPOSITION. *If $\mathfrak{A}$ is a $C^*$-algebra, $\mathscr{B}$ is a $C^*$-subalgebra of $\mathfrak{A}$, and $B \in \mathscr{B}$, then* $\mathrm{sp}_{\mathfrak{A}}(B) = \mathrm{sp}_{\mathscr{B}}(B)$.

*Proof.* As noted above, $\mathrm{sp}_{\mathfrak{A}}(B) \subseteq \mathrm{sp}_{\mathscr{B}}(B)$. In order to establish the reverse inclusion, it suffices to prove the following result: if $A \in \mathscr{B}$, and $A$ has an inverse $A^{-1}$ in $\mathfrak{A}$, then $A^{-1} \in \mathscr{B}$.

We consider first the case in which $A$ is self-adjoint. Since $0 \notin \mathrm{sp}_{\mathfrak{A}}(A)$, the equation $f(t) = t^{-1}$ defines a continuous function on $\mathrm{sp}_{\mathfrak{A}}(A)$. By means of the function calculus for $A$ relative to $\mathfrak{A}$, we obtain an element $f(A)$ of $\mathfrak{A}$, and deduce from Proposition 4.1.4 that $f(A) \in \mathscr{B}$. Since $tf(t) = 1$ for each $t$ in $\mathrm{sp}_{\mathfrak{A}}(A)$, it follows from Theorem 4.1.3(i) and (iv) that $Af(A) = I$; so $A^{-1} = f(A) \in \mathscr{B}$.

Consider next a (not necessarily self-adjoint) element $A$ of $\mathscr{B}$ that has an inverse $C$ in $\mathfrak{A}$. Then $A^*$ lies in $\mathscr{B}$ and has inverse $C^*$ in $\mathfrak{A}$. Since $A^*A$ is a self-adjoint element of $\mathscr{B}$, with inverse $CC^*$ in $\mathfrak{A}$, it follows from the preceding paragraph that $CC^* \in \mathscr{B}$. Accordingly, $A^{-1} = C = (CC^*)A^* \in \mathscr{B}$. ∎

In the circumstances considered in Proposition 4.1.5, we can now omit the suffices $\mathfrak{A}$ and $\mathscr{B}$, and denote by $\mathrm{sp}(B)$ the spectrum of $B$ relative to either algebra.

From the preceding proposition, if $A$ is an invertible self-adjoint element of a $C^*$-algebra $\mathfrak{A}$, then $A^{-1}$ is the norm limit of $p_n(A)$, with each $p_n$ a polynomial (for this, in the proposition take for $\mathscr{B}$ the $C^*$-subalgebra generated by $A$ and $I$). This can be strengthened as follows. If $A$ is self-adjoint and has inverse $A^{-1}$ in $\mathfrak{A}$, then there is a sequence $\{p_n\}$ of polynomials *without constant terms* such that $\|p_n(A) - A^{-1}\| \to 0$. To see this, extend $t^{-1}$ on $\mathrm{sp}(A)$ to a continuous function $f$ on an interval containing $\mathrm{sp}(A)$ and 0, so that $f(0) = 0$. From the Weierstrass approximation theorem, $f$ is the norm (uniform) limit of polynomials $q_n$ on this interval. Of course, $q_n(0) \to f(0) = 0$. Thus the polynomials $p_n$ obtained from $q_n$ by omitting the constant terms ($q_n(0)$) tend to $f$ in norm on this interval (hence on $\mathrm{sp}(A)$). From (i) and (ii) of Theorem 4.1.3, $\|p_n(A) - f(A)\| \to 0$, and from (i) and (iv) of that theorem, $f(A) = A^{-1}$.

Our next result is another spectral mapping theorem (compare Theorem 3.3.6).

4.1.6. THEOREM. *If $A$ is a self-adjoint element of a $C^*$-algebra $\mathfrak{A}$, and $f \in C(\mathrm{sp}(A))$, then*

$$\mathrm{sp}(f(A)) = \{f(t) : t \in \mathrm{sp}(A)\}.$$

*Proof.* In view of Propositions 4.1.4 and 4.1.5, we can interpret $\mathrm{sp}(f(A))$ as the spectrum of $f(A)$ relative to the $C^*$-subalgebra

$$\mathfrak{A}(A) = \{g(A): g \in C(\mathrm{sp}(A))\}$$

of $\mathfrak{A}$. Since the function calculus $g \to g(A)$: $C(\mathrm{sp}(A)) \to \mathfrak{A}$ is a * isomorphism from $C(\mathrm{sp}(A))$ onto $\mathfrak{A}(A)$, it follows that $\mathrm{sp}(f(A))$ coincides with the spectrum of $f$ as an element of $C(\mathrm{sp}(A))$; that is, by Remark 3.4.4,

$$\mathrm{sp}(f(A)) = \{f(t): t \in \mathrm{sp}(A)\}. \quad \blacksquare$$

We conclude this section with some further applications of function calculus.

4.1.7. THEOREM. *Each element $A$ of a $C^*$-algebra $\mathfrak{A}$ is a finite linear combination of unitary elements of $\mathfrak{A}$.*

*Proof.* It is sufficient to consider the case in which $A$ is self-adjoint and $\|A\| \leqslant 1$. In these circumstances, $\mathrm{sp}(A)$ is a subset of the interval $[-1, 1]$, and we can define $f$ in $C(\mathrm{sp}(A))$ by $f(t) = t + i\sqrt{1 - t^2}$. Since

$$t = \tfrac{1}{2}[f(t) + \bar{f}(t)], \qquad f(t)\bar{f}(t) = \bar{f}(t)f(t) = 1,$$

for each $t$ in $\mathrm{sp}(A)$, it follows that the element $f(A)$ $(= U)$ of $\mathfrak{A}$ satisfies

$$A = \tfrac{1}{2}(U + U^*), \qquad UU^* = U^*U = I. \quad \blacksquare$$

Suppose that $\mathfrak{A}$ is a $C^*$-algebra, $A = A^* \in \mathfrak{A}$, and $f$ is a continuous complex-valued function whose domain includes $\mathrm{sp}(A)$. We denote by $f(A)$ the element of $\mathfrak{A}$ that, in the function calculus for $A$, corresponds to the restriction $f|\mathrm{sp}(A)$. This convention is used in Theorem 4.1.8(ii), where we refer to $f(\varphi(A))$ although the domain of $f$ may be strictly larger than $\mathrm{sp}(\varphi(A))$.

4.1.8. THEOREM. *Suppose that $\mathfrak{A}$ and $\mathscr{B}$ are $C^*$-algebras and $\varphi$ is a * homomorphism from $\mathfrak{A}$ into $\mathscr{B}$.*

(i) *For each $A$ in $\mathfrak{A}$, $\mathrm{sp}(\varphi(A)) \subseteq \mathrm{sp}(A)$ and $\|\varphi(A)\| \leqslant \|A\|$; in particular, $\varphi$ is continuous.*

(ii) *If $A$ is a self-adjoint element of $\mathfrak{A}$ and $f \in C(\mathrm{sp}(A))$, then $\varphi(f(A)) = f(\varphi(A))$.*

(iii) *If $\varphi$ is a * isomorphism, then $\|\varphi(A)\| = \|A\|$ and $\mathrm{sp}(\varphi(A)) = \mathrm{sp}(A)$ for each $A$ in $\mathfrak{A}$, and $\varphi(\mathfrak{A})$ is a $C^*$-subalgebra of $\mathscr{B}$.*

*Proof.* (i) The unit elements of $\mathfrak{A}$ and $\mathscr{B}$ will both be denoted by $I$, since the context in each case indicates which one is intended. With $A$ in $\mathfrak{A}$, we prove first that $\mathrm{sp}(\varphi(A)) \subseteq \mathrm{sp}(A)$. For this, if $a \notin \mathrm{sp}(A)$, $aI - A$ has an inverse $S$ in $\mathfrak{A}$; since $\varphi(I) = I$, $aI - \varphi(A)$ has inverse $\varphi(S)$ in $\mathscr{B}$, so $a \notin \mathrm{sp}(\varphi(A))$; hence $\mathrm{sp}(\varphi(A)) \subseteq \mathrm{sp}(A)$.

With $A$ in $\mathfrak{A}$, it results from Proposition 4.1.1(i) that

$$\|A\|^2 = \|A^*A\| = r(A^*A), \tag{1}$$
$$\|\varphi(A)\|^2 = \|\varphi(A)^*\varphi(A)\| = \|\varphi(A^*A)\| = r(\varphi(A^*A)).$$

Since $\mathrm{sp}(\varphi(A^*A)) \subseteq \mathrm{sp}(A^*A)$, we have $r(\varphi(A^*A)) \leqslant r(A^*A)$, and therefore $\|\varphi(A)\| \leqslant \|A\|$.

(ii) If $\{p_n\}$ is a sequence of polynomials tending to $f$ uniformly on $\mathrm{sp}(A)$ (hence, from (i), on $\mathrm{sp}(\varphi(A))$), then $\varphi(p_n(A)) \to \varphi(f(A))$ and $p_n(\varphi(A)) \to f(\varphi(A))$. Since $\varphi(p_n(A)) = p_n(\varphi(A))$ for each $n$ (because $\varphi$ is a homomorphism), (ii) follows.

(iii) Suppose now that $\varphi$ is a * isomorphism. With $B$ self-adjoint in $\mathfrak{A}$, it follows from (i) that $\mathrm{sp}(\varphi(B)) \subseteq \mathrm{sp}(B)$. If strict inclusion occurs, there is a non-zero element $f$ of $C(\mathrm{sp}(B))$, whose restriction to $\mathrm{sp}(\varphi(B))$ is identically zero. From part (ii) of the theorem, we have

$$f(B) \neq 0, \qquad \varphi(f(B)) = f(\varphi(B)) = 0,$$

contrary to the assumption that $\varphi$ is one-to-one. Hence $\mathrm{sp}(\varphi(B)) = \mathrm{sp}(B)$, and so $r(\varphi(B)) = r(B)$, for each self-adjoint $B$ in $\mathfrak{A}$.

With $A$ in $\mathfrak{A}$ and $B = A^*A$, it follows from the preceding paragraph and (1) that

$$\|A\|^2 = r(A^*A) = r(\varphi(A^*A)) = \|\varphi(A)\|^2, \qquad \|\varphi(A)\| = \|A\|.$$

Since $\mathfrak{A}$ is a complete metric space and $\varphi: \mathfrak{A} \to \mathscr{B}$ is an isometry, the * subalgebra $\varphi(\mathfrak{A})$ of $\mathscr{B}$ is closed, contains $I$, and is therefore a $C^*$-subalgebra of $\mathscr{B}$. By Proposition 4.1.5, the spectrum of $\varphi(A)$ in $\mathscr{B}$ is the same as its spectrum in $\varphi(\mathfrak{A})$; and $\mathrm{sp}(A) = \mathrm{sp}_{\varphi(\mathfrak{A})}(\varphi(A))$, since $\varphi$ is an isomorphism from $\mathfrak{A}$ onto $\varphi(\mathfrak{A})$. ∎

The following result strengthens the final conclusion of Theorem 4.1.8(iii).

4.1.9. THEOREM. *If $\mathfrak{A}$ and $\mathscr{B}$ are $C^*$-algebras and $\varphi$ is a * homomorphism from $\mathfrak{A}$ into $\mathscr{B}$, then $\varphi(\mathfrak{A})$ is a $C^*$-subalgebra of $\mathscr{B}$.*

*Proof.* Since $\varphi(\mathfrak{A})$ is a * subalgebra of $\mathscr{B}$ (containing $I$), it suffices to show that $\varphi(\mathfrak{A})$ is closed in $\mathscr{B}$. Accordingly we must prove that, if $B \in \mathscr{B}$ and $\|B - \varphi(A_n)\| \to 0$ for some sequence $\{A_n\}$ of elements of $\mathfrak{A}$, then $B \in \varphi(\mathfrak{A})$. By expressing $B, A_1, A_2, \ldots$ in terms of their real and imaginary parts, we reduce to the case in which $B$ and the $A_n$'s are self-adjoint. Upon passing to a subsequence of $\{A_n\}$, we may suppose also that

$$\|\varphi(A_{n+1}) - \varphi(A_n)\| < 2^{-n} \qquad (n = 1, 2, \ldots).$$

Let $f_n$ be a continuous function on $\mathbb{R}$, with values in the real interval $[-2^{-n}, 2^{-n}]$, such that $f_n(t) = t$ when $|t| \leqslant 2^{-n}$. From Theorem 4.1.8(ii), and since $f_n$ restricts to the identity mapping on $\mathrm{sp}(\varphi(A_{n+1}) - \varphi(A_n))$, we have

$$\varphi(A_{n+1}) - \varphi(A_n) = f_n(\varphi(A_{n+1} - A_n)) = \varphi(f_n(A_{n+1} - A_n)).$$

Since $\|f_n(A_{n+1} - A_n)\| \leqslant 2^{-n}$, the series $A_1 + \sum_{n=1}^{\infty} f_n(A_{n+1} - A_n)$ converges to an element $A$ of $\mathfrak{A}$; and by continuity of $\varphi$ (Theorem 4.1.8(i)),

$$\begin{aligned}\varphi(A) &= \lim_{m\to\infty} \left\{\varphi(A_1) + \sum_{n=1}^{m-1} \varphi(f_n(A_{n+1} - A_n))\right\} \\ &= \lim_{m\to\infty} \left\{\varphi(A_1) + \sum_{n=1}^{m-1} [\varphi(A_{n+1}) - \varphi(A_n)]\right\} \\ &= \lim_{m\to\infty} \varphi(A_m) = B.\end{aligned}$$

Thus $B \in \varphi(\mathfrak{A})$. ■

At later stages (Corollary 4.2.10 and Theorem 10.1.7) we shall show that a closed two-sided ideal $\mathscr{K}$ in a $C^*$-algebra $\mathfrak{A}$ is automatically self-adjoint, and that the quotient algebra $\mathfrak{A}/\mathscr{K}$ is a $C^*$-algebra. With the aid of the latter result, Theorem 4.1.9 becomes a simple consequence of Theorem 4.1.8(iii) (see the proof of Corollary 10.1.8).

## 4.2. Order structure

In Section 2.4 we introduced a concept of positivity for operators acting on a Hilbert space, and a partial order relation on the set of self-adjoint operators. In this section we study the corresponding notions for elements of a $C^*$-algebra.

We recall that a bounded linear operator $A$, acting on a Hilbert space $\mathscr{H}$, is said to be positive if $\langle Ax, x\rangle \geqslant 0$ for each $x$ in $\mathscr{H}$. By Proposition 2.4.6(i) and Theorem 3.2.14(ii), a positive operator $A$ is self-adjoint, and its spectrum is a subset of the non-negative half-line $\mathbb{R}^+ = \{t \in \mathbb{R} : t \geqslant 0\}$. Conversely, if $A$ is a self-adjoint element of $\mathscr{B}(\mathscr{H})$, and $\mathrm{sp}(A) \subseteq \mathbb{R}^+$, the equation $f(t) = t^{1/2}$ defines a real-valued continuous function $f$ on $\mathrm{sp}(A)$. By means of the function calculus, for $A$ as a member of the $C^*$-algebra $\mathscr{B}(\mathscr{H})$, we obtain a self-adjoint operator $H$ $(= f(A))$ such that $H^2 = A$. Since

$$\langle Ax, x\rangle = \langle Hx, Hx\rangle \geqslant 0 \qquad (x \in \mathscr{H}),$$

$A$ is a positive operator. Accordingly, the positive operators are precisely those that are self-adjoint and have spectrum contained in $\mathbb{R}^+$.

Motivated by these considerations, we describe an element $A$ of a $C^*$-algebra $\mathfrak{A}$ as *positive* if $A$ is self-adjoint and $\mathrm{sp}(A) \subseteq \mathbb{R}^+$; we denote by $\mathfrak{A}^+$ the set of all positive elements of $\mathfrak{A}$. From the preceding discussion, this definition is consistent with our earlier conventions when $\mathfrak{A} = \mathscr{B}(\mathscr{H})$. If $\mathscr{B}$ is a $C^*$-subalgebra of $\mathfrak{A}$, a self-adjoint element $B$ of $\mathscr{B}$ is positive relative to $\mathscr{B}$ if and only if it is positive relative to $\mathfrak{A}$ (that is, $\mathscr{B}^+ = \mathscr{B} \cap \mathfrak{A}^+$), since it has the same spectrum in $\mathscr{B}$ as in $\mathfrak{A}$. If $\varphi$ is a * homomorphism from $\mathfrak{A}$ into a $C^*$-algebra $\mathscr{C}$ and $A \in \mathfrak{A}^+$, then $\varphi(A) \in \mathscr{C}^+$; for $\varphi(A)$ is self-adjoint, and $\mathrm{sp}(\varphi(A)) \subseteq \mathrm{sp}(A) \subseteq \mathbb{R}^+$, by Theorem 4.1.8(i). From Proposition 4.1.1(ii), $\|A\| \in \mathrm{sp}(A)$ when $A \in \mathfrak{A}^+$.

With $X$ a compact Hausdorff space, and $f$ in the $C^*$-algebra $C(X)$, $f$ is self-adjoint if and only if $f$ is real valued throughout $X$; moreover,

$$\mathrm{sp}(f) = \{f(x) : x \in X\}.$$

Accordingly, $f$ is positive (in the $C^*$-algebra sense just defined) if and only if $f(x) \geqslant 0$ for each $x$ in $X$.

4.2.1. Lemma. *If $A$ is a self-adjoint element of a $C^*$-algebra $\mathfrak{A}$, $a \in \mathbb{R}$, and $a \geqslant \|A\|$, then $A \in \mathfrak{A}^+$ if and only if $\|A - aI\| \leqslant a$.*

*Proof.* Since $\mathrm{sp}(A) \subseteq [-a, a]$, and

$$\|A - aI\| = r(A - aI) = \sup_{t \in \mathrm{sp}(A)} |t - a| = \sup_{t \in \mathrm{sp}(A)} (a - t),$$

it is apparent that $\|A - aI\| \leqslant a$ if and only if $\mathrm{sp}(A) \subseteq \mathbb{R}^+$. ■

Clauses (ii), (iii), and (v) of the following theorem tell us that $\mathfrak{A}^+$ is a (positive) cone (in the sense explained in the discussion preceding Definition 3.4.5), in the real vector space of all self-adjoint elements of $\mathfrak{A}$.

4.2.2. Theorem. *Suppose that $\mathfrak{A}$ is a $C^*$-algebra.*

(i) *$\mathfrak{A}^+$ is closed in $\mathfrak{A}$.*
(ii) *$aA \in \mathfrak{A}^+$ if $A \in \mathfrak{A}^+$ and $a \in \mathbb{R}^+$.*
(iii) *$A + B \in \mathfrak{A}^+$ if $A, B \in \mathfrak{A}^+$.*
(iv) *$AB \in \mathfrak{A}^+$ if $A, B \in \mathfrak{A}^+$ and $AB = BA$.*
(v) *If $A \in \mathfrak{A}^+$ and $-A \in \mathfrak{A}^+$, then $A = 0$.*

*Proof.* (i) From Lemma 4.2.1,

$$\mathfrak{A}^+ = \{A \in \mathfrak{A} : A = A^* \text{ and } \|A - \|A\|I\| \leqslant \|A\|\},$$

whence $\mathfrak{A}^+$ is closed (since the norm is continuous on $\mathfrak{A}$).

(ii) If $A \in \mathfrak{A}^+$ and $a \in \mathbb{R}^+$, then $aA$ is self-adjoint, and

$$\mathrm{sp}(aA) = \{at : t \in \mathrm{sp}(A)\} \subseteq \mathbb{R}^+.$$

(iii) If $A, B \in \mathfrak{A}^+$, it results from Lemma 4.2.1 that

$$\|A - \|A\|I\| \leqslant \|A\|, \qquad \|B - \|B\|I\| \leqslant \|B\|.$$

Thus

$$\|A + B - (\|A\| + \|B\|)I\| \leqslant \|A\| + \|B\|;$$

and from the same lemma (with $a = \|A\| + \|B\| \geqslant \|A + B\|$), it follows that $A + B \in \mathfrak{A}^+$.

(iv) With $A, B$ commuting self-adjoint elements of $\mathfrak{A}^+$, $AB$ is self-adjoint since $(AB)^* = BA = AB$. Since each of $A$, $B$, $AB$ has the same spectrum in $\mathfrak{A}$ as in the commutative $C^*$-subalgebra generated by $\{I, A, B\}$, it follows from Proposition 3.2.10 that

$$\text{sp}(AB) \subseteq \{st : s \in \text{sp}(A), t \in \text{sp}(B)\} \subseteq \mathbb{R}^+.$$

(v) If $A, -A \in \mathfrak{A}^+$, then $A$ is self-adjoint and $\text{sp}(A) \subseteq \mathbb{R}^+ \cap -\mathbb{R}^+ = \{0\}$; so $\|A\| = r(A) = 0$. ■

4.2.3. Proposition. *Suppose that $A$ is a self-adjoint element of a $C^*$-algebra $\mathfrak{A}$, and $f \in C(\text{sp}(A))$.*

(i) *$f(A) \in \mathfrak{A}^+$ if and only if $f(t) \geqslant 0$ for each $t$ in* $\text{sp}(A)$.

(ii) $\|A\|I \pm A \in \mathfrak{A}^+$.

(iii) *$A$ can be expressed in the form $A^+ - A^-$, where $A^+, A^- \in \mathfrak{A}^+$ and $A^+A^- = A^-A^+ = 0$. These conditions determine $A^+$ and $A^-$ uniquely, and* $\|A\| = \max(\|A^+\|, \|A^-\|)$.

*Proof.* (i) By Theorem 4.1.6, $f(A)$ has spectrum $\{f(t) : t \in \text{sp}(A)\}$; so $f$ takes non-negative values throughout $\text{sp}(A)$ if $f(A) \in \mathfrak{A}^+$. Conversely, if $f(t) \geqslant 0$ for each $t$ in $\text{sp}(A)$, then $f(A)$ is self-adjoint (since $f$ is real valued) and has spectrum a subset of $\mathbb{R}^+$.

(ii) With $f$ in $C(\text{sp}(A))$ defined by $f(t) = \|A\| \pm t$, $f$ takes non-negative values throughout $\text{sp}(A)$. By (i), $f(A) \in \mathfrak{A}^+$; that is, $\|A\|I \pm A \in \mathfrak{A}^+$.

(iii) With $u, u^+, u^-$ the continuous real-valued functions defined, for all real $t$, by

$$u(t) = t, \qquad u^+(t) = \max\{t, 0\}, \qquad u^-(t) = \max\{-t, 0\},$$

we have

$$u = u^+ - u^-, \qquad u^+u^- = u^-u^+ = 0.$$

Since $u(A) = A$, we have

$$A = A^+ - A^-, \qquad A^+A^- = A^-A^+ = 0,$$

where $A^+ = u^+(A)$ and $A^- = u^-(A)$; moreover, $A^+, A^- \in \mathfrak{A}^+$, by (i). The supremum norms of $u, u^+$, and $u^-$, as elements of $C(\text{sp}(A))$, satisfy

$$\|u\| = \max\{\|u^+\|, \|u^-\|\},$$

so $\|A\| = \max\{\|A^+\|, \|A^-\|\}$.

To prove the uniqueness clause of (iii), suppose that $A = B - C$, where $B, C \in \mathfrak{A}^+$ and $BC = CB = 0$. Then

$$A^n = B^n + (-C)^n \qquad (n = 1, 2, 3, \ldots),$$

and therefore $p(A) = p(B) + p(-C)$ whenever $p$ is a polynomial with zero constant term. There is a sequence $\{p_n\}$ of such polynomials that converges to $u^+$ uniformly on $\text{sp}(A) \cup \text{sp}(B) \cup \text{sp}(-C)$; and

$$u^+(A) = \lim p_n(A) = \lim[p_n(B) + p_n(-C)] = u^+(B) + u^+(-C).$$

Since

$$u^+(s) = s \quad (s \in \text{sp}(B)), \qquad u^+(t) = 0 \quad (t \in \text{sp}(-C)),$$

we have $u^+(B) = B, u^+(-C) = 0$. Thus

$$B = u^+(A) = A^+, \qquad C = B - A = A^+ - A = A^-. \quad \blacksquare$$

4.2.4. Corollary. *Each element $A$ of a $C^*$-algebra $\mathfrak{A}$ is a linear combination of at most four members of $\mathfrak{A}^+$.*

*Proof.* From Proposition 4.2.3(ii) or (iii), the real and imaginary parts of $A$ can each be expressed as a difference of elements of $\mathfrak{A}^+$. $\blacksquare$

Our next objective, achieved in Theorem 4.2.6, is to give a number of conditions equivalent to positivity for elements of a $C^*$-algebra. For this purpose, we require the following preliminary result.

4.2.5. Lemma. *If $\mathfrak{A}$ is a $C^*$-algebra, $A \in \mathfrak{A}$, and $-A^*A \in \mathfrak{A}^+$, then $A = 0$.*

*Proof.* Let $A = H + iK$, with $H$ and $K$ self-adjoint in $\mathfrak{A}$. Since $\text{sp}(H) \subseteq \mathbb{R}$ and $\text{sp}(H^2) = \{t^2 : t \in \text{sp}(H)\} \subseteq \mathbb{R}^+$, it follows that $H^2$ (and similarly $K^2$) is positive. Since $AA^*$ is self-adjoint, and $\text{sp}(-AA^*) \subseteq \text{sp}(-A^*A) \cup \{0\} \subseteq \mathbb{R}^+$ by Proposition 3.2.8, $-AA^*$ is positive. Now

$$A^*A + AA^* = (H - iK)(H + iK) + (H + iK)(H - iK) = 2H^2 + 2K^2,$$

$$A^*A = 2H^2 + 2K^2 + (-AA^*).$$

Since all three terms on the right-hand side of the last equation are positive, $A^*A$ (as well as $-A^*A$) is positive and so $A^*A = 0$, by Theorem 4.2.2(iii) and (v). Thus $\|A\|^2 = \|A^*A\| = 0$, and $A = 0$. $\blacksquare$

4.2.6. Theorem. *If $\mathfrak{A}$ is a $C^*$-algebra and $A \in \mathfrak{A}$, the following conditions are equivalent:*

(i) $A \in \mathfrak{A}^+$.
(ii) $A = H^2$, *for some $H$ in $\mathfrak{A}^+$.*
(iii) $A = B^*B$, *for some $B$ in $\mathfrak{A}$.*

*When these conditions are satisfied, the element $H$ occurring in* (ii) *is unique.*

*If $\mathscr{H}$ is a Hilbert space and $\mathfrak{A}$ is a $C^*$-subalgebra of $\mathscr{B}(\mathscr{H})$, the preceding three conditions are equivalent to*

(iv) $\langle Ax, x\rangle \geq 0$, *for each $x$ in $\mathscr{H}$.*

*Proof.* If $A \in \mathfrak{A}^+$, the equation $f(t) = t^{1/2}$ defines a non-negative real-valued continuous function $f$ on $\mathrm{sp}(A)$ ($\subseteq \mathbb{R}^+$). With $H$ defined as $f(A)$, $H \in \mathfrak{A}^+$ and $H^2 = A$. This shows that (i) implies (ii), and it is apparent that (ii) implies (iii).

Suppose next that $A = B^*B$, for some $B$ in $\mathfrak{A}$. Since $A$ is self-adjoint, it has the decomposition $A^+ - A^-$ described in Proposition 4.2.3(iii). With $C$ defined as $BA^-$,

$$C^*C = A^-B^*BA^- = A^-(A^+ - A^-)A^- = -(A^-)^3.$$

Since $A^- \in \mathfrak{A}^+$ and $(A^-)^3$ has spectrum $\{t^3 : t \in \mathrm{sp}(A^-)\}$, it follows that $-C^*C = (A^-)^3 \in \mathfrak{A}^+$. From Lemma 4.2.5, $C = 0$; so $(A^-)^3 = 0$ and, since $A^-$ is self-adjoint, $A^- = 0$ by Corollary 4.1.2. Thus $A = A^+ \in \mathfrak{A}^+$, and (iii) implies (i).

Having proved the equivalence of (i)–(iii), we show next that, when $A \in \mathfrak{A}^+$, the element $H$ in (ii) is unique. For this, suppose that $K$ is any element of $\mathfrak{A}^+$ satisfying $K^2 = A$, while (as above) $H = f(A)$, where $f(t) = t^{1/2}$ ($t \in \mathrm{sp}(A)$). Let $\{p_n\}$ be a sequence of polynomials converging to $f$ uniformly on $\mathrm{sp}(A)$, and let $q_n(t) = p_n(t^2)$. Since $\mathrm{sp}(K) \subseteq \mathbb{R}^+$ and

$$\mathrm{sp}(A) = \mathrm{sp}(K^2) = \{t^2 : t \in \mathrm{sp}(K)\},$$

it follows that

$$\lim_{n\to\infty} q_n(t) = \lim_{n\to\infty} p_n(t^2) = f(t^2) = t,$$

uniformly for $t$ in $\mathrm{sp}(K)$. Hence

$$K = \lim_{n\to\infty} q_n(K) = \lim_{n\to\infty} p_n(K^2) = \lim_{n\to\infty} p_n(A) = f(A) = H,$$

and the uniqueness assertion is proved.

With $\mathfrak{A}$ a $C^*$-subalgebra of $\mathscr{B}(\mathscr{H})$, $\mathfrak{A}^+ = \mathfrak{A} \cap \mathscr{B}(\mathscr{H})^+$. Accordingly, in proving the equivalence of (i) and (iv) in this situation, it suffices to consider the case in which $\mathfrak{A} = \mathscr{B}(\mathscr{H})$. The required result then amounts to the following assertion, already proved in the introductory discussion of the present section: if $A \in \mathscr{B}(\mathscr{H})$, then $\langle Ax, x\rangle \geq 0$ for each $x$ in $\mathscr{H}$ if and only if $A = A^*$ and $\mathrm{sp}(A) \subseteq \mathbb{R}^+$. ■

When $A \in \mathfrak{A}^+$, the element $H$ occurring in condition (ii) of Theorem 4.2.6 is called the *positive square root* of $A$, and is denoted by $A^{1/2}$. A similar procedure can be used to introduce an element $A^\alpha$ of $\mathfrak{A}^+$, for other real values of $\alpha$. With $f_\alpha$ defined by $f_\alpha(t) = t^\alpha$, $f_\alpha$ is a continuous non-negative real-valued function on $\mathrm{sp}(A)$ when $\alpha > 0$ (for all real $\alpha$, if $A$ is invertible). Note that $f_\alpha(t)f_\beta(t) = f_{\alpha+\beta}(t)$, $f_1(t) = t$, and $f_0(t) = 1$ when $A$ is invertible, for all $t$ in $\mathrm{sp}(A)$. With $A^\alpha$ defined as $f_\alpha(A)$, we have $A^\alpha \in \mathfrak{A}^+$, $A^\alpha A^\beta = A^{\alpha+\beta}$, $A^1 = A$, and $A^0 = I$ if $A$ is invertible. It follows easily that this definition of $A^\alpha$ agrees with the elementary one when $\alpha$ is an integer; in particular, if $A$ is invertible, its inverse is the positive element $A^{-1}$ $(= f_{-1}(A))$ of $\mathfrak{A}$.

4.2.7. COROLLARY. *If $\mathfrak{A}$ is a $C^*$-algebra, $A \in \mathfrak{A}^+$, and $B \in \mathfrak{A}$, then $B^*AB \in \mathfrak{A}^+$.*

*Proof.* This follows from Theorem 4.2.6, since $B^*AB = (A^{1/2}B)^*A^{1/2}B$. ■

Suppose that $\mathfrak{A}_h$ is the real linear space consisting of all self-adjoint elements of a $C^*$-algebra $\mathfrak{A}$. Since the adjoint operation is norm continuous, $\mathfrak{A}_h$ is closed in $\mathfrak{A}$, and is therefore a real Banach space. From Theorem 4.2.2, it is a partially ordered vector space with a closed positive cone $\mathfrak{A}^+$. In the partial ordering on $\mathfrak{A}_h$, $A \leqslant B$ if and only if $B - A \in \mathfrak{A}^+$; and, of course,

$$\mathfrak{A}^+ = \{A \in \mathfrak{A}_h : A \geqslant 0\}.$$

From Proposition 4.2.3(ii), $-\|A\|I \leqslant A \leqslant \|A\|I$ for each $A$ in $\mathfrak{A}_h$; in particular, therefore, $I$ is an order unit for $\mathfrak{A}_h$. Moreover

$$\|A\| = \inf\{a : a \geqslant 0, -aI \leqslant A \leqslant aI\},$$

since in its function calculus $A$ corresponds to the identity mapping $\iota$ on $\mathrm{sp}(A)$ $(\subseteq \mathbb{R})$, while $I$ corresponds to the constant function 1, whence

$$\begin{aligned}\|A\| = \|\iota\| &= \sup\{|\iota(t)| : t \in \mathrm{sp}(A)\}\\ &= \inf\{a : -a \cdot 1 \leqslant \iota \leqslant a \cdot 1\}.\end{aligned}$$

Just as in the case of real-number inequalities, one can add inequalities between self-adjoint elements of $\mathfrak{A}$ (because sums of positive elements are positive), multiply through by positive scalars (because positive multiples of positive elements are positive), and take limits (because $\mathfrak{A}^+$ is closed in $\mathfrak{A}$), while multiplication by a negative scalar reverses inequalities. Since a product of commuting positive elements is positive, it follows that $AC \leqslant BC$ whenever $A \leqslant B$, $C \in \mathfrak{A}^+$, and *$C$ commutes with both $A$ and $B$.* This last condition is essential, since without it, $AC$ and $BC$ are not self-adjoint. The corresponding non-commutative result, that $C^*AC \leqslant C^*BC$ whenever $A \leqslant B$ and $C \in \mathfrak{A}$, is a consequence of Corollary 4.2.7.

An element $A$ of $\mathfrak{A}^+$ is invertible if and only if $A \geqslant aI$ for some positive real number $a$. Indeed, $A \geqslant aI$ if and only if $A - aI \in \mathfrak{A}^+$, and since $\mathrm{sp}(A - aI) = \{t - a : t \in \mathrm{sp}(A)\}$, this occurs if and only if $\mathrm{sp}(A) \subseteq [a, \infty)$. Since $\mathrm{sp}(A)$ is a compact subset of $\mathbb{R}^+$, $\mathrm{sp}(A) \subseteq [a, \infty)$ for some positive $a$ if and only if $0 \notin \mathrm{sp}(A)$ (equivalently $A$ is invertible).

4.2.8. PROPOSITION. *Suppose that $A$ and $B$ are self-adjoint elements of a $C^*$-algebra $\mathfrak{A}$.*

(i) *If $-B \leqslant A \leqslant B$, then $\|A\| \leqslant \|B\|$.*
(ii) *If $0 \leqslant A \leqslant B$, then $A^{1/2} \leqslant B^{1/2}$.*
(iii) *If $0 \leqslant A \leqslant B$ and $A$ is invertible, then $B$ is invertible and $B^{-1} \leqslant A^{-1}$.*

*Proof.* (i) Since

$$-\|B\|I \leqslant -B \leqslant A \leqslant B \leqslant \|B\|I,$$

(i) follows.

(ii) and (iii) Suppose that $0 \leqslant A \leqslant B$ and $A$ is invertible. Then $A \geqslant aI$, for some positive real number $a$; hence $B \geqslant aI$, and so $B$ is invertible. Moreover

$$0 \leqslant B^{-1/2}AB^{-1/2} \leqslant B^{-1/2}BB^{-1/2} = I,$$

and $\|B^{-1/2}AB^{-1/2}\| \leqslant 1$ by (i). Thus

$$(1) \quad \|A^{1/2}B^{-1/2}\| = \|(A^{1/2}B^{-1/2})^*A^{1/2}B^{-1/2}\|^{1/2} = \|B^{-1/2}AB^{-1/2}\|^{1/2} \leqslant 1.$$

From this,

$$\|A^{1/2}B^{-1}A^{1/2}\| = \|A^{1/2}B^{-1/2}(A^{1/2}B^{-1/2})^*\| \leqslant 1,$$

whence $A^{1/2}B^{-1}A^{1/2} \leqslant I$, and therefore $B^{-1} \leqslant A^{-1/2}IA^{-1/2} = A^{-1}$.

Furthermore, from Proposition 3.2.8 and (1),

$$\begin{aligned}\|B^{-1/4}A^{1/2}B^{-1/4}\| &= r(B^{-1/4}A^{1/2}B^{-1/4}) \\ &= r(A^{1/2}B^{-1/4}B^{-1/4}) \leqslant \|A^{1/2}B^{-1/2}\| \leqslant 1.\end{aligned}$$

Thus $B^{-1/4}A^{1/2}B^{-1/4} \leqslant I$, and $A^{1/2} \leqslant B^{1/4}IB^{1/4} = B^{1/2}$.

This proves (iii), and (ii) in the case in which $A$ is invertible. Given only that $A, B \in \mathfrak{A}$ and $0 \leqslant A \leqslant B$, we have $0 \leqslant A + \varepsilon I \leqslant B + \varepsilon I$ and $A + \varepsilon I$ is invertible, for every positive real number $\varepsilon$. From the preceding argument,

$$(2) \qquad (A + \varepsilon I)^{1/2} \leqslant (B + \varepsilon I)^{1/2}.$$

With $T$ in $\mathfrak{A}^+$, and $f_\varepsilon$ in $C(\mathrm{sp}(T))$ defined by $f_\varepsilon(t) = (t + \varepsilon)^{1/2}$, we have $f_\varepsilon(T) \in \mathfrak{A}^+$ and $[f_\varepsilon(T)]^2 = T + \varepsilon I$. Thus $(T + \varepsilon I)^{1/2} = f_\varepsilon(T)$; since $f_\varepsilon(t) \to t^{1/2}$ as $\varepsilon \to 0$, uniformly on $\mathrm{sp}(T)$, it follows that $\|(T + \varepsilon I)^{1/2} - T^{1/2}\| \to 0$. Accordingly, from Theorem 4.2.2(i), when $\varepsilon \to 0$, (2) gives $A^{1/2} \leqslant B^{1/2}$. ∎

Suppose that $f$ is a continuous real-valued function defined on a subset $S$ of the real line. We say that $f$ is *operator-monotonic increasing* on $S$ if $f(A) \leqslant f(B)$ whenever $A$ and $B$ are self-adjoint elements of a $C^*$-algebra, $A \leqslant B$ and $\mathrm{sp}(A) \cup \mathrm{sp}(B) \subseteq S$. With $g(t) = t^{1/2}$ $(t \geqslant 0)$ and $h(t) = -1/t$ $(t > 0)$, $g$ and $h$ are operator-monotonic increasing on their respective domains of definition, by Proposition 4.2.8(ii) and (iii). It is clear that an operator-monotonic-increasing function on $S$ is monotonic increasing (in the elementary sense) on $S$. However, the converse is false; for example, if $f(t) = t^2$, then $f$ is not operator-monotonic increasing on $\mathbb{R}^+$ (consider

$$\begin{bmatrix} 1 & 0 \\ 0 & 0 \end{bmatrix}, \quad \begin{bmatrix} 2 & 1 \\ 1 & 1 \end{bmatrix}$$

as operators on $\mathbb{C}^2$). For further information on this subject, see [6].

We conclude this section with some results concerning ideals in $C^*$-algebras, proving first that closed one-sided ideals are "positively generated" in a strong sense. Since the involution induces a one-to-one correspondence between left and right ideals, it suffices to formulate this result for left ideals only.

4.2.9. PROPOSITION. *If $\mathscr{K}$ is a closed left ideal in a $C^*$-algebra $\mathfrak{A}$, each element $S$ of $\mathscr{K}$ can be expressed in the form $S = AK$, with $A$ in $\mathfrak{A}$ and $K$ in $\mathscr{K} \cap \mathfrak{A}^+$.*

*Proof.* Note first that, if $T \in \mathscr{K} \cap \mathfrak{A}^+$, then $T^{1/2} \in \mathscr{K} \cap \mathfrak{A}^+$. Indeed, $\mathscr{K}$ contains all positive powers of $T$, and so contains $p(T)$ whenever $p$ is a polynomial with zero constant term. There is a sequence $\{p_n\}$ of such polynomials, with $\{p_n(t)\}$ converging to $t^{1/2}$ uniformly for $t$ in sp$(T)$. Indeed, the Weierstrass approximation theorem asserts that $t^{1/2}$ is the uniform limit of a sequence $\{q_n(t)\}$ of polynomials, and it suffices to take $p_n(t) = q_n(t) - q_n(0)$. Since $\mathscr{K}$ is closed and $p_n(T) \in \mathscr{K}$, it follows that

$$T^{1/2} = \lim p_n(T) \in \mathscr{K}.$$

With $S$ in $\mathscr{K}$, let $H = (S^*S)^{1/2}$ and $K = H^{1/2}$. Then $S^*S \in \mathscr{K} \cap \mathfrak{A}^+$, and thus $H, K \in \mathscr{K} \cap \mathfrak{A}^+$. For $n = 1, 2, \ldots$, define

$$A_n = S(n^{-1}I + H)^{-1/2},$$

so that

$$S = A_n(n^{-1}I + H)^{1/2}. \tag{3}$$

Then

$$\begin{aligned} \|A_m - A_n\| &= \|S[(m^{-1}I + H)^{-1/2} - (n^{-1}I + H)^{-1/2}]\| \\ &= \|[(m^{-1}I + H)^{-1/2} - (n^{-1}I + H)^{-1/2}]S^*S[(m^{-1}I + H)^{-1/2} \\ &\quad - (n^{-1}I + H)^{-1/2}]\|^{1/2} \\ &= \|[(m^{-1}I + H)^{-1/2} - (n^{-1}I + H)^{-1/2}]H^2[(m^{-1}I + H)^{-1/2} \\ &\quad - (n^{-1}I + H)^{-1/2}]\|^{1/2} \\ &= \|[f_{mn}(H)]^2\|^{1/2} = \|f_{mn}(H)\|, \end{aligned}$$

where $f_{mn}$ is the continuous function defined on the positive half line by

$$f_{mn}(t) = t[(m^{-1} + t)^{-1/2} - (n^{-1} + t)^{-1/2}].$$

Thus

$$\|A_m - A_n\| = \sup\{|f_{mn}(t)| : t \in \mathrm{sp}(H)\}. \tag{4}$$

Now

$$|f_{mn}(t)| = \frac{t|\sqrt{n^{-1} + t} - \sqrt{m^{-1} + t}|}{\sqrt{(n^{-1} + t)(m^{-1} + t)}}$$
$$\leqslant |\sqrt{n^{-1} + t} - \sqrt{m^{-1} + t}|,$$

and $\sqrt{n^{-1} + t} \to \sqrt{t}$ as $n \to \infty$, uniformly for $t$ in sp($H$). Thus $f_{mn}(t) \to 0$ as $\min(m, n) \to \infty$, uniformly for $t$ in sp($H$); from (4), $\{A_n\}$ is a Cauchy sequence in $\mathfrak{A}$. With $A$ defined as $\lim A_n$, it now follows from (3) that

$$S = AH^{1/2} = AK. \quad \blacksquare$$

4.2.10. COROLLARY. *Each closed two-sided ideal $\mathscr{K}$ in a $C^*$-algebra $\mathfrak{A}$ is self-adjoint. A closed two-sided ideal $\mathscr{T}$ in $\mathscr{K}$ is a two-sided ideal in $\mathfrak{A}$.*

*Proof.* Since $\mathscr{K}$ is a closed left ideal in $\mathfrak{A}$, each $S$ in $\mathscr{K}$ has the form $S = AK$, with $A$ in $\mathfrak{A}$ and $K (= (S^*S)^{1/4})$ in $\mathscr{K} \cap \mathfrak{A}^+$, by Proposition 4.2.9 and its proof. Since $\mathscr{K}$ is a right ideal in $\mathfrak{A}$, $S^* = KA^* \in \mathscr{K}$, so $\mathscr{K}$ is self-adjoint.

Suppose next that $S \in \mathscr{T}$. Then $S \in \mathscr{K}$, the preceding paragraph shows that $S^* \in \mathscr{K}$; since $\mathscr{T}$ is a left ideal in $\mathscr{K}$, it now follows that $S^*S \in \mathscr{T} \cap \mathfrak{A}^+$. Upon approximating the square-root function by polynomials with zero constant term, as in the proof of Proposition 4.2.9, we deduce that $\mathscr{T} \cap \mathfrak{A}^+$ contains the positive square root of each of its members. Successive application of this shows that $K^{1/2}$ $(= (S^*S)^{1/8}) \in \mathscr{T}$.

With $B$ in $\mathfrak{A}$, $BAK^{1/2}$ and $K^{1/2}B$ are in $\mathscr{K}$, since $K^{1/2} \in \mathscr{K}$ and $\mathscr{K}$ is a two-sided ideal in $\mathfrak{A}$. Since $\mathscr{T}$ is a left ideal in $\mathscr{K}$, and $K^{1/2} \in \mathscr{T}$,

$$BS = BAK = BAK^{1/2}K^{1/2} \in \mathscr{T};$$

so $\mathscr{T}$ is a left ideal in $\mathfrak{A}$. From this, $AK^{1/2} \in \mathscr{T}$; and since $\mathscr{T}$ is a right ideal in $\mathscr{K}$,

$$SB = AKB = AK^{1/2}K^{1/2}B \in \mathscr{T}.$$

This proves that $\mathscr{T}$ is a right (and hence two-sided) ideal in $\mathfrak{A}$. $\blacksquare$

4.2.11. LEMMA. *Suppose that $\mathscr{B}$ is a closed * subalgebra (not necessarily containing the unit element) of a $C^*$-algebra $\mathfrak{A}$, and let*

$$\Lambda = \{B \in \mathscr{B} \cap \mathfrak{A}^+ : \|B\| < 1\}.$$

(i) *If $B_1, B_2 \in \Lambda$, there is an element $B$ of $\Lambda$ such that $B_1 \leqslant B$ and $B_2 \leqslant B$.*

(ii) *If $S \in \mathscr{B}$ and $\varepsilon > 0$, there is an element $B_0$ of $\Lambda$ with the following property: if $B \in \Lambda$ and $B \geqslant B_0$, then $\|S - SB\| < \varepsilon$.*

*Proof.* (i) Given $B_1$ and $B_2$ in $\Lambda$, we can choose $\delta$ ($> 0$) so that $\|(1 + \delta)B_j\| \leqslant 1$ ($j = 1, 2$). For each positive integer $n$, $t \leqslant t^{1/n}$ when $0 \leqslant t \leqslant 1$, and thus $(1 + \delta)B_j \leqslant [(1 + \delta)B_j]^{1/n}$. When $n$ has the form $2^q$ for some positive integer $q$, repeated application of Proposition 4.2.8(ii) shows that

$$[(1 + \delta)(B_1 + B_2)]^{1/n} \geqslant [(1 + \delta)B_j]^{1/n} \geqslant (1 + \delta)B_j.$$

From the foregoing argument, $B_1 \leqslant B$ and $B_2 \leqslant B$, where

$$B = (1 + \delta)^{-1}[(1 + \delta)(B_1 + B_2)]^{1/n}$$

for a positive integer $n$ of the form $2^q$. Now

$$(1 + \delta)(B_1 + B_2) \in \mathscr{B}, \qquad 0 \leqslant (1 + \delta)(B_1 + B_2) \leqslant 2I;$$

the function $t^{1/n}$ is the uniform limit, on the interval $[0, 2]$, of polynomials without constant term. Since $\mathscr{B}$ is closed, it now follows that $B \in \mathscr{B}$; moreover,

$$0 \leqslant B \leqslant (1 + \delta)^{-1} 2^{1/n} I.$$

When $n$ is sufficiently large, $(1 + \delta)^{-1} 2^{1/n} < 1$; and then, $B \in \Lambda$.

(ii) Given $S$ in $\mathscr{B}$ and $\varepsilon$ ($> 0$), let $S = aT$ where $a > \|S\|$ (so that $T \in \mathscr{B}$ and $\|T\| < 1$). Let $B_0$ be the element $(T^*T)^{1/n}$ of $\Lambda$, where $n$ is a positive integer sufficiently large to ensure that

$$t(1 - t^{1/n}) < \varepsilon^2 a^{-2} \qquad (0 \leqslant t \leqslant 1).$$

If $B \in \Lambda$ and $B \geqslant B_0$, we have

$$\|S - SB\| = a\|T(I - B)\| = a\|T(I - B)^2 T^*\|^{1/2}.$$

Now

$$0 \leqslant I - B \leqslant I - B_0 \leqslant I,$$

and thus

$$0 \leqslant (I - B)^2 \leqslant I - B \leqslant I - B_0 = I - (T^*T)^{1/n};$$

so

$$0 \leqslant T(I - B)^2 T^* \leqslant T[I - (T^*T)^{1/n}]T^*.$$

Hence

$$\begin{aligned} \|T(I - B)^2 T^*\| &\leqslant \|T[I - (T^*T)^{1/n}]T^*\| \\ &= r(T[I - (T^*T)^{1/n}]T^*) \\ &= r(T^*T[I - (T^*T)^{1/n}]) \\ &\leqslant \sup\{t(1 - t^{1/n}) : 0 \leqslant t \leqslant 1\} \\ &< \varepsilon^2 a^{-2}, \end{aligned}$$

from Propositions 4.2.8(i), 4.1.1(i), and 3.2.8, together with Theorem 4.1.6 (noting that $\operatorname{sp}(T^*T) \subseteq [0, 1]$). Hence

$$\|S - SB\| = a\|T(I - B)^2 T^*\|^{1/2} < \varepsilon. \quad \blacksquare$$

If $\mathscr{B}$ contains the identity $I$ of $\mathfrak{A}$, both $B$ in (i) and $B_0$ in (ii) of the preceding lemma can be taken to be appropriate positive multiples of $I$. Of course, the interest of the lemma is precisely in the case where $I \notin \mathscr{B}$.

In Lemma 3.2.24, it was found helpful to introduce an "approximate identity" in the Banach algebra $L_1(\mathbb{R})$. Similar devices are useful in studying ideals and * subalgebras (not containing the unit element) of $C^*$-algebras. Let $\mathscr{S}$ be a subset of a $C^*$-algebra $\mathfrak{A}$. We say that a net $(V_\lambda, \lambda \in \Lambda, \geqslant)$ $(= \{V_\lambda\})$ of self-adjoint elements of $\mathscr{S}$ is an *increasing right approximate identity* for $\mathscr{S}$ if $\|S - SV_\lambda\| \to_\lambda 0$, for each $S$ in $\mathscr{S}$, while $0 \leqslant V_\lambda \leqslant V_u \leqslant I$ whenever $\lambda, \mu \in \Lambda$ and $\lambda \leqslant \mu$. With obvious modifications, we formulate similar definitions for increasing, left, or two-sided approximate identities.

Suppose that $\mathscr{B}$ is a closed * subalgebra of a $C^*$-algebra $\mathfrak{A}$. The set $\Lambda$ occurring in Lemma 4.2.11 is directed by the usual partial order relation on self-adjoint elements of $\mathfrak{A}$, by part (i) of that lemma. By defining $V_B$ to be $B$, when $B \in \Lambda$, we obtain an increasing net $\{V_B\}$ of positive elements in the unit ball of $\mathscr{B}$. From Lemma 4.2.11(ii), $\|S - SV_B\| \to_B 0$, for each $S$ in $\mathscr{B}$; and, since $\mathscr{B}$ is self-adjoint, we also have

$$\|S - V_B S\| = \|S^* - S^* V_B\| \underset{B}{\to} 0.$$

Thus $\{V_B\}$ is an increasing two-sided approximate identity for $\mathscr{B}$.

Suppose next that $\mathscr{K}$ is a closed left ideal in $\mathfrak{A}$. With $\mathscr{K}^*$ the closed right ideal $\{S^* : S \in \mathscr{K}\}$, $\mathscr{K} \cap \mathscr{K}^*$ is a closed * subalgebra of $\mathfrak{A}$, and contains the self-adjoint elements of $\mathscr{K}$. By the preceding paragraph, $\mathscr{K} \cap \mathscr{K}^*$ has an increasing two-sided approximate identity $\{V_\lambda\}$. From Proposition 4.2.9, each $S$ in $\mathscr{K}$ has the form $AK$, with $A$ in $\mathfrak{A}$ and $K$ in $\mathscr{K} \cap \mathfrak{A}^+$ $(\subseteq \mathscr{K} \cap \mathscr{K}^*)$. Since

$$\|S - SV_\lambda\| = \|A(K - KV_\lambda)\| \leqslant \|A\| \, \|K - KV_\lambda\| \underset{\lambda}{\to} 0,$$

it follows that $\{V_\lambda\}$ is an increasing right approximate identity for $\mathscr{K}$.

We have now proved the following result.

4.2.12. PROPOSITION. *A closed left ideal in a $C^*$-algebra has an increasing right approximate identity, and a closed * subalgebra has an increasing two-sided approximate identity.*

Of course, the roles of "left" and "right" can be reversed in Proposition 4.2.12. Note, however, that a closed left ideal in a $C^*$-algebra does not always have a left approximate identity (see Exercise 4.6.37).

## 4.3. Positive linear functionals

In this section, we study linear functionals on a subspace of a $C^*$-algebra. In most applications of the results described below, it suffices to consider the case in which the subspace is the whole $C^*$-algebra; but occasionally, the more general setting is required.

We assume throughout that $\mathcal{M}$ is a self-adjoint subspace of a $C^*$-algebra $\mathfrak{A}$, and contains the unit $I$ of $\mathfrak{A}$. The set $\mathcal{M} \cap \mathfrak{A}^+$ of all positive elements of $\mathcal{M}$ is denoted by $\mathcal{M}^+$. If $\mathcal{M} \subseteq \mathcal{B} \subseteq \mathfrak{A}$, where $\mathcal{B}$ is a $C^*$-subalgebra of $\mathfrak{A}$, then $\mathcal{M}^+ = \mathcal{M} \cap \mathfrak{A}^+ = \mathcal{M} \cap \mathcal{B} \cap \mathfrak{A}^+ = \mathcal{M} \cap \mathcal{B}^+$; so $\mathcal{M}^+$ is unchanged if $\mathcal{M}$ is viewed as a subspace of $\mathcal{B}$ instead of $\mathfrak{A}$. Since $\mathcal{M}$ contains the real and imaginary parts of each of its members, while each self-adjoint $A$ in $\mathcal{M}$ is the difference of two elements, $\frac{1}{2}(\|A\|I \pm A)$, of $\mathcal{M}^+$, it follows that $\mathcal{M}$ is the linear span of $\mathcal{M}^+$.

With $\rho$ a linear functional on $\mathcal{M}$, the equation $\rho^*(A) = \overline{\rho(A^*)}$ $(A \in \mathcal{M})$ defines another such functional $\rho^*$. We describe $\rho$ as *hermitian* if $\rho = \rho^*$; that is, if $\rho(A^*) = \overline{\rho(A)}$ for each $A$ in $\mathcal{M}$. (Compare the discussion of Remark 3.4.9, where this construction is applied to the $C^*$-algebra $C(X)$.) By expressing elements of $\mathcal{M}$ in terms of their real and imaginary parts, it follows that $\rho$ is hermitian if and only if $\rho(H) = \rho^*(H)\,(= \overline{\rho(H)})$ whenever $H = H^* \in \mathcal{M}$; so $\rho$ is hermitian if and only if $\rho(H)$ is real, for each self-adjoint $H$ in $\mathcal{M}$. Each linear functional $\rho$ on $\mathcal{M}$ can be expressed, uniquely, in the form $\rho_1 + i\rho_2$, where $\rho_1$ $(= \frac{1}{2}(\rho + \rho^*))$ and $\rho_2$ $(= \frac{1}{2}i(\rho^* - \rho))$ are hermitian.

If $\rho$ is a bounded hermitian functional on $\mathcal{M}$,

$$\|\rho\| = \sup\{\rho(H): H = H^* \in \mathcal{M}, \|H\| \leqslant 1\};$$

that is, the norm of $\rho$ is the same as the norm of its restriction to the (real linear space of) self-adjoint elements of $\mathcal{M}$. Indeed, if $\varepsilon > 0$, we can choose $A$ in the unit ball of $\mathcal{M}$ so that $|\rho(A)| > \|\rho\| - \varepsilon$. For a suitable scalar $a$ with $|a| = 1$,

$$\|\rho\| - \varepsilon < |\rho(A)| = \rho(aA) = \overline{\rho(aA)} = \rho((aA)^*).$$

With $H_0$ the real part of $aA$, $\|H_0\| \leqslant 1$ and $\rho(H_0) > \|\rho\| - \varepsilon$. Thus $\|\rho\| \leqslant \sup\{\rho(H): H = H^* \in \mathcal{M}, \|H\| \leqslant 1\}$, and the reverse inequality is evident.

A linear functional $\rho$ on $\mathcal{M}$ is said to be *positive* if $\rho(A) \geqslant 0$, for each $A$ in $\mathcal{M}^+$; if, further, $\rho(I) = 1$, $\rho$ is described as a *state* of $\mathcal{M}$. A positive linear functional $\rho$ is hermitian; for if $A = A^* \in \mathcal{M}$, then $\rho(\|A\|I \pm A) \geqslant 0$ since $\|A\|I \pm A \in \mathcal{M}^+$, and $\rho(A)$ is real because

$$\rho(A) = \tfrac{1}{2}[\rho(\|A\|I + A) - \rho(\|A\|I - A)].$$

The real vector space $\mathcal{M}_h$, consisting of all self-adjoint elements of $\mathcal{M}$, is a partially ordered vector space, with positive cone $\mathcal{M}^+$ and order unit $I$. A linear functional $\rho$ on $\mathcal{M}$ is hermitian if and only if its restriction $\rho|\mathcal{M}_h$ is a linear functional (of course, real-valued) on $\mathcal{M}_h$; and each linear functional on $\mathcal{M}_h$ extends, uniquely, to a hermitian linear functional on $\mathcal{M}$. Moreover, $\rho$ is positive (or a state of $\mathcal{M}$) if and only if $\rho|\mathcal{M}_h$ is positive (or a state of $\mathcal{M}_h$) in the sense of Definition 3.4.5. The positive linear functionals on $\mathcal{M}$ form a cone $\mathcal{P}$, in the real vector space consisting of all hermitian linear functionals on $\mathcal{M}$ ($\mathcal{P} \cap -\mathcal{P} = \{0\}$, because $\mathcal{M}$ is the linear span of $\mathcal{M}^+$). Hence there is a partial order relation on the hermitian linear functionals; $\rho_1 \leqslant \rho_2$ if and only if $\rho_2 - \rho_1$ is positive.

With $\mathcal{H}$ a Hilbert space and $x$ in $\mathcal{H}$, the equation

$$\omega_x(A) = \langle Ax, x \rangle \qquad (A \in \mathcal{B}(\mathcal{H}))$$

defines a linear functional $\omega_x$ on $\mathcal{B}(\mathcal{H})$. In view of the equivalence of two concepts of positivity for Hilbert space operators (conditions (i) and (iv) in Theorem 4.2.6, with $\mathfrak{A} = \mathcal{B}(\mathcal{H})$), $\omega_x(A) \geqslant 0$ whenever $A \in \mathcal{B}(\mathcal{H})^+$. Since, also, $\omega_x(I) = \|x\|^2$, it follows that $\omega_x$ is a positive linear functional on $\mathcal{B}(\mathcal{H})$, and is a state if $\|x\| = 1$. If $\mathfrak{A}$ is a $C^*$-subalgebra of $\mathcal{B}(\mathcal{H})$, and (as usual) $\mathcal{M}$ is a self-adjoint subspace of $\mathfrak{A}$ that contains $I$, the restriction $\omega_x|\mathcal{M}$ is a positive linear functional on $\mathcal{M}$. The states of $\mathcal{M}$ that arise in this way, from unit vectors in $\mathcal{H}$, are termed *vector states* of $\mathcal{M}$.

4.3.1. PROPOSITION. *If $\rho$ is a positive linear functional on a $C^*$-algebra $\mathfrak{A}$, then*

$$|\rho(B^*A)|^2 \leqslant \rho(A^*A)\rho(B^*B) \qquad (A, B \in \mathfrak{A}).$$

*Proof.* With $A$ in $\mathfrak{A}$, we have $A^*A \in \mathfrak{A}^+$, and therefore $\rho(A^*A) \geqslant 0$. From this, and since $\rho$ is hermitian, the equation

$$\langle A, B \rangle = \rho(B^*A) \qquad (A, B \in \mathfrak{A})$$

defines an inner product $\langle\ ,\ \rangle$ on $\mathfrak{A}$, and we have the Cauchy–Schwarz inequality

$$|\langle A, B \rangle|^2 \leqslant \langle A, A \rangle \langle B, B \rangle;$$

that is, $|\rho(B^*A)|^2 \leqslant \rho(A^*A)\rho(B^*B)$. ∎

We refer to the inequality occurring in Proposition 4.3.1 as the Cauchy–Schwarz inequality for $\rho$. This inequality appeared, in the case of the $C^*$-algebra $C(X)$, at the end of Remark 3.4.9 and, again, in Remark 3.4.12.

4.3.2. THEOREM. *If $\mathcal{M}$ is a self-adjoint subspace of a $C^*$-algebra $\mathfrak{A}$ and contains the unit $I$ of $\mathfrak{A}$, a linear functional $\rho$ on $\mathcal{M}$ is positive if and only if $\rho$ is bounded and $\|\rho\| = \rho(I)$.*

*Proof.* Suppose first that $\rho$ is positive (and therefore hermitian). With $A$ in $\mathcal{M}$, let $a$ be a scalar of modulus 1 such that $a\rho(A) \geqslant 0$, and let $H$ be the real part of $aA$. Then $\|H\| \leqslant \|A\|$,

$$H \leqslant \|H\|I \leqslant \|A\|I, \qquad \|A\|\rho(I) - \rho(H) = \rho(\|A\|I - H) \geqslant 0,$$

and therefore

$$|\rho(A)| = \rho(aA) = \overline{\rho(aA)} = \rho(\bar{a}A^*)$$
$$= \rho(\tfrac{1}{2}(aA + \bar{a}A^*)) = \rho(H) \leqslant \rho(I)\|A\|.$$

This shows that $\rho$ is bounded, with $\|\rho\| \leqslant \rho(I)$; and the reverse inequality is evident.

Conversely, suppose that $\rho$ is bounded and $\|\rho\| = \rho(I)$; it suffices to consider the case in which $\|\rho\| = \rho(I) = 1$. With $A$ in $\mathscr{M}^+$, let $\rho(A) = a + ib$, where $a$ and $b$ are real. In order to prove that $\rho$ is positive, we have to show that $a \geqslant 0$ and $b = 0$. For small positive $s$,

$$\operatorname{sp}(I - sA) = \{1 - st : t \in \operatorname{sp}(A)\} \subseteq [0, 1],$$

since $\operatorname{sp}(A) \subseteq \mathbb{R}^+$; so $\|I - sA\| = r(I - sA) \leqslant 1$. Hence

$$1 - sa \leqslant |1 - s(a + ib)| = |\rho(I - sA)| \leqslant 1,$$

and therefore $a \geqslant 0$. With $B_n$ in $\mathscr{M}$ defined as $A - aI + inbI$, for each positive integer $n$,

$$\|B_n\|^2 = \|B_n^* B_n\| = \|(A - aI)^2 + n^2b^2I\| \leqslant \|A - aI\|^2 + n^2b^2.$$

Hence

$$(n^2 + 2n + 1)b^2 = |\rho(B_n)|^2 \leqslant \|A - aI\|^2 + n^2b^2 \qquad (n = 1, 2, \ldots),$$

and thus $b = 0$. ■

From Theorem 4.3.2, each state $\rho$ of $\mathscr{M}$ is a bounded linear functional on $\mathscr{M}$, with $\|\rho\| = 1$. Accordingly, the set $\mathscr{S}(\mathscr{M})$ of all states of $\mathscr{M}$ is contained in the surface of the unit ball in the Banach dual space $\mathscr{M}^\sharp$. It is convex and weak* closed, since

$$\mathscr{S}(\mathscr{M}) = \{\rho \in \mathscr{M}^\sharp : \rho(I) = 1,\ \rho(A) \geqslant 0\ (A \in \mathscr{M}^+)\},$$

and is therefore weak* compact, by Corollary 1.6.6. It follows that $\mathscr{S}(\mathscr{M})$, with the weak* topology, is a compact Hausdorff space, the *state space* of $\mathscr{M}$.

4.3.3. Proposition. *If $\mathfrak{A}$ is a $C^*$-algebra with unit $I$, $\mathscr{M}$ is a self-adjoint subspace of $\mathfrak{A}$ containing $I$, $A \in \mathscr{M}$, and $a \in \operatorname{sp}(A)$, then there is a state $\rho$ of $\mathscr{M}$ such that $\rho(A) = a$.*

*Proof.* For all complex numbers $b$ and $c$, $ab + c \in \operatorname{sp}(bA + cI)$, and therefore $|ab + c| \leqslant \|bA + cI\|$. Accordingly, the equation $\rho_0(bA + cI) = ab + c$ defines (unambiguously) a linear functional $\rho_0$ on the subspace

$\{bA + cI: b, c \in \mathbb{C}\}$ of $\mathscr{M}$, and $\rho_0(A) = a$, $\rho_0(I) = 1$, $\|\rho_0\| = 1$. By the Hahn–Banach theorem, $\rho_0$ extends to a bounded linear functional $\rho$ on $\mathscr{M}$, with $\|\rho\| = 1$ $(= \rho(I))$. From Theorem 4.3.2, $\rho$ is positive (and is therefore a state); and $\rho(A) = a$. ■

4.3.4. THEOREM. *Suppose that $\mathfrak{A}$ is a $C^*$-algebra with unit $I$, $\mathscr{M}$ is a self-adjoint subspace of $\mathfrak{A}$ containing $I$, and $A \in \mathscr{M}$.*

(i) *If $\rho(A) = 0$, for each state $\rho$ of $\mathscr{M}$, then $A = 0$.*
(ii) *If $\rho(A)$ is real, for each state $\rho$ of $\mathscr{M}$, then $A$ is self-adjoint.*
(iii) *If $\rho(A) \geqslant 0$, for each state $\rho$ of $\mathscr{M}$, then $A \in \mathscr{M}^+$.*
(iv) *If $A$ is normal, there is a state $\rho$ of $\mathscr{M}$ such that $|\rho(A)| = \|A\|$.*

*Proof.* (i) Suppose first that $A$ is self-adjoint and $\rho(A) = 0$ for each state $\rho$ of $\mathscr{M}$. From Proposition 4.3.3, $\mathrm{sp}(A) = \{0\}$, so $\|A\| = r(A) = 0$, $A = 0$.

Next, let $A = H + iK$, with $H$ and $K$ self-adjoint in $\mathscr{M}$. If $\rho(A) = 0$, for each state $\rho$ of $\mathscr{M}$, then $\rho(H) = \rho(K) = 0$, since $\rho(A) = \rho(H) + i\rho(K)$ and $\rho(H)$ and $\rho(K)$ are real. From the preceding paragraph, $H = K = 0$, whence $A = 0$.

(ii) If $\rho(A)$ is real, for each state $\rho$ of $\mathscr{M}$, then

$$\rho(A - A^*) = \rho(A) - \overline{\rho(A)} = 0,$$

and $A - A^* = 0$ by (i).

(iii) If $\rho(A) \geqslant 0$, for each state $\rho$ of $\mathscr{M}$, then $A$ is self-adjoint by (ii), $\mathrm{sp}(A) \subseteq \mathbb{R}^+$ by Proposition 4.3.3, and so $A \in \mathscr{M}^+$.

(iv) If $A$ is normal, $r(A) = \|A\|$, so $\mathrm{sp}(A)$ contains a scalar $a$ such that $|a| = \|A\|$. By Proposition 4.3.3, $a = \rho(A)$, for some state $\rho$ of $\mathscr{M}$, and then $|\rho(A)| = \|A\|$. ■

Our next objective is to prove a non-commutative analogue of the Hahn–Jordan decomposition for linear functionals on $C(X)$ (Proposition 3.4.11 and Remark 3.4.12). We show in Theorem 4.3.6 that every hermitian functional on $\mathscr{M}$ can be expressed (in a unique optimal manner, when $\mathscr{M}$ is the whole of $\mathfrak{A}$) as a difference of positive linear functionals. For this purpose, we require the following lemma, which will be needed again later when we characterize those subsets of the state space $\mathscr{S}(\mathscr{M})$ that retain some of the properties of $\mathscr{S}(\mathscr{M})$ set out in Theorem 4.3.4.

With $\mathscr{K}$ a subset of the Banach dual space $\mathscr{M}^\sharp$, we write $\overline{\mathrm{co}}(\mathscr{K})$ for the weak* closed convex hull of $\mathscr{K}$.

4.3.5. LEMMA. *Suppose that $\mathfrak{A}$ is a $C^*$-algebra with unit $I$, $\mathscr{M}$ is a self-adjoint subspace of $\mathfrak{A}$ containing $I$, and $\mathscr{S}_0$ is a set of states of $\mathscr{M}$. If*

$$\|H\| = \sup\{|\rho(H)| : \rho \in \mathscr{S}_0\},$$

*for each self-adjoint $H$ in $\mathscr{M}$, then $\overline{\mathrm{co}}(\mathscr{S}_0 \cup -\mathscr{S}_0)$ is the set of all hermitian functionals in the unit ball of $\mathscr{M}^\sharp$.*

*Proof.* The set of all hermitian functionals in the unit ball $(\mathcal{M}^{\#})_1$ is convex and weak* closed, and contains $\mathcal{S}_0 \cup -\mathcal{S}_0$; so it contains $\overline{\text{co}}(\mathcal{S}_0 \cup -\mathcal{S}_0)$. We have to show that the two sets coincide. Suppose the contrary, and let $\rho_0$ be a hermitian functional on $\mathcal{M}$, such that $\|\rho_0\| \leqslant 1$, $\rho_0 \notin \overline{\text{co}}(\mathcal{S}_0 \cup -\mathcal{S}_0)$. By the Hahn–Banach theorem (Corollary 1.2.12), and since the weak* continuous linear functionals on $\mathcal{M}^{\#}$ arise from elements of $\mathcal{M}$ (Proposition 1.3.5), there is an $A$ in $\mathcal{M}$, and a real number $a$, such that

$$\text{Re } \rho_0(A) > a, \qquad \text{Re } \rho(A) \leqslant a \qquad (\rho \in \overline{\text{co}}(\mathcal{S}_0 \cup -\mathcal{S}_0)).$$

With $H$ the real part of $A$,

$$\rho(H) = \tfrac{1}{2}[\rho(A) + \rho(A^*)] = \text{Re } \rho(A),$$

for every hermitian functional $\rho$ on $\mathcal{M}$; so

$$\rho_0(H) > a, \qquad \rho(H) \leqslant a \qquad (\rho \in \overline{\text{co}}(\mathcal{S}_0 \cup -\mathcal{S}_0)).$$

Thus $|\rho(H)| \leqslant a$ $(\rho \in \mathcal{S}_0)$, and

$$a < \rho_0(H) \leqslant \|H\| = \sup\{|\rho(H)| : \rho \in \mathcal{S}_0\} \leqslant a,$$

a contradiction. ∎

4.3.6. Theorem. *If $\mathfrak{A}$ is a $C^*$-algebra with unit $I$ and $\mathcal{M}$ is a self-adjoint subspace of $\mathfrak{A}$ containing $I$, each bounded hermitian functional $\rho$ on $\mathcal{M}$ can be expressed in the form $\rho^+ - \rho^-$, where $\rho^+$ and $\rho^-$ are positive linear functionals on $\mathcal{M}$ and $\|\rho\| = \|\rho^+\| + \|\rho^-\|$. If $\mathcal{M}$ is the whole of $\mathfrak{A}$, these conditions determine $\rho^+$ and $\rho^-$ uniquely.*

*Proof.* We may assume that $\|\rho\| = 1$. With $\mathcal{S}$ the state space of $\mathcal{M}$,

$$\|A\| = \sup\{|\tau(A)| : \tau \in \mathcal{S}\}$$

for each self-adjoint $A$ in $\mathcal{M}$, from Theorem 4.3.4(iv) and since $\|\tau\| = 1$ when $\tau \in \mathcal{S}$. By Lemma 4.3.5, $\rho \in \overline{\text{co}}(\mathcal{S} \cup -\mathcal{S})$.

A straightforward calculation shows that the subset

$$\{a\sigma - b\tau : \sigma, \tau \in \mathcal{S}, a, b \in \mathbb{R}^+, a + b = 1\}$$

of $\overline{\text{co}}(\mathcal{S} \cup -\mathcal{S})$ is convex. It contains $\mathcal{S} \cup -\mathcal{S}$, and is weak* compact since it is the range of the continuous mapping

$$(\sigma, \tau, a) \to a\sigma - (1 - a)\tau : \quad \mathcal{S} \times \mathcal{S} \times [0, 1] \to \mathcal{M}^{\#}.$$

Accordingly, it is the whole of $\overline{\text{co}}(\mathcal{S} \cup -\mathcal{S})$. From this, and the preceding paragraph, $\rho$ has the form $a\sigma - b\tau$, with $\rho$ and $\tau$ in $\mathcal{S}$, $a$ and $b$ in $\mathbb{R}^+$ and $a + b = 1$. With $\rho^+$ and $\rho^-$ the positive linear functionals $a\sigma$ and $b\tau$, respectively, $\rho = \rho^+ - \rho^-$ and

$$\|\rho^+\| + \|\rho^-\| = a + b = 1 = \|\rho\|.$$

Suppose now that $\mathscr{M} = \mathfrak{A}$. To prove the uniqueness of the decomposition of $\rho$, we assume that $\rho = \mu - \nu = \mu' - \nu'$, where $\mu$, $\mu'$, $\nu$, $\nu'$ are positive linear functionals on $\mathfrak{A}$ and

$$\|\mu\| + \|\nu\| = \|\mu'\| + \|\nu'\| = \|\rho\| = 1.$$

Given $\varepsilon$ ($> 0$), choose a self-adjoint $H$ in the unit ball of $\mathfrak{A}$ for which $\rho(H) > \|\rho\| - \frac{1}{2}\varepsilon^2$, and let $K = \frac{1}{2}(I - H)$. Then $0 \leqslant K \leqslant I$,

$$\mu(I) + \nu(I) = \|\mu\| + \|\nu\| = \|\rho\|$$
$$< \rho(H) + \tfrac{1}{2}\varepsilon^2 = \mu(H) - \nu(H) + \tfrac{1}{2}\varepsilon^2,$$
$$\mu(I - H) + \nu(I + H) < \tfrac{1}{2}\varepsilon^2, \qquad \mu(K) + \nu(I - K) < \tfrac{1}{4}\varepsilon^2.$$

Since $K, I - K \in \mathfrak{A}^+$, while $\mu$ and $\nu$ are positive linear functionals,

$$0 \leqslant \mu(K) < \tfrac{1}{4}\varepsilon^2, \qquad 0 \leqslant \nu(I - K) < \tfrac{1}{4}\varepsilon^2.$$

With $A$ in $\mathfrak{A}$, the Cauchy–Schwarz inequality gives

$$|\mu(KA)|^2 = |\mu(K^{1/2} \cdot K^{1/2}A)|^2 \leqslant \mu(K)\mu(A^*KA) \leqslant \tfrac{1}{4}\varepsilon^2\|A\|^2,$$
$$|\nu((I - K)A)|^2 \leqslant \nu(I - K)\nu(A^*(I - K)A) \leqslant \tfrac{1}{4}\varepsilon^2\|A\|^2.$$

From this, and a similar argument for $\mu'$ and $\nu'$, we have

$$|\mu(KA)| \leqslant \tfrac{1}{2}\varepsilon\|A\|, \qquad |\mu'(KA)| \leqslant \tfrac{1}{2}\varepsilon\|A\|,$$
$$|\nu((I - K)A)| \leqslant \tfrac{1}{2}\varepsilon\|A\|, \qquad |\nu'((I - K)A)| \leqslant \tfrac{1}{2}\varepsilon\|A\|.$$

Since $\mu - \mu' = \nu - \nu'$,

$$\mu(A) - \mu'(A) = \mu(KA) - \mu'(KA) + \nu((I - K)A) - \nu'((I - K)A),$$

and so $|\mu(A) - \mu'(A)| < 2\varepsilon\|A\|$. Since the last inequality has been proved for each positive $\varepsilon$, it follows that $\mu = \mu'$, whence $\nu = \nu'$. ∎

4.3.7. Corollary. *If $\mathfrak{A}$ is a $C^*$-algebra with unit $I$ and $\mathscr{M}$ is a self-adjoint subspace of $\mathfrak{A}$ containing $I$, each bounded linear functional on $\mathscr{M}$ is a linear combination of at most four states of $\mathscr{M}$.*

*Proof.* Each bounded linear functional $\tau$ on $\mathscr{M}$ has the form $\rho + i\sigma$, with $\rho$ and $\sigma$ bounded hermitian functionals. Hence

$$\tau = \rho^+ - \rho^- + i\sigma^+ - i\sigma^-,$$

and each term on the right-hand side is a scalar multiple of a state of $\mathscr{M}$. ∎

We shall later give an alternative proof of the existence of the decomposition $\rho = \rho^+ - \rho^-$, by reduction to the case in which $\mathfrak{A}$ is abelian, and appeal to Proposition 3.4.11 (see Remark 4.3.12). The uniqueness clause of

Theorem 4.3.6 fails, in general, if one deletes the assumption that $\mathcal{M} = \mathfrak{A}$ (see Exercise 4.6.22).

Since the state space $\mathcal{S}(\mathcal{M})$ of $\mathcal{M}$ is convex and weak* compact, it has extreme points; indeed, by the Krein–Milman theorem, $\mathcal{S}(\mathcal{M})$ is the weak* closed convex hull $\overline{\text{co}}(\mathcal{P}(\mathcal{M}))$ of the set $\mathcal{P}(\mathcal{M})$ of its extreme points. Elements of $\mathcal{P}(\mathcal{M})$ are termed *pure states* of $\mathcal{M}$, and the weak* closure $\mathcal{P}(\mathcal{M})^-$ is called the *pure state space* of $\mathcal{M}$. In general, $\mathcal{P}(\mathcal{M})$ is not a closed subset of $\mathcal{M}^\#$, and the pure state space then has elements that are not pure states.

When $X$ is a compact Hausdorff space, the pure states of the $C^*$-algebra $C(X)$ are precisely the non-zero multiplicative linear functionals (Theorem 3.4.7); and the set $\mathcal{P}$ of pure states is therefore weak* compact (Proposition 3.2.20). Thus $\mathcal{P}$ coincides with the pure state space, in this case.

A linear functional $\rho$ on $\mathcal{M}$ is a pure state if and only if its restriction $\rho|\mathcal{M}_{\text{h}}$, to the partially ordered vector space $\mathcal{M}_{\text{h}}$ consisting of all self-adjoint elements of $\mathcal{M}$, is a pure state of $\mathcal{M}_{\text{h}}$ in the sense of Definition 3.4.5. Indeed, this follows from similar assertions, concerning hermitian linear functionals and states, occurring in the discussion preceding Proposition 4.3.1.

4.3.8. THEOREM. *Suppose that $\mathfrak{A}$ is a $C^*$-algebra with unit $I$, $\mathcal{M}$ is a self-adjoint subspace of $\mathfrak{A}$ that contains $I$, and $A \in \mathcal{M}$.*

(i) *If $\rho(A) = 0$, for each pure state $\rho$ of $\mathcal{M}$, then $A = 0$.*
(ii) *If $\rho(A)$ is real, for each pure state $\rho$ of $\mathcal{M}$, then $A$ is self-adjoint.*
(iii) *If $\rho(A) \geqslant 0$, for each pure state $\rho$ of $\mathcal{M}$, then $A \in \mathcal{M}^+$.*
(iv) *If $A$ is normal, there is a pure state $\rho_0$ of $\mathcal{M}$ such that $|\rho_0(A)| = \|A\|$.*

*Proof.* If $\rho(A) = 0$ (or $\rho(A)$ is real, or $\rho(A) \geqslant 0$) for all $\rho$ in $\mathcal{P}(\mathcal{M})$, then the same is true for all $\rho$ in $\mathcal{S}(\mathcal{M})$, since every state is a weak* limit of convex combinations of pure states. In view of this, the first three parts of the theorem follow, at once, from the corresponding assertions in Theorem 4.3.4.

Suppose now that $A$ is normal. By Theorem 4.3.4(iv), there is a scalar $c$ and a state $\tau$ of $\mathcal{M}$ such that $\tau(A) = c$, $|c| = \|A\|$. Let $\hat{T}$ be the (weak* continuous) linear functional on $\mathcal{M}^\#$ that takes the value $\rho(T)$ at $\rho$, and let $a$ be a complex number of modulus 1 such that $\tau(aA) = |c| = \|A\|$. From Corollary 1.4.4, there is a $\rho_0$ in $\mathcal{P}(\mathcal{M})$ such that

$$\begin{aligned}\|A\| &\geqslant |\rho_0(A)| \geqslant \text{Re}\,\widehat{aA}(\rho_0)\\ &\geqslant \sup\{\text{Re}\,\widehat{aA}(\rho) : \rho \in \mathcal{S}(\mathcal{M})\}\\ &\geqslant \text{Re}\,\widehat{aA}(\tau) = \text{Re}\,\tau(aA) = \|A\|. \quad \blacksquare\end{aligned}$$

4.3.9. THEOREM. *If $\mathfrak{A}$ is a $C^*$-algebra with unit $I$, $\mathcal{M}$ is a self-adjoint subspace of $\mathfrak{A}$ that contains $I$, and $\mathcal{S}_0$ is a subset of the state space $\mathcal{S}(\mathcal{M})$, the following four conditions are equivalent:*

(i) *If* $A \in \mathscr{M}$ *and* $\rho(A) \geqslant 0$ *for each* $\rho$ *in* $\mathscr{S}_0$, *then* $A \in \mathscr{M}^+$.
(ii) $\|H\| = \sup\{|\rho(H)|: \rho \in \mathscr{S}_0\}$, *for each self-adjoint* $H$ *in* $\mathscr{M}$.
(iii) $\overline{\mathrm{co}}(\mathscr{S}_0) = \mathscr{S}(\mathscr{M})$.
(iv) $\mathscr{P}(\mathscr{M}) \subseteq (\mathscr{S}_0)^-$ *the weak* closure of* $\mathscr{S}_0$ *in* $\mathscr{M}^\sharp$.

*Proof.* With $H$ self-adjoint in $\mathscr{M}$, define $a$ ($\leqslant \|H\|$) by

$$a = \sup\{|\rho(H)| : \rho \in \mathscr{S}_0\},$$

and note that

$$\rho(aI \pm H) = a \pm \rho(H) \geqslant 0 \qquad (\rho \in \mathscr{S}_0).$$

If (i) is satisfied, then $aI \pm H \in \mathscr{M}^+$, $-aI \leqslant H \leqslant aI$, hence $\|H\| \leqslant a$, and so $\|H\| = a$. Thus (i) implies (ii).

Suppose next that (ii) is satisfied. With $\mathscr{S}_1$ defined as $\overline{\mathrm{co}}(\mathscr{S}_0)$, and $(\mathscr{M}^\sharp)_1$ the unit ball in $\mathscr{M}^\sharp$, we have $\mathscr{S}_0 \subseteq \mathscr{S}_1 \subseteq \mathscr{S}(\mathscr{M}) \subseteq (\mathscr{M}^\sharp)_1$, so (ii) remains true when $\mathscr{S}_0$ is replaced by $\mathscr{S}_1$. From Lemma 4.3.5, $\overline{\mathrm{co}}(\mathscr{S}_1 \cup -\mathscr{S}_1)$ is the set of all hermitian functionals in $(\mathscr{M}^\sharp)_1$; in particular, $\mathscr{S}(\mathscr{M}) \subseteq \overline{\mathrm{co}}(\mathscr{S}_1 \cup -\mathscr{S}_1)$. The set

$$\{a\sigma - b\tau : \sigma, \tau \in \mathscr{S}_1,\ a, b \in \mathbb{R}^+,\ a + b = 1\}$$

contains $\mathscr{S}_1 \cup -\mathscr{S}_1$, inherits convexity and weak* compactness from $\mathscr{S}_1$, and so coincides with $\overline{\mathrm{co}}(\mathscr{S}_1 \cup -\mathscr{S}_1)$ (compare this with the proof of Theorem 4.3.6). Accordingly, each state $\rho$ of $\mathscr{M}$ has the form $a\sigma - b\tau$, with $\sigma$ and $\tau$ in $\mathscr{S}_1$, $a$ and $b$ in $\mathbb{R}^+$, and $a + b = 1$. Since

$$1 = \rho(I) = a\sigma(I) - b\tau(I) = a - b = 1 - 2b,$$

we have $b = 0$, $a = 1$ and $\rho = \sigma \in \mathscr{S}_1$. Hence $\mathscr{S}(\mathscr{M}) \subseteq \mathscr{S}_1$, and since the reverse inclusion has already been noted, $\mathscr{S}(\mathscr{M}) = \mathscr{S}_1 = \overline{\mathrm{co}}(\mathscr{S}_0)$. Thus (ii) implies (iii).

If $\overline{\mathrm{co}}(\mathscr{S}_0) = \mathscr{S}(\mathscr{M})$, it follows from Theorem 1.4.5 that $\mathscr{P}(\mathscr{M}) \subseteq (\mathscr{S}_0)^-$; so (iii) implies (iv).

Finally, suppose that $\mathscr{P}(\mathscr{M}) \subseteq (\mathscr{S}_0)^-$. If $A \in \mathscr{M}$ and $\rho(A) \geqslant 0$ for each $\rho$ in $\mathscr{S}_0$, the same is true for each $\rho$ in $(\mathscr{S}_0)^-$ (in particular, for each $\rho$ in $\mathscr{P}(\mathscr{M})$), by weak* continuity of the mapping $\rho \to \rho(A)$. By Theorem 4.3.8(iii), $A \in \mathscr{M}^+$, so (iv) implies (i). ■

4.3.10. COROLLARY. *If* $H$ *is a self-adjoint operator acting on a Hilbert space* $\mathscr{H}$, *then*

$$\|H\| = \sup\{|\langle Hx, x\rangle| : x \in \mathscr{H},\ \|x\| = 1\}.$$

*If* $\mathscr{M}$ *is a self-adjoint subspace of* $\mathscr{B}(\mathscr{H})$ *containing* $I$ *and* $\mathscr{S}_0$ *is the set of all vector states of* $\mathscr{M}$, *then* $\mathscr{P}(\mathscr{M}) \subseteq (\mathscr{S}_0)^-$ *and* $\mathscr{S}(\mathscr{M}) = \overline{\mathrm{co}}(\mathscr{S}_0)$.

*Proof.* If $A \in \mathscr{M}$ and $\rho(A) \geqslant 0$ for each $\rho$ in $\mathscr{S}_0$, then $\langle Ax, x\rangle \geqslant 0$ for each $x$ in $\mathscr{H}$, and thus $A \in \mathscr{M}^+$ ($= \mathscr{M} \cap \mathscr{B}(\mathscr{H})^+$) by Theorem 4.2.6. From Theorem

4.3.9, it now follows that $\overline{\text{co}}(\mathscr{S}_0) = \mathscr{S}(\mathscr{M})$, $\mathscr{P}(\mathscr{M}) \subseteq (\mathscr{S}_0)^-$, and

$$\|H\| = \sup\{|\rho(H)|: \rho \in \mathscr{S}_0\} = \sup\{|\langle Hx, x\rangle| : x \in \mathscr{H}, \|x\| = 1\},$$

for each self-adjoint $H$ in $\mathscr{M}$. Since the last conclusion applies, in particular, when $\mathscr{M} = \mathscr{B}(\mathscr{H})$, the corollary is proved. ∎

The formula for $\|H\|$, in Corollary 4.3.10, can also be proved without reference to $C^*$-algebra theory, by combining Lemma 3.2.13 with Proposition 3.2.15.

By a *function representation* of $\mathscr{M}$ on a compact Hausdorff space $X$, we mean a linear mapping $\varphi: A \to \varphi_A$ from $\mathscr{M}$ into the $C^*$-algebra $C(X)$, such that $\varphi_I$ is the constant function with value 1 throughout $X$, and $\varphi_A \in C(X)^+$ if and only if $A \in \mathscr{M}^+$. If, in addition, given any two distinct points $x$ and $y$ in $X$, there is an element $A$ of $\mathscr{M}$ such that $\varphi_A(x) \neq \varphi_A(y)$, we describe $\varphi$ as a *separating* function representation. Two function representations, $\varphi$: $\mathscr{M} \to C(X)$ and $\psi$: $\mathscr{M} \to C(Y)$, are said to be *equivalent* if there is a homeomorphism $f$ from $X$ onto $Y$, such that $\varphi_A(x) = \psi_A(f(x))$, for each $A$ in $\mathscr{M}$ and $x$ in $X$.

If $\varphi: \mathscr{M} \to C(X)$ is a function representation of $\mathscr{M}$, and $x \in X$, the evaluation mapping $\rho_x$: $A \to \varphi_A(x)$ is a state of $\mathscr{M}$, since it is a positive linear functional and $\rho_x(I) = \varphi_I(x) = 1$. Hence

$$|\varphi_A(x)| = |\rho_x(A)| \leqslant \|A\| \qquad (A \in \mathscr{M}, \quad x \in X)$$

and

$$\|\varphi_A\| = \sup_{x \in X} |\varphi_A(x)| \leqslant \|A\| \qquad (A \in \mathscr{M}).$$

For each self-adjoint $H$ in $\mathscr{M}$, $\varphi_H$ is a real-valued function, since $\varphi_H(x) = \rho_x(H)$ and $\rho_x$ is hermitian. With $c$ defined as $\|\varphi_H\|$

$$\varphi_{cI \pm H}(x) = c \pm \varphi_H(x) \geqslant 0 \qquad (x \in X);$$

so $\varphi_{cI \pm H} \in C(X)^+$; hence $cI \pm H \in \mathscr{M}^+$, $-cI \leqslant H \leqslant cI$, and therefore

$$\|H\| \leqslant c = \|\varphi_H\| \leqslant \|H\|.$$

Thus $\varphi$ maps self-adjoint elements of $\mathscr{M}$, isometrically, onto real-valued functions. By expressing an element $A$ of $\mathscr{M}$ in the form $H + iK$, with $H$ and $K$ self-adjoint in $\mathscr{M}$, it follows that $\varphi$ preserves adjoints; moreover, since $\varphi_A = \varphi_H + i\varphi_K$,

$$\|A\| \leqslant \|H\| + \|K\| = \|\varphi_H\| + \|\varphi_K\| \leqslant 2\|\varphi_A\|.$$

Accordingly,

$$\tfrac{1}{2}\|A\| \leqslant \|\varphi_A\| \leqslant \|A\|, \qquad \|\varphi_H\| = \|H\| \qquad (A \in \mathscr{M}, \quad H = H^* \in \mathscr{M}).$$

The set $\mathscr{N} = \{\varphi_A : A \in \mathscr{M}\}$ is a self-adjoint subspace of the $C^*$-algebra $C(X)$, and contains its unit. Since $\varphi$ is a one-to-one bicontinuous linear mapping from the normed space $\mathscr{M}$ onto the normed space $\mathscr{N}$, its Banach adjoint $\varphi^{\#}: \tau \to \varphi^{\#}_{\tau}$ is a bicontinuous linear mapping from $\mathscr{N}^{\#}$ onto $\mathscr{M}^{\#}$. Since $\varphi_A$ is self-adjoint, or positive, if and only if $A$ has the same property, while $\varphi^{\#}_{\tau}(A) = \tau(\varphi_A)$ $(A \in \mathscr{M}, \tau \in \mathscr{N}^{\#})$, it follows that $\varphi^{\#}_{\tau}$ is hermitian, or positive, if and only if the same is true of $\tau$. Since $\varphi$ is isometric on self-adjoint elements of $\mathscr{M}$, while the norm of a hermitian functional is unchanged by restriction to self-adjoint elements, $\varphi^{\#}$ is isometric on the hermitian functionals in $\mathscr{N}^{\#}$.

We now exhibit some canonical function representations associated with a $C^*$-algebra. For each $A$ in $\mathscr{M}$, the equation

$$\hat{A}(\rho) = \rho(A) \qquad (\rho \in \mathscr{S}(\mathscr{M}))$$

defines a continuous complex-valued function $\hat{A}$ on the state space $\mathscr{S}(\mathscr{M})$. If $\mathscr{S}_0$ is a closed subset of $\mathscr{S}(\mathscr{M})$ and contains the pure state space $\mathscr{P}(\mathscr{M})^{-}$, it is apparent that the restriction $\hat{A}|\mathscr{S}_0$ takes non-negative values throughout $\mathscr{S}_0$ if $A \in \mathscr{M}^{+}$, and the converse assertion follows from Theorem 4.3.9. Since

$$\rho(aA + bB) = a\rho(A) + b\rho(B),$$

we have

$$\widehat{(aA+bB)} = a\hat{A} + b\hat{B} \qquad (A, B \in \mathscr{M}, \quad a, b \in \mathbb{C}).$$

If $\rho_1$ and $\rho_2$ are distinct elements of $\mathscr{S}_0$, then we have $\rho_1(A) \neq \rho_2(A)$ for some $A$ in $\mathscr{M}$; that is, $\hat{A}(\rho_1) \neq \hat{A}(\rho_2)$. Moreover, $\hat{I}(\rho) = \rho(I) = 1$, for each $\rho$ in $\mathscr{S}_0$. Accordingly, the mapping

$$A \to \hat{A}|\mathscr{S}_0: \quad \mathscr{M} \to C(\mathscr{S}_0)$$

is a separating function representation of $\mathscr{M}$ on $\mathscr{S}_0$. The following theorem shows that every separating function representation of $\mathscr{M}$ is equivalent to one that arises in this way, by appropriate choice of $\mathscr{S}_0$. The most important function representations of $\mathscr{M}$ are the two "extreme" ones, obtained from the above construction when $\mathscr{S}_0$ is either the state space $\mathscr{S}(\mathscr{M})$ or the pure state space $\mathscr{P}(\mathscr{M})^{-}$.

4.3.11. THEOREM. *If $\mathfrak{A}$ is a $C^*$-algebra with unit $I$, $\mathscr{M}$ is a self-adjoint subspace of $\mathfrak{A}$ containing $I$, $X$ is a compact Hausdorff space, and $\varphi: \mathscr{M} \to C(X)$ is a separating function representation of $\mathscr{M}$, then there is a unique closed subset $\mathscr{S}_0$ of $\mathscr{S}(\mathscr{M})$, such that $\mathscr{P}(\mathscr{M})^{-} \subseteq \mathscr{S}_0$ and $\varphi$ is equivalent to the function representation*

$$A \to \hat{A}|\mathscr{S}_0: \quad \mathscr{M} \to C(\mathscr{S}_0).$$

*Proof.* For each $x$ in $X$, the equation $\rho_x(A) = \varphi_A(x)$ defines a state $\rho_x$ of $\mathscr{M}$. Given two distinct points $x$ and $y$ of $X$, there is an element $A$ of $\mathscr{M}$ for which $\varphi_A(x) \neq \varphi_A(y)$, whence $\rho_x(A) \neq \rho_y(A)$ and so $\rho_x \neq \rho_y$. If $\{x_a\}$ is a convergent net of elements of $X$, with limit $x$, it follows from the continuity of the function

$\varphi_A$ that

$$\rho_{x_a}(A) = \varphi_A(x_a) \underset{a}{\rightarrow} \varphi_A(x) = \rho_x(A) \qquad (A \in \mathscr{M});$$

so $\rho_{x_a} \rightarrow \rho_x$ in the weak* topology.

From the preceding paragraph, the mapping

$$f: x \rightarrow \rho_x: \quad X \rightarrow \mathscr{S}(\mathscr{M})$$

is one-to-one and continuous. Since $X$ is compact, the same is true of its range, $\mathscr{S}_0 = \{\rho_x : x \in X\}$, and $f$ is a homeomorphism from $X$ onto $\mathscr{S}_0$. If $A \in \mathscr{M}$ and $\rho(A) \geqslant 0$ for each $\rho$ in $\mathscr{S}_0$, then

$$\varphi_A(x) = \rho_x(A) \geqslant 0 \qquad (x \in X);$$

so $\varphi_A \in C(X)^+$, and therefore $A \in \mathscr{M}^+$. Since $\mathscr{S}_0$ is closed, it now follows from Theorem 4.3.9 that $\mathscr{P}(\mathscr{M})^- \subseteq \mathscr{S}_0$. Finally,

$$\hat{A}(f(x)) = \hat{A}(\rho_x) = \rho_x(A) = \varphi_A(x) \qquad (x \in X, \quad A \in \mathscr{M}),$$

and therefore $\varphi$ is equivalent to the function representation

$$A \rightarrow \hat{A}|\mathscr{S}_0: \quad \mathscr{M} \rightarrow C(\mathscr{S}_0).$$

To prove the uniqueness clause of the theorem, suppose that $\varphi$ is equivalent also to the function representation

$$A \rightarrow \hat{A}|\mathscr{S}_1: \quad \mathscr{M} \rightarrow C(\mathscr{S}_1),$$

where $\mathscr{S}_1$ is a closed subset of $\mathscr{S}(\mathscr{M})$ containing $\mathscr{P}(\mathscr{M})^-$. Let $g: x \rightarrow g_x$ be a homeomorphism from $X$ onto $\mathscr{S}_1$, such that

$$\varphi_A(x) = \hat{A}(g_x) \qquad (x \in X, \quad A \in \mathscr{M}).$$

Then $g_x$ is a state of $\mathscr{M}$, and

$$g_x(A) = \hat{A}(g_x) = \varphi_A(x) = \rho_x(A) \qquad (A \in \mathscr{M}).$$

Hence $g_x = \rho_x$ for each $x$ in $X$, and

$$\mathscr{S}_1 = \{g_x : x \in X\} = \{\rho_x : x \in X\} = \mathscr{S}_0. \quad \blacksquare$$

4.3.12. REMARK. We illustrate the use of function representations by giving an alternative proof of the existence of a decomposition of a bounded hermitian functional as a difference of positive linear functionals (Theorem 4.3.6), by reduction to the abelian case (Proposition 3.4.11).

With $\varphi: A \rightarrow \varphi_A$ a function representation of $\mathscr{M}$ on a compact Hausdorff space $X$, and $\mathscr{N}$ the subspace $\{\varphi_A : A \in \mathscr{M}\}$ of $C(X)$, we recall that $\varphi$ has a Banach adjoint operator $\varphi^\#: \tau \rightarrow \varphi_\tau^\#$ from $\mathscr{N}^\#$ onto $\mathscr{M}^\#$; that $\varphi_\tau^\#$ is hermitian, or positive, if and only if $\tau$ has the same property; and that $\varphi^\#$ is isometric on hermitian elements of $\mathscr{N}^\#$. In order to show that a bounded hermitian functional $\rho$ on $\mathscr{M}$ can be expressed as $\rho^+ - \rho^-$, where $\rho^+$ and $\rho^-$ are positive linear functionals and $\|\rho^+\| + \|\rho^-\| = \|\rho\|$, it now suffices to prove the corresponding statement for $\mathscr{N}$. With $\rho$ a bounded hermitian functional on $\mathscr{N}$, $\rho$ extends without increase of norm to a bounded linear functional $\tau$ on $C(X)$. We can suppose that $\tau = \tau_1 + i\tau_2$, where $\tau_1, \tau_2$ are hermitian, and $\|\tau_j\| \leqslant \|\tau\|$ $(j = 1, 2)$. Since $\rho$ and the restrictions $\tau_1|\mathscr{N}$ and $\tau_2|\mathscr{N}$ are hermitian, while $\rho = (\tau_1 + i\tau_2)|\mathscr{N}$, it follows that $\rho = \tau_1|\mathscr{N}$. Upon replacing $\tau$ by $\tau_1$, we may assume that $\tau$ is hermitian. By Proposition 3.4.11, $\tau$ can be expressed as $\tau^+ - \tau^-$, where $\tau^+$ and $\tau^-$ are positive linear functionals on $C(X)$, and $\|\tau^+\| + \|\tau^-\| = \|\tau\|$. With $\rho^+$ and $\rho^-$ defined as $\tau^+|\mathscr{N}$ and $\tau^-|\mathscr{N}$, respectively, $\rho^+$ and $\rho^-$ are positive linear functionals on $\mathscr{N}$, and $\rho = \rho^+ - \rho^-$. Moreover,

$$\begin{aligned}\|\rho^+\| + \|\rho^-\| &\leqslant \|\tau^+\| + \|\tau^-\| = \|\tau\| \\ &= \|\rho\| = \|\rho^+ - \rho^-\| \leqslant \|\rho^+\| + \|\rho^-\|,\end{aligned}$$

so $\|\rho^+\| + \|\rho^-\| = \|\rho\|$. ■

We conclude this section with some further results concerning states and pure states.

4.3.13. THEOREM. *Suppose that $\mathfrak{A}$ is a $C^*$-algebra with unit $I$, $\mathscr{M}$ is a self-adjoint subspace of $\mathfrak{A}$ that contains $I$, and $\rho$ is a state of $\mathscr{M}$. For each self-adjoint $H$ in $\mathfrak{A}$, define*

$$l_H = \sup\{\rho(B) : B = B^* \in \mathscr{M},\ B \leqslant H\},$$

$$u_H = \inf\{\rho(B) : B = B^* \in \mathscr{M},\ B \geqslant H\}.$$

(i) $-\|H\| \leqslant l_H \leqslant u_H \leqslant \|H\|$ $(H = H^* \in \mathfrak{A})$.

(ii) *$\rho$ extends to a state $\tau$ of $\mathfrak{A}$. If $c$ is a real number and $H$ is a self-adjoint element of $\mathfrak{A}$, the extended state $\tau$ can be chosen so that $\tau(H) = c$ if and only if $l_H \leqslant c \leqslant u_H$.*

(iii) *$\rho$ extends uniquely to a state $\tau$ of $\mathfrak{A}$ if and only if $l_H = u_H$ for each self-adjoint $H$ in $\mathfrak{A}$.*

(iv) *If $\rho$ is a pure state of $\mathscr{M}$, then $\rho$ extends to a pure state of $\mathfrak{A}$. If, further, $\rho$ has only one extension as a pure state of $\mathfrak{A}$, then $\rho$ has only one extension as a state of $\mathfrak{A}$.*

*Proof.* (i) With $H$ self-adjoint in $\mathfrak{A}$,

$$\pm \|H\| I \in \mathscr{M}, \qquad -\|H\| I \leqslant H \leqslant \|H\| I.$$

If $B = B^* \in \mathscr{M}$ and $B \leqslant H$, then $B \leqslant \|H\| I$ and therefore $\rho(B) \leqslant \|H\| \rho(I) = \|H\|$. Hence the set

$$\{\rho(B) : B = B^* \in \mathscr{M},\ B \leqslant H\}$$

is bounded above by $\|H\|$ and (with $B = -\|H\| I$) contains $-\|H\|$.

Accordingly, its supremum $l_H$ satisfies $-\|H\| \leqslant l_H \leqslant \|H\|$; and a similar argument shows that $-\|H\| \leqslant u_H \leqslant \|H\|$. If $B_1$ and $B_2$ are self-adjoint elements of $\mathscr{M}$ and $B_1 \leqslant H \leqslant B_2$, then $\rho(B_1) \leqslant \rho(B_2)$. By allowing $B_1$ and $B_2$ to vary, subject only to the conditions just stated, it follows that $l_H \leqslant u_H$.

(ii) With $H$ self-adjoint in $\mathfrak{A}$, $l_H \leqslant u_H$ by (i), so we can choose a real number $c$ such that $l_H \leqslant c \leqslant u_H$. We shall prove that $\rho$ has an extension to a state $\tau$ of $\mathfrak{A}$ such that $\tau(H) = c$.

As a first step, we show that the equation

$$\text{(1)} \qquad \tau_0(aH + A) = ac + \rho(A) \qquad (a \in \mathbb{C}, \quad A \in \mathscr{M})$$

defines a positive linear functional $\tau_0$ on the self-adjoint subspace

$$\mathscr{N} = \{aH + A : a \in \mathbb{C},\ A \in \mathscr{M}\}$$

generated by $H$ and $\mathscr{M}$. This is evident when $H \in \mathscr{M}$, since in this case $\mathscr{N} = \mathscr{M}$, $l_H = u_H = \rho(H)$, hence $c = \rho(H)$; and $\tau_0$, as defined by (1), is $\rho$. We assume henceforth that $H \notin \mathscr{M}$, whence each element $T$ of $\mathscr{N}$ is *uniquely* expressible as $aH + A$, with $a$ in $\mathbb{C}$ and $A$ in $\mathscr{M}$. Accordingly, (1) defines a linear functional $\tau_0$ on $\mathscr{N}$, $\tau_0(H) = c$ and $\tau_0$ extends $\rho$; in particular, $\tau_0(I) = \rho(I) = 1$.

In order to show that $\tau_0$ is positive, suppose that $T \in \mathscr{N}^+$, and let $T = aH + A$, as above. Since

$$0 = T^* - T = (\bar{a} - a)H + A^* - A,$$

$(\bar{a} - a)H = A - A^* \in \mathscr{M}$, and it follows that $a$ is real. If $a = 0$, then $T = A \in \mathscr{M}^+$ and $\tau_0(T) = \rho(A) \geqslant 0$. If $a > 0$, then

$$H + a^{-1}A = a^{-1}T \geqslant 0,$$

so $-a^{-1}A \in \mathscr{M}$, $-a^{-1}A \leqslant H$; from the definition of $l_H$, $-a^{-1}\rho(A) \leqslant l_H$ $(\leqslant c)$, and therefore

$$\tau_0(T) = a[c + a^{-1}\rho(A)] \geqslant 0.$$

If $a < 0$, then $-H - a^{-1}A = (-a)^{-1}T \geqslant 0$, so

$$-a^{-1}A \in \mathscr{M}, \qquad -a^{-1}A \geqslant H;$$

from the definition of $u_H$, $-a^{-1}\rho(A) \geqslant u_H$ $(\geqslant c)$, and therefore

$$\tau_0(T) = -a[-c - a^{-1}\rho(A)] \geqslant 0.$$

The preceding enumeration of cases shows that $\tau_0(T) \geqslant 0$ whenever $T \in \mathscr{N}^+$, so $\tau_0$ is a positive linear functional on $\mathscr{N}$.

From Theorem 4.3.2, $\tau_0$ is bounded, and $\|\tau_0\| = \tau_0(I) = 1$. By the Hahn–Banach theorem, $\tau_0$ extends without change of norm to a bounded linear functional $\tau$ on $\mathfrak{A}$; moreover, $\tau$ is positive, again by Theorem 4.3.2, since

$$\tau(I) = \tau_0(I) = 1 = \|\tau_0\| = \|\tau\|.$$

Accordingly, $\tau$ is a state of $\mathfrak{A}$, $\tau(H) = \tau_0(H) = c$, and $\tau$ extends $\rho$.

Conversely, suppose that $\tau_1$ is any state of $\mathfrak{A}$ that extends $\rho$. If $B_1$ and $B_2$ are self-adjoint elements of $\mathscr{M}$, for which $B_1 \leqslant H \leqslant B_2$, then $\tau_1(B_1) \leqslant \tau_1(H) \leqslant \tau_1(B_2)$; that is, $\rho(B_1) \leqslant \tau_1(H) \leqslant \rho(B_2)$. By allowing $B_1$ and $B_2$ to vary, subject only to the restrictions just stated, it follows that $l_H \leqslant \tau_1(H) \leqslant u_H$.

(iii) This is an immediate consequence of (ii).

(iv) Suppose now that $\rho$ is a pure state of $\mathscr{M}$, and define $\mathscr{S}_0$ to be the set

$$\{\tau \in \mathscr{S}(\mathfrak{A}) : \tau(B) = \rho(B)\ (B \in \mathscr{M})\}$$

of all states of $\mathfrak{A}$ that extend $\rho$. Then $\mathscr{S}_0$ is a closed convex subset of $\mathscr{S}(\mathfrak{A})$, is therefore weak* compact, and is non-empty by (ii). From the Krein–Milman theorem, $\mathscr{S}_0 = \overline{\text{co}}(\mathscr{P}_0)$, the closed convex hull of the set $\mathscr{P}_0$ of all extreme points of $\mathscr{S}_0$. Hence $\mathscr{P}_0$ is not empty, and consists of a single element if and only if $\mathscr{S}_0$ has just one element.

It now suffices to show that each $\tau$ in $\mathscr{P}_0$ is a pure state of $\mathfrak{A}$. For this, suppose that $\tau = a\tau_1 + (1 - a)\tau_2$, where $\tau_1, \tau_2 \in \mathscr{S}(\mathfrak{A})$ and $0 < a < 1$. Since the restrictions $\tau_1|\mathscr{M}$, $\tau_2|\mathscr{M}$ are states of $\mathscr{M}$, while $\rho$ is a pure state of $\mathscr{M}$ and

$$\rho = \tau|\mathscr{M} = a(\tau_1|\mathscr{M}) + (1 - a)(\tau_2|\mathscr{M}),$$

it follows that $\tau_1|\mathscr{M} = \tau_2|\mathscr{M} = \rho$. Thus $\tau_1, \tau_2 \in \mathscr{S}_0$; and since $\tau$ $(= a\tau_1 + (1 - a)\tau_2)$ is an extreme point of $\mathscr{S}_0$, $\tau_1 = \tau_2 = \tau$. Hence $\tau$ is an extreme point of $\mathscr{S}(\mathfrak{A})$; that is, $\tau$ is a pure state of $\mathfrak{A}$. ■

The results on extensions of states and pure states, set out in Theorem 4.3.13, remain valid in the context of a partially ordered vector space $\mathscr{V}$ with order unit $I$, and a subspace $\mathscr{M}$ of $\mathscr{V}$ that contains $I$ (see Exercise 4.6.49).

4.3.14. Proposition. *If $\mathfrak{A}$ is a $C^*$-algebra with center $\mathscr{C}$ and $\rho$ is a pure state of $\mathfrak{A}$, then $\rho(AC) = \rho(A)\rho(C)$ for all $A$ in $\mathfrak{A}$ and $C$ in $\mathscr{C}$. Moreover, the restriction $\rho|\mathscr{C}$ is a pure state of $\mathscr{C}$.*

*Proof.* In order to show that $\rho(AC) = \rho(A)\rho(C)$ when $A \in \mathfrak{A}$ and $C \in \mathscr{C}$, it suffices (by linearity) to consider the case in which $0 \leqslant C \leqslant I$. In this case, for each $H$ in $\mathfrak{A}^+$, we have $0 \leqslant HC \leqslant H$, and thus $0 \leqslant \rho(HC) \leqslant \rho(H)$, since $H$ commutes with $C$ (see the discussion following Corollary 4.2.7). Hence the equation $\rho_0(A) = \rho(AC)$ defines a positive linear functional $\rho_0$ on $\mathfrak{A}$, and $\rho_0 \leqslant \rho$. Since $\rho$ is a pure state, so is its restriction $\rho|\mathfrak{A}_h$ to the partially ordered vector space $\mathfrak{A}_h$ of all self-adjoint elements of $\mathfrak{A}$ (see the paragraph preceding Theorem 4.3.8). Since $\rho_0|\mathfrak{A}_h \leqslant \rho|\mathfrak{A}_h$, it follows from Lemma 3.4.6 that $\rho_0|\mathfrak{A}_h = a(\rho|\mathfrak{A}_h)$, for some scalar $a$. Hence $\rho_0 = a\rho$; and

$$\rho(AC) = \rho_0(A) = a\rho(A) = a\rho(I)\rho(A) = \rho_0(I)\rho(A) = \rho(C)\rho(A),$$

for each $A$ in $\mathfrak{A}$.

From the preceding paragraph it follows, in particular, that the non-zero linear functional $\rho|\mathscr{C}$ is multiplicative on $\mathscr{C}$; so the final assertion of the proposition follows from Proposition 4.4.1 (just below). ■

## 4.4. Abelian algebras

With $X$ a compact Hausdorff space, $C(X)$ is an abelian $C^*$-algebra. Our main purpose in this section is to show that every abelian $C^*$-algebra $\mathfrak{A}$ is * isomorphic to one of the form $C(X)$. As a first step, we prove that the pure states of $\mathfrak{A}$ are precisely the multiplicative linear functionals on $\mathfrak{A}$, a fact already noted in Theorem 3.4.7 for the algebra $C(X)$.

4.4.1. PROPOSITION. *A non-zero linear functional $\rho$ on an abelian $C^*$-algebra $\mathfrak{A}$ is a pure state if and only if $\rho(AB) = \rho(A)\rho(B)$ for all $A$ and $B$ in $\mathfrak{A}$.*

*Proof.* The first assertion of Proposition 4.3.14 includes, as a special case, the fact that pure states of an abelian $C^*$-algebra are multiplicative.

Conversely, suppose that $\rho$ is a multiplicative linear functional on $\mathfrak{A}$. By Proposition 3.2.20, $\rho$ is bounded and $\|\rho\| = \rho(I) = 1$, so $\rho$ is a state of $\mathfrak{A}$. In order to prove that $\rho$ is pure, suppose that $\rho = a\rho_1 + b\rho_2$, where $\rho_1, \rho_2 \in \mathscr{S}(\mathfrak{A})$, $a > 0$, $b > 0$, and $a + b = 1$. With $C$ self-adjoint in $\mathfrak{A}$,

$$[\rho_j(C)]^2 = [\rho_j(IC)]^2 \leqslant \rho_j(I)\rho_j(C^2) = \rho_j(C^2) \qquad (j = 1, 2),$$

by the Cauchy–Schwarz inequality. Accordingly,

$$\begin{aligned} 0 &= \rho(C^2) - [\rho(C)]^2 \\ &= a\rho_1(C^2) + b\rho_2(C^2) - [a\rho_1(C) + b\rho_2(C)]^2 \\ &\geqslant a(a+b)[\rho_1(C)]^2 + b(a+b)[\rho_2(C)]^2 - [a\rho_1(C) + b\rho_2(C)]^2 \\ &= ab[\rho_1(C) - \rho_2(C)]^2. \end{aligned}$$

From this, $\rho_1(C) = \rho_2(C)$ for each self-adjoint $C$ in $\mathfrak{A}$; so $\rho_1 = \rho_2$, whence $\rho$ is a pure state. ■

The argument just given shows that a multiplicative linear functional on a (not necessarily abelian) $C^*$-algebra is a pure state. This can also be proved by reduction to the case of algebras of the form $C(X)$ and an application of Theorem 3.4.7. Indeed, if $\rho = (1 - a)\rho_1 + a\rho_2$, we can show that $\rho(A) = \rho_1(A) = \rho_2(A)$, for a given self-adjoint $A$ in $\mathfrak{A}$, by restricting $\rho$, $\rho_1$, and $\rho_2$ to the $C^*$-subalgebra generated by $I$ and $A$, and identifying this algebra with $C(\mathrm{sp}(A))$.

4.4.2. COROLLARY. *The set $\mathscr{P}(\mathfrak{A})$ of pure states of an abelian $C^*$-algebra $\mathfrak{A}$ is a closed subset of the state space $\mathscr{S}(\mathfrak{A})$.*

*Proof.* By Proposition 4.4.1,

$$\mathscr{P}(\mathfrak{A}) = \{\rho \in \mathscr{S}(\mathfrak{A}) : \rho(AB) = \rho(A)\rho(B)\ (A, B \in \mathfrak{A})\}. \quad \blacksquare$$

4.4.3. THEOREM. *Suppose that $\mathfrak{A}$ is an abelian $C^*$-algebra, $\mathscr{P}(\mathfrak{A})$ is the set of all pure states of $\mathfrak{A}$, and for each $A$ in $\mathfrak{A}$, a complex-valued function $\hat{A}$ is defined throughout $\mathscr{P}(\mathfrak{A})$ by $\hat{A}(\rho) = \rho(A)$. Then $\mathscr{P}(\mathfrak{A})$ is a compact Hausdorff space, relative to the weak* topology, and the mapping $A \to \hat{A}$ is a * isomorphism from $\mathfrak{A}$ onto the $C^*$-algebra $C(\mathscr{P}(\mathfrak{A}))$.*

*Proof.* A substantial part of the argument required to prove this theorem is already contained in the discussion of function representations of (not necessarily abelian) $C^*$-algebras, preceding Theorem 4.3.11. Two new features that are special to the abelian case, those set out in Proposition 4.4.1 and Corollary 4.4.2, suffice to complete the proof. For the sake of clarity, the argument is presented below in unified form, even though this involves some repetition of the earlier discussion.

From Corollary 4.4.2, $\mathscr{P}(\mathfrak{A})$ is weak* compact. With $A$ in $\mathfrak{A}$, it is apparent from the definition of the weak* topology that $\hat{A}$ is a continuous complex-valued function on $\mathscr{P}(\mathfrak{A})$. For all $A$ and $B$ in $\mathfrak{A}$, $a$ and $b$ in $\mathbb{C}$, and $\rho$ in $\mathscr{P}(\mathfrak{A})$

$$(\widehat{aA+bB})(\rho) = \rho(aA + bB) = a\rho(A) + b\rho(B) = a\hat{A}(\rho) + b\hat{B}(\rho),$$

$$|\hat{A}(\rho)| = |\rho(A)| \leqslant \|A\|, \qquad \hat{A}^*(\rho) = \rho(A^*) = \overline{\rho(A)} = \overline{\hat{A}(\rho)};$$

and

$$(\widehat{AB})(\rho) = \rho(AB) = \rho(A)\rho(B) = \hat{A}(\rho)\hat{B}(\rho),$$

since $\rho$ is multiplicative, by Proposition 4.4.1. Since $\mathfrak{A}$ is abelian, $A$ is normal, so by Theorem 4.3.8(iv) there is a pure state $\rho_0$ of $\mathfrak{A}$ such that $|\rho_0(A)| = \|A\|$. From this,

$$\|A\| = |\hat{A}(\rho_0)| \leqslant \sup_{\rho \in \mathscr{P}(\mathfrak{A})} |\hat{A}(\rho)| \leqslant \|A\|,$$

$$\|A\| = \sup_{\rho \in \mathscr{P}(\mathfrak{A})} |\hat{A}(\rho)| = \|\hat{A}\|.$$

The function $\hat{I}$ is the unit of $C(\mathscr{P}(\mathfrak{A}))$, since

$$\hat{I}(\rho) = \rho(I) = 1 \qquad (\rho \in \mathscr{P}(\mathfrak{A})).$$

The preceding discussion shows that the mapping $\varphi : A \to \hat{A}$ is an (isometric) * isomorphism from $\mathfrak{A}$ into $C(\mathscr{P}(\mathfrak{A}))$. Its range

$$\hat{\mathfrak{A}} = \{\hat{A} : A \in \mathfrak{A}\}$$

is therefore a * subalgebra of $C(\mathscr{P}(\mathfrak{A}))$, contains the constant functions, and is closed since $\mathfrak{A}$ is complete and $\varphi$ is an isometry. Given distinct pure states $\rho_1$

and $\rho_2$ of $\mathfrak{A}$, we can choose $A$ in $\mathfrak{A}$ so that $\rho_1(A) \neq \rho_2(A)$, equivalently $\hat{A}(\rho_1) \neq \hat{A}(\rho_2)$; so $\hat{\mathfrak{A}}$ separates the points of $\mathscr{P}(\mathfrak{A})$. By the Stone–Weierstrass theorem, $\hat{\mathfrak{A}} = C(\mathscr{P}(\mathfrak{A}))$. ■

With $\mathfrak{A}$ an abelian $C^*$-algebra, the * isomorphism described in Theorem 4.4.3 is just the function representation of $\mathfrak{A}$ on its pure state space, constructed in the discussion preceding Theorem 4.3.11, and is by far the most important example of a function representation of $\mathfrak{A}$. For this reason, we shall frequently refer to it as *the* function representation of $\mathfrak{A}$.

4.4.4. PROPOSITION. *If $A$ is a normal element of a $C^*$-algebra $\mathfrak{A}$, and $a \in \text{sp}(A)$, there is a pure state $\rho$ of $\mathfrak{A}$ such that $\rho(A) = a$.*

*Proof.* The set of all polynomials in $I$, $A$, and $A^*$ is an abelian * subalgebra of $\mathfrak{A}$, and its closure is an abelian $C^*$-subalgebra $\mathscr{A}$. Now $A$ has the same spectrum relative to $\mathfrak{A}$ or $\mathscr{A}$. From Remark 3.2.11 and Proposition 4.4.1, there is a pure state of $\mathscr{A}$ whose value at $A$ is $a$. From Theorem 4.3.13(iv), this pure state of $\mathscr{A}$ extends to a pure state $\rho$ of $\mathfrak{A}$; and $\rho(A) = a$. ■

In the remainder of this section, we make use of the function representation for abelian $C^*$-algebras in establishing the existence and properties of the function calculus associated with a normal element of a (not necessarily abelian) $C^*$-algebra.

4.4.5. THEOREM. *If $A$ is a normal element of a $C^*$-algebra $\mathfrak{A}$, $C(\text{sp}(A))$ is the abelian $C^*$-algebra of all continuous complex-valued functions on* sp$(A)$, *and $\iota$ in $C(\text{sp}(A))$ is defined by $\iota(t) = t$ $(t \in \text{sp}(A))$, then there is a unique * isomorphism $\varphi: C(\text{sp}(A)) \to \mathfrak{A}$ such that $\varphi(\iota) = A$. For each $f$ in $C(\text{sp}(A))$, $\varphi(f)$ is normal, and is the limit of a sequence of polynomials in $I$, $A$, and $A^*$. The set*

$$\{\varphi(f) : f \in C(\text{sp}(A))\}$$

*is an abelian $C^*$-algebra, and is the smallest $C^*$-subalgebra of $\mathfrak{A}$ that contains $A$. Moreover, $A$ is self-adjoint if and only if $\text{sp}(A) \subseteq \mathbb{R}$, positive if and only if $\text{sp}(A) \subseteq \mathbb{R}^+$, unitary if and only if $\text{sp}(A) \subseteq \mathbb{C}_1$ $(= \{t \in \mathbb{C}: |t| = 1\})$, and a projection if and only if $\text{sp}(A) \subseteq \{0, 1\}$.*

*Proof.* Let $\mathscr{A}$ be any abelian $C^*$-subalgebra of $\mathfrak{A}$ that contains $A$; for example, $\mathscr{A}$ could be the closure of the set of all polynomials in $I$, $A$, and $A^*$. From Theorem 4.4.3, there is a compact Hausdorff space $X$ and a * isomorphism $\psi$ from $\mathscr{A}$ onto $C(X)$. With $u$ the element $\psi(A)$ of $C(X)$,

$$\text{sp}(A) = \text{sp}_{\mathscr{A}}(A) = \text{sp}_{C(X)}(\psi(A)) = \text{sp}_{C(X)}(u) = \{u(x) : x \in X\}.$$

For each $f$ in $C(\text{sp}(A))$, the composite function $f \circ u$ is continuous throughout $X$; the mapping $f \to f \circ u$ is a * isomorphism from $C(\text{sp}(A))$ into $C(X)$. From this, and since $\psi^{-1}$: $C(X) \to \mathscr{A}$ $(\subseteq \mathfrak{A})$ is a * isomorphism, the mapping

$\varphi: f \to \psi^{-1}(f \circ u)$ is a * isomorphism from $C(\mathrm{sp}(A))$ into $\mathfrak{A}$; and $\varphi(\iota) = \psi^{-1}(\iota \circ u) = \psi^{-1}(u) = A$.

Suppose also that $\varphi'$: $C(\mathrm{sp}(A)) \to \mathfrak{A}$ is a * isomorphism, and $\varphi'(\iota) = A$. Then $\varphi$ and $\varphi'$ are linear, multiplicative, and adjoint preserving, $\varphi(\iota) = \varphi'(\iota) = A$, and $\varphi(1) = \varphi'(1) = I$, where 1 denotes the unit of $C(\mathrm{sp}(A))$. From this, $\varphi(f) = \varphi'(f)$ whenever $f$ is a polynomial in 1, $\iota$, and the conjugate complex function $\bar{\iota}$. Since polynomials of this type form an everywhere-dense subset of $C(\mathrm{sp}(A))$, by the Stone–Weierstrass theorem, while $\varphi$ and $\varphi'$ are isometric (see Theorem 4.1.8(iii)), it follows that $\varphi(f) = \varphi'(f)$ for each $f$ in $C(\mathrm{sp}(A))$.

Since $C(\mathrm{sp}(A))$ is an abelian $C^*$-algebra, it results from Theorem 4.1.8(iii) that its image $\{\varphi(f): f \in C(\mathrm{sp}(A))\}$ under $\varphi$ is an abelian $C^*$-subalgebra of $\mathfrak{A}$, and contains $A$ $(= \varphi(\iota))$. From this, $\varphi(f)$ is normal for each $f$ in $C(\mathrm{sp}(A))$. Moreover, $\varphi(f)$ is the limit of a sequence of polynomials in $I$, $A$, and $A^*$, since $f$ is the uniform limit on $\mathrm{sp}(A)$ of polynomials in 1, $\iota$, and $\bar{\iota}$. If $\mathscr{B}$ is a $C^*$-subalgebra of $\mathfrak{A}$, and $A \in \mathscr{B}$, then $\mathscr{B}$ contains $I$, $A$, $A^*$, and hence contains all limits of polynomials in $I$, $A$, and $A^*$; so $\varphi(f) \in \mathscr{B}$, for each $f$ in $C(\mathrm{sp}(A))$.

Since $\varphi$ is a * isomorphism and $\varphi(\iota) = A$, it follows that $A$ is self-adjoint (or positive, or unitary, or a projection) if and only if the same is true of the element $\iota$ of $C(\mathrm{sp}(A))$. Now $\iota$ is self-adjoint if and only if it is real valued on $\mathrm{sp}(A)$; that is, if and only if $\mathrm{sp}(A) \subseteq \mathbb{R}$. Similarly $\iota$ is positive if and only if it takes non-negative values throughout $\mathrm{sp}(A)$ (equivalently, $\mathrm{sp}(A) \subseteq \mathbb{R}^+$). Also $\iota$ is unitary (or a projection) if and only if $\iota(t)\overline{\iota(t)} = 1$ (or $[\iota(t)]^2 = \iota(t) = \overline{\iota(t)}$) for all $t$ in $\mathrm{sp}(A)$, and this occurs if and only if $\mathrm{sp}(A) \subseteq \mathbb{C}_1$ (or $\mathrm{sp}(A) \subseteq \{0, 1\}$). ■

The * isomorphism $\varphi: C(\mathrm{sp}(A)) \to \mathfrak{A}$ described in Theorem 4.4.5 is called the *function calculus* for the normal element $A$ of the $C^*$-algebra $\mathfrak{A}$. With $f$ in $C(\mathrm{sp}(A))$, we usually denote by $f(A)$ the element $\varphi(f)$ of $\mathfrak{A}$. Note that, if $f$ has the form

$$f(t) = \sum_{j=0}^{m} \sum_{k=0}^{n} a_{jk} t^j (\bar{t})^k \qquad (t \in \mathrm{sp}(A))$$

(that is, $f = \sum \sum a_{jk}\, \iota^j (\bar{\iota})^k$), then

$$f(A) = \sum_{j=0}^{m} \sum_{k=0}^{n} a_{jk} A^j (A^*)^k.$$

4.4.6. Remark. Suppose that $\mathscr{B}$ is a $C^*$-subalgebra of a $C^*$-algebra $\mathfrak{A}$ and $A$ is a normal element of $\mathscr{B}$. We can consider two function calculi for $A$,

$$f \to f_{\mathfrak{A}}(A): \quad C(\mathrm{sp}_{\mathfrak{A}}(A)) \to \mathfrak{A}$$

relative to $\mathfrak{A}$, and

$$f \to f_{\mathscr{B}}(A): \quad C(\mathrm{sp}_{\mathscr{B}}(A)) \to \mathscr{B},$$

relative to $\mathscr{B}$. Since $\mathrm{sp}_{\mathfrak{A}}(A) = \mathrm{sp}_{\mathscr{B}}(A)$ and $\mathscr{B} \subseteq \mathfrak{A}$, both function calculi can be considered as mappings from $C(\mathrm{sp}(A))$ into $\mathfrak{A}$, and it is clear from the uniqueness clause in Theorem 4.4.5 that they coincide. In this sense, the function calculus for a normal element is independent of the containing $C^*$-algebra. If $A$ is self-adjoint, the function calculus described in Theorem 4.4.5 coincides with the one considered in Theorem 4.1.3, by the uniqueness clause in either of those theorems. ■

Suppose that $A$ is a normal element of a $C^*$-algebra $\mathfrak{A}$ and $f$ is a continuous complex-valued function whose domain of definition includes $\mathrm{sp}(A)$. Just as in the case of self-adjoint elements, we denote by $f(A)$ the element of $\mathfrak{A}$ that, in the function calculus for $A$, corresponds to the restriction $f|\mathrm{sp}(A)$.

4.4.7. Proposition. *If $\mathfrak{A}$ and $\mathscr{B}$ are $C^*$-algebras, $\varphi$ is a * homomorphism from $\mathfrak{A}$ into $\mathscr{B}$, $A$ is a normal element of $\mathfrak{A}$, and $f \in C(\mathrm{sp}(A))$, then $\varphi(A)$ is a normal element of $\mathscr{B}$, $\mathrm{sp}(\varphi(A)) \subseteq \mathrm{sp}(A)$, and $f(\varphi(A)) = \varphi(f(A))$.*

*Proof.* Since

$$\varphi(A)\varphi(A)^* - \varphi(A)^*\varphi(A) = \varphi(AA^* - A^*A) = 0,$$

$\varphi(A)$ is normal. By Theorem 4.1.8(i), $\mathrm{sp}(\varphi(A)) \subseteq \mathrm{sp}(A)$, and the mappings

$$f \to f(\varphi(A)), \qquad f \to \varphi(f(A)): \quad C(\mathrm{sp}(A)) \to \mathscr{B},$$

being * homomorphisms, are both continuous. For $m, n = 0, 1, 2, \ldots,$

$$\varphi(A^m(A^*)^n) = \varphi(A)^m[\varphi(A)^*]^n,$$

so $\varphi(p(A)) = p(\varphi(A))$ whenever $p$ (in $C(\mathrm{sp}(A))$) has the form

$$p(t) = \sum_{j=0}^{m} \sum_{k=0}^{n} a_{jk} t^j (\bar{t})^k.$$

Accordingly, the mapping

$$f \to \|f(\varphi(A)) - \varphi(f(A))\|: \quad C(\mathrm{sp}(A)) \to \mathbb{R}$$

is continuous throughout $C(\mathrm{sp}(A))$, takes the value zero on the everywhere-dense subset consisting of polynomials $p$ of the type just described, and so vanishes throughout $C(\mathrm{sp}(A))$. ■

4.4.8. Theorem. *If $A$ is a normal element of a $C^*$-algebra $\mathfrak{A}$ and $f \in C(\mathrm{sp}(A))$, then*

$$\mathrm{sp}(f(A)) = \{f(t) : t \in \mathrm{sp}(A)\}.$$

*If $g \in C(\mathrm{sp}(f(A)))$, the composite function $g \circ f$ lies in $C(\mathrm{sp}(A))$, and $(g \circ f)(A) = g(f(A))$, where $g(f(A))$ denotes the element of $\mathfrak{A}$ that corresponds to $g$ in the function calculus for the normal element $f(A)$.*

*Proof.* Since the mapping $f \to f(A)$ is a * isomorphism from $C(\mathrm{sp}(A))$ onto a $C^*$-subalgebra $\mathscr{B}$ of $\mathfrak{A}$, we have

$$\mathrm{sp}(f(A)) = \mathrm{sp}_{\mathscr{B}}(f(A)) = \mathrm{sp}_{C(\mathrm{sp}(A))}(f) = \{f(t) : t \in \mathrm{sp}(A)\}.$$

If $g \in C(\mathrm{sp}(f(A)))$, $g$ is continuous on the range of $f$, so $g \circ f$ is continuous throughout $\mathrm{sp}(A)$. The mappings

$$g \to g \circ f\colon \quad C(\mathrm{sp}(f(A))) \to C(\mathrm{sp}(A)),$$
$$h \to h(A)\colon \quad C(\mathrm{sp}(A)) \to \mathfrak{A}$$

are * isomorphisms, and hence so is

$$\psi\colon g \to (g \circ f)(A)\colon \quad C(\mathrm{sp}(f(A))) \to \mathfrak{A}.$$

With $\iota$ the identity mapping on $\mathrm{sp}(f(A))$, $\iota \circ f = f$, so $\psi(\iota) = f(A)$. From the uniqueness clause in Theorem 4.4.5, $\psi$ coincides with the function calculus $g \to g(f(A))$ for $f(A)$; that is,

$$g(f(A)) = \psi(g) = (g \circ f)(A),$$

for each $g$ in $C(\mathrm{sp}(f(A)))$. ∎

4.4.9. Example. If $X$ is a compact Hausdorff space and $\mathfrak{A}$ is the abelian $C^*$-algebra $C(X)$, each $g$ in $\mathfrak{A}$ is normal. We assert that the function calculus for $g$ is given by

$$f(g) = f \circ g \qquad (f \in C(\mathrm{sp}(g))).$$

For this, note first that, since $\mathrm{sp}(g) = \{g(x) : x \in X\}$, the composite function $f \circ g$ is continuous throughout $X$, when $f \in C(\mathrm{sp}(g))$. Accordingly, the mapping

$$\psi : f \to f \circ g$$

is a * isomorphism from $C(\mathrm{sp}(g))$ into $C(X)\,(= \mathfrak{A})$. With $\iota$ the identity mapping on $\mathrm{sp}(g)$, $\iota \circ g$ is $g$, and thus $\psi(\iota) = g$. From the uniqueness clause in Theorem 4.4.5, $\psi$ is the function calculus for $g$; that is, $f(g) = f \circ g$ for each $f$ in $C(\mathrm{sp}(g))$. ∎

With $\mathfrak{A}$ a $C^*$-algebra and $H$ self-adjoint in $\mathfrak{A}$, the equation $f(t) = \exp it$ defines a continuous function $f$ on $\mathrm{sp}(H)$, and we denote the corresponding element $f(H)$ of $\mathfrak{A}$ by $\exp iH$. Since

$$f(t)\overline{f(t)} = \overline{f(t)}f(t) = 1 \qquad (t \in \mathrm{sp}(H)),$$

we have $f(H)f(H)^* = f(H)^*f(H) = I$, so $\exp iH$ is unitary. Since the series $\sum (it)^n/n!$ converges to $f(t)$, uniformly on $\mathrm{sp}(H)$, it follows by considering its partial sums that

$$\exp iH = \sum_{n=0}^{\infty} \frac{(iH)^n}{n!}.$$

In some $C^*$-algebras $\mathfrak{A}$, every unitary element has the form $\exp iH$, with $H$ self-adjoint in $\mathfrak{A}$ (see, for example, Theorem 5.2.5). In a general $C^*$-algebra, there may be unitaries not of this form (see Exercise 4.6.5), but we have the following result.

4.4.10. PROPOSITION. *If $U$ is a unitary element in a $C^*$-algebra $\mathfrak{A}$, and* $\text{sp}(U)$ *is not the whole unit circle,* $U = \exp iH$ *for some self-adjoint $H$ in $\mathfrak{A}$.*

*Proof.* Since $\text{sp}(U)$ is a proper subset of the unit circle, there is a real number $a$ such that

$$\text{sp}(U) \subseteq \{\exp is : a < s < a + 2\pi\}.$$

In order to use the function calculus for $U$, we define a continuous function $f$ on $\text{sp}(U)$ by

$$f(\exp is) = s \qquad (a < s < a + 2\pi).$$

Since $f$ is real valued on $\text{sp}(U)$, and

$$\exp if(t) = t \qquad (t \in \text{sp}(U)),$$

$f(U)$ is a self-adjoint element $H$ of $\mathfrak{A}$; and

$$\exp iH = \exp if(U) = U,$$

by Theorem 4.4.8. ∎

4.4.11. REMARK. If $A$ and $B$ are normal operators whose spectra are contained in the domain of a continuous function $g$, and if $g$ has a continuous inverse function $f$, then $g(A) = g(B)$ if and only if $A = B$; for if $g(A) = g(B)$, then, from Theorem 4.4.8,

$$A = (f \circ g)(A) = f(g(A)) = f(g(B)) = (f \circ g)(B) = B.$$

As an application of this comment, we note that if $A^n = B^n$ with $A$ and $B$ positive operators and $n$ a positive integer, then $A = B$, so that a positive operator has a *unique* positive $n$th root. ∎

*Bibliography*: [4]

## 4.5. States and representations

By a *representation* of a $C^*$-algebra $\mathfrak{A}$ on a Hilbert space $\mathscr{H}$, we mean a * homomorphism $\varphi$ from $\mathfrak{A}$ into $\mathscr{B}(\mathscr{H})$. If, in addition, $\varphi$ is one-to-one (hence, a * isomorphism), it is described as a *faithful* representation. Our main purpose in this section, achieved in Theorem 4.5.6 (the Gelfand–Neumark theorem) is to show that every $C^*$-algebra has a faithful representation on some Hilbert space.

Suppose that $\varphi$ is a representation of a $C^*$-algebra $\mathfrak{A}$ on a Hilbert space $\mathscr{H}$. In view of our convention that * homomorphisms preserve units, and from Theorem 4.1.8, $\varphi(I) = I$, $\|\varphi(A)\| \leqslant \|A\|$ for each $A$ in $\mathfrak{A}$ (whence $\varphi$ is continuous), and $\|\varphi(A)\| = \|A\|$ if $\varphi$ is faithful. The set $\{A \in \mathfrak{A}: \varphi(A) = 0\}$ is a closed two-sided ideal in $\mathfrak{A}$, the *kernel* of $\varphi$. If there is a vector $x$ in $\mathscr{H}$ for which the linear subspace

$$\varphi(\mathfrak{A})x = \{\varphi(A)x : A \in \mathfrak{A}\}$$

is everywhere dense in $\mathscr{H}$, $\varphi$ is described as a *cyclic* representation, and $x$ is termed a *cyclic vector* (or *generating vector*) for $\varphi$. It turns out that there is an intimate connection between states of $\mathfrak{A}$ and cyclic representations; and the proof of the existence of a faithful representation depends on the construction, from states, of an abundance of cyclic representations.

We give a number of examples to illustrate the concepts just introduced. With $\mathscr{H}$ a Hilbert space and $\mathfrak{A}$ a $C^*$-subalgebra of $\mathscr{B}(\mathscr{H})$, the inclusion mapping from $\mathfrak{A}$ into $\mathscr{B}(\mathscr{H})$ is a faithful representation of $\mathfrak{A}$ on $\mathscr{H}$. Suppose that $\mathscr{K}$ is a closed subspace of $\mathscr{H}$, and is invariant under each operator in $\mathfrak{A}$. When $A \in \mathfrak{A}$, the restriction $A|\mathscr{K}$ can be viewed as a bounded linear operator on $\mathscr{K}$, and coincides with the compression of $A$ to $\mathscr{K}$, as described in Section 2.6. Since compression is an adjoint preserving process, it is easily verified that the mapping $A \to A|\mathscr{K} : \mathfrak{A} \to \mathscr{B}(\mathscr{K})$ is a * homomorphism, and is therefore a representation of $\mathfrak{A}$ on $\mathscr{K}$. With $A$ in $\mathfrak{A}$, $\mathscr{K}$ is invariant under both $A$ and $A^*$, and so reduces $A$; equivalently, $A$ commutes with the projection $E$ from $\mathscr{H}$ onto $\mathscr{K}$. Accordingly, the orthogonal complement $\mathscr{K}^\perp$ is invariant under each operator in $\mathfrak{A}$, and gives rise to a representation $A \to A|\mathscr{K}^\perp$ of $\mathfrak{A}$ on $\mathscr{K}^\perp$.

When $\mathfrak{X} \subseteq \mathscr{H}$, $\mathscr{F} \subseteq \mathscr{B}(\mathscr{H})$, and $x \in \mathscr{H}$, as in Section 1.2 we denote by $[\mathfrak{X}]$ the closed subspace of $\mathscr{H}$ generated by $\mathfrak{X}$; we write $\mathscr{F}\mathfrak{X}$ for the set $\{Ax: A \in \mathscr{F}, x \in \mathfrak{X}\}$, and define $\mathscr{F}x$ to be $\mathscr{F}\{x\}$. Since $\mathfrak{A}\mathfrak{X}$ is invariant under each operator in $\mathfrak{A}$, the same is true of the closed subspace $[\mathfrak{A}\mathfrak{X}]$, so the mapping $A \to A|[\mathfrak{A}\mathfrak{X}]$ is a representation of $\mathfrak{A}$ on $[\mathfrak{A}\mathfrak{X}]$. The representation $A \to A|[\mathfrak{A}x]$ of $\mathfrak{A}$ on $[\mathfrak{A}x]$ is cyclic, having $x$ as a cyclic vector.

When $\mathfrak{A}$ is a $C^*$-subalgebra of $\mathscr{B}(\mathscr{H})$, it is evident (as noted above) that $\mathfrak{A}$ has a faithful representation on $\mathscr{H}$, the inclusion mapping from $\mathfrak{A}$ into $\mathscr{B}(\mathscr{H})$. We shall see in Chapter 10 that it is nevertheless important, even in this case, to study *other* representations of $\mathfrak{A}$, on *different* Hilbert spaces. For the present, however, our main concern is to construct representations of a $C^*$-algebra $\mathfrak{A}$ that is not at the outset presented as a self-adjoint algebra of Hilbert space operators. As an example of this type, consider the $C^*$-algebra $L_\infty$ of all essentially bounded complex-valued measurable functions on a $\sigma$-finite measure space $(S, \mathscr{S}, m)$, with pointwise algebraic structure, complex conjugation as involution, and the essential supremum norm. Each $f$ in $L_\infty$ gives rise to a multiplication operator $M_f$ acting on the Hilbert space $L_2$, and it is apparent from the discussion in Example 2.4.11 that the mapping $f \to M_f$ is a faithful representation of $L_\infty$ on $L_2$. When the measure is Lebesgue measure on a compact interval $X$ ($\subseteq \mathbb{R}$), $C(X)$ is a $C^*$-algebra, and the mapping

$$f \to M_f\colon \quad C(X) \to \mathscr{B}(L_2)$$

is a faithful representation of $C(X)$ on $L_2$.

Suppose that $\varphi$ is a representation of a $C^*$-algebra $\mathfrak{A}$ on a Hilbert space $\mathscr{H}$ and $x$ is a unit vector in $\mathscr{H}$. With $\omega_x$ the corresponding vector state of $\mathscr{B}(\mathscr{H})$, the composite function $\omega_x \circ \varphi$ is a state $\rho$ of $\mathfrak{A}$. Indeed, since $\varphi(I) = I$, $\varphi(\mathfrak{A}^+) \subseteq \mathscr{B}(\mathscr{H})^+$, and

$$\rho(A) = \omega_x(\varphi(A)) = \langle \varphi(A)x, x\rangle \qquad (A \in \mathfrak{A}),$$

it is evident that $\rho$ is a positive linear functional on $\mathfrak{A}$, and $\rho(I) = \omega_x(I) = 1$. We prove, in Theorem 4.5.2, that each state of a $C^*$-algebra arises in this way, from a vector state in an appropriate representation. We first need an auxiliary result.

4.5.1. PROPOSITION. *If $\rho$ is a state of a $C^*$-algebra $\mathfrak{A}$, the set*

$$\mathscr{L}_\rho = \{A \in \mathfrak{A} : \rho(A^*A) = 0\}$$

*is a closed left ideal in $\mathfrak{A}$, and $\rho(B^*A) = 0$ whenever $A \in \mathscr{L}_\rho$ and $B \in \mathfrak{A}$. The equation*

$$\langle A + \mathscr{L}_\rho, B + \mathscr{L}_\rho\rangle = \rho(B^*A) \qquad (A, B \in \mathfrak{A})$$

*defines a definite inner product $\langle\ ,\ \rangle$ on the quotient linear space $\mathfrak{A}/\mathscr{L}_\rho$.*

*Proof.* Since $\rho$ is positive (and hence, also, hermitian), we can define an inner product $\langle\ ,\ \rangle_0$ on $\mathfrak{A}$ by

$$\langle A, B\rangle_0 = \rho(B^*A) \qquad (A, B \in \mathfrak{A});$$

and

$$\mathscr{L}_\rho = \{A \in \mathfrak{A} : \langle A, A\rangle_0 = 0\}.$$

From Proposition 2.1.1(ii), $\mathscr{L}_\rho$ is a linear subspace of $\mathfrak{A}$, and the equation

$$\langle A + \mathscr{L}_\rho, B + \mathscr{L}_\rho\rangle = \langle A, B\rangle_0 = \rho(B^*A)$$

defines a definite inner product on $\mathfrak{A}/\mathscr{L}_\rho$. If $A \in \mathscr{L}_\rho$ and $B \in \mathfrak{A}$,

$$|\rho(B^*A)|^2 \leqslant \rho(B^*B)\rho(A^*A) = 0,$$

so $\rho(B^*A) = 0$. Upon replacing $B$ by $B^*BA$, it follows that

$$\rho((BA)^*BA) = \rho((B^*BA)^*A) = 0, \qquad BA \in \mathscr{L}_\rho,$$

whenever $A \in \mathscr{L}_\rho$ and $B \in \mathfrak{A}$. Hence $\mathscr{L}_\rho$ is a left ideal of $\mathfrak{A}$, and is closed since $\rho$ is continuous. ∎

We refer to $\mathscr{L}_\rho$ as the *left kernel* of the state $\rho$.

4.5.2. THEOREM. *If $\rho$ is a state of a $C^*$-algebra $\mathfrak{A}$, there is a cyclic representation $\pi_\rho$ of $\mathfrak{A}$ on a Hilbert space $\mathscr{H}_\rho$, and a unit cyclic vector $x_\rho$ for $\pi_\rho$, such that $\rho = \omega_{x_\rho} \circ \pi_\rho$; that is,*

$$\rho(A) = \langle \pi_\rho(A)x_\rho, x_\rho \rangle \qquad (A \in \mathfrak{A}).$$

*Proof.* With $\mathscr{L}_\rho$ the left kernel of $\rho$, the quotient linear space $\mathfrak{A}/\mathscr{L}_\rho$ is a pre-Hilbert space relative to the definite inner product defined, as in Proposition 4.5.1, by

$$\langle A + \mathscr{L}_\rho, B + \mathscr{L}_\rho \rangle = \rho(B^*A) \qquad (A, B \in \mathfrak{A}).$$

Its completion is a Hilbert space $\mathscr{H}_\rho$.

If $A, B_1, B_2 \in \mathfrak{A}$, and $B_1 + \mathscr{L}_\rho = B_2 + \mathscr{L}_\rho$, then $B_1 - B_2 \in \mathscr{L}_\rho$, $AB_1 - AB_2 \in \mathscr{L}_\rho$ since $\mathscr{L}_\rho$ is a left ideal in $\mathfrak{A}$, and therefore $AB_1 + \mathscr{L}_\rho = AB_2 + \mathscr{L}_\rho$. Accordingly, the equation $\pi(A)(B + \mathscr{L}_\rho) = AB + \mathscr{L}_\rho$ defines, unambiguously, a linear operator $\pi(A)$ acting on the pre-Hilbert space $\mathfrak{A}/\mathscr{L}_\rho$. Now

$$\|A\|^2 I - A^*A = \|A^*A\|I - A^*A \in \mathfrak{A}^+;$$

hence $B^*(\|A\|^2 I - A^*A)B \in \mathfrak{A}^+$, and therefore

$$\begin{aligned} &\|A\|^2\|B + \mathscr{L}_\rho\|^2 - \|\pi(A)(B + \mathscr{L}_\rho)\|^2 \\ &\quad = \|A\|^2\|B + \mathscr{L}_\rho\|^2 - \|AB + \mathscr{L}_\rho\|^2 \\ &\quad = \|A\|^2\langle B + \mathscr{L}_\rho, B + \mathscr{L}_\rho\rangle - \langle AB + \mathscr{L}_\rho, AB + \mathscr{L}_\rho\rangle \\ &\quad = \|A\|^2\rho(B^*B) - \rho(B^*A^*AB) \\ &\quad = \rho(B^*(\|A\|^2 I - A^*A)B) \geqslant 0, \end{aligned}$$

for all $A$ and $B$ in $\mathfrak{A}$. Thus $\pi(A)$ is bounded, with $\|\pi(A)\| \leqslant \|A\|$; and $\pi(A)$ extends by continuity to a bounded linear operator $\pi_\rho(A)$ acting on $\mathscr{H}_\rho$.

Since $\pi(I)$ is the identity operator on $\mathfrak{A}/\mathscr{L}_\rho$, $\pi_\rho(I)$ is the identity operator on $\mathscr{H}_\rho$. When $A$, $B$, $C \in \mathfrak{A}$ and $a, b \in \mathbb{C}$,

$$\begin{aligned} \pi_\rho(aA + bB)(C + \mathscr{L}_\rho) &= (aA + bB)C + \mathscr{L}_\rho \\ &= a(AC + \mathscr{L}_\rho) + b(BC + \mathscr{L}_\rho) \\ &= (a\pi_\rho(A) + b\pi_\rho(B))(C + \mathscr{L}_\rho), \\ \pi_\rho(AB)(C + \mathscr{L}_\rho) &= ABC + \mathscr{L}_\rho = \pi_\rho(A)(BC + \mathscr{L}_\rho) \\ &= \pi_\rho(A)\pi_\rho(B)(C + \mathscr{L}_\rho), \\ \langle \pi_\rho(A)(B + \mathscr{L}_\rho), C + \mathscr{L}_\rho\rangle &= \langle AB + \mathscr{L}_\rho, C + \mathscr{L}_\rho\rangle = \rho(C^*AB) \\ &= \rho((A^*C)^*B) = \langle B + \mathscr{L}_\rho, A^*C + \mathscr{L}_\rho\rangle \\ &= \langle B + \mathscr{L}_\rho, \pi_\rho(A^*)(C + \mathscr{L}_\rho)\rangle. \end{aligned}$$

From these relations, and since $\mathfrak{A}/\mathscr{L}_\rho$ is everywhere dense in $\mathscr{H}_\rho$, it follows that

$$\pi_\rho(aA + bB) = a\pi_\rho(A) + b\pi_\rho(B),$$
$$\pi_\rho(AB) = \pi_\rho(A)\pi_\rho(B),$$
$$\pi_\rho(A)^* = \pi_\rho(A^*).$$

Accordingly, $\pi_\rho$ is a representation of $\mathfrak{A}$ on $\mathscr{H}_\rho$.

With $x_\rho$ the vector $I + \mathscr{L}_\rho$ in $\mathfrak{A}/\mathscr{L}_\rho$,

$$\pi_\rho(A)x_\rho = \pi_\rho(A)(I + \mathscr{L}_\rho) = A + \mathscr{L}_\rho \qquad (A \in \mathfrak{A}).$$

Hence $\pi_\rho(\mathfrak{A})x_\rho$ is the everywhere-dense subset $\mathfrak{A}/\mathscr{L}_\rho$ of $\mathscr{H}_\rho$, and $x_\rho$ is a cyclic vector for $\pi_\rho$. Moreover,

$$\langle \pi_\rho(A)x_\rho, x_\rho \rangle = \langle A + \mathscr{L}_\rho, I + \mathscr{L}_\rho \rangle = \rho(A) \qquad (A \in \mathfrak{A});$$

in particular, $\|x_\rho\|^2 = \rho(I) = 1$. ■

The method used to produce a representation from a state, in the proof of Theorem 4.5.2, is called the *Gelfand–Neumark–Segal construction*, or GNS construction, and provides one of the basic tools of $C^*$-algebra theory. The associated notation will be used frequently, sometimes without comment; when $\rho$ is a state of a $C^*$-algebra $\mathfrak{A}$, the symbols $\mathscr{H}_\rho$, $\pi_\rho$, and $x_\rho$ always bear the meaning attached to them in the theorem. In applications, the properties of $\mathscr{H}_\rho$, $\pi_\rho$, and $x_\rho$, set out in the theorem, are more important than the details of the construction used to produce them. In a sense made precise in the following proposition, the Hilbert space $\mathscr{H}_\rho$, the cyclic representation $\pi_\rho$, and the unit cyclic vector $x_\rho$ are (essentially) uniquely determined by the condition $\rho = \omega_{x_\rho} \circ \pi_\rho$.

4.5.3. Proposition. *Suppose that $\rho$ is a state of a $C^*$-algebra $\mathfrak{A}$ and $\pi$ is a cyclic representation of $\mathfrak{A}$ on a Hilbert space $\mathscr{H}$ such that $\rho = \omega_x \circ \pi$ for some unit cyclic vector $x$ for $\pi$. If $\mathscr{H}_\rho$, $\pi_\rho$, and $x_\rho$ are the Hilbert space, cyclic representation, and unit cyclic vector produced from $\rho$ by the GNS construction, there is an isomorphism $U$ from $\mathscr{H}_\rho$ onto $\mathscr{H}$ such that*

$$x = Ux_\rho, \qquad \pi(A) = U\pi_\rho(A)U^* \qquad (A \in \mathfrak{A}).$$

*Proof.* For each $A$ in $\mathfrak{A}$,

$$\|\pi(A)x\|^2 = \langle \pi(A)x, \pi(A)x \rangle = \langle \pi(A^*A)x, x \rangle$$
$$= \rho(A^*A) = \langle \pi_\rho(A^*A)x_\rho, x_\rho \rangle = \|\pi_\rho(A)x_\rho\|^2.$$

If $A, B \in \mathfrak{A}$ and $\pi_\rho(A)x_\rho = \pi_\rho(B)x_\rho$, it follows from the above equations (with $A - B$ in place of $A$) that $\pi(A)x = \pi(B)x$. Accordingly, the equation $U_0\pi_\rho(A)x_\rho = \pi(A)x$ $(A \in \mathfrak{A})$ defines a norm-preserving linear operator from $\pi_\rho(\mathfrak{A})x_\rho$ onto $\pi(\mathfrak{A})x$. Since $[\pi_\rho(\mathfrak{A})x_\rho] = \mathscr{H}_\rho$ and $[\pi(\mathfrak{A})x] = \mathscr{H}$, $U_0$ extends by continuity to an isomorphism $U$ from $\mathscr{H}_\rho$ onto $\mathscr{H}$, and

$$Ux_\rho = U_0\pi_\rho(I)x_\rho = \pi(I)x = x.$$

With $A$ and $B$ in $\mathfrak{A}$,

$$\begin{aligned} U\pi_\rho(A)\pi_\rho(B)x_\rho &= U\pi_\rho(AB)x_\rho \\ &= \pi(AB)x \\ &= \pi(A)\pi(B)x = \pi(A)U\pi_\rho(B)x_\rho. \end{aligned}$$

Since vectors of the form $\pi_\rho(B)x_\rho$ ($B \in \mathfrak{A}$) form an everywhere-dense subset of $\mathscr{H}_\rho$, it follows that $U\pi_\rho(A) = \pi(A)U$, and thus $\pi(A) = U\pi_\rho(A)U^*$. ■

Suppose that $\mathfrak{A}$ is a $C^*$-algebra and that $\varphi$ and $\psi$ are representations of $\mathfrak{A}$ on Hilbert spaces $\mathscr{H}$ and $\mathscr{K}$, respectively. We say that $\varphi$ and $\psi$ are (*unitarily*) *equivalent* if there is an isomorphism $U$ from $\mathscr{H}$ onto $\mathscr{K}$ such that $\psi(A) = U\varphi(A)U^*$ for each $A$ in $\mathfrak{A}$. If $\rho$ is a state of $\mathfrak{A}$, $\pi$ is a cyclic representation of $\mathfrak{A}$, and $\rho = \omega_x \circ \pi$ for some unit cyclic vector $x$ for $\pi$, it follows from Proposition 4.5.3 that $\pi$ is equivalent to the representation $\pi_\rho$ obtained from $\rho$ by the GNS construction. In addition, the isomorphism $U$ can be chosen so that $Ux_\rho = x$.

4.5.4. COROLLARY. *If $x$ is a unit vector in a Hilbert space $\mathscr{H}$, $\mathfrak{A}$ is a $C^*$-subalgebra of $\mathscr{B}(\mathscr{H})$, and $\rho$ is the vector state $\omega_x|\mathfrak{A}$, the representation $\pi_\rho$ obtained from $\rho$ by the GNS construction is equivalent to the representation $A \to A|[\mathfrak{A}x]$ of $\mathfrak{A}$ on the Hilbert space $[\mathfrak{A}x]$. The isomorphism $U: \mathscr{H}_\rho \to [\mathfrak{A}x]$ that implements this equivalence can be chosen so that $Ux_\rho = x$.*

*Proof.* This follows from Proposition 4.5.3, since $x$ is a unit cyclic vector for the representation $\pi: A \to A|[\mathfrak{A}x]$, and $\rho = \omega_x \circ \pi$. ■

We prove next that the set of all representations of a $C^*$-algebra $\mathfrak{A}$, obtained from (pure) states of $\mathfrak{A}$ by the GNS construction, is large enough to "separate" the elements of $\mathfrak{A}$.

4.5.5. PROPOSITION. *If $A$ is a non-zero element of a $C^*$-algebra $\mathfrak{A}$, there is a pure state $\rho$ of $\mathfrak{A}$ such that $\pi_\rho(A) \neq 0$, where $\pi_\rho$ is the representation obtained from $\rho$ by the GNS construction.*

*Proof.* By Theorem 4.3.8(i) there is a pure state $\rho$ of $\mathfrak{A}$ such that $\rho(A) \neq 0$, equivalently $\langle \pi_\rho(A)x_\rho, x_\rho \rangle \neq 0$, whence $\pi_\rho(A) \neq 0$. ■

In order to complete the proof that every $C^*$-algebra has a faithful representation, we need the concept of a "direct sum" of representations. Suppose that $\mathfrak{A}$ is a $C^*$-algebra, $(\mathscr{H}_b)_{b \in \mathbb{B}}$ is a family of Hilbert spaces, and $\varphi_b$ is a representation of $\mathfrak{A}$ on $\mathscr{H}_b$ for each $b$ in $\mathbb{B}$. When $A \in \mathfrak{A}$, $\|\varphi_b(A)\| \leqslant \|A\|$

($b \in \mathbb{B}$), so the direct sum $\sum_{b \in \mathbb{B}} \oplus \varphi_b(A)$ is a bounded linear operator acting on the Hilbert space $\sum_{b \in \mathbb{B}} \oplus \mathscr{H}_b$. From the results set out at the end of the subsection on *direct sums*, in Section 2.6, it is apparent that the mapping

$$\varphi: A \to \sum \oplus \varphi_b(A)$$

is a representation of $\mathfrak{A}$ on $\sum \oplus \mathscr{H}_b$. We call $\varphi$ the *direct sum* of the family $(\varphi_b)_{b \in \mathbb{B}}$ of representations of $\mathfrak{A}$, and write $\varphi = \sum \oplus \varphi_b$.

4.5.6. THEOREM (The Gelfand–Neumark theorem). *Each $C^*$-algebra has a faithful representation.*

*Proof.* With $\mathfrak{A}$ a $C^*$-algebra, let $\mathscr{S}_0$ be any family of states of $\mathfrak{A}$ that contains all the pure states. Let $\varphi$ be the direct sum of the family $\{\pi_\rho: \rho \in \mathscr{S}_0\}$, where $\pi_\rho$ is the representation obtained from $\rho$ by the GNS construction. If $A \in \mathfrak{A}$ and $\varphi(A) = 0$, then $\pi_\rho(A) = 0$ ($\rho \in \mathscr{S}_0$) since $\varphi(A)$ is $\sum \oplus \pi_\rho(A)$; in particular, $\pi_\rho(A) = 0$ for each pure state $\rho$ of $\mathfrak{A}$, and $A = 0$ by Proposition 4.5.5. Hence $\varphi$ is a faithful representation of $\mathfrak{A}$. ∎

4.5.7. REMARK. If $\varphi$ is a faithful representation of a $C^*$-algebra $\mathfrak{A}$ on a Hilbert space $\mathscr{H}$, then $\varphi$ is isometric and $\varphi(\mathfrak{A})$ is a $C^*$-subalgebra of $\mathscr{B}(\mathscr{H})$, by Theorem 4.1.8(iii). Accordingly, the Gelfand–Neumark theorem can be restated as follows: if $\mathfrak{A}$ is a $C^*$-algebra, there is a Hilbert space $\mathscr{H}$ such that $\mathfrak{A}$ is * isomorphic to a $C^*$-subalgebra of $\mathscr{B}(\mathscr{H})$. ∎

4.5.8. REMARK. Suppose that $\mathfrak{A}$ is a $C^*$-algebra with state space $\mathscr{S}$ and that $\mathscr{P}$ is the set of pure states of $\mathfrak{A}$. In proving the Gelfand–Neumark theorem, we showed that the representation

$$\sum_{\rho \in \mathscr{S}_0} \oplus \pi_\rho$$

of $\mathfrak{A}$ is faithful whenever $\mathscr{P} \subseteq \mathscr{S}_0 \subseteq \mathscr{S}$. When $\mathscr{S}_0 = \mathscr{S}$, we obtain a faithful representation

$$\Phi = \sum_{\rho \in \mathscr{S}} \oplus \pi_\rho,$$

the *universal representation* of $\mathfrak{A}$, which will be studied in more detail in Section 10.1. With $\tau$ a state of $\mathfrak{A}$, $\tau = \omega_x \circ \pi_\tau$ for a suitable unit vector $x$ $(= x_\tau)$ in $\mathscr{H}_\tau$; thus $\tau = \omega_y \circ \Phi$, where $y$ is the vector $\sum_{\rho \in \mathscr{S}} \oplus y_\rho$, in the Hilbert space $\mathscr{H}_\Phi = \sum_{\rho \in \mathscr{S}} \oplus \mathscr{H}_\rho$, defined by $y_\tau = x$, $y_\rho = 0$ $(\rho \neq \tau)$. Accordingly, *each state of $\mathfrak{A}$ has the form $\omega_y \circ \Phi$, with $y$ a unit vector in $\mathscr{H}_\Phi$.* Since the mapping $\tau \to \tau \circ \Phi^{-1}$ carries the state space of $\mathfrak{A}$ onto the state space of the $C^*$-subalgebra $\Phi(\mathfrak{A})$ of $\mathscr{B}(\mathscr{H}_\Phi)$, it now follows that *each state of $\Phi(\mathfrak{A})$ is a vector state.* This last is the most basic fact about the universal representation, from which its other properties will be deduced in Section 10.1.

When $\mathscr{S}_0 = \mathscr{P}$, we obtain another faithful representation,

$$\psi = \sum_{\rho \in \mathscr{P}} \oplus \pi_\rho$$

of $\mathfrak{A}$. While this representation has useful properties, it is more convenient to work with a "reduced" form of $\psi$, known as the *reduced atomic representation*, which will be studied further in Section 10.3. ■

In Section 3.2, we studied *characters* of $\mathbb{R}$ and their relation to the operator algebras $\mathscr{A}_1(\mathbb{R})$ and $\mathfrak{A}_0(\mathbb{R})$. A character $\xi$ of $\mathbb{R}$ may be viewed as a (continuous) homomorphism of $\mathbb{R}$ into the unitary group of the one-dimensional Hilbert space $\mathbb{C}$, where $t$ corresponds to multiplication by $\xi(t)$ on $\mathbb{C}$. There are more general homomorphisms of $\mathbb{R}$ into the group $\mathscr{U}(\mathscr{H})$ of unitary operators on a Hilbert space $\mathscr{H}$. If $\mathscr{U}(\mathscr{H})$ is provided with the strong-operator topology and the homomorphism of $\mathbb{R}$ into $\mathscr{U}(\mathscr{H})$ is continuous, we refer to the homomorphism as a *one-parameter unitary group* (also, as a *unitary representation* of $\mathbb{R}$). The complete analysis of one-parameter unitary groups (Theorem 5.6.36, Stone's theorem) must await our development of the spectral theory of unbounded self-adjoint operators in Section 5.6. In the present discussion, we shall relate one-parameter unitary groups to certain representations of $\mathfrak{A}_0(\mathbb{R})$. For this purpose, it is convenient to extend our concept of "representation on $\mathscr{H}$" to self-adjoint algebras of operators that may not be norm closed and may not contain $I$ (as, for example, $\mathscr{A}_1(\mathbb{R})$). Again we require that our representation be a * homomorphism into $\mathscr{B}(\mathscr{H})$. Those representations for which the union of the range projections of operators in the image is $I$ are said to be *essential*. When the algebra does not contain $I$, the assumption that a representation is essential replaces the requirement that the image of $I$ is the identity operator.

4.5.9. THEOREM. *If $t \to U_t$ is a one-parameter unitary group on $\mathscr{H}$ there is a representation $\varphi$ of $\mathfrak{A}_0(\mathbb{R})$ on $\mathscr{H}$ such that*

$$\langle \varphi(L_f)x, y\rangle = \int f(t)\langle U_t x, y\rangle\, dt \tag{1}$$

*for each $f$ in $L_1(\mathbb{R})$ and $x, y$ in $\mathscr{H}$; and $\varphi$ restricts to an essential representation of $\mathscr{A}_1(\mathbb{R})$. If $\varphi$ is an essential representation of $\mathscr{A}_1(\mathbb{R})$ on $\mathscr{H}$, there is a one-parameter unitary group $t \to U_t$ on $\mathscr{H}$ such that* (1) *is satisfied—in particular, $\varphi$ extends to a representation of $\mathfrak{A}_0(\mathbb{R})$.*

*Proof.* The integral in (1) defines a conjugate-bilinear functional on $\mathscr{H}$ for each $f$ in $L_1(\mathbb{R})$. As

$$\left|\int f(t)\langle U_t x, y\rangle\, dt\right| \leq \|f\|_1\, \|x\|\, \|y\|,$$

from Theorem 2.4.1 there is an operator $\varphi(L_f)$ on $\mathscr{H}$ such that $\|\varphi(L_f)\| \leqslant \|f\|_1$ and (1) holds. The mapping $\varphi$ of $\mathscr{A}_1(\mathbb{R})$ into $\mathscr{B}(\mathscr{H})$ is linear. Since

$$\begin{aligned}
\langle \varphi(L_{f*g})x, y\rangle &= \int (f*g)(t)\langle U_t x, y\rangle\, dt \\
&= \int\left(\int f(s)g(t-s)\, ds\right)\langle U_t x, y\rangle\, dt \\
&= \int f(s)\left(\int g(t-s)\langle U_t x, y\rangle\, dt\right) ds \\
&= \int f(s)\left(\int g(t)\langle U_t x, U_{-s} y\rangle\, dt\right) ds \\
&= \int f(s)\langle \varphi(L_g)x, U_{-s}y\rangle\, ds \\
&= \int f(s)\langle U_s\varphi(L_g)x, y\rangle\, ds \\
&= \langle \varphi(L_f)\varphi(L_g)x, y\rangle;
\end{aligned}$$

we have that $\varphi(L_f L_g) = \varphi(L_{f*g}) = \varphi(L_f)\varphi(L_g)$. Now

$$\begin{aligned}
\langle \varphi(L_f^*)x, y\rangle &= \int f^*(t)\langle U_t x, y\rangle\, dt \\
&= \int \overline{f(-t)}\;\overline{\langle U_{-t}y, x\rangle}\, dt \\
&= \overline{\langle \varphi(L_f)y, x\rangle} = \langle x, \varphi(L_f)y\rangle;
\end{aligned}$$

so that $\varphi(L_f^*) = \varphi(L_f)^*$. Thus $\varphi$ is a representation of $\mathscr{A}_1(\mathbb{R})$ on $\mathscr{H}$.

To prove that $\varphi$ is an essential representation of $\mathscr{A}_1(\mathbb{R})$, we note that if $\langle \varphi(L_f)x, y\rangle = 0$ for all $f$ in $L_1(\mathbb{R})$ and all $x$ in $\mathscr{H}$, then $y = 0$. For such a vector $y$, $\langle U_t x, y\rangle = 0$ for almost every $t$, from (1). Since $t \to U_t$ is strong-operator continuous, $t \to \langle U_t x, y\rangle$ is continuous and vanishes identically. Thus, $\langle x, y\rangle = 0$ for all $x$ in $\mathscr{H}$, $y = 0$, and $\varphi$ is essential on $\mathscr{A}_1(\mathbb{R})$.

We prove next that $\|\varphi(L_f)\| \leqslant \|L_f\|$, from which it will follow that $\varphi$ has a (unique) bounded extension to $\mathfrak{A}_1(\mathbb{R})$ and from $\mathfrak{A}_1(\mathbb{R})$ to $\mathfrak{A}_0(\mathbb{R})$ (by assigning the identity operator to $I$ and extending the resulting mapping linearly). Let $\mathfrak{A}_0$ be the norm closure of the algebra generated by $I$ and $\varphi(\mathscr{A}_1(\mathbb{R}))$. Then $\mathfrak{A}_0$ is an abelian $C^*$-algebra on $\mathscr{H}$. From Theorem 4.3.8(iv), there is a pure state $\rho$ of $\mathfrak{A}_0$

such that $|\rho(\varphi(L_f))| = \|\varphi(L_f)\|$. Proposition 4.4.1 implies that $\rho$ is multiplicative on $\mathfrak{A}_0$. Thus the equation $\rho_1(f) = \rho(\varphi(L_f))$ defines a non-zero multiplicative linear functional $\rho_1$ on $L_1(\mathbb{R})$. From Theorem 3.2.26, there is a real number $r$ such that $\rho_1(f) = \hat{f}(r)$. From Theorem 3.2.27, there is a $\rho_0$ in $\mathcal{M}_0(\mathbb{R})$ such that $\rho_0(L_f) = \hat{f}(r)$. Thus

$$\|\varphi(L_f)\| = |\rho(\varphi(L_f))| = |\rho_1(f)| = |\hat{f}(r)| = |\rho_0(L_f)| \leqslant \|L_f\|,$$

since $\mathcal{M}_0(\mathbb{R})$ is contained in the unit ball of $\mathfrak{A}_0(\mathbb{R})^\#$.

Suppose now that $\varphi$ is an essential representation of $\mathcal{A}_1(\mathbb{R})$ on $\mathcal{H}$. The argument set out in the preceding paragraph shows that $\|\varphi(L_f)\| \leqslant \|L_f\|$ for all $f$ in $L_1(\mathbb{R})$. The heuristic discussion preceding Theorem 3.2.26 suggests that $\varphi(L_{f_r})$ should be $U_r\varphi(L_f)$, for each $f$ in $L_1(\mathbb{R})$, and that we should define $U_r\varphi(L_f)x$ to be $\varphi(L_{f_r})x$ for each $f$ in $L_1(\mathbb{R})$ and $x$ in $\mathcal{H}$. We must show that if

$$\varphi(L_{f_1})x_1 + \cdots + \varphi(L_{f_n})x_n = 0,$$

then

$$\varphi(L_{(f_1)_r})x_1 + \cdots + \varphi(L_{(f_n)_r})x_n = 0.$$

This last will follow if we establish that the norms of the sums on the left-hand sides of these equalities are equal. More generally, we show that the inner products of two such sums and of their "translates" by $r$ (that is, their images under $U_r$) are equal. For this it suffices to note that

$$\langle \varphi(L_{f_r})x, \varphi(L_{g_r})y \rangle = \langle \varphi(L_{g_r^* * f_r})x, y \rangle$$

and that

$$\begin{aligned}(g_r^* * f_r)(s) &= \int g_r^*(t) f_r(s-t)\,dt = \int \overline{g(-t-r)}\, f(s-t-r)\,dt \\ &= \int g^*(t) f(s-t)\,dt = (g^* * f)(s),\end{aligned}$$

so that $g_r^* * f_r = g^* * f$. It follows that $U_r$ is unambiguously defined and extends to a unitary operator on $\mathcal{H}$ (which we denote, again, by $U_r$). Since

$$\begin{aligned}U_{r+s}\varphi(L_f)x &= \varphi(L_{f_{r+s}})x = \varphi(L_{(f_s)_r})x = U_r\varphi(L_{f_s})x \\ &= U_rU_s\varphi(L_f)x,\end{aligned}$$

$r \to U_r$ is a homomorphism of $\mathbb{R}$ into the unitary group of $\mathcal{H}$. To establish the strong-operator continuity of this mapping, it will suffice to prove that, given $x$ in some set of vectors generating a dense linear manifold in $\mathcal{H}$ and a positive $\varepsilon$, there is a positive $\delta$ such that $\|U_t x - x\| < \varepsilon$ when $|t| < \delta$. For our set of vectors, we may choose $\{\varphi(L_f)u\}$ where $u \in \mathcal{H}$ and $f$ is a continuous function on $\mathbb{R}$ with (compact) support in a finite interval (since such functions are dense in $L_1(\mathbb{R})$).

As $f$ is uniformly continuous and its support is a finite interval, there is a positive $\delta$ such that $\|f - f_t\|_1 < \varepsilon/\|u\|$ provided $|t| < \delta$. Writing $x$ for $\varphi(L_f)u$, when $|t| < \delta$, we have that $\|U_t x - x\| = \|\varphi(L_{f_t} - L_f)u\| < \varepsilon$, for $\|\varphi\| \leqslant 1$, from the preceding paragraph; and $t \to U_t$ is strong-operator continuous.

To complete the proof, we must show that $\varphi$, as constructed, satisfies (1). It will suffice to prove (1) with vectors $u$ of the form $\varphi(L_g)x$ in place of $x$, where $g \in L_1(\mathbb{R})$. Now, the mapping $f \to \langle \varphi(L_f)x, y\rangle$ is a bounded linear functional on $L_1(\mathbb{R})$. From Theorem 1.7.8, there is an $h$ in $L_\infty(\mathbb{R})$ such that $\langle \varphi(L_f)x, y\rangle = \int f(r)h(r)\,dr$ for all $f$ in $L_1(\mathbb{R})$. If $u = \varphi(L_g)x$, then

$$\begin{aligned}\langle \varphi(L_f)u, y\rangle &= \langle \varphi(L_{f*g})x, y\rangle \\ &= \int (f*g)(r)h(r)\,dr \\ &= \int\left(\int f(t)g(r-t)\,dt\right)h(r)\,dr \\ &= \int f(t)\left(\int g_t(r)h(r)\,dr\right)dt \\ &= \int f(t)\langle \varphi(L_{g_t})x, y\rangle\,dt \\ &= \int f(t)\langle U_t u, y\rangle\,dt,\end{aligned}$$

where the Fubini theorem applies since $(|g| * |f|)|h| \in L_1(\mathbb{R})$. ∎

*Bibliography*: [4, 19]

## 4.6. Exercises

4.6.1. Suppose that $S_1, S_2, T_1, T_2$, and $A$ are elements of a $C^*$-algebra $\mathfrak{A}$, and

$$0 \leqslant S_1 \leqslant T_1, \qquad 0 \leqslant S_2 \leqslant T_2.$$

Prove that

$$\|S_1^{1/2}A\| \leqslant \|T_1^{1/2}A\|,$$

and deduce that

$$\|S_1^{1/2}AS_2^{1/2}\| \leqslant \|T_1^{1/2}AT_2^{1/2}\|.$$

4.6.2. Suppose that $\mathfrak{A}$ and $\mathscr{B}$ are $C^*$-algebras and $\varphi$ is a * homomorphism from $\mathfrak{A}$ onto $\mathscr{B}$. Suppose that $B, K \in \mathscr{B}$, with $B$ self-adjoint and $K$ positive, and let $V$ be the exponential unitary $\exp iB$. Show that there exist $A$, $H$, $U$ $(= \exp iA)$ in $\mathfrak{A}$, with $A$ self-adjoint and $H$ positive, such that

$$\varphi(A) = B, \qquad \varphi(H) = K, \qquad \varphi(U) = V.$$

[See also Exercises 4.6.3 and 4.6.59.]

4.6.3. Suppose that $\mathbb{D}$ is the unit disk $\{z \in \mathbb{C} : |z| \leqslant 1\}$, $\mathbb{T}$ is its boundary $\{z \in \mathbb{C} : |z| = 1\}$, and $\mathbb{S}$ is the union $\{\exp i\theta : \theta \in \mathbb{R},\ \pi/4 \leqslant |\theta| \leqslant 3\pi/4\}$ of two closed arcs in $\mathbb{T}$. Consider the $C^*$-algebras $C(\mathbb{D})$, $C(\mathbb{T})$, $C(\mathbb{S})$ and the * homomorphisms $\varphi$ (from $C(\mathbb{D})$ onto $C(\mathbb{T})$) and $\psi$ (from $C(\mathbb{T})$ onto $C(\mathbb{S})$) defined by restriction; that is,

$$\varphi(f) = f \mid \mathbb{T}, \qquad \psi(g) = g|\mathbb{S} \qquad (f \in C(\mathbb{D}), \quad g \in C(\mathbb{T})).$$

Find

(i) a unitary element $u$ of $C(\mathbb{T})$ that is not of the form $\varphi(f)$ for any invertible element $f$ of $C(\mathbb{D})$; [*Hint*. Use the fact (from elementary algebraic topology) that there is no continuous mapping of $\mathbb{D}$ onto $\mathbb{T}$ that leaves each point of $\mathbb{T}$ fixed – that is, $\mathbb{T}$ is not a *retract* of $\mathbb{D}$.]

(ii) a projection $q$ in $C(\mathbb{S})$ that is not of the form $\psi(p)$ for any projection $p$ in $C(\mathbb{T})$.

[See also Exercise 4.6.59.]

4.6.4. Determine whether the following assertion is true or false: if $\mathfrak{A}$ and $\mathscr{B}$ are abelian $C^*$-algebras, $\varphi$ is a * homomorphism from $\mathfrak{A}$ onto $\mathscr{B}$, and $B$ is an invertible self-adjoint element of $\mathscr{B}$, there is an invertible self-adjoint element $A$ of $\mathfrak{A}$ such that $\varphi(A) = B$.

4.6.5. Use the results of Exercises 4.6.2 and 4.6.3(i) to provide an example of an abelian $C^*$-algebra $\mathscr{B}$ and a unitary element $V$ of $\mathscr{B}$ that is not of the form $\exp iB$ for any self-adjoint $B$ in $\mathscr{B}$.

4.6.6. Let $\mathscr{U}$ be the (multiplicative) group of all unitary elements in a $C^*$-algebra $\mathfrak{A}$.

(i) Show that, if $U \in \mathscr{U}$ and $\|I - U\| < 2$, then $U = \exp iH$ for some self-adjoint $H$ in $\mathfrak{A}$.

(ii) Show that, if $V, W \in \mathscr{U}$ and $\|V - W\| < 2$, then $V = W \exp iH$ for some self-adjoint $H$ in $\mathfrak{A}$.

(iii) Let $\mathscr{U}_1$ $(\subseteq \mathscr{U})$ be the set of all products of the form

$$(\exp iH_1)(\exp iH_2) \cdots (\exp iH_k),$$

where $\{H_1, H_2, \ldots, H_k\}$ is a finite set of self-adjoint elements of $\mathfrak{A}$. Show that

$\mathscr{U}_1$ is open, closed, and arcwise connected, in the (relative) norm topology on $\mathscr{U}$.

4.6.7. Suppose that $\mathfrak{A}$ is an abelian $C^*$-algebra, $\mathscr{U}$ is its unitary group (considered as a topological group with the norm topology), $\mathscr{U}_I$ is the connected component of $\mathscr{U}$ that contains the identity $I$, and $U \in \mathscr{U}$. Use the results of Exercise 4.6.6(iii) to show that the following three conditions are equivalent.

(i) $U = \exp iA$ for some self-adjoint $A$ in $\mathfrak{A}$.
(ii) $U$ is connected to $I$ by a continuous arc in $\mathscr{U}$.
(iii) $U \in \mathscr{U}_I$.

4.6.8. Show that, in both of the following cases, each unitary element in the $C^*$-algebra $\mathfrak{A}$ has the form $\exp iA$ for some self-adjoint $A$ in $\mathfrak{A}$.

(i) $\mathfrak{A}$ is the $C^*$-algebra $L_\infty$ associated with a $\sigma$-finite measure space.
(ii) $\mathfrak{A}$ is the $C^*$-algebra $C(X)$, where the compact Hausdorff space $X$ is contractible (that is, there is a point $x_0$ in $X$ and a continuous mapping $f: X \times [0, 1] \to X$ such that $f(x, 0) = x$, $f(x, 1) = x_0$ for each $x$ in $X$). [*Hint*. Use the result of Exercise 4.6.7.]

4.6.9. (i) Show that, if $a$, $b$ are complex numbers and $P$, $Q$ are projections with sum $I$ in a $C^*$-algebra, then the function calculus for the normal element $aP + bQ$ $(= N)$ is given by

$$f(N) = f(a)P + f(b)Q.$$

(ii) Suppose that $\mathbb{T}$ is the unit circle $\{z \in \mathbb{C} : |z| = 1\}$ and $\mathscr{M}$ is the algebra of all $2 \times 2$ complex matrices, so that $\mathscr{M}$ becomes a $C^*$-algebra when identified in the usual way with the set of all linear operators acting on the two-dimensional Hilbert space $\mathbb{C}^2$. The Banach space $C(\mathbb{T}, \mathscr{M})$ (see Example 1.7.2) becomes a $C^*$-algebra $\mathfrak{A}$ when products and adjoints (as well as the linear structure) are defined pointwise. Let $E$ and $F$ be the projections in $\mathfrak{A}$ given by

$$E(e^{i\theta}) = \begin{bmatrix} 0 & 0 \\ 0 & 1 \end{bmatrix}, \qquad F(e^{i\theta}) = \frac{1}{2}\begin{bmatrix} 1 - \cos\theta & -\sin\theta \\ -\sin\theta & 1 + \cos\theta \end{bmatrix}$$

$(0 \leqslant \theta \leqslant 2\pi)$, and let $U$ be the unitary element $(\exp i\pi E)(\exp i\pi F)$ of $\mathfrak{A}$. Show that

$$U(e^{i\theta}) = e^{i\theta}P + e^{-i\theta}Q \qquad (0 \leqslant \theta \leqslant 2\pi),$$

where $P$ and $Q$ are the projections in $\mathscr{M}$ given by

$$P = \frac{1}{2}\begin{bmatrix} 1 & -i \\ i & 1 \end{bmatrix}, \qquad Q = \frac{1}{2}\begin{bmatrix} 1 & i \\ -i & 1 \end{bmatrix}.$$

Deduce that $U$, a product of two exponential unitary elements of $\mathfrak{A}$, is not itself an exponential unitary.

4.6.10. Let $\mathfrak{A}$ be a $C^*$-algebra and $\mathscr{B}$ be a norm-closed (though not necessarily self-adjoint) subalgebra of $\mathfrak{A}$. Let $A$ be a self-adjoint element of $\mathfrak{A}$ in $\mathscr{B}$. Show that $\mathrm{sp}_{\mathscr{B}}(A) = \mathrm{sp}_{\mathfrak{A}}(A)$. [*Hint.* See Exercise 3.5.28.]

4.6.11. Suppose that $A$ is a positive element of a $C^*$-algebra $\mathfrak{A}$, $E$ and $F$ are orthogonal projections in $\mathfrak{A}$, and $EAE = 0$. Show that $EAF = 0$.

4.6.12. Suppose that a $C^*$-algebra $\mathfrak{A}$ has a maximal abelian * subalgebra $\mathscr{A}$ that is finite dimensional.

(i) Show that $\mathscr{A}$ is the linear span of a finite orthogonal family $\{E_1, \ldots, E_n\}$ of projections in $\mathfrak{A}$ with sum $I$ (the identity of $\mathfrak{A}$).

(ii) By considering the family $\{E_1, \ldots, E_n, E_jAE_j\}$, where $A = A^* \in \mathfrak{A}$, show that $E_j\mathfrak{A}E_j = \{aE_j : a \in \mathbb{C}\}$ for $j = 1, \ldots, n$.

(iii) Suppose that $j, k \in \{1, \ldots, n\}$ and $j \neq k$. Given $A$ and $B$ in $\mathfrak{A}$, let $a, b, c, d$ be the scalars determined by $E_jA^*E_kAE_j = aE_j$, $E_jA^*E_kBE_j = bE_j$, $E_jB^*E_kBE_j = cE_j$, $E_kBE_jA^*E_k = dE_k$. Prove that $a \geqslant 0, c \geqslant 0$. By considering suitable expressions for $bdE_j, bdE_k$, and $acE_j$, show that $b = d$ and $ac = |b|^2$. Deduce that

$$(sE_kAE_j + tE_kBE_j)^*(sE_kAE_j + tE_kBE_j) = 0$$

for suitable scalars $s$ and $t$ (not both 0). Deduce that $E_k\mathfrak{A}E_j$ is at most one dimensional.

(iv) Prove that $\mathfrak{A}$ is finite dimensional.

4.6.13. Suppose that $\mathfrak{A}$ is an infinite-dimensional $C^*$-algebra. By using the result of Exercise 4.6.12, show that there is an infinite sequence $\{A_1, A_2, \ldots\}$ of non-zero elements of $\mathfrak{A}^+$ such that $A_jA_k = 0$ when $j \neq k$.

4.6.14. By using the result of Exercise 4.6.12, show that, in an infinite-dimensional $C^*$-algebra, there is a positive element with infinite spectrum.

4.6.15. Let $\rho$ be a state of a $C^*$-algebra $\mathfrak{A}$. We say that $\rho$ is *faithful* if $A = 0$ when $A \in \mathfrak{A}^+$ and $\rho(A) = 0$.

(i) Show that $\mathscr{L}_\rho$, the left kernel of $\rho$, is $\{0\}$ when $\rho$ is a faithful state of $\mathfrak{A}$. Deduce that $\mathscr{H}_\rho$ is the completion of $\mathfrak{A}$ relative to the inner product $(A, B) \to \rho(B^*A)$ $(= \langle A, B\rangle)$, and that $\pi_\rho$ is faithful.

(ii) Let $\rho$ be a state of $\mathfrak{A}$ such that $\pi_\rho$ is faithful. Must $\rho$ be faithful?

4.6.16. Let $A$ be a self-adjoint element in a $C^*$-algebra $\mathfrak{A}$. Let $\rho$ be a state of $\mathfrak{A}$ such that $\rho(A^2) = \rho(A)^2$. (We say that $\rho$ is *definite* on $A$ in this case.) Show that $\rho(AB) = \rho(BA) = \rho(A)\rho(B)$ for each $B$ in $A$.

4.6.17. Suppose that $\mathfrak{A}$, $\mathscr{A}$, and $\{E_1, \ldots, E_n\}$ are as in Exercise 4.6.12. Let $\rho_j$ be a state of $\mathfrak{A}$ that extends the state of $\mathscr{A}$ assigning 1 to $E_j$ and 0 to $E_k$ when $k \neq j$. Let $\rho$ be $n^{-1} \sum_{j=1}^{n} \rho_j$.

(i) Show that $\rho$ is a faithful state of $\mathfrak{A}$.

(ii) Choose $A_{jk}$ in $\mathfrak{A}$ such that $(E_j A_{jk} E_k)^*(E_j A_{jk} E_k) = nE_k$ for those $k$ and $j$ for which $E_j \mathfrak{A} E_k$ is one dimensional. Show that the set of $E_j A_{jk} E_k$ forms an orthonormal basis for $\mathscr{H}_\rho$.

4.6.18. Let $\mathscr{H}$ be a Hilbert space of finite dimension $n$, $\{e_1, \ldots, e_n\}$ be an orthonormal basis for $\mathscr{H}$, $E_{jk}$ be the element of $\mathscr{B}(\mathscr{H})$ that maps $e_k$ to $e_j$ and $e_{k'}$ to 0 when $k' \neq k$, and $\rho$ be an element of the Banach dual space $\mathscr{B}(\mathscr{H})^\#$. Define $A$ to be the element of $\mathscr{B}(\mathscr{H})$ with matrix representation $[a_{jk}]$ relative to $\{e_1, \ldots, e_n\}$, where $a_{jk} = \rho(E_{kj})$, and $\tau$ to be the *normalized trace* ($\tau(B) = n^{-1} \sum_{j=1}^{n} b_{jj} = n^{-1}\operatorname{tr}(B)$, where $B$ has matrix $[b_{jk}]$ relative to $\{e_1, \ldots, e_n\}$). Show that:

(i) $\tau$ is a faithful state of $\mathscr{B}(\mathscr{H})$;

(ii) $\rho(B) = \operatorname{tr}(AB)$ for each $B$ in $\mathscr{B}(\mathscr{H})$, and $\rho$ is a state if and only if $0 \leqslant A$ and $\operatorname{tr}(A) = 1$;

(iii) $A$ is independent of the basis $\{e_1, \ldots, e_n\}$;

(iv) $\rho$ is a pure state of $\mathscr{B}(\mathscr{H})$ if and only if $A$ is a projection with one-dimensional range (in which case $\rho = \omega_x$, where $x$ is a unit vector in the range of $A$);

(v) $\rho$ is a faithful state of $\mathscr{B}(\mathscr{H})$ if and only if $\operatorname{tr}(A) = 1$ and $A \geqslant aI$ for some positive scalar $a$.

4.6.19. With the notation of Exercise 4.6.18, suppose $\rho$ is a state and $E$ is the projection in $\mathscr{B}(\mathscr{H})$ with range the range of $A$. Show that the left kernel $\mathscr{L}$ of $\rho$ is $\mathscr{B}(\mathscr{H})(I - E)$.

4.6.20. Suppose that $\mathfrak{A}$ and $\mathscr{B}$ are $C^*$-algebras, and $\varphi$ is a * homomorphism from $\mathfrak{A}$ onto $\mathscr{B}$.

(i) Let $B_1$ and $B_2$ be elements of $\mathscr{B}^+$ such that $B_1 B_2 = 0$. By considering $B_1 - B_2$, show that there exist elements $A_1$ and $A_2$ of $\mathfrak{A}^+$ such that $A_1 A_2 = 0$, $\varphi(A_1) = B_1$, and $\varphi(A_2) = B_2$.

(ii) Suppose that $\{B_1, B_2, B_3, \ldots.\}$ is a sequence of elements of $\mathscr{B}^+$ such that $B_j B_k = 0$ when $j \neq k$. Prove that there is a sequence $\{A_1, A_2, A_3, \ldots\}$ of elements of $\mathfrak{A}^+$ such that $A_j A_k = 0$ when $j \neq k$, and $\varphi(A_j) = B_j$ for each

$j = 1, 2, 3, \ldots$ . [*Hint*. Upon replacing $B_j$ by $b_j B_j$ for a suitable positive scalar $b_j$, we may assume that the series $\sum B_j$ and $\sum B_j^{1/2}$ converge to elements of $\mathscr{B}$. Prove, by induction on $n$, the following statement: there exist elements $A_1, A_2, \ldots, A_n, X_n$ of $\mathfrak{A}^+$ such that $A_j A_k = A_j X_n = 0$ for all $j, k = 1, \ldots, n$ with $j \neq k$, and

$$\varphi(A_1) = B_1, \ldots, \varphi(A_n) = B_n, \varphi(X_n) = B_{n+1} + B_{n+2} + \cdots .]$$

4.6.21. Suppose that $\mathfrak{A}$ and $\mathscr{B}$ are $C^*$-algebras, $\varphi$ is a * homomorphism from $\mathfrak{A}$ onto $\mathscr{B}$, and

$$A \in \mathfrak{A}^+, \qquad B \in \mathscr{B}, \qquad 0 \leqslant B \leqslant \varphi(A).$$

(i) Show that there is a self-adjoint element $R$ of $\mathfrak{A}$ such that $R \leqslant A$ and $\varphi(R) = B$. Deduce that there exist positive elements $S$ and $T$ of $\mathfrak{A}$ such that

$$\varphi(S) = B, \qquad \varphi(T) = 0, \qquad S \leqslant T + A.$$

(ii) For $n = 1, 2, \ldots$, let

$$U_n = S^{1/2}(n^{-1}I + T + A)^{-1}(T + A)^{1/2}A^{1/2}.$$

Show that

$$U_n^* U_n \leqslant A, \qquad \varphi(U_n) = B^{1/2}[n^{-1}I + \varphi(A)]^{-1}\varphi(A).$$

(iii) By using the result of Exercise 4.6.1, show that

$$\begin{aligned} \|U_m - U_n\| &\leqslant \|(n^{-1} - m^{-1})(m^{-1}I + T + A)^{-1}(n^{-1}I + T + A)^{-1}(T + A)^{3/2}\| \\ &\leqslant |n^{-1} - m^{-1}|^{1/2}, \end{aligned}$$

$$\|B^{1/2} - \varphi(U_n)\| \leqslant \|n^{-1}\varphi(A)^{1/2}[n^{-1}I + \varphi(A)]^{-1}\| \leqslant \tfrac{1}{2}n^{-1/2}.$$

(iv) Deduce that the sequence $\{U_n^* U_n\}$ converges to an element $A_0$ of $\mathfrak{A}$ such that

$$0 \leqslant A_0 \leqslant A, \qquad \varphi(A_0) = B.$$

4.6.22. Let $\mathfrak{A}$ be the $C^*$-algebra $l_\infty(\mathbb{A})$, where $\mathbb{A}$ is the set $\{1, 2, 3, 4\}$, and define a self-adjoint subspace $\mathscr{M}$ of $\mathfrak{A}$ (that contains the identity of $\mathfrak{A}$) by

$$\mathscr{M} = \{f \in \mathfrak{A} : f(1) + f(2) = f(3) + f(4)\}.$$

Find a hermitian linear functional $\rho$ on $\mathscr{M}$ that has more than one expression in the form $\rho = \rho_1 - \rho_2$, where $\rho_1$ and $\rho_2$ are positive linear functionals on $\mathscr{M}$ and $\|\rho\| = \|\rho_1\| + \|\rho_2\|$. [See the discussion following Corollary 4.3.7.]

4.6.23. Let $\varphi$ be a * homomorphism of one $C^*$-algebra $\mathfrak{A}$ onto another $C^*$-algebra $\mathscr{B}$, and let $\rho$ be a linear functional on $\mathscr{B}$.

(i) Show that $\rho \circ \varphi$ is a state of $\mathfrak{A}$ if $\rho$ is a state of $\mathscr{B}$.
(ii) Show that $\rho$ is a state of $\mathscr{B}$ if $\rho \circ \varphi$ is a state of $\mathfrak{A}$.
(iii) Show that $\rho \circ \varphi$ is a pure state of $\mathfrak{A}$ if $\rho$ is a pure state of $\mathscr{B}$.
(iv) Show that $\rho$ is a pure state of $\mathscr{B}$ if $\rho \circ \varphi$ is a pure state of $\mathfrak{A}$.

4.6.24. Let $\mathscr{S}$ be the set of states of a $C^*$-algebra $\mathfrak{A}$. With $A$ in $\mathfrak{A}$, let $b$ be $\sup\{|\rho(A)| : \rho \in \mathscr{S}\}$.

(i) Show that there is a pure state $\rho$ of $\mathfrak{A}$ such that $|\rho(A)| = b$.

(ii) When $A$ is a normal element of $\mathfrak{A}$, $b = \|A\|$, from Theorem 4.3.4. Find an example of a $C^*$-algebra $\mathfrak{A}$ and an element $A$ of $\mathfrak{A}$ for which $b < \|A\|$.

4.6.25. Suppose that $U$ is a unitary element of a $C^*$-algebra $\mathfrak{A}$ and $\rho$ is a state of $\mathfrak{A}$. Show that the equation

$$\rho_U(A) = \rho(UAU^*) \qquad (A \in \mathfrak{A})$$

defines a state $\rho_U$ of $\mathfrak{A}$ and that $\rho_U$ is pure if and only if $\rho$ is pure.

4.6.26. Let $\mathfrak{A}$ be a $C^*$-algebra. Show that:

(i) if $\mathfrak{A}$ is abelian, then $\|\rho_1 - \rho_2\| = 2$ whenever $\rho_1$ and $\rho_2$ are distinct pure states of $\mathfrak{A}$;

(ii) if there is a positive real number $\delta$ such that $\|\rho_1 - \rho_2\| \geqslant \delta$ whenever $\rho_1$ and $\rho_2$ are distinct pure states of $\mathfrak{A}$, then $\mathfrak{A}$ is abelian. [*Hint*. Let $H$ be a self-adjoint element of $\mathfrak{A}$, and define $U(t) = \exp itH$ for all real $t$. With the notation of Exercise 4.6.25, show that $\rho = \rho_{U(t)}$ (and, hence, that $\rho(AU(t)) = \rho(U(t)A)$ for all $A$ in $\mathfrak{A}$) whenever $\rho$ is a pure state of $\mathfrak{A}$ and $|t|$ is sufficiently small. Deduce that $\rho(AH - HA) = 0$.]

4.6.27. Suppose that $\mathscr{P}(\mathfrak{A})$ is the set of all pure states of a $C^*$-algebra $\mathfrak{A}$, and

$$\varphi \colon A \to \hat{A} \colon \quad \mathfrak{A} \to C(\mathscr{P}(\mathfrak{A})^-)$$

is the function representation of $\mathfrak{A}$ on its pure state space $\mathscr{P}(\mathfrak{A})^-$ (see the discussion preceding Theorem 4.3.11).

(i) Show that $\varphi(\mathfrak{A}) = C(\mathscr{P}(\mathfrak{A})^-)$ if and only if $\mathfrak{A}$ is abelian. [*Hint*. Use the result of Exercise 4.6.26(ii).]

(ii) Show that $\mathfrak{A}$ is abelian if and only if $\varphi(\mathfrak{A})$ is a subalgebra of $C(\mathscr{P}(\mathfrak{A})^-)$.

4.6.28. Let $\mathfrak{A}$ be a $C^*$-algebra.

(i) Show that if $A \in \mathfrak{A}$ and $\lambda \in \mathrm{sp}(A^*A)$, then $A^*A - \lambda I$ has neither a left nor right inverse.

(ii) Show that the intersection of the maximal left ideals in $\mathfrak{A}$ is $\{0\}$.

4.6.29. Suppose that $\mathfrak{A}$ is a $C^*$-algebra, and each closed left ideal in $\mathfrak{A}$ is a two-sided ideal. Prove that:

(i) each maximal left ideal in $\mathfrak{A}$ is also a maximal right ideal;

(ii) if $\mathscr{I}$ is a maximal left ideal in $\mathfrak{A}$ (and, hence, a two-sided ideal), then each non-zero element of the quotient Banach algebra $\mathfrak{A}/\mathscr{I}$ is invertible;

(iii) each maximal left ideal in $\mathfrak{A}$ is the kernel of a multiplicative linear functional on $\mathfrak{A}$;

(iv) if $A \in \mathfrak{A}$ and $a \in \text{sp}(A)$, then $a = \rho(A)$ for some multiplicative linear functional $\rho$ on $\mathfrak{A}$.

Deduce that $\mathfrak{A}$ is abelian.

4.6.30. Suppose that $\mathfrak{A}$ is a $C^*$-algebra with the following property: if $A \in \mathfrak{A}$ and $A^2 = 0$, then $A = 0$.

(i) For each positive integer $n$, let $f_n : \mathbb{R} \to [0, 1]$ be a continuous function such that $f_n(t) = 0$ when $|t| \leqslant 1/2n$ and $f_n(t) = 1$ when $|t| \geqslant 1/n$. Prove that

$$f_n(A^*A) = f_{2n}(A^*A)f_n(A^*A), \qquad \|A - Af_n(A^*A)\| \leqslant n^{-1/2},$$

$$f_n(A^*A)B = f_n(A^*A)Bf_{2n}(A^*A),$$

for all $A$ and $B$ in $\mathfrak{A}$.

(ii) By using the result of Exercise 4.6.29, show that $\mathfrak{A}$ is abelian.

4.6.31. Suppose that $A$ is a self-adjoint element of a $C^*$-algebra $\mathfrak{A}$ and $\lambda \in \text{sp}(A)$. Show that there is a pure state $\rho$ of $\mathfrak{A}$ such that $\rho(A) = \lambda$ and $\rho$ is definite on $A$ (see Exercise 4.6.16). [*Hint.* Show that $\lambda = \rho_0(A)$ for some pure state $\rho_0$ of the $C^*$-subalgebra of $\mathfrak{A}$ generated by $A$ and $I$, and use Theorem 4.3.13(iv).]

4.6.32. Let $A$ be a self-adjoint element of the $C^*$-algebra $\mathfrak{A}$. Suppose that for each non-zero self-adjoint $B$ in $\mathfrak{A}$ there is a state $\rho$ of $\mathfrak{A}$, definite on $A$, such that $\rho(B) \neq 0$. Show that $A$ lies in the center of $\mathfrak{A}$.

4.6.33. Suppose that $A$ is a self-adjoint element of the center of a $C^*$-algebra $\mathfrak{A}$. Show that for each non-zero self-adjoint element $B$ of $\mathfrak{A}$ there is a state $\rho$ of $\mathfrak{A}$, definite on $A$, such that $\rho(B) \neq 0$.

4.6.34. Let $A, B$, and $C$, be elements of a $C^*$-algebra $\mathfrak{A}$. Suppose that $A$ is self-adjoint, $C$ is in the center of $\mathfrak{A}$, and $AB - BA = C$. Show that $C = 0$.

4.6.35. Suppose that $\mathscr{L}_0$ is a (not necessarily closed) left ideal in a $C^*$-algebra $\mathfrak{A}$. Given a finite subset $F = \{A_1, \ldots, A_n\}$ of $\mathscr{L}_0$, define $H_F$ and $V_F$ in $\mathscr{L}_0$ by

$$H_F = A_1^*A_1 + \cdots + A_n^*A_n, \qquad V_F = H_F(H_F + n^{-1}I)^{-1}.$$

Show that, with the family $\mathscr{F}$ of all finite subsets of $\mathscr{L}_0$ directed by the inclusion relation $\supseteq$, the net $(V_F, F \in \mathscr{F}, \supseteq)$ is an increasing right approximate identity for $\mathscr{L}_0$.

4.6.36. With the notation of Exercise 4.6.35, let $\mathscr{L}$ be the closure of $\mathscr{L}_0$, so that $\mathscr{L}$ is a closed left ideal in $\mathfrak{A}$. Show that $\{V_F\}$ is an increasing right approximate identity for $\mathscr{L}$. Prove also that, if $\mathscr{L}$ is a two-sided ideal in $\mathfrak{A}$, then $\{V_F\}$ is a two-sided approximate identity for $\mathscr{L}$.

4.6.37. Suppose that $\mathscr{H}$ is a Hilbert space with dimension at least 2, $y$ is a unit vector in $\mathscr{H}$, and $\mathscr{L}$ is the closed left ideal in the $C^*$-algebra $\mathscr{B}(\mathscr{H})$ defined by

$$\mathscr{L} = \{A \in \mathscr{B}(\mathscr{H}) : Ay = 0\}.$$

Show that $\mathscr{L}$ has no left approximate identity.

4.6.38. Provide examples, as indicated below, to show that the statements in Corollary 4.2.10 concerning *closed* two-sided ideals in a $C^*$-algebra $\mathfrak{A}$ are not in general valid for two-sided ideals that are not closed, even when $\mathfrak{A}$ is abelian.

(i) Find an ideal that is not self-adjoint in the abelian $C^*$-algebra $C(\mathbb{D})$, where $\mathbb{D}$ is the unit disk $\{z \in \mathbb{C} : |z| \leqslant 1\}$.

(ii) Suppose that $X$ is the unit interval $[0, 1]$, $\mathfrak{A}$ is the abelian $C^*$-algebra $C(X)$, and $\mathscr{K}$ is the ideal $u\mathfrak{A}$ in $\mathfrak{A}$, where $u(t) = t$ $(0 \leqslant t \leqslant 1)$. Find an ideal in $\mathscr{K}$ that is not an ideal in $\mathfrak{A}$.

4.6.39. Let $\mathscr{I}$ be a closed left ideal in a $C^*$-algebra $\mathfrak{A}$. Suppose that $A \in \mathscr{I}$, $B \in \mathscr{I}^+$, $\|B\| \leqslant 1$, and $AA^* \leqslant B^4$. By considering the sequence $\{C_n\}$, where $C_n = (B + n^{-1}I)^{-1}A$, show that $A = BC$ for some $C$ in $\mathscr{I}$ with $\|C\| \leqslant 1$.

4.6.40. Let $\mathscr{I}$ be a closed two-sided ideal in a $C^*$-algebra $\mathfrak{A}$. Suppose that $A_1, A_2, \ldots \in \mathscr{I}$, and $\sum_{n=1}^{\infty} \|A_n\|^2 \leqslant 1$. Show that there exist elements $B, C_1, C_2, \ldots$ of $\mathscr{I}$ such that $B \geqslant 0$, $\|C_n\| \leqslant 1$, and $A_n = BC_n$. [*Hint*. Use the result of Exercise 4.6.39.]

4.6.41. Suppose that $\mathscr{L}$ is a closed left ideal in a $C^*$-algebra $\mathfrak{A}$, and $\mathscr{S} = \mathscr{L}^+$. Prove that

(i) $\mathscr{S}$ is a closed subset of $\mathfrak{A}^+$,
(ii) $A + B \in \mathscr{S}$ whenever $A, B \in \mathscr{S}$,
(iii) $aA \in \mathscr{S}$ whenever $A \in \mathscr{S}$ and $a \geqslant 0$,
(iv) $A \in \mathscr{S}$ whenever $A \in \mathfrak{A}$ and $0 \leqslant A \leqslant B$ for some $B$ in $\mathscr{S}$. [*Hint*. Use the result of Exercise 4.6.39.]

Show also that the closed left ideal $\mathscr{L}$ is a two-sided ideal if and only if

(v) $UAU^* \in \mathscr{S}$ whenever $A \in \mathscr{S}$ and $U$ is a unitary element of $\mathfrak{A}$.

4.6.42. Suppose that a subset $\mathscr{S}$ of a $C^*$-algebra $\mathfrak{A}$ satisfies conditions (i)–(iv) of Exercise 4.6.41. Show that there is a unique closed left ideal $\mathscr{L}$ in $\mathfrak{A}$ such that $\mathscr{L}^+ = \mathscr{S}$, and that $\mathscr{L} = \{A \in \mathfrak{A} : A^*A \in \mathscr{S}\}$.

4.6.43. Suppose $A$ is an element in a $C^*$-algebra $\mathfrak{A}$, $|A| = (A^*A)^{1/2}$, and $V_n = A(A^*A + n^{-1}I)^{-1/2}$ for each positive integer $n$.

(i) Establish the following two inequalities:

$$\|A^*A[I - (A^*A)^{1/2}(A^*A + n^{-1}I)^{-1/2}]^2\| \leqslant n^{-1},$$

$$\| |A|[I - (A^*A)^{1/2}(A^*A + n^{-1}I)^{-1/2}]\| \leqslant n^{-1/2}.$$

Use these inequalities to show that:

(ii) $\|A - V_n|A|\| \to 0$ as $n \to \infty$;
(iii) $\| |A| - V_n^*A\| \to 0$ as $n \to \infty$.

4.6.44. Suppose $\mathscr{L}$ is a closed left ideal in the $C^*$-algebra $\mathfrak{A}$. Show that $A \in \mathscr{L}$ if and only if $(A^*A)^{1/2}(= |A|) \in \mathscr{L}$.

4.6.45. Suppose $A$ is an invertible element of a $C^*$-algebra $\mathfrak{A}$.

(i) Show that $A = UH$ for some unitary element $U$ in $\mathfrak{A}$ and some positive element $H$ in $\mathfrak{A}$.

(ii) Show that the elements $U$ and $H$ of $\mathfrak{A}$ occurring in the ("polar") decomposition of $A$ described in (i) are unique.

4.6.46. For each positive real number $a$, let $f_a : \mathbb{R}^+ \to \mathbb{R}^+$ be the continuous function defined by $f_a(t) = t^a$. Prove that $f_a$ is operator-monotonic increasing if $0 < a \leqslant 1$ but not if $a > 1$. [*Hint*. Let $X$ be the set of all positive real numbers for which $f_a$ is operator-monotonic increasing. Then $1 \in X$, and (from the discussion following Proposition 4.2.8), $\frac{1}{2} \in X$ and $2 \notin X$. Show that $X$ is closed in $\mathbb{R}^+ \backslash \{0\}$. Given $a, b$ in $X$, prove that $ab \in X$ and (by an argument similar to the proof of Proposition 4.2.8) that $\frac{1}{2}(a + b) \in X$.]

4.6.47. Let $\{u_1, u_2\}$ be an orthonormal basis in a two-dimensional Hilbert space $\mathscr{H}$, and let $\{v_1, v_2\}$ be the orthonormal basis given by

$$v_1 = 2^{-1/2}(u_1 + u_2), \qquad v_2 = 2^{-1/2}(u_1 - u_2).$$

Define $A$ and $B$ in $\mathscr{B}(\mathscr{H})^+$ by

$$Ax = \langle x, u_1 \rangle u_1, \qquad Bx = \lambda\langle x, v_1 \rangle v_1 + \mu\langle x, v_2 \rangle v_2 \qquad (x \in \mathscr{H}),$$

where $\lambda$ and $\mu$ are positive real numbers. Show that, for each continuous function $f: \mathbb{R}^+ \to \mathbb{R}$ with $f(0) = 0$,

$$f(A)x = f(1)\langle x, u_1\rangle u_1, \qquad f(B)x = f(\lambda)\langle x, v_1\rangle v_1 + f(\mu)\langle x, v_2\rangle v_2.$$

By considering the matrix of $f(B) - f(A)$, relative to the basis $\{u_1, u_2\}$, show that $f(A) \leqslant f(B)$ if and only if

$$f(\lambda) + f(\mu) \geqslant \max\{2f(1), 0\}, \qquad f(1)[f(\lambda) + f(\mu)] \leqslant 2f(\lambda)f(\mu).$$

By considering the case in which $\lambda = 1 + \varepsilon$ and $\mu = 1 - \varepsilon + 2\varepsilon^2$ for a sufficiently small positive real number $\varepsilon$, show that the function

$$f_a: t \to t^a: \quad \mathbb{R}^+ \to \mathbb{R}$$

is not operator-monotonic increasing when $a > 1$ (thus re-proving a part of the result of Exercise 4.6.46).

4.6.48. Suppose that $\mathscr{V}$ is a partially ordered vector space with positive cone $\mathscr{V}^+$ and with an order unit $I$, and let $\mathscr{S}$ be the set of all states of $\mathscr{V}$. (The relevant definitions are given in the discussion following Remark 3.4.4.) Prove that:

(i) each element of $\mathscr{V}$ can be expressed as the difference of two positive elements;

(ii) each positive linear functional on $\mathscr{V}$ is a non-negative multiple of a state of $\mathscr{V}$;

(iii) the set of all positive linear functionals on $\mathscr{V}$ is a cone in the algebraic dual space $\mathscr{V}'$ of $\mathscr{V}$;

(iv) for each $H$ in $\mathscr{V}$, the subset $\{\rho(H): \rho \in \mathscr{S}\}$ of $\mathbb{R}$ is bounded;

(v) $\mathscr{S}$ is a convex subset of $\mathscr{V}'$ and is compact in the weak topology $\sigma(\mathscr{V}', \mathscr{V})$ obtained by considering $\mathscr{V}$ as a separating family of linear functionals on $\mathscr{V}'$.

4.6.49. With the notation of Exercise 4.6.48, suppose that $\mathscr{M}$ is a subspace of $\mathscr{V}$ that contains $I$, so that $\mathscr{M}$ is a partially ordered vector space with positive cone $\mathscr{M} \cap \mathscr{V}^+$ and $I$ is an order unit for $\mathscr{M}$. Let $\rho_0$ be a positive linear functional on $\mathscr{M}$, and for each $H$ in $\mathscr{V}$ define

$$l_H = \sup\{\rho_0(B) : B \in \mathscr{M}, B \leqslant H\},$$

$$u_H = \inf\{\rho_0(B) : B \in \mathscr{M}, B \geqslant H\}.$$

(i) Prove that $l_H$ and $u_H$ are real numbers satisfying $l_H \leqslant u_H$.

(ii) Show that, if $H \in \mathscr{V}$, $c \in \mathbb{R}$, and $l_H \leqslant c \leqslant u_H$, the equation

$$\rho_1(aH + B) = ac + \rho_0(B) \qquad (a \in \mathbb{R}, \quad B \in \mathscr{M})$$

defines a positive linear functional $\rho_1$ on the subspace

$$\mathscr{M}_1 = \{aH + B : a \in \mathbb{R}, B \in \mathscr{M}\}$$

of $\mathscr{V}$.

(iii) Let $\mathscr{E}$ be the set of all pairs $(\mathscr{N}, \tau)$ in which $\mathscr{N}$ is a subspace of $\mathscr{V}$ that contains $\mathscr{M}$ and $\tau$ is a positive linear functional on $\mathscr{N}$ that extends $\rho_0$. By use of Zorn's lemma, applied to $\mathscr{E}$ with the partial ordering in which "$(\mathscr{N}_1, \tau_1) \leqslant (\mathscr{N}_2, \tau_2)$" means "$\mathscr{N}_1 \subseteq \mathscr{N}_2$ and $\tau_1 = \tau_2 | \mathscr{N}_1$", prove that $\rho_0$ extends to a positive linear functional $\rho$ on $\mathscr{V}$. Prove also that, if $H \in \mathscr{V}$ and $c \in \mathbb{R}$, the extension $\rho$ can be chosen so that $\rho(H) = c$ if and only if $l_H \leqslant c \leqslant u_H$.

(iv) Show that each state of $\mathscr{M}$ extends to a state of $\mathscr{V}$, and each pure state of $\mathscr{M}$ extends to a pure state of $\mathscr{V}$. Prove also that, if a pure state of $\mathscr{M}$ has only one extension as a pure state of $\mathscr{V}$, then it has only one extension as a state of $\mathscr{V}$. [*Hint*. Note the analogy with Theorem 4.3.13.]

4.6.50. With the notation of Exercise 4.6.48, define

$$\|H\|_I = \inf\{a \in \mathbb{R}^+ : -aI \leqslant H \leqslant aI\}$$

for each $H$ in $\mathscr{V}$.

(i) Prove that $\|H\|_I = \sup\{|\rho(H)| : \rho \in \mathscr{S}\}$. [*Hint*. Use the result of Exercise 4.6.49(iii), with $\mathscr{M}$ the subspace $\{aI : a \in \mathbb{R}\}$ of $\mathscr{V}$.]

(ii) Prove that $\|\ \|_I$ is a semi-norm on $\mathscr{V}$.

(iii) Show that, if $\rho$ is a positive linear functional on $\mathscr{V}$, then

$$|\rho(H)| \leqslant \rho(I)\|H\|_I \qquad (H \in \mathscr{V}).$$

(iv) Show that the subset

$$\mathscr{B} = \{a\rho_1 - b\rho_2 : \rho_1, \rho_2 \in \mathscr{S}, a, b \in \mathbb{R}^+, a + b = 1\}$$

of the algebraic dual space $\mathscr{V}'$ is convex, and is compact in the topology $\sigma(\mathscr{V}', \mathscr{V})$. By means of a Hahn–Banach separation theorem, show that

$$\mathscr{B} = \{\tau \in \mathscr{V}' : |\tau(H)| \leqslant \|H\|_I\,(H \in \mathscr{V})\}.$$

(v) Show that a linear functional $\tau$ on $\mathscr{V}$ can be expressed as the difference of two positive linear functionals on $\mathscr{V}$ if and only if there is a real number $k$ such that

$$|\tau(H)| \leqslant k\|H\|_I \qquad (H \in \mathscr{V}).$$

4.6.51. In the partially ordered vector space $C([0, 1], \mathbb{R})$, let $u$ be the order unit defined by $u(t) = 1$ $(0 \leqslant t \leqslant 1)$, and let $\mathscr{M}$ be the subspace (containing $u$) that consists of all polynomials with real coefficients. When $f \in \mathscr{M}$, let $\bar{f}$ be the unique extension of $f$ as a real polynomial defined throughout $\mathbb{R}$. Show that the equation

$$\rho(f) = \tilde{f}(2) \qquad (f \in \mathcal{M})$$

defines a linear functional $\rho$ on $\mathcal{M}$ that cannot be expressed as the difference of two positive linear functionals on $\mathcal{M}$.

4.6.52. Suppose that $\mathscr{V}$ is a partially ordered vector space with positive cone $\mathscr{V}^+$ and with an order unit $I$, and let $\| \ \|_I$ be the semi-norm defined in Exercise 4.6.50. We say that $\mathscr{V}$ is *archimedian* if the following condition is satisfied: if $H \in \mathscr{V}$ and $H \leqslant \varepsilon I$ for every positive real number $\varepsilon$, then $H \leqslant 0$.

(i) Prove that, if $\mathscr{V}$ is archimedian, then $\| \ \|_I$ is a norm on $\mathscr{V}$, $\mathscr{V}^+$ is closed in the associated norm topology on $\mathscr{V}$, and

$$\mathscr{V}^+ = \{H \in \mathscr{V} : \rho(H) \geqslant 0 \text{ for each state } \rho \text{ of } \mathscr{V}\}.$$

(ii) Show that the real vector space $\mathbb{R}^2$ $(= \mathscr{V})$ becomes a partially ordered vector space, and has an order unit $(1, 0)$ $(= I)$, when the positive cone $\mathscr{V}^+$ is defined by

$$\mathscr{V}^+ = \{(x, y) \in \mathbb{R}^2 : x > 0 \text{ or } x = 0 \text{ and } y \geqslant 0\}.$$

Show also that $\mathscr{V}$ is not archimedian, $\| \ \|_I$ is not a norm on $\mathscr{V}$, and $\mathscr{V} \backslash \mathscr{V}^+$ contains elements $H$ such that $\rho(H) \geqslant 0$ for each state $\rho$ of $\mathscr{V}$.

4.6.53. By a *Banach lattice*, we mean a partially ordered vector space $\mathscr{V}$ (with positive cone $\mathscr{V}^+$ and an order unit $I$) that is archimedian, is a lattice with the partial ordering induced by $\mathscr{V}^+$, and is a Banach space with the norm $\| \ \|_I$ defined in Exercise 4.6.50. When $X$ is a compact Hausdorff space, $C(X, \mathbb{R})$ is a Banach lattice; the present exercise shows that every Banach lattice is isomorphic to one of the form $C(X, \mathbb{R})$.

Suppose that $\mathscr{S}$ is the state space of a Banach lattice $\mathscr{V}$ and $\mathscr{P}^-$ is the closure in $\mathscr{S}$ of the set $\mathscr{P}$ of all pure states of $\mathscr{V}$. When $A \in \mathscr{V}$, define a real-valued function $\hat{A}$ on $\mathscr{P}^-$ by $\hat{A}(\rho) = \rho(A)$.

(i) Prove that $\hat{A} \in C(\mathscr{P}^-, \mathbb{R})$ for each $A$ in $\mathscr{V}$.

(ii) Show that the mapping $A \to \hat{A} : \mathscr{V} \to C(\mathscr{P}^-, \mathbb{R})$ is a linear isometry, with range a closed subspace $\mathcal{M}$ of $C(\mathscr{P}^-, \mathscr{R})$ that contains the constant functions and separates the points of $\mathscr{P}^-$. Show also that $A \geqslant 0$ (in $\mathscr{V}$) if and only if $\hat{A} \geqslant 0$ (in $C(\mathscr{P}^-, \mathbb{R})$).

(iii) Show that each pure state $\rho_0$ of $\mathcal{M}$ extends uniquely to a pure state of $C(\mathscr{P}^-, \mathbb{R})$. By using the results of Exercise 4.6.49, deduce that for all $f$ in $C(\mathscr{P}^-, \mathbb{R})$

$$\begin{aligned} f(\rho) &= \inf\{\hat{A}(\rho) : A \in \mathscr{V}, \hat{A} \geqslant f\} \\ &= \sup\{\hat{A}(\rho) : A \in \mathscr{V}, \hat{A} \leqslant f\} \qquad (\rho \in \mathscr{P}). \end{aligned}$$

(iv) Let $A, B \in \mathscr{V}$, and define $C = A \vee B$, $D = A \wedge B$ (where $\vee$, $\wedge$ denote the lattice operations in $\mathscr{V}$). Show that

$$\hat{C}(\rho) = \max\{\hat{A}(\rho), \hat{B}(\rho)\}, \qquad \hat{D}(\rho) = \min\{\hat{A}(\rho), \hat{B}(\rho)\}$$

for all $\rho$ in $\mathscr{P}^-$. [*Hint*. Since $C \geqslant A$ and $C \geqslant B$, we have $\hat{C} \geqslant \hat{A}$ and $\hat{C} \geqslant \hat{B}$; so $\hat{C} \geqslant f$, where $f$ (in $C(\mathscr{P}^-, \mathbb{R})$) is defined by $f(\rho) = \max\{\hat{A}(\rho), \hat{B}(\rho)\}$. Use (iii) to show that we cannot have $\hat{C}(\rho) > f(\rho)$ when $\rho \in \mathscr{P}$.]

(v) Use the result of Exercise 3.5.49 to show that $\mathscr{M} = C(\mathscr{P}^-, \mathbb{R})$ (and deduce that $\mathscr{P} = \mathscr{P}^-$).

4.6.54. Suppose that $\mathfrak{A}$ is a $C^*$-algebra and, with the usual partial ordering, the set $\mathfrak{A}_h$ of all self-adjoint elements of $\mathfrak{A}$ is a lattice. Prove that $\mathfrak{A}$ is abelian. [*Hint*. Use the results of Exercises 4.6.53 and 4.6.27.]

4.6.55. Show that, if $x$ and $y$ are unit vectors in a Hilbert space $\mathscr{H}$ and the corresponding vector states of $\mathscr{B}(\mathscr{H})$ satisfy $\|\omega_x - \omega_y\| < \varepsilon$, then $\|x - cy\| < \varepsilon^{1/2}$ for some complex number $c$ for which $|c| = 1$.

Deduce that if a state $\omega$ of $\mathscr{B}(\mathscr{H})$ is the norm limit of a sequence of vector states of $\mathscr{B}(\mathscr{H})$, then $\omega$ is a vector state. [This is a special case of a result proved in Theorem 7.3.11.]

4.6.56. With the notation of Exercises 1.9.19 and 3.5.4, observe that $l_\infty$ becomes an abelian $C^*$-algebra in which $c$ is a $C^*$-subalgebra and $c_0$ ($\subseteq c$) is a closed ideal when the involution in $l_\infty$ is pointwise complex conjugation (that is, $\{x_n\}^*$ is $\{\bar{x}_n\}$ for each bounded complex sequence $\{x_n\}$). Note also that the equation

$$\rho_0(\{x_n\}) = \lim_{n \to \infty} x_n \qquad (\{x_n\} \in c)$$

defines a pure state $\rho_0$ of $c$.

(i) Let $\rho$ be any pure state of $l_\infty$ that extends $\rho_0$ (Theorem 4.3.13(iv)). Show that $\rho$ is a multiplicative linear functional on $l_\infty$, and that

$$\liminf x_n \leqslant \rho(\{x_n\}) \leqslant \limsup x_n$$

for every bounded real sequence $\{x_n\}$.

(ii) Show that the multiplicative linear functionals $\rho$ satisfying the conditions set out in (i) are precisely the elements of $\beta(\mathbb{N})\backslash\mathbb{N}$ (see Exercises 3.5.5 and 3.5.6).

4.6.57. Let $\mathscr{H}$ be a Hilbert space and let $\mathscr{L}_0$ be the set of all sequences $\{u_1, u_2, \ldots\}$ of elements of $\mathscr{H}$ that are weakly convergent to 0. Let $\rho$ be a pure state of the $C^*$-algebra $l_\infty$, with the properties set out in Exercise 4.6.56(i).

(i) Show that $\mathscr{L}_0$ becomes a complex vector space when $a\{u_n\} + b\{v_n\}$ is defined to be $\{au_n + bv_n\}$ for all $\{u_n\}$ and $\{v_n\}$ in $\mathscr{L}_0$.

(ii) Show that, if $\{u_n\}, \{v_n\} \in \mathscr{L}_0$, then the complex sequences $\{\langle u_n, v_n\rangle\}$, $\{\|u_n\|\}$ are in $l_\infty$. Show also that the equation

$$\langle\{u_n\}, \{v_n\}\rangle_0 = \rho(\{\langle u_n, v_n\rangle\})$$

defines an inner product $\langle\, ,\rangle_0$ on $\mathscr{L}_0$, and the corresponding semi-norm $\|\;\|_0$ on $\mathscr{L}_0$ is given by

$$\|\{u_n\}\|_0 = \rho(\{\|u_n\|\}).$$

(iii) Let $\mathscr{N}_0$ be the subspace $\{\{u_n\} \in \mathscr{L}_0 : \|\{u_n\}\|_0 = 0\}$ of $\mathscr{L}_0$. With $\mathscr{L}_1$ the quotient space $\mathscr{L}_0/\mathscr{N}_0$, let $\langle\, ,\rangle_1$ be the definite inner product on $\mathscr{L}_1$, and $\|\;\|_1$ the corresponding norm, derived (as in Proposition 2.1.1) from $\langle\, ,\rangle_0$. Let $\mathscr{L}$ be the completion of the pre-Hilbert space $\mathscr{L}_0/\mathscr{N}_0$ so obtained, and use the same symbols, $\langle\, ,\rangle_1$ and $\|\;\|_1$, for its inner product and norm. Show that, for each $T$ in $\mathscr{B}(\mathscr{H})$, the equation

$$\pi_0(T)\{u_n\} = \{Tu_n\} \qquad (\{u_n\} \in \mathscr{L}_0)$$

defines a linear operator $\pi_0(T)$ acting on $\mathscr{L}_0$, and

$$\|\pi_0(T)\{u_n\}\|_0 \leqslant \|T\|\,\|\{u_n\}\|_0.$$

Deduce that the mapping

$$\{u_n\} + \mathscr{N}_0 \to \{Tu_n\} + \mathscr{N}_0 : \quad \mathscr{L}_1 \to \mathscr{L}_1$$

is well defined and extends uniquely to a bounded linear operator $\pi(T)$ acting on $\mathscr{L}$ with $\|\pi(T)\| \leqslant \|T\|$.

(iv) Show that the mapping

$$\pi : T \to \pi(T) : \quad \mathscr{B}(\mathscr{H}) \to \mathscr{B}(\mathscr{L})$$

is a representation of $\mathscr{B}(\mathscr{H})$.

(v) Show that the kernel of $\pi$ is the ideal $\mathscr{K}$ consisting of all compact linear operators acting on $\mathscr{H}$. [*Hint*. Use condition (ii) in Exercise 2.8.20 as the defining property of a compact linear operator.]

4.6.58. Suppose that $\mathscr{H}$ is a separable Hilbert space, $\mathscr{K}$ ($\subseteq \mathscr{B}(\mathscr{H})$) is the ideal consisting of all compact linear operators acting on $\mathscr{H}$, and $\{e_q : q \in \mathbb{Q}\}$ is an orthonormal basis of $\mathscr{H}$ indexed by the (countable) set $\mathbb{Q}$ of all rational numbers. Let $\varphi$ be a representation of $\mathscr{B}(\mathscr{H})$ that has kernel $\mathscr{K}$ (see Exercise 4.6.57). For each real number $t$, choose a sequence $\{q(1), q(2), \ldots\}$ of rational numbers (with no repetitions) that converges to $t$, and let $E_t$ be the projection from $\mathscr{H}$ onto the subspace generated by $\{e_{q(1)}, e_{q(2)}, \ldots\}$. Show that:

(i) $\{E_t : t \in \mathbb{R}\}$ is a commuting family of projections such that $E_t \notin \mathscr{K}$, $E_s E_t \in \mathscr{K}$, whenever $s, t \in \mathbb{R}$ and $s \neq t$;

(ii) the Hilbert space on which $\varphi(\mathscr{B}(\mathscr{H}))$ acts is not separable;

(iii) the result of Exercise 4.6.20(ii) cannot be extended to the case in which $\{B_1, B_2, \ldots\}$ is replaced by an uncountable family, even when $\mathfrak{A}$ is abelian.

4.6.59. Suppose that $\{e_0, e_1, e_2, \ldots\}$ is an orthonormal basis in a separable Hilbert space $\mathscr{H}$, and $W$ is the isometric linear operator on $\mathscr{H}$ defined (as in Example 3.2.18) by $We_j = e_{j+1}$ $(j = 0, 1, 2, \ldots)$. Let $\mathscr{K}$ be the ideal in $\mathscr{B}(\mathscr{H})$ that consists of all compact linear operators acting on $\mathscr{H}$, and let $\pi$ be a representation of $\mathscr{B}(\mathscr{H})$ that has kernel $\mathscr{K}$ (see Exercise 4.6.57)..

(i) Show that $W^*W = I$ and $WW^* = I - E_0$, where $E_0$ is the projection from $\mathscr{H}$ onto the one-dimensional subspace containing $e_0$.

(ii) Show that $\pi(W)$ is a unitary operator $U$.

(iii) Show that there is no invertible operator $T$ in $\mathscr{B}(\mathscr{H})$ such that $\pi(T) = U$. [*Hint*. If $K \in \mathscr{K}$ and $T = W + K$, use the relation $T^*W = I + K^*W$ and the result of Exercise 3.5.18(iv) to show that $T$ is not invertible.]

(iv) Show that there is no normal operator $N$ in $\mathscr{B}(\mathscr{H})$ such that $\pi(N) = U$. [*Hint*. Suppose that $K \in \mathscr{K}$ and $W + K$ is a normal operator $N$. Let $\mathscr{M}$ be the null space of $N$, so that $\mathscr{M}$ is also the null space of $N^*$ (Proposition 2.4.6(iii)). Use the relation $W^*N = I + W^*K$ and the properties of compact linear operators to show that $\mathscr{M}$ is finite dimensional. Prove also that $N + E$ $(= W + K + E)$ is normal and one-to-one, where $E$ is the projection from $\mathscr{H}$ onto $\mathscr{M}$. Hence reduce to the case in which $\mathscr{M} = \{0\}$. In this case, use the relation $N^*W = I + K^*W$ and the properties of compact linear operators to show that $N$ is invertible in $\mathscr{B}(\mathscr{H})$, contradicting the conclusion of (iii).]

4.6.60. Suppose that $\mathscr{I}$ is a proper closed two-sided ideal in a $C^*$-algebra $\mathfrak{A}$, $\{V_\lambda\}$ is an increasing two-sided approximate identity for $\mathscr{I}$, and $\varphi\colon \mathfrak{A} \to \mathfrak{A}/\mathscr{I}$ is the quotient mapping from $\mathfrak{A}$ onto the Banach algebra $\mathfrak{A}/\mathscr{I}$ (Proposition 3.1.8). Prove that:

(i) $\mathfrak{A}/\mathscr{I}$ has an involution defined by $\varphi(A)^* = \varphi(A^*)$ $(A \in \mathfrak{A})$;

(ii) the usual quotient norm on $\mathfrak{A}/\mathscr{I}$ satisfies

$$\|\varphi(A)\| = \lim_\lambda \|A - AV_\lambda\| = \lim_\lambda \|A - V_\lambda A\| \qquad (A \in \mathfrak{A});$$

(iii) with the quotient norm and the involution defined in (i), $\mathfrak{A}/\mathscr{I}$ is a $C^*$-algebra. [This important result will be proved by another method in Theorem 10.1.7.]

4.6.61. Show that, if $\mathscr{I}$ is a proper closed two-sided ideal in a $C^*$-algebra $\mathfrak{A}$, then there is a representation of $\mathfrak{A}$ that has kernel $\mathscr{I}$.

4.6.62. Suppose that $\mathfrak{A}$ and $\mathscr{B}$ are $C^*$-algebras, $\varphi$ is a * homomorphism from $\mathfrak{A}$ into $\mathscr{B}$, and $\mathscr{I}$ is a closed two-sided ideal in $\mathfrak{A}$.

(i) By adapting the proof of Theorem 4.1.9, show that $\varphi(\mathscr{I})$ is closed in $\mathscr{B}$.

(ii) Let $\mathscr{C}$ be the $C^*$-subalgebra $\{cI + S : c \in \mathbb{C}, S \in \mathscr{I}\}$ of $\mathfrak{A}$, and let $\mathscr{K}$ be the kernel of the * homomorphism $\varphi \mid \mathscr{C} : \mathscr{C} \to \mathscr{B}$. By considering the induced * isomorphism $\psi$ from the $C^*$-algebra $\mathscr{C}/\mathscr{K}$ into $\mathscr{B}$, give a second proof that $\varphi(\mathscr{I})$ is closed in $\mathscr{B}$.

4.6.63. Suppose that $\mathscr{I}$ and $\mathscr{J}$ are closed two-sided ideals in a $C^*$-algebra $\mathfrak{A}$. By using the results of Exercises 4.6.60(iii) and 4.6.62, show that the ideal $\mathscr{I} + \mathscr{J}$ is closed in $\mathfrak{A}$.

4.6.64. Suppose that $\mathscr{I}$ and $\mathscr{J}$ are closed two-sided ideals in a $C^*$-algebra $\mathfrak{A}$, and $A = B + C \in \mathfrak{A}^+$, where $B \in \mathscr{I}$ and $C \in \mathscr{J}$.

(i) Show that $A = S + T$ for suitably chosen self-adjoint elements $S$ of $\mathscr{I}$ and $T$ of $\mathscr{J}$.

(ii) Suppose that $\varepsilon > 0$, and define

$$H = |S| + |T| + \varepsilon I, \qquad D = A^{1/2}H^{-1/2},$$

$$S_1 = D|S|D^*, \qquad T_1 = D|T|D^*,$$

where $|S|$ and $|T|$ denote the positive square roots of $S^2$ and $T^2$, respectively. Prove that $D^*D \leqslant I$, and deduce that

$$A - \varepsilon I \leqslant S_1 + T_1 \leqslant A.$$

Prove also that

$$S_1 \in \mathscr{I}^+, \qquad T_1 \in \mathscr{J}^+, \qquad \|S_1\| \leqslant \|A\|, \qquad \|T_1\| \leqslant \|A\|,$$

and $0 \leqslant A_1 \leqslant \varepsilon I$, where

$$A_1 = A - S_1 - T_1 = (S - S_1) + (T - T_1) \in (\mathscr{I} + \mathscr{J})^+.$$

(iii) By repeated application of the result of (ii), show that $A$ can be expressed in the form $X + Y$, with $X$ in $\mathscr{I}^+$ and $Y$ in $\mathscr{J}^+$. [This exercise shows that $(\mathscr{I} + \mathscr{J})^+ = \mathscr{I}^+ + \mathscr{J}^+$ when $\mathscr{I}$ and $\mathscr{J}$ are closed two-sided ideals in a $C^*$-algebra.]

4.6.65. Suppose that $\mathfrak{A}$ is a $C^*$-algebra and $\delta : \mathfrak{A} \to \mathfrak{A}$ is a linear mapping such that

$$\delta(AB) = A\delta(B) + \delta(A)B \qquad (A, B \in \mathfrak{A}).$$

(Such a mapping $\delta$ is called a *derivation of* $\mathfrak{A}$.) Let $\mathscr{I}$ be the set of all elements $A$

in $\mathfrak{A}$ for which the linear mapping

$$T \to \delta(AT) \; : \; \mathfrak{A} \to \mathfrak{A}$$

is continuous.

(i) Show that if $A \in \mathfrak{A}$, then $A \in \mathscr{I}$ if and only if the linear mapping

$$T \to A\delta(T) \; : \; \mathfrak{A} \to \mathfrak{A}$$

is continuous.

(ii) Show that $\mathscr{I}$ is a closed two-sided ideal in $\mathfrak{A}$.

(iii) Show that the restriction $\delta | \mathscr{I}$ is continuous. [*Hint.* If $\delta | \mathscr{I}$ is discontinuous, there is a sequence $\{A_1, A_2, \ldots\}$ in $\mathscr{I}$ such that $\sum \|A_j\|^2 < 1$ and $\|\delta(A_j)\| \to \infty$. Use the result of Exercise 4.6.40 to obtain a contradiction.]

(iv) Show that the quotient $C^*$-algebra $\mathfrak{A}/\mathscr{I}$ is finite dimensional. [*Hint.* Suppose the contrary, and deduce from Exercises 4.6.13 and 4.6.20(ii) that the unit ball of $\mathfrak{A}$ contains a sequence $\{S_1, S_2, \ldots\}$ of positive elements not in $\mathscr{I}$ such that $S_j S_k = 0$ when $j \neq k$. Prove that there is a sequence $\{T_1, T_2, \ldots\}$ in $\mathfrak{A}$ such that

$$\|T_j\| \leqslant 2^{-j}, \qquad \|\delta(S_j^2 T_j)\| \geqslant j + \|\delta(S_j)\| \qquad (j = 1, 2, \ldots).$$

Obtain a contradiction by considering $S_j \delta(C)$, where $C = \sum S_j T_j$.]

(v) Deduce that $\delta : \mathfrak{A} \to \mathfrak{A}$ is continuous.

4.6.66. Suppose that $\mathfrak{A}$ is a $C^*$-algebra and $\mathscr{X}$ is a Banach space. We describe $\mathscr{X}$ as a *Banach $\mathfrak{A}$-module* if there are bounded bilinear mappings

$$(A, x) \to Ax, \qquad (A, x) \to xA \; : \; \mathfrak{A} \times \mathscr{X} \to \mathscr{X}$$

such that $Ix = xI = x$ for each $x$ in $\mathscr{X}$, and the associative law holds for each type of triple product $A_1 A_2 x$, $A_1 x A_2$, $x A_1 A_2$. By a *derivation* from $\mathfrak{A}$ into a Banach $\mathfrak{A}$-module $\mathscr{X}$, we mean a linear mapping $\delta : \mathfrak{A} \to \mathscr{X}$ such that

$$\delta(AB) = A\delta(B) + \delta(A)B \qquad (A, B \in \mathfrak{A}).$$

Adapt the program set out in Exercise 4.6.65 to prove that every derivation from a $C^*$-algebra $\mathfrak{A}$ into a Banach $\mathfrak{A}$-module is continuous.

4.6.67. Show that the set $\mathscr{P}$ of pure states of $\mathscr{B}(\mathscr{H})$ is weak* closed when $\mathscr{H}$ is finite dimensional.

4.6.68. Let $\mathscr{H}$ be a Hilbert space. Show that each vector state $\omega_x$ of $\mathscr{B}(\mathscr{H})$ is pure.

4.6.69. Suppose $\mathscr{H}$ is an infinite-dimensional Hilbert space, $\mathscr{K}$ is the ideal of compact operators in $\mathscr{B}(\mathscr{H})$, $\mathscr{P}$ is the set of pure states of $\mathscr{B}(\mathscr{H})$, and $\mathscr{S}_0$ is the set of vector states of $\mathscr{B}(\mathscr{H})$.

(i) Use Exercises 4.6.57 and 4.6.23 to show that there is a pure state $\rho$ of $\mathscr{B}(\mathscr{H})$ that is 0 on $\mathscr{K}$.

(ii) With $a$ in $[0, 1]$, $x$ a unit vector in $\mathscr{H}$, and $\rho$ in $\mathscr{P}^-$ and 0 on $\mathscr{K}$, let $\omega$ be the state $a\omega_x + (1 - a)\rho$ of $\mathscr{B}(\mathscr{H})$. Show that $\omega$ is in $\mathscr{P}^-$, the pure state space of $\mathscr{B}(\mathscr{H})$, and that $\omega$ is in $\mathscr{S}_0^-$. [*Hint*. Use Corollary 4.3.10 to approximate $\rho$ by a vector state $\omega_{y'}$ of $\mathscr{B}(\mathscr{H})$. With $A_1, \ldots, A_n$ self-adjoint operators in $(\mathscr{B}(\mathscr{H}))_1$ and $E$ the projection with range $[x, A_1x, \ldots, A_nx]$, estimate $|(\omega - \omega_z)(A_j)|$ where $z = a^{1/2}x + (1 - a)^{1/2}y$ and $y = \|(I - E)y'\|^{-1}(I - E)y'$.]

(iii) Conclude that $\mathscr{P}$ is not weak* closed (that is, $\mathscr{P} \neq \mathscr{P}^-$) when $\mathscr{H}$ is infinite dimensional.

4.6.70. Let $\mathfrak{A}$ be a $C^*$-algebra and $\mathscr{B}$ be a self-adjoint subalgebra of $\mathfrak{A}$ containing $I$. Suppose that for each pair $\rho_1, \rho_2$ of distinct states of $\mathfrak{A}$ there is a $B$ in $\mathscr{B}$ such that $\rho_1(B) \neq \rho_2(B)$ (that is, $\mathscr{B}$ *separates* the states of $\mathfrak{A}$). Show that $\mathscr{B}$ is norm dense in $\mathfrak{A}$.

# CHAPTER 5

# ELEMENTARY VON NEUMANN ALGEBRA THEORY

Those $C^*$-algebras (von Neumann algebras) that are strong-operator closed in their action on some Hilbert space play a fundamental role in the subject. Historically they were the first class of such operator algebras introduced. Their study will occupy us in this and the following four chapters. In the present chapter we develop the elements of the subject.

The strengthened closure assumption on the algebra entails significant structural changes. On the technical level, the strong-operator closed algebras abound in projections; while the general $C^*$-algebra may contain no projections other than 0 and $I$. In a less technical (and deeper) sense, the passage from the general to the strong-operator closed $C^*$-algebra corresponds to the passage from the algebra of continuous functions to the algebra of (bounded) measurable functions. This correspondence can be made precise in the commutative case and lends force to the interpretation of the theory of von Neumann algebras as "non-commutative measure theory."

## 5.1. The weak- and strong-operator topologies

Recall that the strong-operator topology on $\mathscr{B}(\mathscr{H})$ has a base of neighborhoods of an operator $T_0$ consisting of sets of the type

$$V(T_0 : x_1, \ldots, x_m ; \varepsilon) = \{T \in \mathscr{B}(\mathscr{H}) : \|(T - T_0)x_j\| < \varepsilon \, (j = 1, \ldots, m)\},$$

where $x_1, \ldots, x_m$ are in $\mathscr{H}$ and $\varepsilon$ is positive. Thus the net $\{T_j\}$ is strong-operator convergent to $T_0$ if and only if $\{\|(T_j - T_0)x\|\}$ converges to 0 for each $x$ in $\mathscr{H}$, that is, if and only if the net $\{T_j x\}$ of vectors in $\mathscr{H}$ converges to $T_0 x$ for each $x$ in $\mathscr{H}$. (See the discussion following Proposition 2.5.8 and the comments in Remark 2.5.9.)

Another topology on $\mathscr{B}(\mathscr{H})$ will be important for us.

5.1.1. Definition. The *weak-operator topology* on $\mathscr{B}(\mathscr{H})$ is the weak topology on $\mathscr{B}(\mathscr{H})$ (in the sense described in Section 1.3) induced by the family

$\mathscr{F}_w$ of linear functionals $\omega_{x,y}$: $\mathscr{B}(\mathscr{H}) \to \mathbb{C}$ defined by the equation

$$\omega_{x,y}(A) = \langle Ax, y \rangle \qquad (x, y \in \mathscr{H}, \quad A \in \mathscr{B}(\mathscr{H})). \quad \blacksquare$$

If $\omega_{x,y}(A) = 0$ for all $x$ and $y$ in $\mathscr{H}$, then $A = 0$, whence $\mathscr{F}_w$ is a separating family of linear functionals for $\mathscr{B}(\mathscr{H})$. It follows that the weak-operator topology on $\mathscr{B}(\mathscr{H})$ is a locally convex topology determined by semi-norms $|\omega_{x,y}(A)|$. The family of sets of the form

$$V(T_0 : \omega_{x_1,y_1}, \ldots, \omega_{x_m,y_m}; \varepsilon)$$
$$= \{T \in \mathscr{B}(\mathscr{H}) : |\langle (T - T_0)x_j, y_j \rangle| < \varepsilon \, (j = 1, \ldots, m)\},$$

where $\varepsilon$ is positive and $x_1, \ldots, x_m$, $y_1, \ldots, y_m$ are in $\mathscr{H}$, constitutes a base of convex (open) neighborhoods of $T_0$ in the weak-operator topology. Since $|\langle (T - T_0)x, y \rangle| < \varepsilon$ when $\|(T - T_0)x\| < \varepsilon(1 + \|y\|)^{-1}$, each open set relative to the weak-operator topology is open relative to the strong-operator topology. Hence the weak-operator topology is weaker (coarser) than the strong-operator topology. (See Exercise 5.7.2 where it is noted that this relation is "strict.") As a consequence, the requirement that a subset of $\mathscr{B}(\mathscr{H})$ be strong-operator closed is less stringent than the requirement that it be weak-operator closed. An important exception to this occurs in the class of convex sets of operators.

5.1.2. THEOREM. *The weak- and strong-operator closures of a convex subset $\mathscr{K}$ of $\mathscr{B}(\mathscr{H})$ coincide.*

*Proof.* An operator in the strong-operator closure of $\mathscr{K}$ is in the weak-operator closure of $\mathscr{K}$. Suppose $A$, in the weak-operator closure of $\mathscr{K}$, and vectors $x_1, \ldots, x_n$ in $\mathscr{H}$ are given. Let $\tilde{\mathscr{H}}$ be the direct sum $\mathscr{H} \oplus \cdots \oplus \mathscr{H}$ of $\mathscr{H}$ with itself $n$ times. For $T$ in $\mathscr{B}(\mathscr{H})$, let $\tilde{T}(y_1, \ldots, y_n)$ be $(Ty_1, \ldots, Ty_n)$ (that is, $\tilde{T} = T \oplus \cdots \oplus T$). Then $\{\tilde{T} : T \text{ in } \mathscr{K}\}$ is a convex subset $\tilde{\mathscr{K}}$ of $\mathscr{B}(\tilde{\mathscr{H}})$; and $\tilde{\mathscr{K}}\tilde{x}$ is a convex subset of $\tilde{\mathscr{H}}$, where $\tilde{x} = (x_1, \ldots, x_n)$. As $\tilde{A}$ is in the weak-operator closure of $\tilde{\mathscr{K}}$, $\tilde{A}\tilde{x}$ is in the weak closure of $\tilde{\mathscr{K}}\tilde{x}$ (in $\tilde{\mathscr{H}}$). From Theorem 1.3.4, $\tilde{A}\tilde{x}$ is in the norm closure of $\tilde{\mathscr{K}}\tilde{x}$ (in $\tilde{\mathscr{H}}$). Thus for some $K$ in $\mathscr{K}$, $\|Kx_j - Ax_j\|$ is small for each $j$ in $\{1, \ldots, n\}$. It follows that $A$ is in the strong-operator closure of $\mathscr{K}$ and that the weak- and strong-operator closures of $\mathscr{K}$ coincide. $\blacksquare$

By polarization (see 2.4(3)) the span of the functionals $\omega_{x,x}$ $(= \omega_x)$ coincides with the span of $\mathscr{F}_w$, so that the semi-norms defined by $|\langle Ax, x \rangle|$ determine the weak-operator topology on $\mathscr{B}(\mathscr{H})$. In fact, restricted to $(\mathscr{B}(\mathscr{H}))_1$, the unit ball in $\mathscr{B}(\mathscr{H})$ (or, equally, any bounded subset of $\mathscr{B}(\mathscr{H})$), the weak-operator topology is determined by the semi-norms $|\langle Ax_j, x_k \rangle|$ where $(x_j)$ spans a dense linear manifold in $\mathscr{H}$. For this, note that $|\langle Ax, x \rangle|$ is small (with $A$ in $(\mathscr{B}(\mathscr{H}))_1$) provided $|\langle Ay, y \rangle|$ is small with $y$ (in the span of $(x_j)$) sufficiently near $x$.

Since $\langle BAx, y\rangle = \langle Ax, B^*y\rangle$, the mappings $A \to BA$ and $A \to AB$ ($B \in \mathscr{B}(\mathscr{H})$) of $\mathscr{B}(\mathscr{H})$ into $\mathscr{B}(\mathscr{H})$ are weak-operator continuous. That is, left and right multiplication by $B$ are weak-operator continuous. We note, too, from Theorem 1.3.1, that each weak-operator continuous linear functional on $\mathscr{B}(\mathscr{H})$ lies in the linear span of $\mathscr{F}_{\text{w}}$.

Another useful aspect of the weak-operator topology resides in a special compactness property it possesses.

5.1.3. THEOREM. *The unit ball* $(\mathscr{B}(\mathscr{H}))_1$ *of* $\mathscr{B}(\mathscr{H})$ *is weak-operator compact.*

*Proof.* Let $\mathbb{D}_{x,y}$ be the closed disk of radius $\|x\| \cdot \|y\|$ in the plane $\mathbb{C}$ of complex numbers. The mapping which assigns to each $T$ in $(\mathscr{B}(\mathscr{H}))_1$ the point $\{\langle Tx, y\rangle : x, y \text{ in } \mathscr{H}\}$ of $\prod_{x,y} \mathbb{D}_{x,y}$ is a homeomorphism of $(\mathscr{B}(\mathscr{H}))_1$, with the weak-operator topology, onto its image $X$ in the topology induced on $X$ by the product topology on $\prod_{x,y} \mathbb{D}_{x,y}$ (from the very definition of these topologies). As $\prod_{x,y} \mathbb{D}_{x,y}$ is a compact Hausdorff space in the product topology (Tychonoff's theorem), $X$ is compact if it is closed. If $b$ is a point in the closure of $X$ and $x_1$, $y_1$, $x_2$, $y_2$ are elements of $\mathscr{H}$, then, for each positive number $\varepsilon$, there is a $T$ in $(\mathscr{B}(\mathscr{H}))_1$ such that each of

$$|a \cdot b(x_j, y_k) - a\langle Tx_j, y_k\rangle|, \qquad |b(x_j, y_k) - \langle Tx_j, y_k\rangle|,$$

$$|b(ax_1 + x_2, y_j) - \langle T(ax_1 + x_2), y_j\rangle|, \qquad |b(x_j, ay_1 + y_2) - \langle Tx_j, ay_1 + y_2\rangle|$$

is less than $\varepsilon$, where $j, k = 1, 2$. It follows that

$$|b(ax_1 + x_2, y_1) - a \cdot b(x_1, y_1) - b(x_2, y_1)| < 3\varepsilon$$

and

$$|b(x_1, ay_1 + y_2) - \bar{a} \cdot b(x_1, y_1) - b(x_1, y_2)| < 3\varepsilon.$$

Thus $b(ax_1 + x_2, y_1) = a \cdot b(x_1, y_1) + b(x_2, y_1)$ and $b(x_1, ay_1 + y_2) = \bar{a} \cdot b(x_1, y_1) + b(x_1, y_2)$. In addition $|b(x, y)| \leqslant \|x\| \cdot \|y\|$, since $b(x, y) \in \mathbb{D}_{x,y}$. Hence $b$ is a conjugate-bilinear functional on $\mathscr{H}$ bounded by 1. From the Riesz representation (Theorem 2.4.1) of such bilinear functionals, there is an operator $T_0$ in $(\mathscr{B}(\mathscr{H}))_1$ such that $b(x, y) = \langle T_0 x, y\rangle$ for all $x$ and $y$ in $\mathscr{H}$. Thus $b \in X$, $X$ is closed, $X$ is compact, and $(\mathscr{B}(\mathscr{H}))_1$ is weak-operator compact. ■

The weak-operator topology and the Riesz representation of bounded conjugate-bilinear functionals on a Hilbert space appear once again in establishing a key order-topological property of $\mathscr{B}(\mathscr{H})$. It concerns nets $\{H_a, \mathbb{A}, \leqslant\}$ of self-adjoint operators $H_a$ for which the operator-ordering and the partial ordering of the directed index set $\mathbb{A}$ agree (that is, $H_a \leqslant H_{a'}$ if $a \leqslant a'$). We say that such nets are *monotone increasing* (*decreasing* if the operator-ordering reverses the ordering of $\mathbb{A}$). Although it will prove useful to have the result that follows for nets, for present purposes the simpler circumstances of sequences would suffice.

5.1.4. LEMMA. *If $\{H_a\}$ is a monotone increasing net of self-adjoint operators on the Hilbert space $\mathscr{H}$ and $H_a \leqslant kI$ for all $a$, then $\{H_a\}$ is strong-operator convergent to a self-adjoint operator $H$, and $H$ is the least upper bound of $\{H_a\}$.*

*Proof.* Since the convergence of $\{H_a\}$ and that of $\{H_a, a \geqslant a_0\}$ are equivalent, we may assume that $\{H_a\}$ is bounded below (by $H_{a0}$) as well as above. Thus $-\|H_{a0}\|I \leqslant H_a \leqslant kI$, and $\{H_a\}$ is a bounded set of operators. From the weak-operator compactness of a closed ball in $\mathscr{B}(\mathscr{H})$ (Theorem 5.1.3), some subnet $\{H_{a'}\}$ of $\{H_a\}$ is weak-operator convergent to an operator $H$ in $\mathscr{B}(\mathscr{H})$.

As $\{H_a\}$ is monotone increasing, $\langle H_{a'}x, x\rangle \geqslant \langle H_{a_1}x, x\rangle$ when $a' \geqslant a_1$. Thus $H \geqslant H_{a_1}$ for all $a_1$, since $\langle H_{a_1}x, x\rangle \leqslant \lim_{a'}\langle H_{a'}x, x\rangle = \langle Hx, x\rangle$. If $a \geqslant a'$, then $0 \leqslant H - H_a \leqslant H - H_{a'}$; and

$$0 \leqslant \langle (H - H_a)x, x\rangle = \|(H - H_a)^{1/2}x\|^2 \leqslant \langle (H - H_{a'})x, x\rangle \to 0.$$

Hence $\{(H - H_a)^{1/2}\}$ is strong-operator convergent to 0. The strong-operator continuity of multiplication on bounded sets of operators (see Remark 2.5.10) allows us to conclude that $\{H - H_a\}$ is strong-operator convergent to 0.

We have noted that $H$ is an upper bound for $\{H_a\}$. If $K \geqslant H_a$ for all $a$, then $\langle Kx, x\rangle \geqslant \langle H_a x, x\rangle \to_a \langle Hx, x\rangle$. Hence $\langle Kx, x\rangle \geqslant \langle Hx, x\rangle$ for all $x$ in $\mathscr{H}$; and $K \geqslant H$. It follows that $H$ is the least upper bound of $\{H_a\}$. ■

If $\{K_a\}$ is a family of positive operators acting on $\mathscr{H}$, the net of finite subsums of $\sum K_a$ is monotone increasing. From the preceding lemma, this net is strong-operator convergent (equivalently, the family $\{K_a\}$ is summable, in the sense of the discussion preceding Proposition 1.2.19, in the strong-operator topology) if the finite subsums are bounded above.

5.1.5. LEMMA. *If $A$ is a bounded operator on the Hilbert space $\mathscr{H}$ and $0 \leqslant A \leqslant I$, then $\{A^{1/n}\}$ is a monotone increasing sequence of operators whose strong-operator limit is the projection on the closure of the range of $A$.*

*Proof.* Passing to the function algebra representing $\mathfrak{A}(A)$, the $C^*$-subalgebra of $\mathfrak{A}$ generated by $A$ and $I$ (see Theorem 4.1.3), $\{A^{1/n}\}$ is seen to be a monotone increasing sequence bounded above by $I$. From Lemma 5.1.4, $\{A^{1/n}\}$ has a strong-operator limit $E$, which is the least upper bound of $\{A^{1/n}\}$. Hence $\{A^{2/n}\}$ has $E^2$ as strong-operator limit. But $\{A^{1/n}\}$ $(= \{A^{2/2n}\})$ is a subsequence of $\{A^{2/n}\}$, so that $E = E^2$, and $E$ is a projection.

Applying the Stone–Weierstrass theorem in the function algebra representing $\mathfrak{A}(A)$, we see that $A^{1/n}$ is the norm limit of polynomials, without constant term, in $A$. Thus $A^{1/n}x = 0$ if $Ax = 0$; and $Ex = 0$ when $Ax = 0$. On the other hand, if $Ex = 0$, then $0 = \langle Ex, x\rangle \geqslant \langle A^{1/n}x, x\rangle = \|A^{1/2n}x\|^2 \geqslant 0$, so that $A^{1/2n}x = 0$ and $Ax = 0$. It follows that the (self-adjoint) operators $E$ and $A$ have the same null space. Hence $E = R(A)$ (see Proposition 2.5.13). ■

A $C^*$-algebra $\mathscr{R}$, acting on a Hilbert space $\mathscr{H}$, that is closed in the weak-operator topology and contains $I$ is said to be a *von Neumann algebra*. If the center of $\mathscr{R}$ consists of scalar multiples of $I$, we say that $\mathscr{R}$ is a *factor*.

5.1.6. EXAMPLE. In Section 4.1 we noted that the algebra $\mathscr{A}$ of multiplications by essentially bounded (measurable) functions on $L_2(S, \mathscr{S}, m)$ $(= L_2)$ is a $C^*$-algebra. Since $M_f M_g = M_{fg} = M_g M_f$, $\mathscr{A}$ is abelian. Let $e_n$ be the characteristic function of the subset $S_n$ of finite measure in $S$, where the sets $S_n$ are so chosen that they are mutually disjoint with union $S$. Then $M_{e_n}$ is a projection and $\sum M_{e_n} = \vee M_{e_n} = I$ (see Proposition 2.5.8 and Example 2.5.12).

If $T$ is a bounded operator on $L_2$ commuting with $\mathscr{A}$ and $Te_n = f_n$, then (by a measure-theoretic argument similar to that used in proving 2.4(13)) $f_n$ is essentially bounded; for

$$TM_{e_n}(g) = T(ge_n) = M_g T(e_n) = f_n g \; (= M_{f_n}(g)),$$

for each essentially bounded $g$ in $L_2$. Since $TM_{e_n}$ and $M_{f_n}$ are bounded and the essentially bounded functions are dense in $L_2$, $TM_{e_n} = M_{f_n}$. From Example 2.4.11, $\|M_{f_n}\|$ is the essential bound of $f_n$. As $\|M_{f_n}\| \leqslant \|T\|$, the function $f$ defined by $f|S_n = f_n|S_n$ is essentially bounded. Note for this that $f_n = T(e_n) = T(e_n \cdot e_n) = e_n T(e_n) = e_n \cdot f_n$. Thus, $M_f M_{e_n} = M_{fe_n} = M_{f_n} = TM_{e_n}$ for each $n$. Since $\sum M_{e_n} = I$, $M_f = T$.

It follows that $\mathscr{A}$ is not a proper subset of a commuting family of bounded operators on $L_2$. We say that $\mathscr{A}$ is *maximal abelian* in this case. Since $\{T: \langle (TA - AT)x, y\rangle = 0\}$ is a weak-operator closed set and $\mathscr{A}$ is the intersection, over all $A$ in $\mathscr{A}$ and all $x$ and $y$ in $L_2$, of these sets (from the facts we established above); $\mathscr{A}$ is weak-operator closed. In the terminology just introduced, $\mathscr{A}$ is a von Neumann algebra.

We now specialize to the case where $S$ is the interval $[0, 1]$ and $m$ is Lebesgue measure, so that the set $\{M_f : f \in C(S)\}$ is a $C^*$-subalgebra $\mathfrak{A}$ of $\mathscr{A}$. We assert that $\mathfrak{A}$ has weak-operator closure $\mathfrak{A}$ (illustrating the fact that the passage from a $C^*$-algebra of operators to the von Neumann algebra it generates is analogous to the transition from continuous functions to bounded measurable functions). For this, note that each $f$ in $L_\infty$ is the limit almost everywhere of a sequence $\{f_n\}$ of continuous functions such that $\|f_n\|_\infty \leqslant \|f\|_\infty$ for each $n$; and $M_f$ is the strong- (and hence weak-)operator limit of $\{M_{f_n}\}$, by the final paragraph of Remark 2.5.12. ■

5.1.7. EXAMPLE. The algebra $\mathscr{B}(\mathscr{H})$ of all bounded operators on a Hilbert space is an example of a factor. The projections $E_0$ with one-dimensional range are *minimal projections* in $\mathscr{B}(\mathscr{H})$ (contain no other non-zero projections in $\mathscr{B}(\mathscr{H})$). We shall note, at a later point (Theorem 6.6.1), that the property of having a (single) minimal projection characterizes $\mathscr{B}(\mathscr{H})$ among the factors (up to * isomorphism). On the other hand, if $\{E_n\}$ is a family of minimal projections in $\mathscr{B}(\mathscr{H})$ whose ranges correspond to an orthonormal basis $\{x_n\}$ for $\mathscr{H}$ (that is, the range of $E_n$ is spanned by $x_n$), then $\{E_n\}$ is an orthogonal family with sum $I$. ■

Although we imposed the requirement that a von Neumann algebra contain $I$ (for convenience and for simplicity of statements of theorems), this requirement entails no essential restriction. The result that follows permits us to consider the action of a weak-operator closed algebra on $\mathscr{H}$, stable under the adjoint operation, on the range of its maximal projection – where it becomes a von Neumann algebra in our sense.

5.1.8. PROPOSITION. *If $\mathfrak{A}$ is a weak-operator closed self-adjoint algebra of operators acting on the Hilbert space $\mathscr{H}$, the union and intersection of each family of projections in $\mathfrak{A}$ lie in $\mathfrak{A}$. There is a projection $P$ in $\mathfrak{A}$, larger than all other projections in $\mathfrak{A}$, such that $PA = AP = A$ for all $A$ in $\mathfrak{A}$.*

*Proof.* Since $R(T^*) = R(T^*T)$, from Proposition 2.5.13, once we note that the range projection of each positive self-adjoint operator in $\mathfrak{A}$ lies in $\mathfrak{A}$, the same is true for each operator in $\mathfrak{A}$. Now Lemma 5.1.5 assures us that $R(A)$ is the strong- (hence, weak-)operator limit of $\{A^{1/n}\}$, when $0 \leqslant A \leqslant I$. With $A$ in $\mathfrak{A}$, $A^{1/n} \in \mathfrak{A}$, since $\mathfrak{A}$ is norm closed. Thus $R(A) \in \mathfrak{A}$. It follows that $R(T) \in \mathfrak{A}$ for each $T$ in $\mathfrak{A}$.

If $E$ and $F$ are projections, $R(E + F) = E \vee F$ (see Proposition 2.5.14). Thus $E \vee F \in \mathfrak{A}$, if $E$ and $F$ are in $\mathfrak{A}$. It follows that finite unions of projections in $\mathfrak{A}$ lie in $\mathfrak{A}$. If $\{E_a\}$ is a family of projections in $\mathfrak{A}$, the unions of finite subfamilies lie in $\mathfrak{A}$ and form a monotone increasing net bounded above by $I$. From Lemma 5.1.4, this net has a least upper bound $E$ which is its strong-operator limit. Thus $E \in \mathfrak{A}$; and, from Remark 2.5.9, $E = \bigvee_a E_a$. Again, from 2.5(4) (applied in the Hilbert space $P(\mathscr{H})$), $\bigwedge_a E_a = P - \bigvee_a(P - E_a)$, where $P$ is the union of all projections in $\mathfrak{A}$. From the foregoing, $P \in \mathfrak{A}$ and $\bigwedge_a E_a \in \mathfrak{A}$.

As $P$ contains the range projection of each $A$ in $\mathfrak{A}$, $PA = A$ and $PA^* = A^*$. Thus $PA = AP = A$; and $P$ is a multiplicative unit for $\mathfrak{A}$. ■

## 5.2. Spectral theory for bounded operators

We use the special lattice-theoretic properties of abelian von Neumann algebras to identify important features of their representing function systems.

These features are used, in turn, to develop the classical form of the spectral theorem for bounded self-adjoint operators. We treat the case of unbounded operators in Section 5.6. In the case of finite-dimensional Hilbert spaces, the results of this section reduce to the familiar diagonalization of self-adjoint (and normal) matrices relative to orthonormal bases.

5.2.1. THEOREM. *If $\mathscr{A}$ is an abelian von Neumann algebra, then $\mathscr{A} \cong C(X)$, where $X$ is an extremely disconnected compact Hausdorff space.*

*Proof.* From Theorem 4.4.3, $\mathscr{A} \cong C(X)$ for some compact Hausdorff space $X$. Since the isomorphism transports the operator order on the set of self-adjoint operators in $\mathscr{A}$ to the pointwise order on the set of real-valued functions in $C(X)$, each increasing net $\{f_a\}$ in $C(X)$, bounded above by $k$, corresponds to an increasing net $\{A_a\}$ of self-adjoint operators in $\mathscr{A}$, bounded above by $kI$. From Lemma 5.1.4, $\{A_a\}$ has a least upper bound $A$ in $\mathscr{A}$. There is an $f$ in $C(X)$, corresponding to $A$, which is the least upper bound of $\{f_a\}$. If $\{f_a\}$ is an arbitrary family of real-valued functions in $C(X)$ bounded above by $k$, the least upper bounds of finite subsets of $\{f_a\}$ form an increasing net, bounded above by $k$, whose least upper bound (in $C(X)$) is a least upper bound (in $C(X)$) for $\{f_a\}$. From Theorem 3.4.16, $X$ is extremely disconnected. ■

Our next result describes the *spectral resolution* of a (bounded) self-adjoint operator.

5.2.2. THEOREM. *If $A$ is a self-adjoint operator acting on a Hilbert space $\mathscr{H}$ and $\mathscr{A}$ is an abelian von Neumann algebra containing $A$, there is a family $\{E_\lambda\}$ of projections, indexed by $\mathbb{R}$, in $\mathscr{A}$ such that*

(i) $E_\lambda = 0$ *if* $\lambda < -\|A\|$, *and* $E_\lambda = I$ *if* $\|A\| \leqslant \lambda$;
(ii) $E_\lambda \leqslant E_{\lambda'}$ *if* $\lambda \leqslant \lambda'$;
(iii) $E_\lambda = \bigwedge_{\lambda' > \lambda} E_{\lambda'}$;
(iv) $AE_\lambda \leqslant \lambda E_\lambda$ *and* $\lambda(I - E_\lambda) \leqslant A(I - E_\lambda)$ *for each* $\lambda$;
(v) $A = \int_{-\|A\|}^{\|A\|} \lambda \, dE_\lambda$ *in the sense of norm convergence of approximating Riemann sums; and $A$ is the norm limit of finite linear combinations with coefficients in* sp($A$) *of orthogonal projections* $E_{\lambda'} - E_\lambda$.

*With $\mathscr{A}$ isomorphic to $C(X)$ and $X$ an extremely disconnected compact Hausdorff space, if $f$ and $e_\lambda$ in $C(X)$ correspond to $A$ and $E_\lambda$ in $\mathscr{A}$, then $e_\lambda$ is the characteristic function of the largest clopen subset $X_\lambda$ on which $f$ takes values not exceeding $\lambda$.*

*Proof.* From Theorem 5.2.1, we have that $\mathscr{A}$ is isomorphic to $C(X)$ for some extremely disconnected compact Hausdorff space $X$. If $f$ corresponds to $A$ and $X_\lambda = X \backslash f^{-1}((\lambda, \infty))^-$, then $X_\lambda$ is a clopen subset of $X$ on which $f$ takes values not exceeding $\lambda$. If $Y$ is another clopen subset of $X$ on which $f$ takes values not exceeding $\lambda$, then $Y \subseteq X \backslash f^{-1}((\lambda, \infty))$ so that $f^{-1}((\lambda, \infty)) \subseteq X \backslash Y$. As $Y$ is open, $X \backslash Y$ is closed; and $f^{-1}((\lambda, \infty))^{-} \subseteq X \backslash Y$. Thus $Y \subseteq X_\lambda$; and $X_\lambda$ is the *largest* clopen set in $X$ on which $f$ takes values not exceeding $\lambda$. If $e_\lambda$ is the characteristic function of $X_\lambda$, $e_\lambda \in C(X)$ since $X_\lambda$ is clopen; and $E_\lambda$ in $\mathscr{A}$ corresponding to $e_\lambda$ is a projection. Conditions (i), (ii), and (iv), in the statement of this theorem, are apparent from the definition of $X_\lambda$ and the properties of the isomorphism of $\mathscr{A}$ with $C(X)$. (For the second inequality in (iv), we make special use of the continuity of $f$ to conclude that $f$ takes values greater than or equal to $\lambda$ on $f^{-1}((\lambda, \infty))^{-}$.)

To prove (iii), note, first, that $e_\lambda$ is the greatest lower bound in $C(X)$ of $\{e_{\lambda'} : \lambda' > \lambda\}$; for if $h$ is a (positive) lower bound in $C(X)$ and $h(p) \neq 0$, then $h$ is not 0 at each point of the clopen set $h^{-1}((\frac{1}{2}h(p), 2h(p)))^{-}$ $(= Y)$. Since $h \leqslant e_{\lambda'}$ if $\lambda < \lambda'$; $Y \subseteq X_{\lambda'}$. Thus $Y$ is a clopen set on which $f$ takes values not exceeding $\lambda'$, for each $\lambda'$ greater than $\lambda$ – that is, $f$ takes values not exceeding $\lambda$ on $Y$. From our characterization of $X_\lambda$ as the *largest* such clopen set, $Y \subseteq X_\lambda$ and $h \leqslant e_\lambda$. From (ii), $e_\lambda$ is a lower bound for $\{e_{\lambda'} : \lambda' > \lambda\}$, so that $e_\lambda$ is the greatest lower bound of this set in $C(X)$. It follows that $E_\lambda$ is the greatest lower bound of $\{E_{\lambda'} : \lambda' > \lambda\}$ in $\mathscr{A}$. From Proposition 5.1.8, $\bigwedge_{\lambda' > \lambda} E_{\lambda'} \in \mathscr{A}$. From Corollary 2.5.7, $\bigwedge_{\lambda' > \lambda} E_{\lambda'}$ is the greatest lower bound in $\mathscr{B}(\mathscr{H})$ of $\{E_{\lambda'} : \lambda' > \lambda\}$. Thus $E_\lambda = \bigwedge_{\lambda' > \lambda} E_{\lambda'}$.

To prove (v), choose $\lambda_0$ less than $-\|A\|$ and let $\{\lambda_0, \lambda_1, \ldots, \lambda_n\}$ be a partition of $[\lambda_0, \|A\|]$ (so that $\lambda_n = \|A\|$). If $[\lambda_{j-1}, \lambda_j] \cap \mathrm{sp}(A) \neq \varnothing$, let $\lambda_j'$ be a point of this intersection – otherwise, let $\lambda_j'$ be $\lambda_{j-1}$. If this intersection is empty, $f^{-1}([\lambda_{j-1}, \lambda_j]) = \varnothing$, since $\mathrm{sp}(A)$ is the range of $f$. Thus

$$f^{-1}((\lambda_{j-1}, \infty)) = f^{-1}((\lambda_j, \infty))$$

and $e_{\lambda_{j-1}} = e_{\lambda_j}$. It follows that $\sum_{j=1}^{n} \lambda_j' \, (e_{\lambda_j} - e_{\lambda_{j-1}})$ $(= h)$ is a linear combination of mutually "orthogonal" characteristic functions $e_{\lambda'} - e_\lambda$ with coefficients in $\mathrm{sp}(A)$. Now each $p$ in $X$ lies in exactly one set $X_{\lambda_j} \backslash X_{\lambda_{j-1}}$ $(= Y_j)$, $j = 1, \ldots, n$, since $X_{\lambda_0} = \varnothing$ and $X_{\lambda_n} = X$. If $p \in Y_j$, then $h(p) = \lambda_j'$ and $\lambda_{j-1} \leqslant f(p) \leqslant \lambda_j$. Hence $\|f - h\| \leqslant \max_j\{|\lambda_j - \lambda_{j-1}|\}$, and (v) follows. ∎

A family $\{E_\lambda\}$ of projections indexed by $\mathbb{R}$, satisfying

(i′) $\bigwedge_{\lambda \in \mathbb{R}} E_\lambda = 0$ and $\bigvee_{\lambda \in \mathbb{R}} E_\lambda = I$

and (ii), (iii) of Theorem 5.2.2 is said to be a *resolution of the identity*. Since (i) of Theorem 5.2.2 guarantees (i′) above, the family $\{E_\lambda\}$ determined in the argument of Theorem 5.2.2 is a resolution of the identity. If there is a constant $a$ such that $E_\lambda = 0$ when $\lambda < -a$ and $E_\lambda = I$ when $a < \lambda$ (as there is in the case of Theorem 5.2.2), we say that $\{E_\lambda\}$ is a *bounded* resolution of the identity – otherwise we say that $\{E_\lambda\}$ is an *unbounded* resolution of the identity. At this point, we have a resolution of the identity for $A$ in each abelian von Neumann algebra containing $A$. In Theorem 5.2.3, we show that a resolution of the identity satisfying either (iv) or (v) of Theorem 5.2.2 is the resolution of the identity for $A$ in the abelian von Neumann algebra generated by $A$ and $I$; so that we may speak of *the* resolution of the identity for $A$ (or *the spectral resolution of* $A$).

5.2.3. THEOREM. *If $\{F_\lambda\}$ is a resolution of the identity and $A$ is a (bounded) self-adjoint operator such that $AF_\lambda \leqslant \lambda F_\lambda$ and $\lambda(I - F_\lambda) \leqslant A(I - F_\lambda)$ for each $\lambda$, or if $A = \int_{-a}^{a} \lambda\, dF_\lambda$ for each a exceeding some $a_0$, then $\{F_\lambda\}$ is the resolution of the identity for $A$ in $\mathscr{A}_0$, the abelian von Neumann algebra generated by $A$ and $I$.*

*Proof.* The fact that $AF_\lambda$ is self-adjoint is implicit in the assumption that $AF_\lambda \leqslant \lambda F_\lambda$. Thus $A$ commutes with each $F_\lambda$. Since $F_\lambda \leqslant F_{\lambda'}$ when $\lambda \leqslant \lambda'$, $\{F_\lambda\}$ is an abelian family. Let $\mathscr{A}$ be an abelian von Neumann algebra that contains $A$ and $\{F_\lambda\}$ and $X$ be the extremely disconnected compact Hausdorff space such that $\mathscr{A} \cong C(X)$. If $\{E_\lambda\}$ is the resolution of the identity for $A$ in $\mathscr{A}$, $f$ in $C(X)$ corresponds to $A$, and $e_\lambda$ in $C(X)$ corresponds to $E_\lambda$, then $e_\lambda$ is the characteristic function of $X_\lambda$, the largest clopen set in $X$ on which $f$ takes values not exceeding $\lambda$. If $f_\lambda$ in $C(X)$ corresponds to $F_\lambda$, then $f_\lambda$ is the characteristic function of a clopen set $Y_\lambda$ on which $f$ takes values not exceeding $\lambda$, since $f \cdot f_\lambda \leqslant \lambda f_\lambda$. Thus $Y_\lambda \subseteq X_\lambda$. As $F_\lambda = \wedge_{\lambda' > \lambda} F_{\lambda'}$, $Y_\lambda$ is the largest clopen set in $X$ contained in $\bigcap_{\lambda' > \lambda} Y_{\lambda'}$. Now $\lambda' \leqslant f(p)$ if $p \in X \backslash Y_{\lambda'}$, since $\lambda'(I - F_{\lambda'}) \leqslant A(I - F_{\lambda'})$, so that $X \backslash Y_{\lambda'} \subseteq f^{-1}((\lambda, \infty))^-$ when $\lambda' > \lambda$. Thus $X_\lambda \subseteq Y_{\lambda'}$ when $\lambda' > \lambda$; and $X_\lambda$ is a clopen set contained in $\bigcap_{\lambda' > \lambda} Y_{\lambda'}$. Since $Y_\lambda$ is the largest such clopen set, $X_\lambda \subseteq Y_\lambda$. Hence $X_\lambda = Y_\lambda$ and $E_\lambda = F_\lambda$. The resolution of the identity for $A$ in $\mathscr{A}_0$ satisfies (iv) of Theorem 5.2.2 and $\mathscr{A}_0 \subseteq \mathscr{A}$. From what we have just proved, that resolution coincides with $\{E_\lambda\}$ (and $\{F_\lambda\}$).

Suppose, now, that $A = \int_{-a}^{a} \lambda\, dF_\lambda$ for each $a$ exceeding some $a_0$. If, for such an $a$, $\lambda \in [-a, a]$ and $\{\lambda_0, \ldots, \lambda_n\}$ is a partition of $[-a, a]$, with $\lambda$ as some $\lambda_k$, such that $(B =) \sum_{j=1}^{n} \lambda_j'(F_{\lambda_j} - F_{\lambda_{j-1}})$ is close (in norm) to $A$; then $\|AF_\lambda - BF_\lambda\|$ is small and

$$BF_\lambda = \sum_{j=1}^{k} \lambda_j'(F_{\lambda_j} - F_{\lambda_{j-1}}) \leqslant \sum_{j=1}^{k} \lambda_k(F_{\lambda_j} - F_{\lambda_{j-1}}) = \lambda(F_\lambda - F_{-a}) \leqslant \lambda F_\lambda.$$

Thus $AF_\lambda \leqslant \lambda F_\lambda$. At the same time, $\|A(I - F_\lambda) - B(I - F_\lambda)\|$ is small and

$$B(I - F_\lambda) = \sum_{j=k+1}^{n} \lambda_j'(F_{\lambda_j} - F_{\lambda_{j-1}}) \geqslant \sum_{j=k+1}^{n} \lambda_k(F_{\lambda_j} - F_{\lambda_{j-1}}) = \lambda(F_a - F_\lambda).$$

Thus $A(I - F_\lambda) \geqslant \lambda(F_a - F_\lambda)$ for each $a$ greater than $a_0$. Letting $a$ tend to $+\infty$, $F_a$ tends to $I$ in the strong-operator topology, so that $A(I - F_\lambda) \geqslant \lambda(I - F_\lambda)$. From the first part of this proof, $F_\lambda = E_\lambda$ for each $\lambda$. ■

In Theorem 5.2.4 we start with a bounded resolution of the identity and construct a bounded self-adjoint operator whose spectral resolution is the given resolution of the identity.

5.2.4. THEOREM. *If $\{E_\lambda\}$ is a bounded resolution of the identity on a Hilbert space $\mathscr{H}$, then $\int_{-a}^{a} \lambda\, dE_\lambda$ converges to a self-adjoint operator $A$ on $\mathscr{H}$ such that $\|A\| \leqslant a$ and for which $\{E_\lambda\}$ is the spectral resolution, where $E_\lambda = 0$ if $\lambda \leqslant -a$ and $E_\lambda = I$ if $a \leqslant \lambda$.*

*Proof.* If $\{\lambda_0, \ldots, \lambda_n\}$ $(= \mathscr{P})$ and $\{\mu_0, \ldots, \mu_m\}$ $(= \mathscr{Q})$ are partitions of $[-a, a]$, $|\mathscr{P}|$ and $|\mathscr{Q}|$ are the lengths of their largest subintervals, and $\{\gamma_0, \ldots, \gamma_r\}$ is their common refinement, then

$$\left\| \sum_{j=1}^{n} \lambda_j'(E_{\lambda_j} - E_{\lambda_{j-1}}) - \sum_{k=1}^{r} \gamma_k'(E_{\gamma_k} - E_{\gamma_{k-1}}) \right\| \leqslant |\mathscr{P}|$$

and

$$\left\| \sum_{j=1}^{m} \mu_j'(E_{\mu_j} - E_{\mu_{j-1}}) - \sum_{k=1}^{r} \gamma_k'(E_{\gamma_k} - E_{\gamma_{k-1}}) \right\| \leqslant |\mathscr{Q}|,$$

so that

$$\left\| \sum_{j=1}^{n} \lambda_j'(E_{\lambda_j} - E_{\lambda_{j-1}}) - \sum_{k=1}^{m} \mu_k'(E_{\mu_k} - E_{\mu_{k-1}}) \right\| \leqslant |\mathscr{P}| + |\mathscr{Q}|.$$

Thus the family of approximating Riemann sums to $\int_{-a}^{a} \lambda\, dE_\lambda$, indexed by their corresponding partition of $[-a, a]$ and the set of these partitions partially ordered (and directed) by refinement, forms a Cauchy net in the norm topology on $\mathscr{B}(\mathscr{H})$. Since $\mathscr{B}(\mathscr{H})$ is complete in its norm topology, this net converges in norm to a bounded self-adjoint operator on $\mathscr{H}$. From Theorem 5.2.3, $\{E_\lambda\}$ is the spectral resolution of $A$. Passing to $C(X)$, where $\mathscr{A} \cong C(X)$ and $\mathscr{A}$ is an abelian von Neumann algebra containing $A$, we see that the conditions, $E_\lambda = 0$ if $\lambda \leqslant -a$ and $E_\lambda = I$ if $a \leqslant \lambda$, imply that the function in $C(X)$ representing $A$ has range in $[-a, a]$. Thus $\|A\| \leqslant a$. ∎

We studied unitary operators in $C^*$-algebras in Section 4.4, and noted, there, that $\exp iH$ is a unitary element in each $C^*$-algebra containing the self-adjoint element $H$. We remarked, in the discussion preceding Proposition 4.4.10, that not each unitary element of a $C^*$-algebra has this form. In essence, the possibility of finding "log $U$" in the $C^*$-algebra generated by $U$ (an algebra of *continuous* functions) may be blocked by topological (homotopy) considerations. This is not the case in the von Neumann algebra generated by $U$—where the topological obstructions vanish before the (essentially measure-theoretic) constructions available in von Neumann algebras. We prove this von Neumann algebra analogue to Proposition 4.4.10 in the theorem that follows.

5.2.5. THEOREM. *If $U$ is a unitary operator acting on the Hilbert space $\mathscr{H}$ and $\mathscr{A}$ is the (abelian) von Neumann algebra generated by $U$ and $U^*$, there is a* *positive operator $H$ in $\mathscr{A}$ such that $\|H\| \leqslant 2\pi$ and $U = \exp iH$. In addition, $U$ is the norm limit of finite linear combinations of mutually orthogonal projections in $\mathscr{A}$ with coefficients in* $\mathrm{sp}(U)$.

*Proof.* From Theorem 5.2.1, $\mathscr{A} \cong C(X)$ with $X$ an extremely disconnected compact Hausdorff space. If $u$ in $C(X)$ corresponds to $U$, then $\bar{u}$ corresponds to $U^*$; and $|u|^2 = 1$. Let $X_\lambda$ be the complement of the closure of the set of points at which the values of $u$ do not lie in $\{\exp i\lambda' : \lambda' \text{ in } [0, \lambda]\}$ $(= C_\lambda)$, for $\lambda$ in $[0, 2\pi)$. Arguing as in the proof of Theorem 5.2.2, $X_\lambda$ is the largest clopen set on which $u$ takes values in $C_\lambda$. Let $e_\lambda$ be 0 if $\lambda < 0$, 1 if $2\pi \leqslant \lambda$; and let $e_\lambda$ be the characteristic function of $X_\lambda$ for $\lambda$ in $[0, 2\pi)$. Then $e_\lambda$ is the greatest lower bound of $\{e_{\lambda'} : \lambda < \lambda'\}$ if $\lambda < 0$ or $\lambda \geqslant 2\pi$. As $e_\lambda \leqslant e_{\lambda'}$ when $\lambda \leqslant \lambda'$, $e_\lambda$ is a lower bound of $\{e_{\lambda'} : \lambda < \lambda'\}$ for all $\lambda$. To see that $e_\lambda$ is the greatest lower bound when $\lambda \in [0, 2\pi)$, note that each clopen subset $\mathcal{O}$ of $\bigcap_{\lambda' > \lambda} X_{\lambda'}$ is contained in $X_\lambda$ (for $u$ takes values on $\mathcal{O}$ in each $C_{\lambda'}$, with $\lambda'$ exceeding $\lambda$, so that $u$ takes values on $\mathcal{O}$ in $C_\lambda$). As in Theorem 5.2.2, the projections $E_\lambda$ in $\mathscr{A}$ corresponding to $e_\lambda$ give rise to a (bounded) resolution of the identity $\{E_\lambda\}$.

From Theorem 5.2.4, $\int \lambda \, dE_\lambda$ converges (in norm) to a self-adjoint operator $H$ in $\mathscr{A}$. Let $h$ be the function in $C(X)$ corresponding to $H$. Letting $X_\lambda$ be $\varnothing$ when $\lambda < 0$ and $X$ when $\lambda \geqslant 2\pi$, $X_\lambda$ is the largest clopen set on which $h$ takes values not exceeding $\lambda$. The range of $h$ is contained in $[0, 2\pi]$ so that $H$ is positive and $\|H\| \leqslant 2\pi$. Note, too, that $h$ cannot take the value $2\pi$ at each point of a non-null clopen set $\mathcal{O}_0$; for otherwise $\mathcal{O}_0$ is disjoint from $\bigcup_{\lambda' < 2\pi} X_{\lambda'}$. But then $u(p_0) \neq 1$ for some $p_0$ in $\mathcal{O}_0$ (otherwise $\mathcal{O}_0 \subseteq X_0$). By continuity of $u$, there is a clopen subset $\mathcal{O}_1$ of $\mathcal{O}_0$ containing $p_0$ and there is a $\lambda_0$ in $(0, 2\pi)$ such that $u(q) \in C_{\lambda_0}$ for each $q$ in $\mathcal{O}_1$. Thus $\mathcal{O}_1 \subseteq X_{\lambda_0}$ contrary to the choice of $\mathcal{O}_0$ disjoint from $X_{\lambda_0}$. With this information, we can now see that $X_\lambda$ is the largest clopen set on which $\exp ih$ takes values in $C_\lambda$ $(\lambda \in [0, 2\pi))$ – whence $\exp ih = u$, and $\exp iH = U$. Indeed, if $\mathcal{O}$ is a clopen set such that $\exp ih(p) \in C_\lambda$ for each $p$ in $\mathcal{O}$, then either $h(p) \in [0, \lambda]$ or $h(p) = 2\pi$ for each $p$ in $\mathcal{O}$. If $h(p) = 2\pi$ for some $p$ in $\mathcal{O}$, then, by continuity of $h$, there is a clopen subset of $\mathcal{O}$ containing $p$ on which $h$ takes values near $2\pi$ – in particular, not in $[0, \lambda]$, since $\lambda < 2\pi$. By choice of $\mathcal{O}$, then, $h$ takes the value $2\pi$ on this entire clopen subset – contrary to what we have just proved. Thus $h(p) \in [0, \lambda]$ for all $p$ in $\mathcal{O}$, and $\mathcal{O} \subseteq X_\lambda$.

If $\sum_{j=1}^n \lambda_j'(E_{\lambda_j} - E_{\lambda_{j-1}})$ is close to $H$ in norm, then

$$\sum_{j=1}^n (\exp i\lambda_j')(E_{\lambda_j} - E_{\lambda_{j-1}})$$

is close to $\exp iH$ $(= U)$ in norm. From Theorem 5.2.2, we can choose $\lambda_j'$ in $\mathrm{sp}(H)$ if $E_{\lambda_j} \neq E_{\lambda_{j-1}}$. With this choice, $\exp i\lambda_j' \in \mathrm{sp}(U)$ if $E_{\lambda_j} \neq E_{\lambda_{j-1}}$. ∎

In Example 5.1.6 we noted that the multiplication algebra $\mathscr{A}$ of a $\sigma$-finite measure space $(S, \mathscr{S}, m)$ is an abelian von Neumann algebra. If $f$ is a real-valued essentially bounded measurable function on $S$, $M_f$ is a (bounded) self-adjoint operator on $L_2$. With $E_\lambda$ the projection corresponding to multiplication by the characteristic function of the set $S_\lambda$ on which $f$ takes values not exceeding $\lambda$, $\{E_\lambda\}$ is the spectral resolution of $M_f$. The key observation needed for this is the fact that $\bigcap_{\lambda' > \lambda} S_{\lambda'} = S_\lambda$ (and, thus, $\bigwedge_{\lambda' > \lambda} E_{\lambda'} = E_\lambda$).

In Theorem 5.2.6 we describe a simultaneous spectral resolution for a commuting family of operators forming an abelian $C^*$-algebra. In this case, the (joint) spectrum of the family is the pure state space. Somewhat more precisely, we describe the spectral resolution of a *representation* of an abelian $C^*$-algebra.

5.2.6. THEOREM. *If $X$ is a compact Hausdorff space, $\mathscr{H}$ is a Hilbert space, and $\varphi$ is a representation of $C(X)$ on $\mathscr{H}$, then, to each Borel subset $S$ of $X$ there corresponds a projection $E(S)$ such that*

(i) *$E(S) \in \mathscr{A}$, the strong-operator closure of $\varphi(C(X))$;*

(ii) *$E(S) = \bigwedge\{E(\mathscr{O}) : S \subseteq \mathscr{O}, \mathscr{O}$ open$\}$;*

(iii) *$E(\bigcup_{n=1}^{\infty} S_n) = \sum_{n=1}^{\infty} E(S_n)$ for each countable family $\{S_n\}$ of mutually disjoint Borel subsets of $X$, in particular, $E(S_n)E(S_m) = 0$ if $n \neq m$, and $E(\varnothing) = 0$;*

(iv) *$E(X_0) = I$, for $X_0$ a Borel subset of $X$, if the span of the ranges of those $\varphi(f)$ such that $f \in C(X)$ and $f$ vanishes on $X \setminus X_0$ is dense in $\mathscr{H}$;*

(v) *for each $x$ in $\mathscr{H}$, $S \to \langle E(S)x, x\rangle$ is a regular Borel measure, $\mu_x$, and, for $f$ in $C(X)$,*

$$\langle \varphi(f)x, x\rangle = \int_X f(p)\, d\mu_x(p).$$

*Proof.* If $\mathscr{O}$ is an open subset of $X$ and $f$ in $C(X)$ has range in $[0, 1]$ and vanishes on $X \setminus \mathscr{O}$, then $0 \leqslant \varphi(f) \leqslant I$. Thus $\varphi(\mathscr{F}(\mathscr{O}))$ has a least upper bound $E(\mathscr{O})$ in the abelian von Neumann algebra $\mathscr{A}$, where $\mathscr{F}(\mathscr{O})$ is the set (directed by its natural order) of such functions $f$. As $\{f^2 : f \in \mathscr{F}(\mathscr{O})\} = \mathscr{F}(\mathscr{O})$, $E(\mathscr{O})$ is a projection. With $S$ a Borel subset of $X$, let $E(S)$ be $\bigwedge\{E(\mathscr{O}) : S \subseteq \mathscr{O}, \mathscr{O}$ open$\}$. If $x$ is a unit vector in $\mathscr{H}$, $f \to \langle \varphi(f)x, x\rangle$ is a state of $C(X)$. From the Riesz representation of such functionals (see the discussion preceding Lemma 1.7.7), there is a regular Borel measure $\mu_x$ on $X$ such that $\langle \varphi(f)x, x\rangle = \int_X f(p)\, d\mu_x(p)$. By (inner) regularity of $\mu_x$, given an open set $\mathscr{O}$, there is a compact subset $\mathscr{K}$ of $\mathscr{O}$ such that $\mu_x(\mathscr{K})$ is close to $\mu_x(\mathscr{O})$. Since $X$ is a normal space, there is a continuous function $f$ on $X$ with range in $[0, 1]$, vanishing outside $\mathscr{O}$, and 1 on $\mathscr{K}$. Then

$$\mu_x(\mathscr{K}) \leqslant \int f(p)\, d\mu_x(p) = \langle \varphi(f)x, x\rangle \leqslant \langle E(\mathscr{O})x, x\rangle.$$

It follows that $\mu_x(\mathcal{O}) \leqslant \langle E(\mathcal{O})x, x\rangle$. From the definition of $E(\mathcal{O})$, $\langle E(\mathcal{O})x, x\rangle \leqslant \mu_x(\mathcal{O})$, so that $\langle E(\mathcal{O})x, x\rangle = \mu_x(\mathcal{O})$. From (outer) regularity of $\mu_x$, we have that

$$\begin{aligned}\mu_x(S) &= \inf\{\mu_x(\mathcal{O}) : S \subseteq \mathcal{O},\ \mathcal{O} \text{ open}\}\\ &= \inf\{\langle E(\mathcal{O})x, x\rangle : S \subseteq \mathcal{O},\ \mathcal{O} \text{ open}\} = \langle E(S)x, x\rangle\end{aligned}$$

for each Borel set $S$. If, now, $\{S_n\}$ is a family of disjoint Borel subsets of $X$, then

$$\left\langle E\left(\bigcup_{n=1}^{\infty} S_n\right)x, x\right\rangle = \mu_x\left(\bigcup_{n=1}^{\infty} S_n\right) = \sum_{n=1}^{\infty} \mu_x(S_n) = \sum_{n=1}^{\infty} \langle E(S_n)x, x\rangle.$$

In particular, if $x$ is a unit vector in the range of one of the $E(S_n)$, say, $E(S_1)$, then

$$1 \geqslant \left\langle E\left(\bigcup_{n=1}^{\infty} S_n\right)x, x\right\rangle = \sum_{n=1}^{\infty} \langle E(S_n)x, x\rangle \geqslant \langle E(S_1)x, x\rangle = 1.$$

Since $0 \leqslant \langle E(S_n)x, x\rangle$ for all $n$, $\langle E(S_n)x, x\rangle = 0$ unless $n = 1$. It follows that $E(S_n)E(S_m) = 0$ if $n \neq m$ and that $E(\bigcup_{n=1}^{\infty} S_n) = \sum_{n=1}^{\infty} E(S_n)$.

With $X_0$ as in (iv), if $\mathcal{O}$ is an open set containing $X_0$ and $f$ (real-valued) in $C(X)$ vanishes on $X\backslash\mathcal{O}$, then the range projection of $\varphi(f_0)$ is a subprojection of $E(\mathcal{O})$, where $f_0 = \|f\|^{-1}|f|$. But $\varphi(f_0)$ and $\varphi(f)$ have the same range projection (for $\varphi(|f|) = \varphi(f_+) + \varphi(f_-)$, $\varphi(f) = \varphi(f_+) - \varphi(f_-)$, and $f_+f_- = 0$ – see Remark 3.4.9). Thus $E(\mathcal{O})$ contains the range projection of the image of each function in $C(X)$ vanishing on $X\backslash X_0$; and, by assumption, $E(\mathcal{O}) = I$. Hence $E(X_0) = I$. ∎

We apply this theorem to the important special case of $\mathbb{R}$-essential representations of $C(\{\mathbb{R}, \infty\})$ (that is, representations essential on the ideal of functions in $C(\{\mathbb{R}, \infty\})$ vanishing at $\infty$, where, as usual, $\{\mathbb{R}, \infty\}$ denotes the one-point compactification of $\mathbb{R}$).

5.2.7. COROLLARY. *Each $\mathbb{R}$-essential representation $\varphi_0$ of $C(\{\mathbb{R}, \infty\})$ corresponds to a (possibly unbounded) resolution of the identity $\{E_\lambda\}$ such that, for each $f$ in $C(\{\mathbb{R}, \infty\})$ with $f(\infty) = 0$,*

$$\langle \varphi_0(f)x, x\rangle = \int_{\mathbb{R}} f(\lambda)\, d\langle E_\lambda x, x\rangle.$$

*Proof.* From Theorem 5.2.6, there is a projection-valued measure $S \to E(S)$ on $\{\mathbb{R}, \infty\}$ such that $\langle \varphi_0(f)x, x\rangle = \int_{\mathbb{R}} f(\lambda)\, d\mu_x(\lambda)$, where $f \in C(\{\mathbb{R}, \infty\})$, $f(\infty) = 0$, and $\mu_x(S) = \langle E(S)x, x\rangle$. If $E_\lambda = I - E((\lambda, \infty))$, then $\{E_\lambda\}$ is a (possibly unbounded) spectral resolution of the identity. To see this, note that $E((\lambda, \infty)) \leqslant E((\lambda', \infty))$, when $\lambda' \leqslant \lambda$, so that $E_{\lambda'} \leqslant E_\lambda$, in this case. Since $\varphi_0$ is $\mathbb{R}$-essential, from (iv) of Theorem 5.2.6, we have that $E(\mathbb{R}) = I$. As

$(m, \infty) = \cup_{n \geqslant m}(n, n+1]$, $E((m, \infty)) = \sum_{n=m}^{\infty} E((n, n+1])$, and $\wedge_m E((m, \infty)) = 0$. It follows that

$$I = I - \bigwedge_{\lambda} E((\lambda, \infty)) = \bigvee_{\lambda}(I - E((\lambda, \infty))) = \bigvee_{\lambda} E_\lambda .$$

At the same time,

$$I = E(\mathbb{R}) = \sum_{n=-\infty}^{\infty} E((n, n+1]) = \bigvee_{m=-\infty}^{\infty} E((m, \infty)),$$

so that

$$0 = I - \bigvee_{m=-\infty}^{\infty} E((m, \infty)) = \bigwedge_{m=-\infty}^{\infty} (I - E((m, \infty))) = \bigwedge_{m=-\infty}^{\infty} E_m$$

and

$$0 \leqslant \bigwedge_{\lambda} E_\lambda \leqslant \bigwedge_{m=-\infty}^{\infty} E_m = 0.$$

Let $\lambda(n)$ be $\lambda + n^{-1}$. Then

$$E((\lambda, \infty)) = E((\lambda + 1, \infty)) + \sum_{n=1}^{\infty} E((\lambda(n+1), \lambda(n)]),$$

so that

$$\bigwedge_{n=1}^{\infty} E_{\lambda(n)} \leqslant E_{\lambda+1} \wedge \left[ \bigwedge_{n=1}^{\infty} (I - E((\lambda(n+1), \lambda(n)])) \right] = I - E((\lambda, \infty)) = E_\lambda .$$

Since

$$E_\lambda \leqslant \bigwedge_{\lambda < \lambda'} E_{\lambda'} \leqslant \bigwedge_{n=1}^{\infty} E_{\lambda(n)} \leqslant E_\lambda ,$$

it follows that $E_\lambda = \wedge_{\lambda < \lambda'} E_{\lambda'}$ and $\{E_\lambda\}$ is a resolution of the identity.

We have

$$E((\lambda, \lambda']) = E((\lambda, \infty)) - E((\lambda', \infty)) = E_{\lambda'} - E_\lambda ,$$

so that

$$\mu_x((\lambda, \lambda']) = \langle E_{\lambda'} x, x \rangle - \langle E_\lambda x, x \rangle .$$

Combining this with (v) of Theorem 5.2.6, we have

$$\langle \varphi_0(f)x, x \rangle = \int_{\mathbb{R}} f(\lambda)\, d\mu_x(\lambda) = \int_{\mathbb{R}} f(\lambda)\, d\langle E_\lambda x, x \rangle \tag{1}$$

when $f \in C(\{\mathbb{R}, \infty\})$ and $f$ vanishes outside a finite interval. Since such $f$ are norm dense in the set of those functions in $C(\{\mathbb{R}, \infty\})$ vanishing at $\infty$, (1) is valid for all $f$ in this set. ■

Using Theorem 5.2.6, we can produce the spectral resolution by measure-theoretic means. If $A$ is a (bounded) self-adjoint or normal operator on $\mathscr{H}$ and $\mathfrak{A}$ is the (abelian) $C^*$-algebra generated by $A$ (and $A^*$) and $I$, then the mapping $f \to f(A)$ provides us with a representation of $C(X)$ on $\mathscr{H}$, where $X = \{\mathbb{R}, \infty\}$ if $A$ is self-adjoint and $X = \{\mathbb{C}, \infty\}$, the one-point compactification of $\mathbb{C}$, if $A$ is normal. From Theorem 5.2.6, there is a *projection-valued measure* assigning a projection $E(S)$ on $\mathscr{H}$ to a Borel subset $S$ of $X$ and such that, for each $f$ in $C(X)$,

$$\langle f(A)x, x\rangle = \int_X f(p)\, d\mu_x(p),$$

where $\mu_x(S) = \langle E(S)x, x\rangle$. If $\mathcal{O}$ is an open set disjoint from sp$(A)$ and $f$ in $C(X)$ is in $\mathscr{F}(\mathcal{O})$ (see the notation of Theorem 5.2.6), then $f(A) = 0$, so that $E(\mathcal{O}) = 0$ and $\mu_x(\mathcal{O}) = 0$ for each $x$ in $\mathscr{H}$. We write

$$\langle f(A)x, x\rangle = \int_{\text{sp}(A)} f(p)\, d\mu_x(p),$$

and speak of $S \to E(S)$ as the spectral measure for $A$. In case $A$ is self-adjoint, Corollary 5.2.7 shows us how to pass from this spectral measure for $A$ to a spectral resolution $\{E_\lambda\}$. From the proof of that corollary, we have that $E_\lambda = I - E((\lambda, \infty))$. As just noted, $E((\lambda, \infty)) = 0$ if $(\lambda, \infty)$ is disjoint from sp$(A)$. Thus $E_\lambda = I$ if $\lambda \geqslant \|A\|$. At the same time, $I - E((\lambda, \infty)) = E(X \backslash (\lambda, \infty)) = 0$ if $\lambda < -\|A\|$, so that $E_\lambda = 0$ when $\lambda < -\|A\|$. Thus $\{E_\lambda\}$ is a bounded resolution of the identity in this case, and

$$\langle f(A)x, x\rangle = \int_{-\|A\|}^{\|A\|} f(\lambda)\, d\langle E_\lambda x, x\rangle \tag{2}$$

(at first, for each $f$ in $C(X)$ vanishing at $\infty$, but then for each $f$ continuous on sp$(A)$ since each such agrees on sp$(A)$ with some function vanishing at $\infty$). In particular

$$\langle Ax, x\rangle = \int_{-\|A\|}^{\|A\|} \lambda\, d\langle E_\lambda x, x\rangle$$

for each $x$ in $\mathscr{H}$, so that $AE_\lambda \leqslant \lambda E_\lambda$ and $\lambda(I - E_\lambda) \leqslant A(I - E_\lambda)$. It follows (from Theorem 5.2.3) that $\{E_\lambda\}$ is the spectral resolution of $A$.

Since $\mu_x(S) = \langle E(S)x, x\rangle$ for each Borel subset $S$ of $X$ and $x$ in $\mathscr{H}$ and

$$\langle f(A)x, x\rangle = \int_X f(p)\, d\mu_x(p),$$

polarization of $\langle E(S)x, y\rangle$ allows us to define a complex (Radon) measure $\mu_{x,y}$ on $X$ as a linear combination of positive measures $\mu_z$ and

$$\langle f(A)x, y\rangle = \int_X f(p)\, d\mu_{x,y}(p) \tag{3}$$

for each $f$ in $C(X)$ and all $x$, $y$ in $\mathscr{H}$. If $A$ is self-adjoint, (3) amounts to the formula,

$$\langle f(A)x, y\rangle = \int_{-\|A\|}^{\|A\|} f(\lambda)\, d\langle E_\lambda x, y\rangle,$$

which "polarizes" to (2). If $A$ is normal, (3) provides us with the possibility of defining $g(A)$ when $g$ is a bounded Borel function on $\mathbb{C}$. Note that $|\int g(p)\, d\mu_{x,y}(p)| \leqslant \|x\|\,\|y\| \sup|g(p)|$ (this is apparent from (3) when $g \in C(X)$, and extends by a measure-theoretic argument to the case where $g$ is a bounded Borel function). Thus

$$(x, y) \to \int_X g(p)\, d\mu_{x,y}(p)$$

is a bounded conjugate-bilinear functional on $\mathscr{H}$ with bound not exceeding $\sup|g(p)|$ and, so, corresponds to an operator $g(A)$ satisfying

$$\langle g(A)x, y\rangle = \int_X g(p)\, d\mu_{x,y}(p), \qquad \|g(A)\| \leqslant \sup_{p \in X} |g(p)|, \tag{4}$$

where $\mu_{x,y}(S) = \langle E(S)x, y\rangle$ for each Borel subset $S$ of $X$ and all $x$, $y$ in $\mathscr{H}$. Again, if $T$ in $\mathscr{B}(\mathscr{H})$ commutes with $\mathfrak{A}$ (that is, with $A$ and $A^*$), then, from (3), $\mu_{Tx,y} = \mu_{x,T^*y}$, so that $\langle g(A)Tx, y\rangle = \langle Tg(A)x, y\rangle$. Thus $g(A)T = Tg(A)$. We have

$$\begin{aligned}\langle \bar{g}(A)x, y\rangle &= \int_X \overline{g(p)}\, d\mu_{x,y}(p) = \overline{\int_X g(p)\, d\mu_{y,x}(p)}\\ &= \overline{\langle g(A)y, x\rangle} = \langle g(A)^*x, y\rangle\end{aligned}$$

for all $x$ and $y$ in $\mathscr{H}$. Thus

$$\bar{g}(A) = g(A)^* \tag{5}$$

for each bounded Borel function $g$ on $\mathbb{C}$ and each normal $A$. Since $A$ and $A^*$ commute with $\mathfrak{A}$, $g(A)$ and $\bar{g}(A)$ commute with $\mathfrak{A}$ and, hence, with each other. Thus $g(A)$ is normal.

We proceed, now, to establish the other properties of a (bounded) Borel function calculus for (bounded) normal operators. If $g$ and $h$ are bounded Borel functions on $\mathbb{C}$ (or $\mathbb{R}$ if $A$ is self-adjoint), then, for each $x$ in $\mathscr{H}$,

$$\langle (ag + h)(A)x, x\rangle = \int_X (ag(p) + h(p))\, d\mu_x(p) = \langle (ag(A) + h(A))x, x\rangle,$$

so that

$$(ag + h)(A) = ag(A) + h(A). \tag{6}$$

If $g$ is the characteristic function of a Borel subset $S$ of $X$, then, for each $x$ in $\mathscr{H}$,

$$\langle g(A)x, x\rangle = \int_X g(p)\, d\mu_x(p) = \mu_x(S) = \langle E(S)x, x\rangle.$$

Thus $g(A) = E(S)$; and, in particular $g^2(A) = g(A) = g(A)^2$. If $h$ is the characteristic function of a Borel set disjoint from $S$, then, from Theorem 5.2.6(iii), $0 = g(A)h(A) = (g \cdot h)(A)$. It follows, now, that $g^2(A) = g(A)^2$, so that $(g + h)^2(A) = [g(A) + h(A)]^2$ and

$$(7) \qquad (g \cdot h)(A) = g(A) \cdot h(A),$$

when $g$ and $h$ are finite linear combinations of characteristic functions of disjoint Borel subsets of $X$ ("step functions"). Since each bounded Borel function is a uniform (norm) limit of such step functions (and $\|g(A)\| \leqslant \|g\|$ for each bounded Borel function $g$), we have (7) for arbitrary bounded Borel functions $g$ and $h$.

The identities (5), (6), and (7) that we have established thus for assure us that the rule

$$(8) \qquad (g \circ h)(A) = g(h(A))$$

holds when $g$ is a polynomial (in $z$ and $\bar{z}$) and $h$ is an arbitrary bounded Borel function. Using the Stone–Weierstrass theorem (3.4.14) to approximate a continuous function uniformly on a closed disk in $\mathbb{C}$ containing the range of $h$ by a polynomial (in $z$ and $\bar{z}$), it follows that (8) is valid for each continuous function $g$ and each bounded Borel function $h$. Since $g(h(A)) = \bar{g}(h(A)) = g(h(A))^*$ if $g$ is real-valued and $g(h(A)) \geqslant 0$ if $g \geqslant 0$, we have that $\{g_n(h(A))\}$ is an increasing sequence of self-adjoint operators when $\{g_n\}$ is an increasing sequence of bounded Borel functions. If each $g_n$ is continuous and tends pointwise to a bounded Borel function $g$, then $g_n(h(A)) = (g_n \circ h)(A)$ and $\{g_n \circ h\}$ is an increasing sequence tending pointwise to $g \circ h$. It will be useful for us to note that if $\{f_n\}$ is an increasing sequence of bounded Borel functions tending pointwise to the bounded Borel function $f$, then $\{f_n(A)\}$ is an increasing sequence of self-adjoint operators with least upper bound $f(A)$ (and a similar conclusion holds for decreasing sequences). We say that the mapping $f \to f(A)$ with this monotone sequential convergence property is *σ-normal.* To prove this, choose $x$ in $\mathscr{H}$ and note that

$$\langle f_n(A)x, x\rangle = \int_X f_n(p)\, d\mu_x(p) \to \int_X f(p)\, d\mu_x(p) = \langle f(A)x, x\rangle,$$

from the monotone convergence theorem. Thus, in the case of the continuous $g_n$, we have (8) for their limit $g$. In particular, we have (8) for $g$ the characteristic function of an open set $\mathcal{O}$ in $\mathbb{C}$ (and, so, for a closed set as well). To construct the sequence $\{g_n\}$ for $g$, express $\mathcal{O}$ as a countable union of open disks $\mathcal{O}_j$ (with radius $r_j$). Let $f_{jn}$ be a continuous function on $\mathbb{C}$ with range in $[0, 1]$, vanishing outside $\mathcal{O}_j$, and 1 on the closed disk with the same center as $\mathcal{O}_j$ and radius $(n-1)r_j/n$. Then $f_{1n} \vee f_{2n} \vee \cdots \vee f_{nn}$ will serve as $g_n$.

Let $\mathcal{F}$ be the family of Borel sets $S$ whose characteristic function $g$ satisfies (8) for all bounded Borel functions $h$. We have just seen that $\mathcal{F}$ contains all open and all closed sets. From the properties we have established for the mapping $f \to f(A)$ (in particular, $\sigma$-normality) we see that $\mathcal{F}$ is a $\sigma$-algebra. Hence $\mathcal{F}$ is the family of all Borel sets; and (8) holds for all bounded Borel functions $h$, when $g$ is the characteristic function of a Borel set. As a consequence, (8) is valid for each step function $g$ and then, by passing to (norm) limits, (8) follows for each pair of bounded Borel functions $g$ and $h$.

We summarize this discussion in the theorem that follows.

5.2.8. THEOREM. *If $A$ is a (bounded) normal operator on the complex Hilbert space $\mathcal{H}$ the * homomorphism $f \to f(A)$ of $C(X)$ into the $C^*$-algebra $\mathfrak{A}$ generated by $A$, $A^*$, and $I$, where $X$ is the one-point compactification of $\mathbb{C}$, extends to a $\sigma$-normal * homomorphism $g \to g(A)$ of the algebra $\mathcal{B}$ of bounded Borel functions $g$ on $\mathbb{C}$ into the abelian von Neumann algebra $\mathcal{A}$ consisting of operators commuting with each operator commuting with $\mathfrak{A}$. If $g$ in $\mathcal{B}$ vanishes on* $\mathrm{sp}(A)$, *then $g(A) = 0$. With $g$ and $h$ in $\mathcal{B}$, $\bar{g}(A) = g(A)^*$ and $(g \circ h)(A) = g(h(A))$. Letting $S$ be a Borel subset of $X$, $g$ be its characteristic function, and $E(S)$ be $g(A)$, the mapping $S \to E(S)$ is a projection-valued measure on $X$. Moreover*

$$\|f(A)\| \leqslant \sup\{|f(a)| : a \in \mathrm{sp}(A)\}$$

*and*

$$\langle f(A)x, x\rangle = \int_X f(p)\, d\mu_x(p) = \int_{\mathrm{sp}(A)} f(p)\, d\mu_x(p)$$

*for each $f$ in $\mathcal{B}$, where $\mu_x(S) = \langle E(S)x, x\rangle$. If $A$ is self-adjoint its spectral resolution is $\{E_\lambda\}$, where $E_\lambda = I - E((\lambda, \infty))$.*

If $\mathcal{I}$ is the ideal of functions in $\mathcal{B}$ vanishing on $\mathrm{sp}(A)$, then $\mathcal{I}$ is the kernel of the homomorphism of $\mathcal{B}$ onto $\mathcal{B}(\mathrm{sp}(A))$ obtained by restricting a function in $\mathcal{B}$ to $\mathrm{sp}(A)$. Thus $\mathcal{B}/\mathcal{I} \cong \mathcal{B}(\mathrm{sp}(A))$. As noted in Theorem 5.2.8, the kernel of the ($\sigma$-normal) homomorphism, $g \to g(A)$, of $\mathcal{B}$ into $\mathcal{A}$, contains $\mathcal{I}$. Thus the mapping, $g + \mathcal{I} \to g(A)$, gives rise to a homomorphism of $\mathcal{B}(\mathrm{sp}(A))$ into $\mathcal{A}$. If $g \in \mathcal{B}(\mathrm{sp}(A))$ and we define $\tilde{g}$ to be $g$ on $\mathrm{sp}(A)$ and 0 on the complement of $\mathrm{sp}(A)$, then $\tilde{g} \in \mathcal{B}$ and $\tilde{g}(A)$ is the image of $g$ under the homomorphism described. We write $g(A)$ for this image. From this same observation, we see that $g \to g(A)$ is a $\sigma$-normal homomorphism of $\mathcal{B}(\mathrm{sp}(A))$ into $\mathcal{A}$.

In the theorem that follows, we prove that our bounded Borel function calculus is unique. (Compare Theorem 4.4.5 and Remark 4.4.6.) The uniqueness is stated in terms of $\mathscr{B}(\mathrm{sp}(A))$, although the preceding paragraph applies to any Borel subset of $\mathbb{C}$ containing $\mathrm{sp}(A)$ and the following result (and argument) apply to each bounded Borel subset of $\mathbb{C}$ containing $\mathrm{sp}(A)$.

5.2.9. THEOREM. *If $A$ is a (bounded) normal operator on a complex Hilbert space $\mathscr{H}$, $\mathscr{B}(\mathrm{sp}(A))$ is the algebra of (complex-valued) bounded Borel functions on $\mathrm{sp}(A)$, $\varphi$ is a $\sigma$-normal homomorphism of $\mathscr{B}(\mathrm{sp}(A))$ into an abelian von Neumann algebra $\mathscr{A}$, $\varphi(1) = I$, and $\varphi(\iota) = A$, where $\iota(a) = a$ for each $a$ in $\mathrm{sp}(A)$, then $\varphi$ maps $\mathscr{B}(\mathrm{sp}(A))$ into $\mathscr{A}_0$, the abelian von Neumann algebra generated by $A$, $A^*$, and $I$, and $\varphi(g) = g(A)$ for each $g$ in $\mathscr{B}(\mathrm{sp}(A))$.*

*Proof.* With complex conjugation as involution (* operation) and $\sup\{|g(a)| : a \in \mathrm{sp}(A)\}$ as $\|g\|$, $\mathscr{B}(\mathrm{sp}(A))$ is a $C^*$-algebra; for $\|\bar{g} \cdot g\| = \|g\|^2$. If $g$ is real-valued and $a \in \mathbb{C}\backslash\mathbb{R}$, then $g - a1$ has an inverse $h$ in $\mathscr{B}(\mathrm{sp}(A))$. Since $I = [\varphi(g) - aI] \cdot \varphi(h)$, $a \notin \mathrm{sp}(\varphi(g))$. The elements of $\mathscr{A}$ are normal, so that $\varphi(g)$ is self-adjoint. Thus $\varphi$ is a * homomorphism of the $C^*$-algebra $\mathscr{B}(\mathrm{sp}(A))$ into $\mathscr{A}$. From Theorem 4.1.8(i), $\varphi$ is order preserving and does not increase norm.

Applying Theorem 4.4.5 to $\varphi$ restricted to $C(\mathrm{sp}(A))$, we have that $\varphi(f) = f(A)$ for each $f$ in $C(\mathrm{sp}(A))$ and $f(A) \in \mathscr{A}_0$. Since $\varphi$ and the mapping $g \to g(A)$ are $\sigma$-normal, we have, as in the argument proving (8), that $\varphi(g) = g(A)$ $(\in \mathscr{A}_0)$ when $g$ is the characteristic function of an open set. If $\mathscr{F}$ is the family of Borel sets whose characteristic function $g$ satisfy $\varphi(g) = g(A)$, then, again, by $\sigma$-normality, $\mathscr{F}$ contains the union of each countable subfamily. Moreover, since $\varphi(1) = I$, $\mathscr{F}$ contains the complement of each set in $\mathscr{F}$. Thus $\mathscr{F}$ coincides with the family of all Borel subsets of $\mathrm{sp}(A)$; and $\varphi(g) = g(A)$ $(\in \mathscr{A}_0)$ for the characteristic function $g$ of such a set. Since $\varphi$ is linear and norm continuous, $\|g(A)\| \leq \|g\|$, and the step functions are norm dense in $\mathscr{B}(\mathrm{sp}(A))$, it follows, now, that $\varphi(g) = g(A)$ $(\in \mathscr{A}_0)$ for each $g$ in $\mathscr{B}(\mathrm{sp}(A))$. ∎

Lemma 5.2.10 provides the foundation for developing the Borel function calculus in purely topological–function-theoretic terms. The approach of Theorem 5.2.8, stemming from Theorem 5.2.6, is essentially measure-theoretic, while the construction of the spectral resolution $\{E_\lambda\}$ in Theorem 5.2.2 is topological and function-theoretic. We shall take the latter path, and refer to the following results, when we treat the function calculus for unbounded normal operators (Section 5.6).

Recall that a subset of a topological space $X$ is *nowhere dense* (in $X$) if its closure has empty interior and that a subset is *meager* (or *of the first category*) in $X$ if it is a countable union of sets nowhere dense in $X$. (See [K: p. 201].) A subset of a nowhere-dense set is nowhere dense, so that a subset of a meager set is meager. A countable union of meager sets is meager.

5.2.10. LEMMA. *If $X$ is an extremely disconnected compact Hausdorff space, each Borel subset of $X$ differs from a (unique) clopen set by a meager set.*

*Each bounded Borel function $g$ on $X$ differs from a (unique) continuous function $f$ on a meager set. The mapping that assigns $f$ to $g$ is a (conjugation-preserving, $\sigma$-normal) homomorphism of $\mathscr{B}(X)$, the algebra of bounded Borel functions, onto $C(X)$ with kernel consisting of those functions vanishing outside a meager set.*

*Proof.* Let $\mathscr{F}$ be the family of subsets of $X$ that differ from a clopen set by a meager set. If $S \in \mathscr{F}$ and $X_0$ is a clopen set such that $(S \backslash X_0) \cup (X_0 \backslash S)$ is meager, then $X \backslash S$ and $X \backslash X_0$ differ by this same set. As $X \backslash X_0$ is clopen, $X \backslash S \in \mathscr{F}$. Each open set $\mathcal{O}$ lies in $\mathscr{F}$, since $\mathcal{O}^-$ is clopen and $\mathcal{O}^- \backslash \mathcal{O}$ is nowhere dense. If $S_j \in \mathscr{F}$ for $j = 1, 2, \ldots$ and $X_j$ is a clopen set such that $(S_j \backslash X_j) \cup (X_j \backslash S_j)$ $(= M_j)$ is meager, then

$$\left[\left(\bigcup_{j=1}^{\infty} S_j\right) \Big\backslash \left(\bigcup_{j=1}^{\infty} X_j\right)\right] \cup \left[\left(\bigcup_{j=1}^{\infty} X_j\right) \Big\backslash \left(\bigcup_{j=1}^{\infty} S_j\right)\right] \subseteq \bigcup_{j=1}^{\infty} M_j.$$

As $\bigcup_{j=1}^{\infty} M_j$ is meager and $\bigcup_{j=1}^{\infty} X_j$ is open, $\bigcup_{j=1}^{\infty} S_j \in \mathscr{F}$. Hence $\mathscr{F}$ contains the $\sigma$-algebra generated by the open subsets of $X$; that is, $\mathscr{F}$ contains the Borel subsets of $X$.

The Baire category theorem [K: p. 200, Theorem 34] assures us that the complement of a meager set is dense in $X$, so that two continuous functions agree on the complement of a meager set only if they are equal. Thus there is at most one continuous function agreeing with a given bounded Borel function on the complement of a meager set. If $S$ is a Borel subset of $X$, $g$ is its characteristic function, $X_0$ is a clopen subset of $X$ such that $(X_0 \backslash S) \cup (S \backslash X_0)$ $(= M)$ is meager and $f$ is the characteristic function of $X_0$, then $f$ is continuous and $g - f$ is 0 on $X \backslash M$. We see, from this and the preceding "uniqueness" remark, that there is at most one clopen set differing from $S$ by a meager set. At the same time, we see that a finite linear combination of characteristic functions of (disjoint) Borel subsets of $X$ (step functions) agrees with a (unique) continuous function on the complement of a meager set. Since the step functions are (supremum-)norm dense in $\mathscr{B}(X)$; if $g$ is in $\mathscr{B}(X)$, there is a sequence $\{g_n\}$ of step functions such that $\|g - g_n\| \to 0$. Let $\{f_n\}$ be a sequence of continuous functions such that $f_n$ and $g_n$ agree on the complement of a meager set $M_n$. Then $\|f_n - f_m\| \leqslant \|g_n - g_m\|$, since $f_n - f_m$ and $g_n - g_m$ agree on the complement of $M_n \cup M_m$, a dense set (so that $|(f_n - f_m)(p)| \leqslant \|g_n - g_m\|$ for each $p$ in this dense set). Thus $\{f_n\}$ is a Cauchy sequence and converges in norm to some $f$ in $C(X)$. As $\{g_n\}$ tends to $g$ and $\{f_n\}$ tends to $f$ pointwise, $f$ and $g$ agree on the complement of $\bigcup_{j=1}^{\infty} M_j$, a meager set.

If $g_1$ and $g_2$ in $\mathscr{B}(X)$ differ from $f_1$ and $f_2$ in $C(X)$ on the meager sets $M_1$ and $M_2$, then $\bar{g}_1$, $ag_1 + g_2$ and $g_1 g_2$ differ from $\bar{f}_1$, $af_1 + f_2$ and $f_1 f_2$ on a subset of $M_1 \cup M_2$. Thus the assignment to $g$ in $\mathscr{B}(X)$ of the unique $f$ in $C(X)$ differing from $g$ on a meager set is a (conjugate-preserving) homomorphism of $\mathscr{B}(X)$ onto $C(X)$. Of course $g$ corresponds to 0 if and only if $g$ vanishes outside a meager set. If $\{g_n\}$ is a monotone increasing sequence of bounded Borel functions on $X$ tending pointwise to the bounded Borel function $g$ and $f_n$ in $C(X)$ differs from $g_n$ on the meager set $M_n$, then $f_n(p) \leqslant f_{n+1}(p)$ for $p$ in $X \backslash (M_n \cup M_{n+1})$, a dense set, so that $f_n \leqslant f_{n+1}$. Moreover $f_n \leqslant f$, where $f$ in $C(X)$ differs from $g$ on the meager set $M_0$. Now $\{f_n(p)\}$ is $\{g_n(p)\}$ and tends to $f(p)\ (= g(p))$ for $p$ in $X \backslash (\cup_{j=0}^{\infty} M_j)$. Thus $f$ is the least upper bound in $C(X)$ of $\{f_n\}$; and our homomorphism of $\mathscr{B}(X)$ onto $C(X)$ is $\sigma$-normal. ■

5.2.11. Corollary. *If $\mathscr{O}$ is an open dense subset of the extremely disconnected compact Hausdorff space $X$ and $f$ is a continuous bounded function defined on $\mathscr{O}$, then there is a (unique) continuous function $h$ extending $f$ from $\mathscr{O}$ to $X$.*

*Proof.* Let $g(p)$ be $f(p)$ if $p \in \mathscr{O}$ and 0 if $p \in X \backslash \mathscr{O}$. Then $g$ is a bounded Borel function on $X$ and differs from a (unique) function $h$ in $C(X)$ on a meager subset of $X$. If $h(p) \neq f(p)$ for some $p$ in $\mathscr{O}$, then, by continuity of $h - f$ on $\mathscr{O}, f$ and hence $g$ differ from $h$ on a non-empty clopen subset of $\mathscr{O}$, contradicting the choice of $h$. Thus $h$ is a continuous extension of $f$ from $\mathscr{O}$ to $X$. ■

5.2.12. Remark. Lemma 5.2.10 provides us with another means of proving Theorem 5.2.5. In the notation of the proof of Theorem 5.2.5, define $h_0(p)$ to be that $\lambda$ in $[0, 2\pi)$ such that $\exp i\lambda = u(p)$. Then $h_0$ is continuous on $X \backslash u^{-1}(1)$ and $h_0$ is 0 on $u^{-1}(1)$. Hence $h_0 \in \mathscr{B}(X)$ and there is a function $h$ in $C(X)$ agreeing with $h_0$ on the complement of a meager subset $M$ of $X$. Thus $\exp ih$ and $\exp ih_0\ (= u)$ agree on the dense subset $X \backslash M$. Since $\exp ih$ and $u$ are continuous, $\exp ih = u$ and $\exp iH = U$, where $H$ in $\mathscr{A}$ corresponds to $h$. ■

5.2.13. Remark. If $\mathscr{A}$ is an abelian von Neumann algebra, $\mathscr{A} \cong C(X)$, and $A$ in $\mathscr{A}$ is represented by $f$ in $C(X)$; the mapping $g \to g \circ f$ is a $\sigma$-normal homomorphism of $\mathscr{B}(\mathrm{sp}(A))$ into $\mathscr{B}(X)$. Composing this mapping with the $\sigma$-normal homomorphism of $\mathscr{B}(X)$ onto $C(X)$ (described in Lemma 5.2.10) and then with the ($\sigma$-normal) isomorphism of $C(X)$ onto $\mathscr{A}$ yields a $\sigma$-normal homomorphism $\varphi$ of $\mathscr{B}(\mathrm{sp}(A))$ into $\mathscr{A}$ carrying 1 onto $I$ and the identity transform $\iota$ on $\mathrm{sp}(A)$ onto $A$. Theorem 5.2.9 applies and $\varphi(g) = g(A)$ for each $g$ in $\mathscr{B}(\mathrm{sp}(A))$. The process of forming $\varphi$ recaptures the bounded Borel function calculus by topological–function-theoretic means. At the same time, we see that $g(A)$, obtained from $g \circ f$ with $f$ representing $A$ in $C(X)$, is independent of the abelian von Neumann algebra $\mathscr{A}$ containing $A$ (and its isomorphic $C(X)$) we use. ■

We reinforce the preceding "independence" comment in the proposition that follows. We say that a * homomorphism $\psi$ of one von Neumann algebra into another is $\sigma$-normal when $\psi$ maps the least upper bound of each increasing sequence of self-adjoint operators bounded above in the first algebra onto the least upper bound of the image sequence. (Compare Proposition 4.4.7.)

5.2.14. PROPOSITION. *If $\varphi$ is a $\sigma$-normal homomorphism of one von Neumann algebra $\mathscr{R}_1$ into another $\mathscr{R}_2$ mapping $I$ onto $I$, then $\varphi(f(A)) = f(\varphi(A))$ for each normal $A$ in $\mathscr{R}_1$ and each bounded Borel function $f$ on* sp$(A)$.

*Proof.* From Proposition 4.4.7, $\mathrm{sp}(\varphi(A)) \subseteq \mathrm{sp}(A)$. If $\mathscr{A}_0$ is the abelian von Neumann algebra generated by $A$, $A^*$, and $I$, then $\mathscr{A}_0 \subseteq \mathscr{R}_1$ and $\varphi(\mathscr{A}_0)$ is contained in an abelian von Neumann subalgebra $\mathscr{A}$ of $\mathscr{R}_2$ containing $\varphi(A)$. The mapping that assigns $\varphi(h(A))$ to $h$ in $\mathscr{B}(\mathrm{sp}(A))$ is a $\sigma$-normal homomorphism. Composing this mapping with the $\sigma$-normal homomorphism of $\mathscr{B}(\mathrm{sp}(\varphi(A)))$ into $\mathscr{B}(\mathrm{sp}(A))$ that assigns to each $h$ the function equal to it on $\mathrm{sp}(\varphi(A))$ and 0 on $\mathrm{sp}(A)\backslash\mathrm{sp}(\varphi(A))$ $(= Y)$ yields a $\sigma$-normal homomorphism $\psi$ of $\mathscr{B}(\mathrm{sp}(\varphi(A)))$ into $\mathscr{A}$. Let $g$ be the characteristic function of $Y$ (an open subset of $\mathrm{sp}(A)$). Let $\mathscr{O}$ be an open subset of $\mathbb{C}$ such that $Y = \mathscr{O} \cap \mathrm{sp}(A)$. As noted in the proof of Theorem 5.2.8, the characteristic function of $\mathscr{O}$ is the pointwise limit of an increasing sequence of positive continuous functions on $\mathscr{O}$. The sequence $\{f_n\}$ of restrictions of these functions to $\mathrm{sp}(A)$ is increasing and tends pointwise to $g$. Moreover, each $f_n$ vanishes on $\mathrm{sp}(\varphi(A))$ so that $f_n(\varphi(A)) = 0$. From Proposition 4.4.7, $0 = f_n(\varphi(A)) = \varphi(f_n(A))$. Thus, by $\sigma$-normality of $\varphi$ and the mapping $h \to h(A)$ of $\mathscr{B}(\mathrm{sp}(A))$ into $\mathscr{A}_0$, $\varphi(g(A)) = 0$. For any $h$ in $\mathscr{B}(\mathrm{sp}(A))$ vanishing on $\mathrm{sp}(\varphi(A))$, we have $g \cdot h = h$, so that $g(A)h(A) = h(A)$ and $\varphi(h(A)) = 0$. It follows that, with $\iota$ and $\iota_0$ the identity transforms on $\mathrm{sp}(A)$ and $\mathrm{sp}(\varphi(A))$,

$$\varphi(A) = \varphi(\iota(A)) = \varphi([\iota \cdot (1 - g)](A)) = \psi(\iota_0)$$

and $\psi$ maps the constant function 1 on $\mathrm{sp}(\varphi(A))$ to $\varphi((1 - g)(A))$, which is $\varphi(I)$. From Theorem 5.2.9, $\psi(h) = h(\varphi(A))$ for each $h$ in $\mathscr{B}(\mathrm{sp}(\varphi(A)))$. With $f$ in $\mathscr{B}(\mathrm{sp}(A))$ and $f_0$ its restriction to $\mathrm{sp}(\varphi(A))$, $f(\varphi(A)) = f_0(\varphi(A))$ (in essence, by definition), so that

$$\varphi(f(A)) = \varphi([f \cdot (1 - g)](A)] = \psi(f_0) = f_0(\varphi(A)) = f(\varphi(A)). \quad \blacksquare$$

5.2.15. REMARK. Again, we recapture the composite function rule, $(f \circ g)(A) = f(g(A))$, for $f$ and $g$ bounded Borel functions on $\mathbb{C}$. The mapping $f \to (f \circ g)(A)$ of $\mathscr{B}(\mathbb{D})$ is a $\sigma$-normal homomorphism mapping the identity transform onto $g(A)$ and 1 onto $I$, where $\mathbb{D}$ is a disk containing $\mathrm{sp}(g(A)) \cup \mathrm{range}(g)$. It follows from Theorem 5.2.9 that $(f \circ g)(A) = f(g(A))$. $\blacksquare$

*Bibliography*: [7, 17, 18, 23]

## 5.3. Two fundamental approximation theorems

If $\mathscr{F}$ is a family of bounded operators on the Hilbert space $\mathscr{H}$, $\mathscr{F}'$ will denote those bounded operators on $\mathscr{H}$ commuting with all operators in $\mathscr{F}$. The set $\mathscr{F}'$ is called *the commutant of* $\mathscr{F}$. Note that $\mathscr{F}'$ is weak-operator closed.

5.3.1. THEOREM (Double commutant). *If $\mathfrak{A}$ is a self-adjoint algebra of operators containing the identity operator and acting on the Hilbert space $\mathscr{H}$, then the weak- and strong-operator closures of $\mathfrak{A}$ coincide with $(\mathfrak{A}')'$.*

*Proof.* Since $\mathfrak{A}$ is convex, Theorem 5.1.2 assures us that the weak- and strong-operator closures of $\mathfrak{A}$ coincide. We shall conclude this, once again, in the present case, by showing that $\mathfrak{A}''$ is the strong-operator closure of $\mathfrak{A}$. Since $\mathfrak{A} \subseteq \mathfrak{A}''$ and $(\mathfrak{A}')'$ is weak- (and strong-)operator closed, both the weak- and strong-operator closures of $\mathfrak{A}$ are contained in $\mathfrak{A}''$.

Suppose $T \in \mathfrak{A}''$ and $x_1, \ldots, x_n$ are given in $\mathscr{H}$. We want to find $T_0$ in $\mathfrak{A}$ such that $\|(T - T_0)x_j\| < 1$ for each $j$. Let $\tilde{\mathscr{H}}$ be the $n$-fold direct sum of $\mathscr{H}$ with itself. With $A$ in $\mathscr{B}(\mathscr{H})$, let $\tilde{A}$ be $A \oplus \cdots \oplus A$ (so that $\tilde{A}(y_1, \ldots, y_n) = (Ay_1, \ldots, Ay_n)$). If $\tilde{x} = (x_1, \ldots, x_n)$ and $\tilde{\mathfrak{A}} = \{\tilde{A} : A \in \mathfrak{A}\}$, then $\tilde{\mathfrak{A}}$ is a self-adjoint algebra of operators on $\tilde{\mathscr{H}}$, and $[\tilde{\mathfrak{A}}\tilde{x}]$ is invariant under $\tilde{\mathfrak{A}}$. From the final paragraph of Section 2.6, *Subspaces*, the orthogonal projection $\bar{E}$ with range $[\tilde{\mathfrak{A}}\tilde{x}]$ commutes with $\tilde{\mathfrak{A}}$ (that is, $\bar{E} \in \tilde{\mathfrak{A}}'$). If we show that $\tilde{T} \in \tilde{\mathfrak{A}}''$, then $\tilde{T}$ commutes with $\bar{E}$, and the range of $\bar{E}$ is stable under $\tilde{T}$. In this case, $\tilde{T}\tilde{x} = (Tx_1, \ldots, Tx_n)$ is in this range, since $\tilde{x} = (Ix_1, \ldots, Ix_n) \in \bar{E}(\tilde{\mathscr{H}})$. As $\tilde{\mathfrak{A}}\tilde{x}$ is dense in $[\tilde{\mathfrak{A}}\tilde{x}]$, there is a $T_0$ in $\mathfrak{A}$ such that $\|(T - T_0)x_j\| < 1$ for all $j$.

It remains to prove that $\tilde{T} \in \tilde{\mathfrak{A}}''$. In Section 2.6, *Matrix representations*, we observed that operators in $\mathscr{B}(\tilde{\mathscr{H}})$ can be represented as $n \times n$ matrices with entries from $\mathscr{B}(\mathscr{H})$. In this representation, $\tilde{T}$ has each diagonal entry $T$ and all others 0. At the same time, simple matrix calculations show that $\tilde{\mathfrak{A}}'$ consists of those matrices with all entries in $\mathfrak{A}'$, and $\tilde{\mathfrak{A}}''$ consists of those matrices with a single operator from $\mathfrak{A}''$ at all diagonal entries and 0 at all other entries. Since $T$ is assumed to be in $\mathfrak{A}''$, $\tilde{T} \in \tilde{\mathfrak{A}}''$. ■

There are several easy consequences of the double commutant theorem. If $\mathscr{F} \subseteq \mathscr{B}(\mathscr{H})$ and $\mathscr{F}^* = \{A^* : A \in \mathscr{F}\}$, then $\{\mathscr{F} \cup \mathscr{F}^*\}''$ is the von Neumann algebra generated by $\mathscr{F}$. If $\mathscr{M}$ is a factor, $\{\mathscr{M} \cup \mathscr{M}'\}' = \mathscr{M} \cap \mathscr{M}'$, which is the set of scalars. Thus $\{\mathscr{M} \cup \mathscr{M}'\}'' = \mathscr{B}(\mathscr{H})$. More generally, $\{\mathscr{R} \cup \mathscr{R}'\}''$ is $\mathscr{C}'$, where $\mathscr{C}$ is the center of the von Neumann algebra $\mathscr{R}$; for $\{\mathscr{R} \cup \mathscr{R}'\}' = \mathscr{R} \cap \mathscr{R}' = \mathscr{C}$. At the same time, we see that $\mathscr{C}$ is the center of $\mathscr{R}'$. It follows that $\mathscr{R}'$ is a factor if $\mathscr{R}$ is a factor.

As another simple consequence of the double commutant theorem, we may conclude that a von Neumann algebra $\mathscr{R}$ on $\mathscr{H}$ is $\mathscr{B}(\mathscr{H})$ if each projection in $\mathscr{R}'$ is either 0 or $I$. For this, we note that, from Theorem 5.2.2(v), each von Neumann algebra is the norm closure of the linear span of its projections. Thus $\mathscr{R}'$ is $\{\lambda I\}$, and, under the present assumption, $\mathscr{R}$ $(= \mathscr{R}'') = \mathscr{B}(\mathscr{H})$.

Our second approximation result (the Kaplansky density theorem) requires for its proof certain facts about strong-operator continuity of some classes of functions. These continuity results are interesting and useful in their own right. The most basic of them, strong-operator continuity of multiplication on bounded subsets of $\mathscr{B}(\mathscr{H})$, was established in Remark 2.5.10.

5.3.2. PROPOSITION. *Each continuous real- or complex-valued function $f$ (on $\mathbb{R}$ or $\mathbb{C}$) is strong-operator continuous on bounded sets of (self-adjoint or normal) operators on the Hilbert space $\mathscr{H}$.*

*Proof.* We may suppose that the bounded set of normal operators under consideration is contained in the ball of radius $r$. If $A_0$ is a normal operator in this ball, $\varepsilon > 0$, and $x_1, \ldots, x_n$ is a set of vectors in $\mathscr{H}$, we want to find vectors $y_1, \ldots, y_m$ and a positive $\delta$ such that $\|(f(A) - f(A_0))x_k\| < \varepsilon$ if $\|(A - A_0)y_j\| < \delta$, $A$ is normal, and $\|A\| \leqslant r$. If we can accomplish this with $x_1, \ldots, x_n$ replaced by a single vector $x_0$, then we can do it for $x_1, \ldots, x_n$—increasing the set of $y$'s successively. Replace $\varepsilon$ by $\varepsilon\|x_0\|^{-1}$. If

$$\left\|(f(A) - f(A_0))\frac{x_0}{\|x_0\|}\right\| < \frac{\varepsilon}{\|x_0\|},$$

then $\|(f(A) - f(A_0))x_0\| < \varepsilon$. Thus we may assume that $\|x_0\| = 1$.

From the Weierstrass approximation theorem (see Remark 3.4.15), there is a polynomial $p$ (in $z$ and $\bar{z}$) such that $\|f - p\|_{\mathbb{C}_r} < \varepsilon/3$, where $\mathbb{C}_r$ is the closed disk in $\mathbb{C}$ with center 0 and radius $r$. Using the strong-operator continuity of multiplication on bounded sets of operators and of the adjoint operation on the set of normal operators (see Remark 2.5.10), we can find vectors $y_1, \ldots, y_m$ and a positive $\delta$ such that $\|(p(A) - p(A_0))x_0\| < \varepsilon/3$ if $\|(A - A_0)y_j\| < \delta$, $A$ is normal, and $\|A\| \leqslant r$ (where $\bar{z}$ is replaced by $A^*$ in determining $p(A)$). In this case,

$$\begin{aligned}\|(f(A) - f(A_0))x_0\| &\leqslant \|(f(A) - p(A))x_0\| + \|(p(A) - p(A_0))x_0\| \\ &\quad + \|(p(A_0) - f(A_0))x_0\| \\ &\leqslant \|f(A) - p(A)\| + \|p(A_0) - f(A_0)\| + \varepsilon/3 \\ &\leqslant 2\|f - p\|_{\mathbb{C}_r} + \varepsilon/3 < \varepsilon.\end{aligned}$$

To conclude that $\|f(A) - p(A)\|$ and $\|f(A_0) - p(A_0)\|$ are majorized by $\|f - p\|_{\mathbb{C}_r}$, in the preceding inequality, we pass to the function representation of the (abelian) $C^*$-algebra generated by $A$ and $A^*$. This representation can be viewed as taking place on $\mathrm{sp}(A)$ ($\subseteq \mathbb{C}_r$) with $A$ represented by $z$, $A^*$ by $\bar{z}$, $f(A)$ by $f|\mathrm{sp}(A)$, and $p(A)$ by $p|\mathrm{sp}(A)$. Since the representation is an isometry, $\|f(A) - p(A)\| = \|f - p\|_{\mathrm{sp}(A)}$ (and, similarly, $\|f(A_0) - p(A_0)\| = \|f - p\|_{\mathrm{sp}(A_0)}$). ■

Presently we shall prove a stong-operator continuity result applicable to all (rather than bounded sets of) self-adjoint operators. An important transform from self-adjoint operators to unitary operators is needed for this – the Cayley transform. It assigns to a self-adjoint operator $H$ the (unitary) operator

$(H - iI)(H + iI)^{-1}$ $(= U(H))$. Since

$$U(H)U(H)^* = U(H)^*U(H) = (H - iI)^{-1}(H + iI)(H - iI)(H + iI)^{-1}$$
$$= (H^2 + I)(H^2 + I)^{-1} = I,$$

$U(H)$ is unitary. Now $-i(z + 1)(z - 1)^{-1}$ is a real number $t$, provided $|z| = 1$ and $z \neq 1$. In this case, $z = (t - i)(t + i)^{-1}$. Similarly, $-i(U + I)(U - I)^{-1}$ $(= H)$ is defined if $1 \notin \text{sp}(U)$; and $H$ is self-adjoint if $U$ is unitary. Note for this that

$$(U^* + I)(U - I) = U - U^* = -(U^* - I)(U + I),$$

so that

$$H^* = i(U^* + I)(U^* - I)^{-1} = -i(U + I)(U - I)^{-1} = H.$$

Analogous computations yield $U(H) = U$ and $-i(U(H) + I)(U(H) - I)^{-1} = H$. (These identities result, as well, from the corresponding facts for real and complex numbers through the use of the function representation for $\mathfrak{A}(H)$.)

5.3.3. Lemma. *The Cayley transform is strong-operator continuous.*

*Proof.* We have

$$(H + iI)(U(H) - U(H_0))(H_0 + iI) = 2i(H - H_0),$$
$$\|(U(H) - U(H_0))x_0\| = 2\|(H + iI)^{-1}(H - H_0)(H_0 + iI)^{-1}x_0\|$$
$$\leqslant 2\|(H - H_0)(H_0 + iI)^{-1}x_0\|$$

(noting that $\|(H + iI)^{-1}\| \leqslant 1$, from the function representation of $\mathfrak{A}(H)$). The strong-operator continuity of $H \to U(H)$ follows from this inequality. ∎

5.3.4. Theorem. *If $h$ is a continuous real-valued function vanishing at $\infty$ on $\mathbb{R}$, then $h$ is strong-operator continuous on the set of self-adjoint operators.*

*Proof.* Let $f(z)$ be $h(-i(z + 1)(z - 1)^{-1})$ when $z \neq 1$ and $|z| = 1$. If we define $f(1)$ to be 0, $f$ is a continuous function on the unit circle in $\mathbb{C}$ with values in $\mathbb{R}$, since $h$ vanishes at $\infty$ on $\mathbb{R}$. From the discussion preceding Lemma 5.3.3, $f(U(H)) = h(H)$ for each self-adjoint $H$. Since $f$ is continuous on the unit circle in $\mathbb{C}$, $f$ gives rise to a strong-operator continuous function on the (bounded) set of unitary operators, from Proposition 5.3.2. As $h$ is the composition of the Cayley transform and $f$, applying Lemma 5.3.3, we see that $h$ is strong-operator continuous. ∎

We write $\mathfrak{A}^-$ for the strong- (equivalently, weak-)operator closure of an algebra $\mathfrak{A}$ and (much less frequently used) $\mathfrak{A}^=$ for its norm closure.

5.3.5. THEOREM (Kaplansky density theorem). *If* $\mathfrak{A}$ *is a self-adjoint algebra of operators, then each A in* $(\mathfrak{A}^-)_1$, *the unit ball of* $\mathfrak{A}^-$, *lies in* $(\mathfrak{A})_1^-$, *the strong-operator closure of the unit ball* $(\mathfrak{A})_1$ *of* $\mathfrak{A}$. *If H is a self-adjoint operator in* $(\mathfrak{A}^-)_1$, *then H is in the strong-operator closure of the set of self-adjoint operators in* $(\mathfrak{A})_1$.

*Proof.* If $H$ is a self-adjoint operator in $\mathfrak{A}^-$ and $\{T_a\}$ is a net of operators in $\mathfrak{A}$ weak-operator convergent to $H$, then $\{\frac{1}{2}(T_a + T_a^*)\}$ is a net of self-adjoint operators in $\mathfrak{A}$ weak-operator convergent to $H$. Thus $H$ is in the weak-operator closure of the set of self-adjoint operators in $\mathfrak{A}$. Since this set is convex; from Theorem 5.1.2, $H$ is in its strong-operator closure.

Suppose, now, that $H$ is a self-adjoint operator in $(\mathfrak{A}^-)_1$ and $\{H_a\}$ is a net of self-adjoint operators in $\mathfrak{A}$ strong-operator convergent to $H$. If $f(t)$ is $t$, for $t$ in $[-1,1]$, and $t^{-1}$ for $t$ not in $[-1,1]$, then $f$ is real-valued, continuous, and vanishes at $\infty$, on $\mathbb{R}$. From Theorem 5.3.4, $f$ gives rise to a strong-operator continuous function; and $\{f(H_a)\}$ is strong-operator convergent to $f(H)$. Since $f$ is the identity mapping on $\text{sp}(H)$, $f(H) = H$. Since $f$ has bound 1, $\|f(H_a)\| \leqslant 1$. Thus $H$ is in the strong-operator closure of the set of self-adjoint elements in $(\mathfrak{A}^=)_1$, the unit ball of the norm closure $\mathfrak{A}^=$ of $\mathfrak{A}$. But each self-adjoint element in $(\mathfrak{A}^=)_1$ is a norm limit (hence, strong-operator limit) of self-adjoint elements in $(\mathfrak{A})_1$. Thus $H$ is the strong-operator limit of such elements.

If $T \in (\mathfrak{A}^-)_1$, then $T'$, the $2 \times 2$ matrix with entries 0 on the diagonal and $T$ and $T^*$ at the other positions, acting on $\mathscr{H} \oplus \mathscr{H}$, is self-adjoint and in the unit ball of $\mathfrak{A}_2^-$, where $\mathfrak{A}_2$ is the algebra of $2 \times 2$ matrices with entries from $\mathfrak{A}$. From what we have proved to this point, $T'$ lies in the strong-operator closure of the unit ball of $\mathfrak{A}_2$. Thus each entry in $T'$, in particular $T$, is in the strong-operator closure of the set of corresponding entries of elements in the unit ball of $\mathfrak{A}_2$ – and these entries are all in the unit ball of $\mathfrak{A}$. ∎

Note that we do not assume that $I \in \mathfrak{A}$ in Theorem 5.3.5.

5.3.6. COROLLARY. *If* $\mathfrak{A}$ *is a self-adjoint algebra of operators acting on the Hilbert space* $\mathscr{H}$ *and H is a positive operator in* $(\mathfrak{A}^-)_1$, *then H is in the strong-operator closure of* $(\mathfrak{A}^+)_1$.

*Proof.* Since $H \geqslant 0$, $H = K^2$, with $K$ a positive operator in $(\mathfrak{A}^-)_1$. From Theorem 5.3.5, $K$ is the strong-operator limit of a net $\{K_a\}$ of self-adjoint operators in $(\mathfrak{A})_1$. By strong-operator continuity of multiplication on the unit ball, $\{K_a^2\}$ is strong-operator convergent to $K^2$ $(= H)$. Since $K_a$ is self-adjoint, $K_a^2 \in (\mathfrak{A}^+)_1$. ∎

5.3.7. COROLLARY. *If* $\mathfrak{A}$ *is a C*-algebra acting on the Hilbert space* $\mathscr{H}$ *and U is a unitary operator in* $\mathfrak{A}^-$, *then U is in the strong-operator closure of the set of unitary operators in* $\mathfrak{A}$.

*Proof.* From Theorem 5.2.5, $U = \exp iH$, with $H$ a self-adjoint operator in $\mathfrak{A}^-$. Using Theorem 5.3.5 (with the easy extension of its assertion to a ball of arbitrary radius), we have that $H$ is the strong-operator limit of $\{H_a\}$, with $H_a$ a self-adjoint operator in the ball of radius $\|H\|$ in $\mathfrak{A}$. Since $t \to \exp it$ is continuous, $\{\exp iH_a\}$ is strong-operator convergent to $\exp iH$ $(= U)$. ∎

In each of the statements of the results related to the Kaplansky density theorem, except Corollary 5.3.7, we did not have to assume that $\mathfrak{A}$ is norm closed. This assumption is needed in Corollary 5.3.7 (see Exercise 5.7.34). The Kaplansky density theorem, as well as the double commutant theorem, are used so often in what follows that we shall generally make use of them without mention or citation.

*Bibliography*: [9, 11]

## 5.4. Irreducible algebras – an application

A family $\mathscr{F}$ of bounded operators on a Hilbert space $\mathscr{H}$ is said to act *topologically irreducibly* when (0) and $\mathscr{H}$ are the only (closed) stable subspaces under $\mathscr{F}$. If (0) and $\mathscr{H}$ are the only linear manifolds (not necessarily closed) in $\mathscr{H}$ stable under $\mathscr{F}$, we say that $\mathscr{F}$ acts *algebraically irreducibly*.

5.4.1. Theorem. *If $\mathscr{F}$ is a self-adjoint family of bounded operators acting on the Hilbert space $\mathscr{H}$, then $\mathscr{F}$ acts topologically irreducibly on $\mathscr{H}$ if and only if $\mathscr{F}'$ consists of scalars or, equivalently, $\mathscr{F}'' = \mathscr{B}(\mathscr{H})$.*

*Proof.* Note, first, that if $\mathscr{F}'$ consists of scalars, $\mathscr{F}'' = \mathscr{B}(\mathscr{H})$. If $\mathscr{F}'' = \mathscr{B}(\mathscr{H})$, then $\mathscr{F}'''$ consists of scalars. However, $\mathscr{F} \subseteq \mathscr{F}''$, so that $\mathscr{F}''' \subseteq \mathscr{F}'$; and $\mathscr{F}'$ commutes with $(\mathscr{F}')'$, so that $\mathscr{F}' \subseteq \mathscr{F}'''$. Thus $\mathscr{F}' = \mathscr{F}'''$, and $\mathscr{F}'$ consists of scalars.

With $\mathscr{F}$ a self-adjoint family of bounded operators, $\mathscr{F}'$ is a von Neumann algebra. From Theorem 5.2.2(v), then, $\mathscr{F}'$ consists of scalars if and only if each projection in $\mathscr{F}'$ is either 0 or $I$. Since $\mathscr{F}$ is self-adjoint, a projection lies in $\mathscr{F}'$ if and only if its range is stable under $\mathscr{F}$ (see Section 2.6, *Subspaces*). Thus $\mathscr{F}$ acts topologically irreducibly if and only if $\mathscr{F}'$ consists of scalars. ∎

Making essential use of the Kaplansky density theorem, we shall prove (Corollary 5.4.4) that topological and algebraic irreducibility are the same for a $C^*$-algebra. Toward this end, we need the following result.

5.4.2. Lemma. *If $\{x_1, \ldots, x_n\}$ is an orthonormal set in the Hilbert space $\mathscr{H}$ and $z_1, \ldots, z_n$ are vectors in the ball of radius $r$ and center 0 in $\mathscr{H}$, there is a $B$ in $\mathscr{B}(\mathscr{H})$ such that $\|B\| \leqslant (2n)^{1/2}r$ and $Bx_j = z_j$ for all $j$. If $Ax_j = z_j$ for some self-adjoint operator $A$, then $B$ may be chosen self-adjoint.*

*Proof.* Let $E$ be the orthogonal projection of $\mathscr{H}$ onto the (finite-dimensional) subspace spanned by $\{x_1, \ldots, x_n\}$. If $Tx$ is $\sum_{j=1}^{n} \langle Ex, x_j \rangle z_j$, then

$TE = T$ and

$$\|Tx\| \leqslant r \sum_{j=1}^{n} |\langle Ex, x_j \rangle| \leqslant r\left( \sum_{j=1}^{n} |\langle Ex, x_j \rangle|^2 \right)^{1/2} \left( \sum_{j=1}^{n} 1 \right)^{1/2} \leqslant n^{1/2} r \|Ex\|$$
$$\leqslant n^{1/2} r \|x\|.$$

Thus $\|T\| \leqslant n^{1/2} r$.

If $Ax_j = z_j$ for some self-adjoint $A$, then $ET = EAE$, so that $ET$ is self-adjoint. Thus $B$ is self-adjoint, where $B = T + T^*(I - E) = ET + (I - E)T + T^*(I - E)$, and $Bx_j = Tx_j = z_j$. Since $T(I - E) = 0 = (I - E)T^*$,

$$\|BB^*\| = \|B\|^2 = \|TT^* + T^*(I - E)T\| \leqslant 2\|TT^*\| = 2\|T\|^2 \leqslant 2nr^2. \quad \blacksquare$$

5.4.3. Theorem. *If the $C^*$-algebra $\mathfrak{A}$ acts topologically irreducibly on the Hilbert space $\mathscr{H}$, $\{y_1, \ldots, y_n\}$ is a set of vectors and $\{x_1, \ldots, x_n\}$ is a linearly independent set of vectors in $\mathscr{H}$, there is an $A$ in $\mathfrak{A}$ such that $Ax_j = y_j$. If $Bx_j = y_j$ for some self-adjoint operator $B$, then $A$ can be chosen self-adjoint.*

*Proof.* Replacing $\{x_1, \ldots, x_n\}$ by an orthonormal basis for the space they span and $\{y_1, \ldots, y_n\}$ by the transform of this basis under the mapping taking $x_j$ to $y_j$, we may assume that $\{x_1, \ldots, x_n\}$ is an orthonormal set. With $B_0$ in $\mathscr{B}(\mathscr{H})$ such that $B_0 x_j = y_j$, choose $A_0$ in $\mathfrak{A}$ such that $\|A_0 x_j - B_0 x_j\| = \|A_0 x_j - y_j\| \leqslant [2(2n)^{1/2}]^{-1}$. For the choice of $A_0$, we use the fact that $\mathfrak{A}^- = \mathfrak{A}'' = \mathscr{B}(\mathscr{H})$ (Theorem 5.4.1). Choose $B_1$ such that $B_1 x_j = y_j - A_0 x_j$, with $\|B_1\| \leqslant \frac{1}{2}$ (using Lemma 5.4.2). Note that $A_0$ can be chosen self-adjoint if $B_0$ can, since the self-adjoint operators in $\mathfrak{A}$ are strong-operator dense in those of its strong-operator closure, $\mathscr{B}(\mathscr{H})$. In this case, $B_1$ can be chosen self-adjoint, from Lemma 5.4.2. Theorem 5.3.5 (Kaplansky density) provides us with an operator $A_1$ in $\mathfrak{A}$ (self-adjoint if $B_0$ is self-adjoint) such that $\|A_1\| \leqslant \frac{1}{2}$ and $\|A_1 x_j - B_1 x_j\| \leqslant [4(2n)^{1/2}]^{-1}$.

Suppose, now, that $B_k$ has been constructed so that $\|B_k\| \leqslant 2^{-k}$, $B_k x_j = y_j - A_0 x_j - A_1 x_j - \cdots - A_{k-1} x_j$, and $B_k$ is self-adjoint if $B_0$ is. Choose $A_k$ in $\mathfrak{A}$ (self-adjoint if $B_k$ is) such that $\|A_k\| \leqslant 2^{-k}$ and $\|A_k x_j - B_k x_j\| \leqslant [2^{k+1}(2n)^{1/2}]^{-1}$ (employing Kaplansky density for this choice). From Lemma 5.4.2, there is a $B_{k+1}$ with $\|B_{k+1}\| \leqslant 2^{-(k+1)}$ and $B_{k+1} x_j = y_j - A_0 x_j - \cdots - A_k x_j$ (self-adjoint if $A_k$ is). The sum $\sum_{k=0}^{\infty} A_k$ converges in norm to an operator $A$ in $\mathfrak{A}$ (self-adjoint if $B_0$ is); and

$$y_j - Ax_j = y_j - \sum_{k=0}^{\infty} A_k x_j = \lim_k (y_j - A_0 x_j - \cdots - A_k x_j)$$
$$= \lim_k B_{k+1} x_j = 0. \quad \blacksquare$$

In the preceding proof, we could have chosen $A_0$ such that $\|A_0 x_j - y_j\| \leqslant [2(2n)^{1/2}]^{-1}\varepsilon$ and $\|A_0\| \leqslant \|B_0\|$, $B_1$ such that $\|B_1\| \leqslant \varepsilon/2$, $A_1$ such that

$\|A_1\| \leqslant \varepsilon/2$, and, finally, $A_k$ such that $\|A_k\| \leqslant 2^{-k}\varepsilon$. With these choices, $\|A\| \leqslant \sum_{k=0}^{\infty} \|A_k\| \leqslant \|B_0\| + \varepsilon$ for each preassigned positive $\varepsilon$, though, in this process, $A$ depends on the given $\varepsilon$. A function calculus argument (see Exercise 5.7.41) proves that $A$ may be chosen so that $\|A\| \leqslant \|B_0\|$.

The property of $\mathfrak{A}$ described in the first sentence of Theorem 5.4.3 is referred to as *transitivity*. It is an immediate consequence of the transitivity of $\mathfrak{A}$ that it acts algebraically irreducibly.

5.4.4. COROLLARY. *If the $C^*$-algebra $\mathfrak{A}$ acts topologically irreducibly on the Hilbert space $\mathscr{H}$, then it acts algebraically irreducibly.*

Theorem 5.4.3 asserts a form of "self-adjoint transitivity" for $\mathfrak{A}$ as well as "general transitivity." In Theorem 5.4.5 we establish "unitary transitivity" for $\mathfrak{A}$ – a result we shall need in the deeper study of states (see Section 10.2). We need no longer distinguish topological and algebraic irreducibility for $C^*$-algebras. We shall speak of $C^*$-algebras "acting irreducibly."

5.4.5. THEOREM. *If $\mathfrak{A}$ is a $C^*$-algebra acting irreducibly on $\mathscr{H}$ and $V$ is a unitary operator such that $Vx_1 = y_1, \ldots, Vx_n = y_n$, then there is a self-adjoint operator $H$ in $\mathfrak{A}$ such that $Ux_1 = y_1, \ldots, Ux_n = y_n$, where $U = \exp iH$.*

*Proof.* Since two operators that agree on a basis for the space generated by $x_1, \ldots, x_n$ agree on the space, we can assume that $\{x_1, \ldots, x_n\}$ is an orthonormal set. The same is then the case for $\{y_1, \ldots, y_n\}$. Let $\{x_1, \ldots, x_m\}$ and $\{y_1, \ldots, y_m\}$ be extensions of $\{x_1, \ldots, x_n\}$ and $\{y_1, \ldots, y_n\}$ to orthonormal bases of the space generated by $x_1, \ldots, x_n, y_1, \ldots, y_n$. Define $V_0$ on this space by $V_0 x_j = y_j$ for all $j$. We may replace $V$ by $V_0$ for the purposes of finding $U$ of the statement of this theorem. We may replace $\{x_1, \ldots, x_m\}$ by an orthonormal basis that diagonalizes $V_0$. Hence, we may assume that $V_0 x_j = c_j x_j$ for all $j$, with $|c_j| = 1$. Let $a_1, \ldots, a_m$ be real numbers such that $c_j = \exp ia_j$. From Theorem 5.4.3, there is a self-adjoint $H$ in $\mathfrak{A}$ such that $Hx_j = a_j x_j$. If $U = \exp iH$, then $U$ is a unitary operator in $\mathfrak{A}$, $U$ is a norm limit of polynomials in $H$, and $Ux_j = c_j x_j$. ■

*Bibliography*: [5, 8]

## 5.5. Projection techniques and constructs

In this section we study some topics involving projections and constructs with von Neumann algebras related to them.

*Central carriers.* If $A$ is in the von Neumann algebra $\mathscr{R}$ and $\{P_a\}$ is a family of central projections in $\mathscr{R}$ (that is, each $P_a$ is in the center $\mathscr{C}$ of $\mathscr{R}$) such that $P_a A = 0$ for all $a$, then $PA = 0$, where $P = \bigvee_a P_a$. Note, for this, that $Ax$ is orthogonal to the range of $P_a$ for all $a$, and, hence, to the union of these spaces. Thus $PAx = 0$ for all $x$ and $PA = 0$.

5.5.1. Definition. The *central carrier* $C_A$ of an operator $A$ in a von Neumann algebra $\mathscr{R}$ is the projection $I - P$, where $P$ is the union of all central projections $P_a$ in $\mathscr{R}$ such that $P_aA = 0$. ■

Since the center $\mathscr{C}$ of $\mathscr{R}$ is $\mathscr{R} \cap \mathscr{R}'$, $\mathscr{C}$ is a von Neumann algebra. From Proposition 5.1.8, $P$ and hence $C_A$ are in $\mathscr{C}$. As we noted, $PA = 0$, so that $C_AA = A$. We could equally well have defined $C_A$ as the intersection of all central projections $Q$ such that $QA = A$. The context will usually make clear the von Neumann algebra relative to which a central carrier is formed. It may, however, require clarification, in which case some phrase such as "relative to $\mathscr{R}$" will appear. That the von Neumann algebra plays a role in determining central carriers can be illustrated by identifying the central carrier of a projection $E$ (different from 0 and $I$) relative to the algebra of all bounded operators and relative to the von Neumann algebra generated by $E$ and $I$. In the first case the central carrier is $I$ and in the second it is $E$.

5.5.2. Proposition. *The central carrier of an operator $A$ in a von Neumann algebra $\mathscr{R}$ acting on a Hilbert space $\mathscr{H}$ has range $[\mathscr{R}A(\mathscr{H})]$.*

*Proof.* Since $C_AA = A$, $A(\mathscr{H})$ is contained in the range of $C_A$. Since $C_A$ commutes with $\mathscr{R}$, and $\mathscr{R}$ is a self-adjoint family of bounded operators on $\mathscr{H}$, the range of $C_A$ is stable under $\mathscr{R}$. Thus $[\mathscr{R}A(\mathscr{H})]$ is contained in the range of $C_A$.

The projection $Q$ with range $[\mathscr{R}A(\mathscr{H})]$ commutes with $\mathscr{R}$ and $\mathscr{R}'$, since $[\mathscr{R}A(\mathscr{H})]$ is stable under $\mathscr{R}$ and $\mathscr{R}'$. Thus $Q$ is in the center of $\mathscr{R}$ (see the discussion following Theorem 5.3.1). As $[\mathscr{R}A(\mathscr{H})]$ contains the range of $A$, $QA = A$. Hence $C_A \leqslant Q$ (for $(I - Q)A = 0$, so that $I - Q \leqslant I - C_A$). From the earlier discussion, $Q \leqslant C_A$, so that $Q = C_A$. ■

5.5.3. Proposition. *If $\{E_a\}$ is a family of projections in a von Neumann algebra $\mathscr{R}$ and $E = \bigvee_a E_a$, then $C_E = \bigvee_a C_{E_a}$. If $Q$ is a central projection in $\mathscr{R}$, then $QC_A = C_{QA}$ for each $A$ in $\mathscr{R}$.*

*Proof.* Since $E_a \leqslant E \leqslant C_E$, $C_{E_a} \leqslant C_E$; and $\bigvee_a C_{E_a} \leqslant C_E$. If $P = C_E - \bigvee_a C_{E_a}$, then $PE_a = 0$ for each $a$, so that $PE = 0$, and $0 = PC_E = P$. Since $\mathscr{R}QA(\mathscr{H}) = Q\mathscr{R}A(\mathscr{H})$, it follows from Proposition 5.5.2 that $QC_A = C_{QA}$. ■

The analogous identity for intersections of projections is not valid. If $\{x_n\}$ is an orthonormal basis for a (separable) Hilbert space $\mathscr{H}$ and $E_n$ is the projection with range spanned by $\{x_j : j \geqslant n\}$, then $\bigwedge_n E_n = 0$. If central carriers are formed relative to $\mathscr{B}(\mathscr{H})$, then $C_{E_n} = I$ and $C_0 = 0$. Thus $\bigwedge_n C_{E_n} = I \neq C_0$.

5.5.4. Theorem. *If $\mathscr{R}$ is a von Neumann algebra with center $\mathscr{C}$ acting on the Hilbert space $\mathscr{H}$, then $\sum_{j=1}^n A_jA_j' = 0$, with $A_j$ in $\mathscr{R}$ and $A_j'$ in $\mathscr{R}'$, if and only if there are operators $C_{jk}, j, k$ in $\{1, \ldots, n\}$, in $\mathscr{C}$ such that $\sum_{j=1}^n A_jC_{jk} = 0$ for $k$ in*

*$\{1,\ldots,n\}$ and $\sum_{k=1}^{n} C_{jk}A'_k = A'_j$ for $j$ in $\{1,\ldots,n\}$. In particular, $AA' = 0$, with $A$ in $\mathscr{R}$ and $A'$ in $\mathscr{R}'$, if and only if $C_A C_{A'} = 0$.*

*Proof.* Given operators $C_{jk}$ with the stated properties, we have

$$\sum_{j=1}^{n} A_j A'_j = \sum_{j=1}^{n}\sum_{k=1}^{n} A_j C_{jk} A'_k = \sum_{k=1}^{n}\left(\sum_{j=1}^{n} A_j C_{jk}\right)A'_k = 0.$$

Suppose, now, that $\sum_{j=1}^{n} A_j A'_j = 0$. Let $\mathscr{R}'_n$ be the algebra of $n \times n$ matrices with entries from $\mathscr{R}'$ acting on the $n$-fold direct sum of $\mathscr{H}$ with itself. Then $\mathscr{R}'_n$ has commutant consisting of those $n \times n$ matrices with some element of $\mathscr{R}$ repeated at each diagonal position and 0 at each off-diagonal position. Let $\tilde{A}$ be the $n \times n$ matrix whose first row has the entries $A_1, \ldots, A_n$ and all of whose other entries are 0. Let $[C_{jk}]$ be the union of all projections $\tilde{E}'$ in $\mathscr{R}'_n$ such that $\tilde{A}\tilde{E}' = 0$. Then $\tilde{A}\tilde{P} = 0$, where $\tilde{P} = [C_{jk}]$.

Let $\tilde{F}'$ be the projection with entry the projection $F'$ in $\mathscr{R}'$ at each diagonal position and 0 at all other positions of an $n \times n$ matrix. Since $A_j F' = F' A_j$ for all $j$, $\tilde{A}\tilde{F}'\tilde{P} = \tilde{F}'\tilde{A}\tilde{P} = 0$. It follows from this and the construction of $\tilde{P}$ that the range of $\tilde{P}$ contains that of $\tilde{F}'\tilde{P}$. Hence

$$\tilde{F}'\tilde{P} = \tilde{P}\tilde{F}'\tilde{P} = (\tilde{P}\tilde{F}'\tilde{P})^* = \tilde{P}\tilde{F}';$$

and $C_{jk}$ commutes with each projection $F'$ in $\mathscr{R}'$. From Theorem 5.2.2, $C_{jk} \in \mathscr{C}$.

Since $\tilde{A}\tilde{P} = 0$, matrix multiplication yields the equations $\sum_{j=1}^{n} A_j C_{jk} = 0$ for $k$ in $\{1,\ldots,n\}$. If $\tilde{A}'$ is the $n \times n$ matrix with entries $A'_1, \ldots, A'_n$ in the first column and 0 at all other entries, $\tilde{A}\tilde{A}' = 0$. Thus $\tilde{A}$ annihilates the range projection of $\tilde{A}'$ and $\tilde{P}\tilde{A}' = \tilde{A}'$. Again, matrix multiplication yields $\sum_{k=1}^{n} C_{jk} A'_k = A'_j$ for $j$ in $\{1,\ldots,n\}$.

If $AA' = 0$ with $A$ in $\mathscr{R}$ and $A'$ in $\mathscr{R}'$, applying the result just proved, there is a central operator $Q$ such that $0 = AQ = QA$ and $QA' = A'$. Moreover, the $1 \times 1$ matrix with $Q$ as entry is the projection $\tilde{P}$. Thus $Q$ may be chosen as a projection in $\mathscr{C}$. Since $QA = 0$, $Q \leqslant I - C_A$. As $QA' = A'$, $C_{A'} \leqslant Q$. Thus $C_A C_{A'} = 0$. Conversely if $C_A C_{A'} = 0$, then $AA' = C_A A C_{A'} A' = A C_A C_{A'} A' = 0$. ■

Theorem 5.5.4 says, in effect, that the algebra generated by $\mathscr{R}$ and $\mathscr{R}'$ is isomorphic to the (algebraic) tensor product of $\mathscr{R}$ and $\mathscr{R}'$ as modules over $\mathscr{C}$. In particular, when $\mathscr{R}$ is a factor, $\mathscr{C}$ is the scalars; and the algebra generated by $\mathscr{R}$ and $\mathscr{R}'$ is the more standard tensor product of algebras over the field of scalars.

*Some constructions.* If $\mathscr{R}$ is a von Neumann algebra acting on a Hilbert space $\mathscr{H}$ and $E'$ is a projection in $\mathscr{R}'$, the range of $E'$ is stable under $\mathscr{R}$. We noted in Section 4.5 that the restriction of $\mathscr{R}$ to this range is a representation of $\mathscr{R}$. The image of this representation, a self-adjoint algebra of operators on $E'(\mathscr{H})$, is * isomorphic to $\mathscr{R}E'$ (and will often be denoted by $\mathscr{R}E'$). In the discussion that follows, we shall prove that $\mathscr{R}E'$, acting on $E'(\mathscr{H})$, is a von Neumann algebra; and we shall identify its center and commutant.

5.5.5. PROPOSITION. *If $\mathscr{R}$ is a von Neumann algebra acting on the Hilbert space $\mathscr{H}$ and $E'$ is a projection in $\mathscr{R}'$, then the mapping $AE' \to AC_{E'}$ is a * isomorphism of $\mathscr{R}E'$ onto $\mathscr{R}C_{E'}$.*

*Proof.* Since $(aA + B)E' = aAE' + BE'$, $ABE' = AE'BE'$, $(aA + B)C_{E'} = aAC_{E'} + BC_{E'}$, and $ABC_{E'} = AC_{E'}BC_{E'}$; the mapping $AE' \to AC_{E'}$ is a * homomorphism of $\mathscr{R}E'$ onto $\mathscr{R}C_{E'}$ once we note that it is single-valued ("well defined"). The same operator in $\mathscr{R}E'$ may appear as $AE'$ and $BE'$ for some $A$ and $B$ in $\mathscr{R}$. In this case, we wish to conclude that $AC_{E'} = BC_{E'}$. Equivalently, if $TE' = 0$ for $T$ in $\mathscr{R}$, we wish to conclude that $TC_{E'} = 0$. But Theorem 5.5.4 tells us that $C_T C_{E'} = 0$ when $TE' = 0$, so that $0 = C_T C_{E'} T = C_T T C_{E'} = TC_{E'}$. At the same time, we see that $TE'$ is 0 if $TC_{E'} = 0$ (for, then, $0 = TC_{E'}E' = TE'$). Thus the mapping $AE' \to AC_{E'}$ is a * isomorphism (clearly onto $\mathscr{R}C_{E'}$). ■

5.5.6. PROPOSITION. *If $\mathscr{R}$ is a von Neumann algebra with center $\mathscr{C}$, acting on the Hilbert space $\mathscr{H}$, and $E'$ is a projection in $\mathscr{R}'$, then $\mathscr{R}E'$, acting on $E'(\mathscr{H})$, is a von Neumann algebra with center $\mathscr{C}E'$ and commutant $E'\mathscr{R}'E'$.*

*Proof.* The mapping $A \to AE'$ of $\mathscr{R}$ onto $\mathscr{R}E'$ is weak-operator continuous. From Theorem 5.1.3, $(\mathscr{B}(\mathscr{H}))_1$ is weak-operator compact so that its intersection $(\mathscr{R})_1$ with the weak-operator closed $\mathscr{R}$ is weak-operator compact. Thus $(\mathscr{R})_1 E'$ is weak-operator compact (hence closed). From the Kaplansky density theorem, $(\mathscr{R}E')_1$ is dense in the unit ball of the weak-operator closure of $\mathscr{R}E'$. We show that $(\mathscr{R}E')_1 = (\mathscr{R})_1 E'$, from which $(\mathscr{R}E')_1$ is the unit ball of the weak-operator closure of $\mathscr{R}E'$; and $\mathscr{R}E'$ is weak-operator closed. Choose $AE'$ in $(\mathscr{R}E')_1$, and observe that $AE' = AC_{E'}E'$. With $T_n$ in $\mathscr{R}$, if $\|T_n C_{E'} - T\| \to 0$, then $\|T_n C_{E'}^2 - TC_{E'}\| = \|T_n C_{E'} - TC_{E'}\| \to 0$. Thus $T \in \mathscr{R}$ and $T = TC_{E'} \in \mathscr{R}C_{E'}$. Thus $\mathscr{R}C_{E'}$ is a $C^*$-algebra (norm-closed with unit $C_{E'}$). From Proposition 5.5.5, $TC_{E'} \to TE'$ is a * isomorphism of $\mathscr{R}C_{E'}$ onto $\mathscr{R}E'$. Since a * isomorphism of a $C^*$-algebra is an isometry (see Theorem 4.1.8(iii)), $\|AC_{E'}\| = \|AE'\| \leqslant 1$. Thus $AE' = AC_{E'}E' \in (\mathscr{R})_1 E'$, from which $(\mathscr{R}E')_1 \subseteq (\mathscr{R})_1 E'$. The reverse inclusion is immediate; and $(\mathscr{R})_1 E' = (\mathscr{R}E')_1$.

If $T$ is in the center of $\mathscr{R}C_{E'}$, then, in particular, $T = TC_{E'}$. Thus $0 = TA(I - C_{E'}) = A(I - C_{E'})T$, for $A$ in $\mathscr{R}$. As $A = AC_{E'} + A(I - C_{E'})$ and $AC_{E'} \in \mathscr{R}C_{E'}$, $TA = AT$. It follows that $T \in \mathscr{C}$, so that $T \in \mathscr{C}C_{E'}$. Of course $\mathscr{C}C_{E'}$ commutes with $\mathscr{R}C_{E'}$. Combining these conclusions, $\mathscr{C}C_{E'}$ is the center of $\mathscr{R}C_{E'}$. Hence $\mathscr{C}E'$ is the center of $\mathscr{R}E'$.

If $T'$ in $\mathscr{B}(E'(\mathscr{H}))$ commutes with $\mathscr{R}E'$, then, denoting by $T'$ again the operator on $\mathscr{H}$ that is 0 on $(I - E')(\mathscr{H})$ and $T'$ on $E'(\mathscr{H})$, with $T$ in $\mathscr{R}$, $T' = E'T'E'$ and $T'T = T'E'T = TE'T' = TT'$. Thus $T' \in \mathscr{R}'$, so that $T' \in E'\mathscr{R}'E'$. Of course $E'\mathscr{R}'E'$ commutes with $\mathscr{R}E'$. It follows that $E'\mathscr{R}'E'$ is the commutant of $\mathscr{R}E'$ on $E'(\mathscr{H})$. ■

5.5.7. COROLLARY. *If $\mathscr{R}$ is a von Neumann algebra acting on the Hilbert space $\mathscr{H}$ and $E$ is a projection in $\mathscr{R}$, then $E\mathscr{R}E$ acting on $E(\mathscr{H})$ is a von Neumann algebra with commutant $\mathscr{R}'E$.*

*Proof.* From Proposition 5.5.6, $E\mathscr{R}E$ acting on $E(\mathscr{H})$ is the commutant of $\mathscr{R}'E$, and $\mathscr{R}'E$ is a von Neumann algebra. Thus $E\mathscr{R}E$ is a von Neumann algebra and its commutant in $\mathscr{B}(E(\mathscr{H}))$ is $\mathscr{R}'E$. ∎

If $\{\mathscr{H}_a\}$ is a family of Hilbert spaces and $\{\mathscr{R}_a\}$ is a family of von Neumann algebras such that $\mathscr{R}_a$ acts on $\mathscr{H}_a$, we can form $\sum \oplus \mathscr{H}_a (= \mathscr{H})$ and the family $\mathscr{R}$, of all bounded operators on $\mathscr{H}$ of the form $\sum \oplus A_a$, where $A_a \in \mathscr{R}_a$ and $\{\|A_a\|\}$ is bounded (see Section 2.6, *Direct sums*, for a discussion of direct sums of Hilbert spaces and operators). It is clear that $\mathscr{R}$ is a self-adjoint algebra of operators on $\mathscr{H}$. Suppose that $A$ is in the strong-operator closure of $\mathscr{R}$. We denote by $\mathscr{H}_a$ again the subspace of $\mathscr{H}$ consisting of those vectors all of whose components, except possibly the component in $\mathscr{H}_a$, are 0. Each operator in $\mathscr{R}$ commutes with $P_a$, the orthogonal projection of $\mathscr{H}$ onto $\mathscr{H}_a$, so that the same is true of $A$. Now $AP_a$ is in the strong-operator closure of $\mathscr{R}P_a$ since right multiplication by $P_a$ is strong-operator continuous. Hence $A_a$, the restriction of $AP_a$ to $\mathscr{H}_a$, is in $\mathscr{R}_a$; and $A = \sum \oplus A_a \in \mathscr{R}$. It follows that $\mathscr{R}$ is a von Neumann algebra. We call $\mathscr{R}$ the *direct sum* of $\{\mathscr{R}_a\}$ and denote it by $\sum \oplus \mathscr{R}_a$.

Since $P_a \in \mathscr{R}$, each $A'$ in $\mathscr{R}'$ commutes with $P_a$. Thus $A' = \sum \oplus A'_a$, where $A'_a$ is the restriction of $A'P_a$ to $\mathscr{H}_a$. Hence $\mathscr{R}' \subseteq \sum \oplus \mathscr{R}'_a$. As the reverse inclusion is apparent, we see that $\mathscr{R}' = \sum \oplus \mathscr{R}'_a$.

*Cyclicity, separation, and countable decomposability.* We have noted and used the fact that the range of a projection in a von Neumann algebra is stable under the action of the commutant. For certain of these projections, the image of a single vector acted on by the commutant is dense in the range. Such projections play an important role in the theory. In the discussion that follows, we study their properties. We begin by noting that the projection $E$ with range $[\mathscr{R}'Y]$ is in $\mathscr{R}$ $(= \mathscr{R}'')$, where $\mathscr{R}$ is a von Neumann algebra and $Y$ is a set of vectors, since $[\mathscr{R}'Y]$ is stable under the self-adjoint family $\mathscr{R}'$.

5.5.8. DEFINITION. A projection $E$ in a von Neumann algebra $\mathscr{R}$ acting on the Hilbert space $\mathscr{H}$ is said to be *cyclic in* $\mathscr{R}$ (or *under* $\mathscr{R}'$) when its range is $[\mathscr{R}'x]$ for some vector $x$. In this case, $x$ is said to be a *generating vector for* $E$ (under $\mathscr{R}'$). If $[\mathscr{R}'x] = \mathscr{H}$, we say that $x$ is a *generating vector under* $\mathscr{R}'$. More generally, we say that a family $Y$ of vectors is *generating for* $\mathscr{R}'$ when $[\mathscr{R}'Y] = \mathscr{H}$, and that $Y$ is *generating for* $E$ when $[\mathscr{R}'Y]$ is the range of $E$. If $A$, in $\mathscr{R}$, is 0 when $Ay = 0$ for all $y$ in $Y$, we say that $Y$ is *separating* for $\mathscr{R}$. In particular, we speak of a vector $y$, such that $A = 0$ if $Ay = 0$ and $A \in \mathscr{R}$, as a *separating vector* for $\mathscr{R}$. ∎

5.5.9. PROPOSITION. *If $E$ is a cyclic projection in the von Neumann algebra $\mathscr{R}$ with generating vector $x$ and $F$ is a projection in $\mathscr{R}$ such that $F \leqslant E$, then $F$ is* *cyclic in $\mathscr{R}$ with generating vector $Fx$. Each projection in $\mathscr{R}$ is the union of an orthogonal family of cyclic projections in $\mathscr{R}$.*

*Proof.* Note that $[A'Fx: A' \in \mathscr{R}'] = [FA'x: A' \in \mathscr{R}']$. Since $F$ is continuous and $\{A'x: A' \in \mathscr{R}'\}$ is dense in $E(\mathscr{H})$, $\{FA'x: A' \in \mathscr{R}'\}$ is dense in $FE(\mathscr{H})$ $(= F(\mathscr{H}))$. Thus $F$ is cyclic in $\mathscr{R}$ and $Fx$ is a generating vector for $F$.

Suppose $E$ is an arbitrary projection in $\mathscr{R}$. If $E$ is 0, then $E$ is cyclic in $\mathscr{R}$ with generating vector 0. If $E \neq 0$ and $x$ is some non-zero vector in its range, then $[\mathscr{R}'x]$ is the range of a cyclic projection $E_0$. Since the ranges of $E_0$ and $E$ are stable under the self-adjoint family $\mathscr{R}'$; $E_0 \leqslant E$ and $E_0 \in \mathscr{R}'' = \mathscr{R}$. Thus $E$ has a non-zero cyclic subprojection if $E \neq 0$. The set of orthogonal families of non-zero cyclic subprojections of $E$ is non-empty, and the union of each totally ordered subset is an upper bound for that subset under inclusion ordering. Zorn's lemma guarantees the existence of a maximal orthogonal family $\{E_a\}$ of non-zero cyclic subprojections of $E$. If $E - \bigvee_a E_a$ is not 0, it contains a non-zero cyclic subprojection $E_0$. Adjoining $E_0$ to $\{E_a\}$ contradicts the maximality of $\{E_a\}$. Thus $E$ is the union of the orthogonal family $\{E_a\}$ of non-zero cyclic projections. ■

The simple Zorn's-lemma argument employed at the end of the preceding proof will be needed (with minor modifications) frequently. For the most part, all that will appear will be the imperative, "Let $\{E_a\}$ be a maximal orthogonal family of cyclic subprojections."

5.5.10. PROPOSITION. *If $\{Q_n\}$ is a countable, orthogonal family of central projections in a von Neumann algebra $\mathscr{R}$ and $\{E_n\}$ is a family of cyclic projections in $\mathscr{R}$ such that $E_n \leqslant Q_n$, then $\sum E_n$ is cyclic in $\mathscr{R}$.*

*Proof.* If $x_n$ is a generating unit vector for $E_n$ under $\mathscr{R}'$, then $\{x_n\}$ is an orthonormal set, so that $\sum n^{-1}x_n$ converges to a vector $x$. As $[\mathscr{R}'x_n] = [\mathscr{R}'Q_n x] \subseteq [\mathscr{R}'x]$, the range of $\sum E_n$ is contained in $[\mathscr{R}'x]$. Since the range of $\sum E_n$ is stable under $\mathscr{R}'$ and $x$ is in that range, it coincides with $[\mathscr{R}'x]$. That is, $x$ is a generating vector for $\sum E_n$, and $\sum E_n$ is cyclic in $\mathscr{R}$. ■

5.5.11. PROPOSITION. *If $\mathscr{R}$ is a von Neumann algebra acting on the Hilbert space $\mathscr{H}$, a subset $Y$ of $\mathscr{H}$ is generating for $\mathscr{R}$ if and only if it is separating for $\mathscr{R}'$.*

*Proof.* Suppose $Y$ is generating for $\mathscr{R}$, $A' \in \mathscr{R}'$, and $A'y = 0$ for all $y$ in $Y$. Then $0 = AA'y = A'Ay$ for all $A$ in $\mathscr{R}$ and $y$ in $Y$. As $\{Ay: y \in Y, A \in \mathscr{R}\}$ spans $\mathscr{H}$, $A' = 0$. Thus $Y$ is separating for $\mathscr{R}'$.

If $Y$ is not generating for $\mathscr{R}$, then $[\mathscr{R}Y]$ is the range of a projection $E'$ in $\mathscr{R}'$, different from $I$. Thus $I - E' \neq 0$, and $(I - E')y = 0$ for all $y$ in $Y$ (since $y$ is in the range of $E'$). Hence $Y$ is not separating for $\mathscr{R}'$ in this case. ■

The special case of the preceding result in which $Y$ consists of a single vector is the one used most frequently.

5.5.12. Corollary. *A vector is generating for a von Neumann algebra if and only if it is separating for the commutant.*

5.5.13. Proposition. *If $\mathscr{R}$ is a von Neumann algebra and $E$ and $E'$ are projections in $\mathscr{R}$ and $\mathscr{R}'$ with ranges $[\mathscr{R}'x]$ and $[\mathscr{R}x]$, respectively, then $C_E = C_{E'}$.*

*Proof.* From Proposition 5.5.2, $C_E$ and $C_{E'}$ have ranges $[\mathscr{R}\mathscr{R}'x]$ and $[\mathscr{R}'\mathscr{R}x]$, respectively. Since $\mathscr{R}$ and $\mathscr{R}'$ commute, these subspaces coincide; and $C_E = C_{E'}$. ∎

5.5.14. Definition. A projection $E$ in a von Neumann algebra $\mathscr{R}$ is said to be *countably decomposable* relative to $\mathscr{R}$ when each orthogonal family of non-zero subprojections of $E$ in $\mathscr{R}$ is countable. When $I$ is countably decomposable relative to $\mathscr{R}$, we say that $\mathscr{R}$ is countably decomposable. ∎

The term "$\sigma$-finite" is often used in place of "countably decomposable." The von Neumann algebra $\mathscr{R}$ relative to which countable decomposability is asserted for a projection $E$ is important, as can be seen by taking $E$ to be $I$ and $\mathscr{R}$ to be, first, $\mathscr{B}(\mathscr{H})$ with $\mathscr{H}$ non-separable, then to be $\{aI\}$. The (minimal) projections corresponding to an orthonormal basis for $\mathscr{H}$ form an uncountable orthogonal family of subprojections of $I$, so that $I$ is not countably decomposable relative to $\mathscr{B}(\mathscr{H})$; but $I$ is countably decomposable relative to $\{aI\}$. Despite the necessity for caution when a projection is claimed to be countably decomposable, reference to $\mathscr{R}$ will be omitted when no confusion can arise.

If $\mathscr{H}$ is a separable Hilbert space, each orthogonal family of projections is countable, so that $\mathscr{B}(\mathscr{H})$ and each von Neumann algebra on $\mathscr{H}$ are countably decomposable.

5.5.15. Proposition. *If $E$ is a cyclic projection in a von Neumann algebra $\mathscr{R}$, then $E$ is countably decomposable.*

*Proof.* If $x$ is a unit generating vector for $E$ under $\mathscr{R}'$ and $\{E_a\}$ is an orthogonal family of non-zero subprojections of $E$ in $\mathscr{R}$, then, from Proposition 5.5.9, $E_a x$ is a generating vector for $E_a$. Since $E_a \neq 0$, $E_a x \neq 0$. From Bessel's inequality (Remark 2.5.17), $\sum_a \|E_a x\|^2 \leqslant \|x\|^2 = 1$. If the set of indices $a$ is uncountable, there is a positive integer $n$ such that $1/n \leqslant \|E_a x\|$ for an infinite set of indices $a$. In this case, $\sum_a \|E_a x\|^2$ is not finite. Thus $\{E_a\}$ is countable. It follows that $E$ is countably decomposable. ∎

5.5.16. Proposition. *A central projection $P$ in a von Neumann algebra $\mathscr{R}$ is the central carrier of a cyclic projection in $\mathscr{R}$ if and only if $P$ is countably decomposable relative to the center $\mathscr{C}$ of $\mathscr{R}$.*

*Proof.* Suppose $P = C_E$ with $E$ a cyclic projection in $\mathscr{R}$, and $x$ is a unit generating vector for $E$. If $\{P_a\}$ is an orthogonal family of central subprojections of $P$, then $\sum_a \|P_a x\|^2 \leqslant \|x\|^2 = 1$. Thus (as in the proof of Proposition 5.5.15) $P_a x = 0$ for all but a countable set of indices. If $P_a x = 0$, then $0 = A'P_a x = P_a A'x$ for each $A'$ in $\mathscr{R}'$. As $[\mathscr{R}'x]$ is the range of $E$, $P_a E = 0$. From Theorem 5.5.4, $0 = P_a C_E = P_a P = P_a$. Thus $P$ is countably decomposable relative to $\mathscr{C}$.

Suppose now that $P$ is countably decomposable relative to $\mathscr{C}$. Let $\{E_a\}$ be a family of non-zero projections cyclic in $\mathscr{R}$ maximal with respect to the property that their central carriers form an orthogonal family of subprojections of $P$. By hypothesis $\{C_{E_a}\}$ and consequently $\{E_a\}$ are countable. From Proposition 5.5.10, $\sum E_a$ is cyclic in $\mathscr{R}$; and, from Proposition 5.5.3, its central carrier is $\bigvee_a C_{E_a}$. If $P - \bigvee_a C_{E_a} \neq 0$, it contains a non-zero projection $E_0$ cyclic in $\mathscr{R}$. Since $C_{E_0}$ is orthogonal to each $C_{E_a}$, adjoining $E_0$ to $\{E_a\}$ contradicts the maximality of $\{E_a\}$. Thus $P = \bigvee_a C_{E_a}$; and $\sum E_a$ is a cyclic projection in $\mathscr{R}$ with central carrier $P$. ∎

An important consequence of the preceding result deals with the case where $\mathscr{R}$ is abelian.

5.5.17. COROLLARY. *If $\mathscr{A}$ is a countably decomposable abelian von Neumann algebra acting on the Hilbert space $\mathscr{H}$, $\mathscr{A}$ has a separating vector. If $\mathscr{A}$ is maximal abelian, in addition, the separating vector is generating for $\mathscr{A}$.*

*Proof.* Since $\mathscr{A}$ is its own center and $I$ is countably decomposable relative to $\mathscr{A}$, $I$ is (the central carrier of) a cyclic projection in $\mathscr{A}$. With $x$ a generating vector for $\mathscr{A}'$, $x$ is separating for $\mathscr{A}$ $(= \mathscr{A}'')$, from Corollary 5.5.12. If $\mathscr{A}$ is maximal abelian, $x$ is generating for $\mathscr{A}$ $(= \mathscr{A}')$ as well. ∎

A partial converse to the preceding result states that if an abelian von Neumann algebra has a generating vector then it is maximal abelian. While an ad-hoc argument could be given to establish that converse, at this point it is best to defer further discussion to Section 7.2 (see Corollary 7.2.16), where suitable techniques are developed. An illustration of the situation discussed in Corollary 5.5.17 is provided by the example (see Example 5.1.6) of the multiplication algebra of the unit interval under Lebesgue measure. In this case the algebra is maximal abelian; and the constant function 1 is a separating (and generating) vector for it.

5.5.18. PROPOSITION. *If $\mathscr{R}$ is a countably decomposable von Neumann algebra acting on the Hilbert space $\mathscr{H}$, there is a central projection cyclic in $\mathscr{R}$ whose orthogonal complement is cyclic in $\mathscr{R}'$.*

*Proof.* Let $\{x_n\}$ be a set of unit vectors in $\mathscr{H}$ maximal with respect to the property that $\{E_n\}$ and $\{E'_n\}$ are orthogonal families of projections, where $E_n$

has range $[\mathscr{R}'x_n]$ and $E'_n$ has range $[\mathscr{R}x_n]$. Since $\{E_n\}$ is an orthogonal family of projections in $\mathscr{R}$, it is countable. We may assume that the index $n$ is a positive integer. As $\{x_n\}$ is an orthonormal set, $\sum n^{-1}x_n$ converges to a vector $x$ in $\mathscr{H}$. If $E$ is $\sum E_n$ and $E'$ is $\sum E'_n$, then $E$ and $E'$ are cyclic projections in $\mathscr{R}$ and $\mathscr{R}'$, respectively; and $x$ is a generating vector for each. Of course, $x$ is in the range of both $E$ and $E'$. In addition $[\mathscr{R}'x_n] = [\mathscr{R}'E'_nx] \subseteq [\mathscr{R}'x]$. Thus $[\mathscr{R}'x]$ is the range of $E$. Symmetrically, $[\mathscr{R}x]$ is the range of $E'$.

If $(I - E)(I - E') \neq 0$, a unit vector $x_0$ in the range of $(I - E)(I - E')$ will generate cyclic projections $E_0$ in $\mathscr{R}$ and $E'_0$ in $\mathscr{R}'$ orthogonal to $\{E_n\}$ and $\{E'_n\}$, respectively. Adjoining $x_0$ to $\{x_n\}$ contradicts the maximality of $\{x_n\}$. Thus $(I - E)(I - E') = 0$; and, from Theorem 5.5.4, $C_{I-E}C_{I-E'} = 0$. Since $I - C_{I-E} \leqslant E$, $I - C_{I-E}$ is cyclic in $\mathscr{R}$ (from Proposition 5.5.9). Similarly $I - C_{I-E'}$ is cyclic in $\mathscr{R}'$. As $C_{I-E} \leqslant I - C_{I-E'}$, $C_{I-E}$ is cyclic in $\mathscr{R}'$. Thus $I - C_{I-E}$ is a central projection cyclic in $\mathscr{R}$ whose orthogonal complement, $C_{I-E}$, is cyclic in $\mathscr{R}'$. ■

5.5.19. PROPOSITION. *If $E$ is the union of a countable family $\{E_n\}$ of cyclic projections in a von Neumann algebra $\mathscr{R}$, then $E$ is countably decomposable in $\mathscr{R}$.*

*Proof.* Let $x_n$ be a unit generating vector for $E_n$ under $\mathscr{R}'$. Let $\{F_a : a \in \mathbb{A}\}$ be an orthogonal family of non-zero subprojections of $E$ in $\mathscr{R}$; and let $\mathscr{S}_n$ be $\{a : F_a x_n \neq 0\}$. If $a$ is not in $\mathscr{S}_n$, $F_a x_n = 0$; and $(0) = [\mathscr{R}'F_a x_n] = [F_a \mathscr{R}'x_n]$. Since $F_a \leqslant \bigvee_n E_n$, $F_a$ does not annihilate the range $[\mathscr{R}'x_n]$ of each $E_n$. Thus $\cup \mathscr{S}_n = \mathbb{A}$. On the other hand, $\sum_{a\in\mathbb{A}} \|F_a x_n\|^2 \leqslant \|x_n\|^2 = 1$, so that $F_a x_n \neq 0$ for at most a countable number of elements $a$ of $\mathbb{A}$ – that is, $\mathscr{S}_n$ is countable. If $\aleph$ is the cardinal number of $\mathbb{A}$, $\aleph \leqslant \aleph_0 \cdot \aleph_0 = \aleph_0$ (see the final paragraph of the proof of Theorem 2.2.10), since $\mathbb{A} = \cup \mathscr{S}_n$. ■

## 5.6. Unbounded operators and abelian von Neumann algebras

In this section we study the spectral theory of unbounded self-adjoint and normal operators. We associate an (unbounded) spectral resolution with each (unbounded) self-adjoint operator (compare Theorem 5.2.2 and the discussion following it). We extend the function calculus (both continuous and bounded Borel) to (unbounded) normal operators (compare Theorem 5.2.8).

We begin with a discussion (really, a continuation of Example 5.1.6) that details the relation between unbounded self-adjoint (and normal) operators and the multiplication algebra of a measure space. If we are prepared to confine attention to, say, the case of separable Hilbert space, this discussion combined with some of the general theory of abelian von Neumann algebras to be developed in Chapter 6 contains all that we need in dealing with unbounded self-adjoint (and normal) operators.

If $g$ is a (complex) measurable function (finite almost everywhere) on $S$ – now, without the restriction that it be essentially bounded – multiplication by $g$ will not yield an everywhere-defined operator on $L_2(S)$, for many of the products will not lie in $L_2(S)$. Enough functions $f$ will have product $fg$ in $L_2(S)$, however, to form a dense linear submanifold $\mathscr{D}$ of $L_2(S)$ and constitute a (dense) domain for an (unbounded) multiplication operator $M_g$. To see this, let $E_n$ be the (bounded) multiplication operator corresponding to the characteristic function of the (measurable) set on which $|g| \leqslant n$. Since $g$ is finite almost everywhere, $\{E_n\}$ is an increasing sequence of projections with union $I$. The union $\mathscr{D}_0$ of the ranges of the $E_n$ is a dense linear submanifold of $L_2(S)$ contained in $\mathscr{D}$. A measure-theoretic argument shows that $M_g$ is closed with $\mathscr{D}_0$ as a core. In fact, if $\{f_n\}$ is a sequence in $\mathscr{D}$ converging in $L_2(S)$ to $f$ and $\{gf_n\}$ converges in $L_2(S)$ to $h$, then, passing to subsequences, we may assume that $\{f_n\}$ and $\{gf_n\}$ converge almost everywhere to $f$ and $h$, respectively. But, then, $\{gf_n\}$ converges almost everywhere to $gf$, so that $gf$ and $h$ are equal almost everywhere. Thus $gf \in L_2(S), f \in \mathscr{D}, h = M_g(f)$, and $M_g$ is closed. With $f_0$ in $\mathscr{D}$, $\{E_n f_0\}$ converges to $f_0$ and $\{M_g E_n f_0\} = \{E_n M_g f_0\}$ converges to $M_g f_0$. Now $E_n f_0 \in \mathscr{D}_0$, so that $\mathscr{D}_0$ is a core for $M_g$. Note that $M_g E_n$ is bounded and that its bound does not exceed $n$. Using the lemma that follows, we see that $M_g$ is an (unbounded) self-adjoint operator when $g$ is real-valued, since $M_g E_n$ is a bounded self-adjoint operator in that case.

5.6.1. LEMMA. *If $\{E_n\}$ is an increasing sequence of projections on the Hilbert space $\mathscr{H}$ and $A_0$ is a linear operator with dense domain $\bigcup_{n=1}^{\infty} E_n(\mathscr{H})$ $(= \mathscr{D}_0)$ such that $A_0 E_n$ is a bounded self-adjoint operator on $\mathscr{H}$, then $A_0$ is preclosed and its closure is self-adjoint. If $A$ is closed with core $\mathscr{D}_0$ and $AE_n$ is a bounded self-adjoint operator, $A$ is self-adjoint.*

*Proof.* With $x$ and $y$ in $\mathscr{D}_0$, there is an $m$ such that

$$\langle A_0 x, y\rangle = \langle A_0 E_m x, y\rangle = \langle x, A_0 E_m y\rangle = \langle x, A_0 y\rangle.$$

Thus $y \in \mathscr{D}(A_0^*)$ and $A_0^*$ is densely defined. From Theorem 2.7.8(ii), $A_0$ is preclosed. With $A$ the closure of $A_0$, it remains to prove the last assertion of this lemma.

With $x$ and $y$ in $\mathscr{D}(A)$, we can choose sequences $\{x_n\}$ and $\{y_n\}$ in $\mathscr{D}_0$ converging to $x$ and $y$, respectively, such that $\{Ax_n\}$ and $\{Ay_n\}$ converge to $Ax$ and $Ay$. Then

$$\langle Ax_n, y_n\rangle = \langle AE_m x_n, y_n\rangle = \langle x_n, AE_m y_n\rangle = \langle x_n, Ay_n\rangle.$$

Now $\langle Ax_n, y_n\rangle$ tends to $\langle Ax, y\rangle$ and $\langle x_n, Ay_n\rangle$ tends to $\langle x, Ay\rangle$ as $n$ tends to infinity. Thus $\langle Ax, y\rangle = \langle x, Ay\rangle$; and $A$ is symmetric. Note that $(A \pm iI)E_n$ has range $E_n(\mathscr{H})$ since $AE_n$ is bounded and self-adjoint (so that $AE_n \pm iE_n$ has a bounded inverse on $E_n(\mathscr{H})$). Thus $A \pm iI$ has a dense range. From Lemma 2.7.9, this range is $\mathscr{H}$; and, from Proposition 2.7.10, $A$ is self-adjoint. ∎

If $M_g$ is unbounded, we cannot expect it to belong to the multiplication algebra $\mathscr{A}$ of the measure space $(S, \mathscr{S}, m)$. Nonetheless, there are various ways in which $M_g$ behaves as if it were in $\mathscr{A}$ – for example, $M_g$ is unchanged when it is "transformed" by a unitary operator $U$ commuting with $\mathscr{A}$. In this case (see Example 5.1.6), $U \in \mathscr{A}$, so that $U = M_u$ where $u$ is a bounded measurable function on $S$ with modulus 1 almost everywhere (see Example 2.4.11). With $f$ in $\mathscr{D}(M_g)$, $guf \in L_2(S)$; while, if $guh \in L_2(S)$, then $gh \in L_2(S)$ and $h \in \mathscr{D}(M_g)$. Thus $U$ transforms $\mathscr{D}(M_g)$ onto itself. Moreover

$$(U^*M_gU)(f) = \bar{u}guf = |u|^2 gf = gf.$$

Thus $U^*M_gU = M_g$.

The fact that $M_g$ "commutes" with all unitary operators commuting with $\mathscr{A}$ in conjunction with Theorem 4.1.7 and the double commutant theorem (5.3.1) (from which it follows that a bounded operator having this property lies in $\mathscr{A}$) provides us with an indication of the extent to which $M_g$ "belongs" to $\mathscr{A}$. We formalize this property in the definition that follows.

5.6.2. Definition. We say that a closed densely defined operator $T$ is *affiliated* with a von Neumann algebra $\mathscr{R}$ (and write $T \eta \mathscr{R}$) when $U^*TU = T$ for each unitary operator $U$ commuting with $\mathscr{R}$. ■

Note that the equality, $U^*TU = T$, of the preceding definition is to be understood in the strict sense that $U^*TU$ and $T$ have the same domain and (formal) equality holds for the transforms of vectors in that domain. As far as the domains are concerned, the effect is that $U$ transforms $\mathscr{D}(T)$ onto itself.

5.6.3. Remark. If $T$ is a closed densely defined operator with core $\mathscr{D}_0$ and $U^*TUx = Tx$ for each $x$ in $\mathscr{D}_0$ and each unitary operator $U$ commuting with a von Neumann algebra $\mathscr{R}$, then $T \eta \mathscr{R}$. To see this, note that, with $y$ in $\mathscr{D}(T)$, there is a sequence $\{y_n\}$ in $\mathscr{D}_0$ such that $y_n \to y$ and $Ty_n \to Ty$ (since $\mathscr{D}_0$ is a core for $T$). Now $Uy_n \to Uy$ and $TUy_n = UTy_n \to UTy$. Since $T$ is closed, $Uy \in \mathscr{D}(T)$ and $TUy = UTy$. Thus $\mathscr{D}(T) \subseteq U^*(\mathscr{D}(T))$. Applied to $U^*$, we have $\mathscr{D}(T) \subseteq U(\mathscr{D}(T))$, so that $U(\mathscr{D}(T)) = \mathscr{D}(T)$. Hence $\mathscr{D}(U^*TU) = \mathscr{D}(T)$ and $U^*TUy = Ty$ for each $y$ in $\mathscr{D}(T)$. ■

Our discussion to this point establishes that $M_g$ is a closed operator (self-adjoint, when $g$ is real-valued) affiliated with the multiplication algebra $\mathscr{A}$. Conversely, if $A$ is a closed (unbounded) operator affiliated with $\mathscr{A}$, then $A$ has the form $M_g$ (and $g$ is real-valued when $A$ is self-adjoint). We prove this in the discussion that follows.

Since each $B_0$ in $\mathscr{A}$ is a linear combination of four unitary operators in $\mathscr{A}$ (see Theorem 4.1.7) and $UA \subseteq AU$ for each such unitary operator (as $\mathscr{A} \subseteq \mathscr{A}'$ and $A \eta \mathscr{A}$), we have $B_0A \subseteq AB_0$. In particular, if $E$ is the projection corresponding to multiplication by the characteristic function of a measurable subset $S_0$ of $S$, $EA \subseteq AE$, so that $Ef \in \mathscr{D}(A)$ if $f \in \mathscr{D}(A)$.

Let $\mathscr{D}_1$ be the set of essentially bounded functions in $\mathscr{D}(A)$. If $f \in \mathscr{D}(A)$ and $E_n$ is multiplication by the characteristic function of $\{x: |f(x)| \leqslant n\}$, $\{E_n\}$ is an ascending sequence of projections in $\mathscr{A}$ tending to $I$ in the strong-operator topology (since $f$ is finite almost everywhere). From the foregoing $E_n f \in \mathscr{D}_1$, $\{E_n f\}$ tends to $f$ and $AE_n f = E_n Af \to Af$ as $n \to \infty$. Thus $\mathscr{D}_1$ is a core for $A$.

If $f$ and $g$ are in $\mathscr{D}_1$, then

$$fAg = M_f Ag = AM_f g = A(fg) = AM_g f = M_g Af = gAf. \tag{1}$$

Let $\{S_n\}$ be a family of mutually disjoint sets of finite positive measure with union $S$. If $x_n$ is the characteristic function of $S_n$, there is a sequence $\{f_{n_j}\}$ of elements of $\mathscr{D}_1$ tending to $x_n$. If

$$S_n^0 = \{s : s \in S_n, f_{n_j}(s) = 0 \text{ for all } j\}$$

and $x$ is the characteristic function of $S_n^0$, then $0 = M_x f_{n_j} \to_j M_x x_n = x$, so that $S_n^0$ has measure 0. Let $g(s)$ be $[(Af_{n_j})(s)][f_{n_j}(s)]^{-1}$, where $s \in S_n \backslash S_n^0$ and $j$ is the least integer such that $f_{n_j}(s) \neq 0$. Then $g$ is a measurable function on $S$ defined almost everywhere. For each $f$ in $\mathscr{D}_1$, $f_{n_j}(s)(Af)(s) = f(s)(Af_{n_j})(s)$ for all $n$ and $j$ except on a set of measure 0, from (1). Thus $(Af)(s) = g(s)f(s)$ almost everywhere. We have noted that $M_g$ is closed and affiliated with $\mathscr{A}$. Since $M_g$ is an extension of the restriction of the closed operator $A$ to the core $\mathscr{D}_1$ for $A$, we have that $A \subseteq M_g$. As we have noted earlier, the set of functions $h$ in $L_2$ vanishing on the complement of sets $\{s: s \in S, |g(s)| \leqslant m\}$ forms a core for $M_g$. With $h$ such a function, let $\{f_n\}$ be a sequence of functions in $\mathscr{D}_1$ tending to $h$. If $x$ is the characteristic function of the set of points at which $h$ does not vanish, we may replace $f_n$ by $xf_n$. Changing notation, we may assume that each $f_n$ vanishes when $h$ does. In this case, since $M_g M_x$ is bounded,

$$Af_n = M_g M_x f_n \to M_g M_x h = M_g h.$$

As $A$ is closed, $h \in \mathscr{D}(A)$ and $Ah = M_g h$. It follows that $A = M_g$.

If $A$ is self-adjoint, $M_{gx}$ is a bounded self-adjoint operator, so that $gx$ is real-valued almost everywhere. Hence $g$ is real-valued almost everywhere. Again, as with bounded multiplication operators, the projections $E_\lambda$, corresponding to multiplication by the characteristic function of the set where $g$ does not exceed $\lambda$, form a spectral resolution $\{E_\lambda\}$ of $A$ (see the discussion following Theorem 5.2.5). In this case, if $A$ is, in fact, unbounded, there will be no non-negative real number $a$ (replacing $\|A\|$ when $A$ is bounded) such that $E_\lambda = 0$ if $\lambda < -a$ and $E_\lambda = I$ if $a < \lambda$; and we speak of $\{E_\lambda\}$ as an *unbounded* resolution of the identity (see the discussion preceding Theorem 5.2.3).

We summarize the foregoing conclusions in the theorem that follows.

5.6.4. THEOREM. *If $(S, \mathscr{S}, m)$ is a $\sigma$-finite measure space and $\mathscr{A}$ is its multiplication algebra acting on $L_2(S)$, then $A$ is a closed densely defined* *operator affiliated with $\mathscr{A}$ if and only if $A = M_g$ for some measurable function $g$ finite almost everywhere on $S$. In this case, $A$ is self-adjoint if and only if $g$ is real-valued almost everywhere.*

An unbounded self-adjoint operator $A$ can be associated ("affiliated") with an abelian von Neumann algebra. As in the case of a bounded self-adjoint operator (see Theorem 5.2.2), we can use Theorem 5.2.1 to locate a function on an extremely disconnected compact Hausdorff space that "represents" $A$. As might be expected, this function is neither everywhere defined nor bounded. We will be able to use this representing function (as in Theorem 5.2.2) to find a resolution of the identity (unbounded) for $A$. (See the discussion of resolutions following Theorem 5.2.2.) In preparation for this analysis, in the definition that follows we describe the functions that appear.

5.6.5. DEFINITION. If $X$ is an extremely disconnected compact Hausdorff space a *normal function* on $X$ is a continuous complex-valued function $f$ defined on an open dense subset $X \setminus Z$ of $X$ such that $\lim_{q \to p} |f(q)| = \infty$ for each $p$ in $Z$ (where $q \in X \setminus Z$). A *self-adjoint function* on $X$ is a real-valued normal function on $X$. If $f$ is self-adjoint and defined on $X \setminus Z$, we denote by $Z_+$ those points $p$ of $Z$ such that $\lim_{q \to p} f(q) = +\infty$ and by $Z_-$ those $p$ in $Z$ such that $\lim_{q \to p} f(q) = -\infty$. We denote by $\mathscr{N}(X)$ and $\mathscr{S}(X)$ the sets of normal and self-adjoint functions on $X$. ■

It is one of the many surprising properties of extremely disconnected compact Hausdorff spaces (and their associated constructs) that $Z = Z_+ \cup Z_-$ (that is, there are no points near which an $f$ in $\mathscr{S}(X)$ takes arbitrarily large positive and negative values). This follows from the fundamental property of such spaces that disjoint open sets $\mathscr{O}_1$ and $\mathscr{O}_2$ have disjoint closures ($X \setminus \mathscr{O}_1$ is closed and, thus, contains $\mathscr{O}_2^-$, a clopen set, so that $X \setminus \mathscr{O}_2^-$ is closed and contains $\mathscr{O}_1^-$). To see this, note that the sets

$$\{q : q \in X \setminus Z, f(q) > 1\}, \qquad \{q : q \in X \setminus Z, f(q) < -1\}$$

are disjoint open sets (since $f$ is continuous on $X \setminus Z$ and $X \setminus Z$ is open in $X$), and that a point near which $f$ takes arbitrarily large positive and negative values would have to lie in both of their closures.

Clearly $|f|$ is normal on $X$ if $f$ is normal. The following simple lemma will prove useful to us.

5.6.6. LEMMA. *If $f$ and $g$ are normal functions on $X$ defined on $X \setminus Z$ and $X \setminus Z'$, respectively, and $f(q) = g(q)$ for each $q$ in some dense subset $S$ of $X \setminus (Z \cup Z')$, then $Z = Z'$ and $f = g$.*

*Proof.* Since $X \setminus (Z \cup Z')$ is dense in $X$, $S$ is dense in $X$. If $p \in Z'$ and $q$ in $S$ is near $p$, then $|g(q)|$ $(= |f(q)|)$ is large so that $p \in Z$. Thus $Z' \subseteq Z$ and, symmetrically, $Z \subseteq Z'$. Hence $f - g$ is defined and continuous on $X \backslash Z$ and 0 on the dense subset $S$. It follows that $f = g$. ■

5.6.7. Lemma. *If $A$ is a self-adjoint operator acting on a Hilbert space $\mathscr{H}$, $A$ is affiliated with some abelian von Neumann algebra. If $A \eta \mathscr{A}$ and $\mathscr{A}$ is isomorphic to $C(X)$, with $X$ an extremely disconnected compact Hausdorff space, there is a unique self-adjoint function $h$ on $X$ such that $h \hat{\cdot} e$ is in $C(X)$ and represents $AE$ when $E$ is a projection in $\mathscr{A}$ such that $AE$ is a bounded everywhere-defined operator, where $e$ in $C(X)$ corresponds to $E$, and $(h \hat{\cdot} e)(p)$ is $h(p)$ if $e(p) = 1$ and 0 otherwise. There is a resolution of the identity $\{E_\lambda\}$ in $\mathscr{A}$ such that $\bigcup_{n=1}^{\infty} F_n(\mathscr{H})$ is a core for $A$, where $F_n = E_n - E_{-n}$, and $Ax = \int_{-n}^{n} \lambda \, dE_\lambda x$ for each $x$ in $F_n(\mathscr{H})$ and all $n$, in the sense of norm convergence of approximating Riemann sums.*

*Proof.* From Proposition 2.7.10 and Remark 2.7.11, $A + iI$ and $A - iI$ have range $\mathscr{H}$, null space (0), and inverses $T_+$ and $T_-$ that are everywhere defined with bound not exceeding 1. Note that

$$\begin{aligned}\langle T_+(A + iI)x, (A - iI)y \rangle &= \langle x, (A - iI)y \rangle = \langle (A + iI)x, y \rangle \\ &= \langle (A + iI)x, T_-(A - iI)y \rangle,\end{aligned}$$

when $x$ and $y$ are in $\mathscr{D}(A)$, since $A$ is self-adjoint. Thus $T_- = T_+^*$. (Recall that $A \pm iI$ have range $\mathscr{H}$, so that $(A + iI)x$ and $(A - iI)y$ represent arbitrary vectors in $\mathscr{H}$.) Again, since $A \pm iI$ have range $\mathscr{H}$, we can represent an arbitrary vector as $(A - iI)(A + iI)x$, where $x \in \mathscr{D}(A)$ and $Ax \in \mathscr{D}(A)$. In this case,

$$(A - iI)(A + iI)x = (A^2 + I)x = (A + iI)(A - iI)x,$$

and $T_+T_- = T_-T_+$. Since $T_- = T_+^*$, $T_+$ is normal. Let $\mathscr{A}$ be an abelian von Neumann algebra containing $I$, $T_+$, and $T_-$. If $U$ is a unitary operator in $\mathscr{A}'$, for each $x$ in $\mathscr{D}(A)$, $Ux = UT_+(A + iI)x = T_+U(A + iI)x$ so that $(A + iI)Ux = U(A + iI)x$; and $U^{-1}(A + iI)U = A + iI$. Thus $U^{-1}AU = A$ and $A \eta \mathscr{A}$. In particular $A \eta \mathscr{A}_0$, where $\mathscr{A}_0$ is the (abelian) von Neumann algebra generated by $I$, $T_+$, and $T_-$.

From Theorem 5.2.1, $\mathscr{A} \cong C(X)$, where $X$ is an extremely disconnected compact Hausdorff space. Let $g_+$ and $g_-$ be the functions in $C(X)$ corresponding to $T_+$ and $T_-$. Let $h_+$ and $h_-$ be the functions defined as the reciprocals of $g_+$ and $g_-$, respectively, at those points where $g_+$ and $g_-$ do not vanish. Then $h_+$ and $h_-$ are continuous where they are defined on $X$, as is $\frac{1}{2}(h_+ + h_-)\,(= h)$. In a formal sense, $h$ is the function that corresponds to $A$. We shall see that $h$ is real-valued and find the spectral resolution of $A$ by subjecting $h$ to the same spectral analysis as was performed on the functions of $C(X)$ in proving Theorem 5.2.2.

Since $T_+$ and $T_-$ are adjoints of one another, $g_+$ and $g_-$ are complex conjugates of one another (in particular, they vanish at the same points of $X$).

Thus $h_+$ and $h_-$ are complex conjugates of one another; and $h$ is real-valued. The set $Z$ on which $g_+$ (and $g_-$) vanishes is closed (since $g_+$ is continuous) and nowhere dense; for if it contains a non-null open set it contains its closure, a non-null clopen set. The projection corresponding to this non-null clopen set would have product with $T_+$ (and $T_-$) equal to 0 – contradicting the fact that $T_+$ (and $T_-$) have null space (0). Thus each point $p$ in $Z$ is a limit of points $q$ in $X \setminus Z$ (at which $h$ is defined).

For each $y$ in $\mathscr{H}$,

$$AT_+T_-y = (A + iI - iI)T_+T_-y = T_-y - iT_+T_-y,$$

so that $AT_+T_- = T_- - iT_+T_-$. Similarly $AT_-T_+ = T_+ + iT_+T_-$. Hence

$$2iT_+T_- = T_- - T_+ \tag{2}$$

and

$$AT_+T_- = \tfrac{1}{2}(T_+ + T_-). \tag{3}$$

It follows from (2) that $(h(q) + i)^{-1} = g_+(q)$ (and that $(h(q) - i)^{-1} = g_-(q)$) for $q$ in $X \setminus Z$. Hence, for each $q$ in $X \setminus Z$ near $p$ (in $Z$), $g_+(q)$ is near 0 and $|h(q)|$ is large. Thus $h$ is a self-adjoint function on $X$.

Let $\mathcal{O}_\lambda$ be $U_\lambda \cup Z_+$, where $U_\lambda$ is the set of points of $X \setminus Z$ at which $h$ exceeds $\lambda$ and $Z_+$ has the meaning explained in Definition 5.6.5. We show that $\mathcal{O}_\lambda$ is open. Since $h$ is continuous on $X \setminus Z$ and $X \setminus Z$ is open in $X$, $U_\lambda$ is open in $X$. If $p \in Z_+$, there is an open set $\mathcal{O}$ containing $p$ such that if $q \in \mathcal{O} \cap (X \setminus Z)$, then $h(q) > 0$ – and since $|h(q)|$ is large for $q$ (in $X \setminus Z$) near $p$, we may choose $\mathcal{O}$ such that $h(q) > \lambda$ for $q$ in $\mathcal{O} \cap (X \setminus Z)$. In this case, $\mathcal{O} \cap (X \setminus Z) \subseteq U_\lambda \subseteq \mathcal{O}_\lambda$. If there were a $p'$ in $\mathcal{O} \cap Z$ with $p'$ in $Z_-$, then, from the definition of $Z_-$, there would be a $q$ in $\mathcal{O}$ with $h(q)$ strictly negative – contradicting the choice of $\mathcal{O}$. Thus $\mathcal{O} \cap Z \subseteq Z_+$; $\mathcal{O} \subseteq \mathcal{O}_\lambda$; and $\mathcal{O}_\lambda$ is open, as asserted. Let $X_\lambda$ be $X \setminus \mathcal{O}_\lambda^-$. Again, as in the proof of Theorem 5.2.2, $X_\lambda$ contains each clopen set $Y$ such that $h(p) \leqslant \lambda$ for all $p$ in $Y$. (For points $p$ of $Z_-$, where "$h(p) = -\infty$," we write "$h(p) \leqslant \lambda$" as well, so that $Y$ may contain points of $Z_-$.) Indeed, if $q \in \mathcal{O}_\lambda$, then $q \in X \setminus Y$, a closed set, so that $\mathcal{O}_\lambda^- \subseteq X \setminus Y$ and $Y \subseteq X \setminus \mathcal{O}_\lambda^- = X_\lambda$. At the same time, $X_\lambda$ is such a clopen set, for if $p \in X_\lambda \cap (X \setminus Z)$, then, since $p \notin U_\lambda$, $h(p) \leqslant \lambda$. If $p \in X_\lambda \cap Z$, then $p \in Z \setminus Z_+$ $(= Z_-)$ and $h(p) \leqslant \lambda$ (in the extended sense). Thus $X_\lambda$ is the *largest* clopen set on which $h(p) \leqslant \lambda$; and $X_\lambda \cap Z = Z_-$.

We proceed now as in the proof of Theorem 5.2.2. Let $e_\lambda$ be the characteristic function of $X_\lambda$ and $E_\lambda$ be the projection in $\mathscr{A}$ corresponding to $e_\lambda$. In this case, $\{E_\lambda\}$ satisfies $E_\lambda \leqslant E_{\lambda'}$ if $\lambda \leqslant \lambda'$ and $E_\lambda = \bigwedge_{\lambda < \lambda'} E_{\lambda'}$. Since $Z$ is nowhere dense, $\bigvee_\lambda e_\lambda = 1$ and $\bigwedge_\lambda e_\lambda = 0$, so that $\bigvee_\lambda E_\lambda = I$ and $\bigwedge_\lambda E_\lambda = 0$. That is, we have constructed a resolution of the identity $\{E_\lambda\}$ (and this resolution is unbounded if $h \notin C(X)$). Let $F$ be $E_b - E_a$, where $a < b$. Then $e_b - e_a$ $(= f)$, the characteristic function of $X_b \setminus X_a$, corresponds to $F$. Since $X_b \cap Z = X_a \cap Z$ $(= Z_-)$, we have that $X_b \setminus X_a \subseteq X \setminus Z$; and $g_+(p)g_-(p) \neq 0$

when $f(p) = 1$. For $p$ in $X \backslash Z$,

$$h(p) = \left(\frac{g_+ + g_-}{2g_+g_-}\right)(p), \tag{4}$$

by choice of $h$. Moreover, there is a positive function $k$ in $C(X)$ such that $kg_+g_- = f$ and $kf = k$ (since $g_+g_-$ is continuous and vanishes nowhere on the *clopen* set $X_b \backslash X_a$). If $K$ in $\mathscr{A}$ corresponds to $k$, then

$$KT_+T_- = F. \tag{5}$$

From our information about $X_\lambda$, if $p \in X_b \backslash X_a$, then $p \notin Z$ and $a \leqslant h(p) \leqslant b$. Thus from (4),

$$ag_+g_-f \leqslant \tfrac{1}{2}(g_+ + g_-)f \leqslant bg_+g_-f,$$

and

$$akg_+g_-f = af \leqslant \tfrac{1}{2}(g_+ + g_-)kf = \tfrac{1}{2}(g_+ + g_-)k \leqslant bkg_+g_-f = bf.$$

Thus

$$aF \leqslant \tfrac{1}{2}(T_+ + T_-)K \leqslant bF. \tag{6}$$

Combining (3), (5), and (6), we have

$$aF \leqslant AF \leqslant bF. \tag{7}$$

It follows that $AF$ is bounded; and from (3), (4), and (5), the corresponding element of $C(X)$ is $h \hat{\cdot} f$. With $E$ a projection in $\mathscr{A}$ such that $AE \in \mathscr{B}(\mathscr{H})$ and $U$ a unitary operator in $\mathscr{A}'$, $U^{-1}AEU = U^{-1}AUE = AE$, so that $AE \in \mathscr{A}$. Let $f_n$ in $C(X)$ correspond to $F_n$ and $Y_n$ be the (clopen) subset of $X$ on which $f_n$ takes the value 1. Since $\vee_{n=1}^{\infty} F_n = I$, $\{Y_n\}$ has union dense in $X$. If $e$ in $C(X)$ corresponds to $E$, then (replacing $F$, above, by $F_n$) $(h \hat{\cdot} f_n)e$ corresponds to $AF_nE$ $(= AEF_n)$. Suppose $e(p) = 1$. There is some $n$ and a $q$ in $Y_n$ near $p$. In this case, $((h \hat{\cdot} f_n)e)(q) = h(q)$ and $|h(q)| \leqslant \|AEF_n\| \leqslant \|AE\|$. Hence $p \notin Z$, and $|h(p)| \leqslant \|AE\|$. Thus $h \hat{\cdot} e \in C(X)$. Now $(h \hat{\cdot} f_n)e = h \hat{\cdot} (ef_n) = (h \hat{\cdot} e)f_n$. If $\tilde{h}$ in $C(X)$ represents $AE$, then $\tilde{h}f_n$ represents $AEF_n$. Thus $\tilde{h}f_n = (h \hat{\cdot} e)f_n$ for all $n$ and $\tilde{h} = h \hat{\cdot} e$.

Since $(2F_n - I)A(2F_n - I) = A$, we have $F_nA \subseteq AF_n$. Thus, with $x$ in $\mathscr{D}(A)$, $F_nx \to x$ and $AF_nx = F_nAx \to Ax$. Hence $\cup_{n=1}^{\infty} F_n(\mathscr{H})$ is a core for $A$. Since $(h \hat{\cdot} e)ee_\lambda \leqslant \lambda ee_\lambda$ and $\lambda(e - ee_\lambda) \leqslant (h \hat{\cdot} e)(e - ee_\lambda)$, we have, from Theorem 5.2.3, that $\{EE_\lambda | E(\mathscr{H})\}$, a resolution of the identity on $E(\mathscr{H})$, is the resolution of the identity for $AE|E(\mathscr{H})$. With $F_n$ in place of $E$, Theorem 5.2.2(v) (applied to $AF_n|F_n(\mathscr{H})$ and the resolution $\{F_nE_\lambda|F_n(\mathscr{H})\}$) yields the equality, $Ax = \int_{-n}^{n} \lambda \, dE_\lambda x$, for $x$ in $F_n(\mathscr{H})$ and all $n$. ■

If $x \in \mathscr{D}(A)$,

$$\int_{-n}^{n} \lambda \, dE_\lambda x = \int_{-n}^{n} \lambda \, dE_\lambda F_nx = AF_nx \to Ax.$$

Interpreted as an improper integral, we write

$$(8) \qquad Ax = \int_{-\infty}^{+\infty} \lambda \, dE_\lambda x$$

when $x \in \mathscr{D}(A)$.

In the circumstances set out in Lemma 5.6.7, we say that $h$ (in $\mathscr{S}(X)$) *represents* $A$ ($\eta\, \mathscr{A}$).

5.6.8. LEMMA. *If $\mathscr{A}$ is an abelian von Neumann algebra acting on the Hilbert space $\mathscr{H}$ and $\mathscr{A}$ is isomorphic to $C(X)$ with $X$ an extremely disconnected compact Hausdorff space, then each $h$ in $\mathscr{S}(X)$ represents some self-adjoint operator $A$ affiliated with $\mathscr{A}$.*

*Proof.* From the proof of Lemma 5.6.7, $h$ determines a resolution of the identity $\{E_\lambda\}$ in $\mathscr{A}$ and $h \,\hat{\cdot}\, f_n \in C(X)$, where $f_n = e_n - e_{-n}$ with $e_n$ in $C(X)$ representing $E_n$. Let $A_n$ in $\mathscr{A}$ correspond to $h \,\hat{\cdot}\, f_n$. Noting that $(h \,\hat{\cdot}\, f_m) f_n = h \,\hat{\cdot}\, f_n$ when $n \leqslant m$, we have $A_m F_n = A_n$, in this case, where $F_m$ in $\mathscr{A}$ corresponds to $f_m$. Thus, defining $A_0 x$ to be $A_n x$ when $x \in F_n(\mathscr{H})$, $A_0$ is a linear transformation with domain $\bigcup_{n=1}^{\infty} F_n(\mathscr{H})$ $(= \mathscr{D}_0)$. From Lemma 5.6.1, $A_0$ is preclosed and its closure $A$ is self-adjoint with core $\mathscr{D}_0$. If $U$ is a unitary operator in $\mathscr{A}'$ and $x_n \in F_n(\mathscr{H})$, then $Ux_n \in F_n(\mathscr{H})$ so that $U^{-1}AUx_n = U^{-1}A_nUx_n = A_n x_n = Ax_n$. From Remark 5.6.3, $A \,\eta\, \mathscr{A}$.

If $\tilde{h}$ in $\mathscr{S}(x)$ represents $A$, then, from Lemma 5.6.7, $\tilde{h} \,\hat{\cdot}\, f_n$ represents $AF_n$ $(= A_n)$. Thus $h \,\hat{\cdot}\, f_n = \tilde{h} \,\hat{\cdot}\, f_n$ for each $n$. From Lemma 5.6.6, $h = \tilde{h}$, since $h$ and $\tilde{h}$ agree on dense subsets of $X$. Thus $h$ represents $A$. ∎

5.6.9. LEMMA. *If $\{E_\lambda\}$ is a resolution of the identity on a Hilbert space $\mathscr{H}$ and $\mathscr{A}$ is an abelian von Neumann algebra containing $\{E_\lambda\}$, there is a self-adjoint $A$ affiliated with $\mathscr{A}$ such that*

$$(9) \qquad Ax = \int_{-n}^{n} \lambda \, dE_\lambda x,$$

*for each $x$ in $F_n(\mathscr{H})$ and all $n$, where $F_n = E_n - E_{-n}$; and $\{E_\lambda\}$ is the resolution of the identity for $A$ (as constructed in Lemma 5.6.7).*

*Proof.* Suppose $\mathscr{A}$ is isomorphic to $C(X)$ with $X$ an extremely disconnected compact Hausdorff space. Let $e_\lambda$ in $C(X)$ correspond to $E_\lambda$ and let $X_\lambda$ be the clopen subset of $X$ on which $e_\lambda$ takes the value 1. Let $Z_-$ be $\bigcap_\lambda X_\lambda$ and $Z_+$ be $X \backslash (\bigcup_\lambda X_\lambda)$. Then $Z_-$ and $Z_+$ are closed subsets of $X$. Both are nowhere dense in $X$ since $\bigwedge_\lambda E_\lambda = 0$ and $\bigvee_\lambda E_\lambda = I$. Their union $Z$ is a closed nowhere-dense subset of $X$. If $p \in X \backslash Z$, let $h(p) = \inf\{\lambda : p \in X_\lambda\}$. Given a positive $\varepsilon$ and a point $q_0$ of $X \backslash Z$ at which $h$ takes the value $\lambda$, if $q \in X_{\lambda+\varepsilon} \backslash X_{\lambda-\varepsilon}$, then $|h(q) - h(q_0)| \leqslant \varepsilon$. Thus $h$ is continuous on $X \backslash Z$. By definition, $h$ tends to $+\infty$ at points of $Z_+$ and to $-\infty$ at points of $Z_-$. Thus $h \in \mathscr{S}(X)$; and, from Lemma 5.6.8, $h$ corresponds to a self-adjoint operator $A$ affiliated with $\mathscr{A}$. We note that $\{E_\lambda\}$ is the resolution of the identity for $A$, so that (9) holds, by identifying $X_\lambda$ as the largest clopen set on which $h$ takes values not exceeding $\lambda$. If $Y$ is another such clopen set, $e$ is its characteristic function, and $E$ in $\mathscr{A}$ corresponds to $e$, then $Y \subseteq X_{\lambda'}$ for each $\lambda'$ exceeding $\lambda$. Thus $E \leqslant \bigwedge_{\lambda' > \lambda} E_{\lambda'} = E_\lambda$ and $Y \subseteq X_\lambda$. ■

5.6.10. Lemma. *If $A$ is a closed operator on the Hilbert space $\mathscr{H}$, $\{E_\lambda\}$ is a resolution of the identity on $\mathscr{H}$, $\bigcup_{n=1}^{\infty} F_n(\mathscr{H})$ $(= \mathscr{D}_0)$ is a core for $A$, where $F_n = E_n - E_{-n}$, and*

$$Ax = \int_{-n}^{n} \lambda \, dE_\lambda x \tag{10}$$

*for each $x$ in $F_n(\mathscr{H})$ and all $n$, then $A$ is self-adjoint and $\{E_\lambda\}$ is the resolution of the identity for $A$.*

*Proof.* From (10), $AF_n$ is bounded, everywhere defined, and is the strong-operator limit of finite real-linear combinations of $\{E_\lambda\}$. Thus $E_\lambda AF_n = AF_n E_\lambda$ and $AF_n$ is self-adjoint. From Lemma 5.6.1, $A$ is self-adjoint. If $x \in \mathscr{D}(A)$, there are sequences $\{n_j\}$ (tending to $\infty$) and $\{x_j\}$ such that $x_j = F_{n_j} x_j \to x$ and $Ax_j \to Ax$, since $\mathscr{D}_0$ is a core for $A$. For each $n$,

$$F_n Ax = \lim_j F_n AF_{n_j} x_j = \lim_j AF_{n_j} F_n x_j = \lim_j AF_n x_j = AF_n x$$

so that $F_n A \subseteq AF_n$ for all $n$. At the same time $AF_n E_\lambda x = E_\lambda AF_n x \to E_\lambda Ax$ and $F_n E_\lambda x \to E_\lambda x$. Since $A$ is closed, $E_\lambda x \in \mathscr{D}(A)$ and $AE_\lambda x = E_\lambda Ax$. Thus $E_\lambda A \subseteq AE_\lambda$ and $(2E_\lambda - I)A(2E_\lambda - I) = A$. It follows that $\{E_\lambda\}$ commutes with $T_+$ and $T_-$. Let $\mathscr{A}$ be the (abelian) von Neumann algebra generated by $\{E_\lambda\}$, $T_+$, and $T_-$. As noted in Lemma 5.6.7, $A \,\eta\, \mathscr{A}$. From Lemma 5.6.9, there is a self-adjoint operator $\tilde{A}$ affiliated with $\mathscr{A}$ such that $\tilde{A}x = \int_{-n}^{n} \lambda \, dE_\lambda x$ for each $x$ in $F_n(\mathscr{H})$ and all $n$. Since $\mathscr{D}_0$ is a core for both $A$ and $\tilde{A}$ on which they agree, $\tilde{A} = A$ and $\{E_\lambda\}$ is the resolution for $A$. ■

5.6.11. Remark. The (abelian) von Neumann algebra $\mathscr{A}_0$ generated by $T_+$ and $T_-$ is the *smallest* von Neumann algebra with which (the self-adjoint operator) $A$ is affiliated. (See the proof of Lemma 5.6.7 for the main part of this.) We refer to $\mathscr{A}_0$ as *the von Neumann algebra generated by $A$*. ■

We assemble the foregoing results in the theorem that follows.

5.6.12. Theorem. *If $\mathscr{A}$ is an abelian von Neumann algebra acting on a Hilbert space $\mathscr{H}$, $\mathscr{S}(\mathscr{A})$ is the family of self-adjoint operators affiliated with $\mathscr{A}$, $\mathscr{A}$ is isomorphic to $C(X)$ with $X$ an extremely disconnected compact Hausdorff space, and $\mathscr{S}(X)$ is the family of self-adjoint functions on $X$, then:*

(i) *there is a one-to-one mapping* $\varphi$ *of* $\mathscr{S}(\mathscr{A})$ *onto* $\mathscr{S}(X)$ *extending the isomorphism of* $\mathscr{A}$ *with* $C(X)$ *for which* $\varphi(A)\hat{\cdot}\, e$ *corresponds to* $AE$ *for each projection* $E$ *in* $\mathscr{A}$ *with* $AE$ *in* $\mathscr{A}$, *where* $e$ *in* $C(X)$ *corresponds to* $E$ *and* $(\varphi(A)\hat{\cdot}\, e)(p)$ *is* $(\varphi(A))(p)$ *or* 0 *according as* $e(p)$ *is* 1 *or* 0;

(ii) *there is a resolution of the identity* $\{E_\lambda\}$ *in the abelian von Neumann subalgebra* $\mathscr{A}_0$ *of* $\mathscr{A}$ *generated by an* $A$ *in* $\mathscr{S}(\mathscr{A})$ *such that*

$$Ax = \int_{-n}^{n} \lambda \, dE_\lambda x \tag{11}$$

*for each* $x$ *in* $F_n(\mathscr{H})$ *and all* $n$, *where* $F_n = E_n - E_{-n}$, *and* $\bigcup_{n=1}^{\infty} F_n(\mathscr{H})$ *is a core for* $A$;

(iii) *if* $\{E'_\lambda\}$ *is a resolution of the identity on* $\mathscr{H}$ *such that* $Ax = \int_{-n}^{n} \lambda \, dE'_\lambda x$ *for each* $x$ *in* $F'_n(\mathscr{H})$ *and all* $n$, *and* $\bigcup_{n=1}^{\infty} F'_n(\mathscr{H})$ *is a core for* $A$, *where* $F'_n = E'_n - E'_{-n}$, *then* $E_\lambda = E'_\lambda$ *for all* $\lambda$;

(iv) *if* $\{E_\lambda\}$ *is a resolution of the identity in* $\mathscr{A}$ *there is an* $A$ *in* $\mathscr{S}(\mathscr{A})$ *for which* (11) *holds*;

(v) *if* $e_\lambda$ *in* $C(X)$ *corresponds to* $E_\lambda$ *and* $X_\lambda$ *is the clopen set on which* $e_\lambda$ *takes the value* 1, *then* $X_\lambda$ *is the largest clopen subset of* $X$ *on which* $\varphi(A)$ *takes values not exceeding* $\lambda$ *(in the extended sense).*

It will prove convenient to have the scope of our study broadened to include unbounded "normal" operators as well as self-adjoint operators. We say that a closed densely defined operator $A$ is *normal* when the two self-adjoint operators $A^*A$ and $AA^*$ $(= A^{**}A^*)$ are equal. A spectral theory and function calculus for such operators will make it possible for us to apply complex function theory techniques (see, for example, Stone's theorem (5.6.36) and Section 9.2).

We begin with some preparatory material concerning extensions of unbounded operators. The following simple facts are easily verified.

$$\text{If } A \subseteq B \text{ and } C \subseteq D, \text{ then } A + C \subseteq B + D. \tag{12}$$

$$\text{If } A \subseteq B, \text{ then } CA \subseteq CB \text{ and } AC \subseteq BC. \tag{13}$$

$$(A + B)C = AC + BC, \qquad CA + CB \subseteq C(A + B). \tag{14}$$

In connection with the last assertion of (14), note that we do not have equality in general. This is illustrated by a densely (but not, everywhere-)defined $C$ and $A = I$, $B = -I$. In this case, $C(A + B)$ is 0 but $CA + CB$ is $0|\mathscr{D}(C)$. It follows from these rules that if $CA \subseteq AC$ for each $C$ in some family $\mathscr{F}$, then $TA \subseteq AT$ for each sum $T$ of products of operators in $\mathscr{F}$. We cannot speak of the "algebra" generated by $\mathscr{F}$, for, as we have just noted, a distributive law fails. However, if $\mathscr{F}$ consists of everywhere-defined operators (in particular, of operators in $\mathscr{B}(\mathscr{H})$), we can speak of this algebra.

We may add to (12), (13), (14) another easily proved rule.

(15) If $\{T_a\}$ is a net of operators in $\mathscr{B}(\mathscr{H})$ tending to $T$ in the strong-operator topology and $T_aA \subseteq BT_a$ for each $a$, where $B$ is closed, then $TA \subseteq BT$.

To see this, suppose $x \in \mathscr{D}(A)$. Then $T_ax \in \mathscr{D}(B)$, and $BT_ax = T_aAx \to TAx$. Now, $T_ax \to Tx$. As $B$ is closed $Tx \in \mathscr{D}(B)$ and $BTx = TAx$, from which (15) follows.

Combining the results of this discussion, we have the following lemma.

5.6.13. LEMMA. *If $A$ is a closed operator acting on the Hilbert space $\mathscr{H}$ and $CA \subseteq AC$ for each $C$ in a self-adjoint subset $\mathscr{F}$ of $\mathscr{B}(\mathscr{H})$, then $TA \subseteq AT$ for each $T$ in the von Neumann algebra generated by $\mathscr{F}$.*

If $A$ is a closed operator and $E$ is a projection on $\mathscr{H}$ such that $EA \subseteq AE$ and $AE$ is a bounded everywhere-defined operator on $\mathscr{H}$, we say that $E$ is a *bounding* projection for $A$. If $\{E_n\}$ is an increasing sequence of projections each of which is bounding for $A$ and $\vee_{n=1}^{\infty} E_n = I$, we say that $\{E_n\}$ is a *bounding sequence* for $A$.

5.6.14. LEMMA. *If $E$ is a bounding projection for a closed densely defined operator $A$ on the Hilbert space $\mathscr{H}$, then $E$ is bounding for $A^*$, $A^*A$, and $AA^*$; and $(AE)^* = A^*E$. If $\{E_n\}$ is a bounding sequence for $A$, then $\cup_{n=1}^{\infty} E_n(\mathscr{H})$ is a core for each of $A$, $A^*$, $A^*A$, and $AA^*$.*

*Proof.* Note that $EA$ is preclosed, densely defined, and bounded, since $EA \subseteq AE$ and $AE$ is bounded. Thus $EA$ has closure $AE$ and $(EA)^* = (AE)^*$ from Theorem 2.7.8(i). If $x \in E(\mathscr{H})$ and $y \in \mathscr{D}(A)$, then $\langle Ay, x\rangle = \langle y, (EA)^*x\rangle$, so that $x \in \mathscr{D}(A^*)$ and $A^*x = (EA)^*x$. It follows that $A^*E = (EA)^*E$. But $(I - E)\overline{EA} = 0$ so that $(EA)^* = (\overline{EA})^* = (EA)^*E = A^*E$. Now $EA^* \subseteq (AE)^* = (EA)^* = A^*E$; and $E$ is bounding for $A^*$. Since $EA^*A \subseteq A^*AE$ $(= A^*EAE)$, $E$ is bounding for $A^*A$ and, similarly, for $AA^*$.

It follows that $\{E_n\}$ is a bounding sequence for $A^*$, $A^*A$, and $AA^*$ if it is for $A$. If $x \in \mathscr{D}(A)$, then $E_nx \to x$, $E_nx \in \mathscr{D}(A)$, and $AE_nx = E_nAx \to Ax$. Thus $\cup_{n=1}^{\infty} E_n(\mathscr{D}(A))$ is a core for $A$. Since $E_n(\mathscr{H}) \subseteq \mathscr{D}(A)$, $\cup_{n=1}^{\infty} E_n(\mathscr{H})$ is a core for $A$ (and $A^*$, $A^*A$, $AA^*$ as well). ■

Recall that the statement "$A \,\eta\, \mathscr{A}$" includes the assumption that $A$ is closed and densely defined.

5.6.15. THEOREM. *If $\mathscr{A}$ is an abelian von Neumann algebra acting on the Hilbert space $\mathscr{H}$ and $A, B \,\eta\, \mathscr{A}$, then:*

(i) *each finite set of operators affiliated with $\mathscr{A}$ has a common bounding sequence in $\mathscr{A}$;*

(ii) *$A + B$ is densely defined and preclosed and its closure $A \hat{+} B \,\eta\, \mathscr{A}$;*

(iii) *$A \cdot B$ is densely defined and preclosed and its closure $A \hat{\cdot} B \,\eta\, \mathscr{A}$;*

(iv) *$A \hat{\cdot} B = B \hat{\cdot} A$ and $A^*A = AA^*$ $(= A^* \hat{\cdot} A)$;*

(v) *$(aA \hat{+} B)^* = \bar{a}A^* \hat{+} B^*$;*

(vi) *$(A \hat{\cdot} B)^* = B^* \hat{\cdot} A^*$;*

(vii) *if $A \subseteq B$, then $A = B$; if $A$ is symmetric, $A = A^*$;*

(viii) *the family $\mathscr{N}(\mathscr{A})$ of operators affiliated with $\mathscr{A}$ forms a commutative * algebra (with unit $I$) under the operations of addition $\hat{+}$ and multiplication $\hat{\cdot}$ described in* (ii) *and* (iii).

*Proof.* Throughout this argument, $U$ denotes a unitary operator in $\mathscr{A}'$. Since $U^*AU = A$, we have $U^*A^*U = A^*$; and $A^* \,\eta\, \mathscr{A}$. At the same time $U^*A^*AU = A^*A$ and $A^*A \,\eta\, \mathscr{A}$. If $E$ is a projection in $\mathscr{A}$, $(2E - I)$ is a unitary operator in $\mathscr{A}$ $(\subseteq \mathscr{A}')$; so that $(2E - I)A(2E - I) = A$. Thus $EA \subseteq AE$. From Theorem 2.7.8(v), $A^*A$ is self-adjoint. Let $\{E_\lambda\}$ be its spectral resolution and let $F_n$ be $E_n - E_{-n}$. From Theorem 5.6.12(ii), $E_\lambda \in \mathscr{A}$. As $A^*AF_n$ is bounded and everywhere defined, $AF_n$ is everywhere defined and closed, since $A$ is closed and $F_n$ is bounded. The closed graph theorem (1.8.6) tells us that $AF_n$ is bounded. (This follows directly, as well, since $\|AF_nx\|^2 = \langle F_nx, A^*AF_nx\rangle \leqslant \|A^*AF_n\|\,\|x\|^2$.) As $\{F_n\}$ is an increasing sequence of projections in $\mathscr{A}$ with least upper bound $I$ and $F_nA \subseteq AF_n$, if $x \in \mathscr{D}(A)$, $F_nx \to x$ and $AF_nx = F_nAx \to Ax$. Thus $\bigcup_{n=1}^\infty F_n(\mathscr{H})$ is a core for $A$ and $\{F_n\}$ is a bounding sequence in $\mathscr{A}$ for $A$.

Suppose $\{E_n\}$ is a bounding sequence in $\mathscr{A}$ for $\{A_j\}$, $j = 1, \ldots, m - 1$ and $\{F_n\}$ is a bounding sequence in $\mathscr{A}$ for $A_m$, where $A_j \in \mathscr{A}$. Then $\{E_nF_n\}$ is a bounding sequence in $\mathscr{A}$ for $A_1, \ldots, A_m$. In particular, $\bigcup_{n=1}^\infty E_nF_n(\mathscr{H})$ is a common core for $A_1, \ldots, A_m$. It follows that both $A + B$ and $A^* + B^*$ are densely defined. But $A^* + B^* \subseteq (A + B)^*$, so that $(A + B)^*$ is densely defined and $A + B$ is preclosed (see Theorem 2.7.8(ii)).

If $\{E_n\}$ is a bounding sequence in $\mathscr{A}$ for $A$, $B$, $A^*$, and $B^*$, then $E_nAB \subseteq AE_nB \subseteq ABE_n$ and $AE_nBE_n \subseteq ABE_n$. As $AE_n$ and $BE_n$ are bounded and defined everywhere, $AE_nBE_n = ABE_n$. Thus $\{E_n\}$ is a bounding sequence for $AB$ and, similarly, for $BA$ and $B^*A^*$. In particular $B^*A^*$ is densely defined. As $B^*A^* \subseteq (AB)^*$, $(AB)^*$ is densely defined and $AB$ is preclosed. At the same time, $ABE_n = AE_nBE_n = BE_nAE_n = BAE_n$. Thus $A \hat{\cdot} B$ and $B \hat{\cdot} A$ agree on their common core $\bigcup_{n=1}^\infty E_n(\mathscr{H})$; and $A \hat{\cdot} B = B \hat{\cdot} A$. As $A^*A$ and $AA^*$ are self-adjoint, $A^*A = A^* \hat{\cdot} A = A \hat{\cdot} A^* = AA^*$. If $x \in \mathscr{D}(A) \cap \mathscr{D}(B)$ $(= \mathscr{D}(A + B))$, $Ux \in \mathscr{D}(A + B)$ and $U^*x \in \mathscr{D}(A + B)$. Thus $U(\mathscr{D}(A + B)) = \mathscr{D}(A + B)$ and $U^*(A + B)U = A + B$. It follows that $U^*(A \hat{+} B)U = A \hat{+} B$ and $A \hat{+} B \,\eta\, \mathscr{A}$. If $y \in \mathscr{D}(AB)$, then $y \in \mathscr{D}(B)$ and $By \in \mathscr{D}(A)$. Thus $Uy \in \mathscr{D}(B)$ and $BUy = UBy \in \mathscr{D}(A)$. It follows that $Uy \in \mathscr{D}(AB)$. Since $U^*y \in \mathscr{D}(AB)$, $U(\mathscr{D}(AB)) = \mathscr{D}(AB)$. As $U^*ABUy = ABy$, $U^*A \hat{\cdot} BU = A \hat{\cdot} B$ and $A \hat{\cdot} B \,\eta\, \mathscr{A}$.

With $\{E_n\}$ bounding for $A$ and $A^*$, $E_nA^* \subseteq A^*E_n$ and $E_nA^* \subseteq (AE_n)^*$. Thus $A^*E_n$ and $(AE_n)^*$ are bounded, everywhere-defined extensions of the same densely defined operator $E_nA^*$. It follows that $(AE_n)^* = A^*E_n$ (or we may cite Lemma 5.6.14). Suppose that $\{E_n\}$ is bounding for $B$, $B^*$, $aA \mathbin{\hat{+}} B$, $(aA \mathbin{\hat{+}} B)^*$, $A \mathbin{\hat{\cdot}} B$, $(A \mathbin{\hat{\cdot}} B)^*$, and $A^* \mathbin{\hat{\cdot}} B^*$ $(= B^* \mathbin{\hat{\cdot}} A^*)$ as well as for $A$ and $A^*$. Then, from the foregoing,

$$\begin{aligned}(\bar{a}A^* \mathbin{\hat{+}} B^*)E_n = \bar{a}A^*E_n + B^*E_n = \bar{a}(AE_n)^* + (BE_n)^* &= ((aA \mathbin{\hat{+}} B)E_n)^* \\ &= (aA \mathbin{\hat{+}} B)^*E_n\end{aligned}$$

and

$$\begin{aligned}(A \mathbin{\hat{\cdot}} B)^*E_n = ((A \mathbin{\hat{\cdot}} B)E_n)^* = (AE_nBE_n)^* = (BE_n)^*(AE_n)^* &= B^*E_nA^*E_n \\ &= (B^* \mathbin{\hat{\cdot}} A^*)E_n.\end{aligned}$$

Since $(aA \mathbin{\hat{+}} B)^*$ and $\bar{a}A^* \mathbin{\hat{+}} B^*$ agree on their common core $\bigcup_{n=1}^{\infty} E_n(\mathscr{H})$, they are equal. Similarly $(A \mathbin{\hat{\cdot}} B)^* = B^* \mathbin{\hat{\cdot}} A^*$ $(= A^* \mathbin{\hat{\cdot}} B^*)$.

If $A \subseteq B$ and $\{E_n\}$ is a bounding sequence in $\mathscr{A}$ for both $A$ and $B$, then $AE_n \subseteq BE_n$ so that $AE_n = BE_n$. Thus $A$ and $B$ agree on their common core $\bigcup_{n=1}^{\infty} E_n(\mathscr{H})$. Hence $A = B$. If $A$ is symmetric, $A \subseteq A^*$ and, from the preceding conclusion, $A = A^*$.

It is routine to verify identities such as

$$(A \mathbin{\hat{\cdot}} B) \mathbin{\hat{\cdot}} C = A \mathbin{\hat{\cdot}} (B \mathbin{\hat{\cdot}} C),$$

by choosing a common bounding sequence for all operators involved. Thus (viii) follows. ■

As noted in (iv) of the preceding theorem, $A^*A = AA^*$ for each $A$ affiliated with an abelian von Neumann algebra $\mathscr{A}$. By analogy with the case of bounded operators, we expect normal operators to be affiliated with abelian von Neumann algebras. With the aid of the lemmas that follow, we shall prove this. We conclude from this that the multiplication operators corresponding to unbounded (complex-valued) measurable functions (finite almost everywhere) are normal (compare Theorem 5.6.4). Our first lemma is an analogue to Lemma 5.6.1.

5.6.16. LEMMA. *If $\{F_n\}$ is a bounding sequence for the closed operator $A$ on the Hilbert space $\mathscr{H}$ and $AF_n$ is normal for each $n$, then $A$ is normal.*

*Proof.* From Lemma 5.6.14, $(AF_n)^* = A^*F_n$, so that

$$A^*AF_n = A^*F_nAF_n = (AF_n)^*AF_n = AF_n(AF_n)^* = AF_nA^*F_n = AA^*F_n.$$

Thus the self-adjoint operators $A^*A$ and $AA^*$ agree on $\bigcup_{n=1}^{\infty} F_n(\mathscr{H})$, a core for each of them. Thus $A^*A = AA^*$. ■

5.6.17. LEMMA. *If $BA \subseteq AB$ and $\mathscr{D}(A) \subseteq \mathscr{D}(B)$, where $A$ is a self-adjoint operator and $B$ is a closed operator on the Hilbert space $\mathscr{H}$, then $E_\lambda B \subseteq BE_\lambda$ for each $E_\lambda$ in the spectral resolution $\{E_\lambda\}$ of $A$.*

*Proof.* We note that $B(A + iI) = BA + iB$ under the present assumptions. For this, observe that, from (14), $BA + iB \subseteq B(A + iI)$. Suppose $x \in \mathscr{D}(B(A + iI))$. Then $x \in \mathscr{D}(A)$ and $Ax + ix \in \mathscr{D}(B)$. By assumption $x \in \mathscr{D}(A) \subseteq \mathscr{D}(B)$ so that $Ax \in \mathscr{D}(B)$, as well. Thus $x \in \mathscr{D}(BA + iB)$ and $B(A + iI)x = BAx + iBx$. Hence $B(A + iI) \subseteq BA + iB$ and the stated equality follows. Similarly $B(A - iI) = BA - iB$.

Let $T_+$ and $T_-$ be the (bounded, everywhere-defined) inverses to $A + iI$ and $A - iI$, respectively. Then, from (12)–(14) and the preceding paragraph,

$$T_+B = T_+B(A + iI)T_+ = T_+(BA + iB)T_+ \subseteq T_+(AB + iB)T_+$$
$$= T_+(A + iI)BT_+ \subseteq BT_+.$$

Similarly $T_-B \subseteq BT_-$. From the proof of Lemma 5.6.7, $T_+ = T_-^*$ so that Lemma 5.6.13 applies; and $TB \subseteq BT$ for each $T$ in the von Neumann algebra $\mathscr{A}$ generated by $T_+$ and $T_-$. In particular $E_\lambda B \subseteq BE_\lambda$ for each $\lambda$. ∎

5.6.18. THEOREM. *An operator $A$ is normal if and only if it is affiliated with an abelian von Neumann algebra. If $A$ is normal, there is a smallest von Neumann algebra $\mathscr{A}_0$ such that $A \,\eta\, \mathscr{A}_0$. The algebra $\mathscr{A}_0$ is abelian.*

*Proof.* From Theorem 5.6.15(iv), each operator affiliated with an abelian von Neumann algebra is normal. Assume, now, that $A$ is normal. Since $AA^*A = A^*AA$ and $\mathscr{D}(A^*A) \subseteq \mathscr{D}(A)$, Lemma 5.6.17 applies. Thus $E_\lambda A \subseteq AE_\lambda$ for each $\lambda$, where $\{E_\lambda\}$ is the spectral resolution of $A^*A$; and $F_nA \subseteq AF_n$ for each $n$, where $F_n = E_n - E_{-n}$. In the same way, $A^*A^*A = A^*AA^*$ and $\mathscr{D}(A^*A) = \mathscr{D}(AA^*) \subseteq \mathscr{D}(A^*)$, so that $F_nA^* \subseteq A^*F_n$ for each $n$. As in the proof of Theorem 5.6.15, $AF_n$ and $A^*F_n$ are bounded since $A^*AF_n$ $(= AA^*F_n)$ is. Moreover, $F_nA^* \subseteq (AF_n)^*$, so that both $(AF_n)^*$ and $A^*F_n$ are bounded extensions of the densely defined $F_nA^*$. Thus $(AF_n)^* = A^*F_n$ (and $(A^*F_n)^* = AF_n$). Note, too, that $AF_nAF_m \subseteq AAF_n$ and $AF_mAF_n \subseteq AAF_n$, when $n \leqslant m$. Since $AF_nAF_m$ and $AF_mAF_n$ are everywhere defined, $AF_nAF_m = AAF_n = AF_mAF_n$. At the same time, $A^*F_mAF_n = A^*AF_n = AA^*F_n = AF_nA^*F_m$. Thus $\{F_n, AF_n, A^*F_n : n = 1, 2, \ldots\}$ generates an abelian von Neumann algebra $\mathscr{A}_0$. Since $\vee_{n=1}^\infty F_n = I$ and $F_nA \subseteq AF_n$, $\cup_{n=1}^\infty F_n(\mathscr{H})$ $(= \mathscr{D}_0)$ is a core for $A$. If $U$ is a unitary operator in $\mathscr{A}_0'$ and $x \in \mathscr{D}_0$, $AUx = AUF_nx = AF_nUx = UAF_nx = UAx$ (for some $n$). From Remark 5.6.3, $A \,\eta\, \mathscr{A}_0$ (and $A^* \,\eta\, \mathscr{A}_0$). If $A \,\eta\, \mathscr{R}$, then $A^* \,\eta\, \mathscr{R}$ and $A^*A \,\eta\, \mathscr{R}$. From Remark 5.6.11, $A^*A$ generates an abelian von Neumann algebra $\mathscr{A}_1$ contained in $\mathscr{R}$. Thus $F_n \in \mathscr{R}$ so that $AF_n$, $A^*F_n$ are in $\mathscr{R}$; and $\mathscr{A}_0 \subseteq \mathscr{R}$. ∎

We refer to $\mathscr{A}_0$ as *the von Neumann algebra generated by* (the normal operator) $A$.

5.6.19. THEOREM. *If $\mathscr{A}$ is an abelian von Neumann algebra, $\varphi$ is an isomorphism of $\mathscr{A}$ onto $C(X)$ where $X$ is a compact Hausdorff space, and $A \,\eta\, \mathscr{A}$,* *there is a unique normal function $\varphi(A)$ on $X$ such that $\varphi(AE) = \varphi(A) \hat{\cdot} \varphi(E)$ when $E$ is a bounding projection in $\mathscr{A}$ for $A$ where $(\varphi(A) \hat{\cdot} \varphi(E))(p)$ is $\varphi(A)(p)\varphi(E)(p)$ if $\varphi(A)(p)$ is defined and* 0 *otherwise. If $\mathscr{N}(X)$ is the family of normal functions on $X$ and $f$, $g$ are in $\mathscr{N}(X)$, there are unique normal functions $\bar{f}$, $af$, $f \hat{+} g$, and $f \hat{\cdot} g$ such that $\bar{f}(p) = \overline{f(p)}$, $(af)(p) = af(p)$, $(f \hat{+} g)(p) = f(p) + g(p)$, and $(f \hat{\cdot} g)(p) = f(p)g(p)$, when $f$ and $g$ are defined at $p$. Endowed with the operations $f \to \bar{f}$, $(a, f) \to af$, $(f, g) \to f \hat{+} g$ and $(f, g) \to f \hat{\cdot} g$, $\mathscr{N}(X)$ is an associative, commutative algebra with unit* 1 *and involution $f \to \bar{f}$. The mapping $\varphi$, as extended, is a * isomorphism of $\mathscr{N}(\mathscr{A})$ onto $\mathscr{N}(X)$.*

*Proof.* We show first that $\varphi(A)$, as described, is unique. From Theorem 5.6.15, $A$ has a bounding sequence $\{E_n\}$ in $\mathscr{A}$. If $f$ and $g$ are normal functions with the properties ascribed to $\varphi(A)$ and both $f$ and $g$ are defined at $p$, then

$$f(p)\varphi(E_n)(p) = \varphi(AE_n)(p) = g(p)\varphi(E_n)(p)$$

for each $n$. Thus if $\varphi(E_n)(p) = 1$ for some $n$, $f(p) = g(p)$. As $\{E_n\}$ is monotone increasing to $I$, $f$ and $g$ agree on a dense subset of $X$; and, from Lemma 5.6.6, $f = g$.

If $A \in \mathscr{S}(\mathscr{A})$, the notation of Theorem 5.6.12(i) agrees with the present notation and the function associated with $A$ there has the properties required of $\varphi(A)$. Thus $\varphi$ is defined on $\mathscr{S}(\mathscr{A})$.

If $A$ and $B$ are in $\mathscr{S}(\mathscr{A})$, from Theorem 5.6.15(i), we can choose a bounding sequence $\{E_n\}$ for both $A$ and $B$. Then $AE_n$, $BE_n$, $(A + B)E_n$, and $ABE_n$ are in $\mathscr{A}$. Moreover $(A \hat{+} B)E_n = (A + B)E_n = AE_n + BE_n$ and $(A \hat{\cdot} B)E_n = ABE_n = AE_nBE_n$, so that

$$\begin{aligned} \varphi(A \hat{+} B)(p)\varphi(E_n)(p) &= \varphi((A \hat{+} B)E_n)(p) \\ &= \varphi(A)(p)\varphi(E_n)(p) + \varphi(B)(p)\varphi(E_n)(p) \end{aligned}$$

and

$$\varphi(A \hat{\cdot} B)(p)\varphi(E_n)(p) = \varphi((A \hat{\cdot} B)E_n)(p) = \varphi(A)(p)\varphi(B)(p)\varphi(E_n)(p)$$

when $\varphi(A)$, $\varphi(B)$, $\varphi(A \hat{+} B)$, and $\varphi(A \hat{\cdot} B)$ are defined at $p$. Since $\{E_n\}$ is monotone increasing to $I$, $\varphi(A \hat{+} B)$ and $\varphi(A) + \varphi(B)$ agree on a dense subset of $X$ as do $\varphi(A \hat{\cdot} B)$ and $\varphi(A)\varphi(B)$. Thus $\varphi(A \hat{+} B)$ and $\varphi(A \hat{\cdot} B)$ are finite, when both $\varphi(A)$ and $\varphi(B)$ are defined, and, by continuity, are normal extensions of $\varphi(A) + \varphi(B)$ and $\varphi(A)\varphi(B)$, respectively. Define $\varphi(A) \hat{+} \varphi(B)$ and $\varphi(A) \hat{\cdot} \varphi(B)$ to be these normal extensions. Since each $h$ in $\mathscr{S}(X)$ corresponds to some $A$ in $\mathscr{S}(\mathscr{A})$, from Theorem 5.6.12, the operations $\hat{+}$ and $\hat{\cdot}$ apply to all functions in $\mathscr{S}(X)$. It is easy to verify (by repeated use of Lemma 5.6.6) that $\mathscr{S}(X)$ endowed with these operations (and the indicated multiplication by real scalars) is an associative, commutative algebra (over $\mathbb{R}$) with unit 1. From the preceding discussion, $\varphi$ is an isomorphism of $\mathscr{S}(\mathscr{A})$ onto $\mathscr{S}(X)$.

If $A \in \mathcal{N}(\mathcal{A})$, then $A = A_1 \hat{+} iA_2$, where $A_1$ $(=\frac{1}{2}(A \hat{+} A^*))$ and $A_2$ $(= -\frac{1}{2}i(A \hat{+} -A^*))$ are in $\mathcal{S}(\mathcal{A})$. If $\varphi(A_1)$ and $\varphi(A_2)$ are defined on $X\backslash Z_1$ and $X\backslash Z_2$, respectively, then $\varphi(A_1) + i\varphi(A_2)$ is defined on $X\backslash(Z_1 \cup Z_2)$. If $p \in Z_1$, then $|\varphi(A_1)(q)|$ and hence $|\varphi(A_1)(q) + i\varphi(A_2)(q)|$ are large at points $q$ of $X\backslash(Z_1 \cup Z_2)$ near $p$. A similar comment applies to $p$ in $Z_2$, so that $\varphi(A_1)+i\varphi(A_2)$ defined on $X\backslash(Z_1 \cup Z_2)$ is an element, $\varphi(A_1) \hat{+} i\varphi(A_2)$ $(= \varphi(A))$, of $\mathcal{N}(X)$.

If $h$ in $\mathcal{N}(X)$ is defined on $X\backslash Z$, then $h^{-1}(0)$ is a closed set in $X$. To see this, note that $h^{-1}(0)$ is closed in $X\backslash Z$ by continuity of $h$ on $X\backslash Z$, and $|h|$ is large at points of $X\backslash Z$ near a point $p$ of $Z$ so that $p \notin h^{-1}(0)^-$. Thus the interior $X_0$ of $h^{-1}(0)$ is a clopen subset of $X$ and the function $g$, defined as 1 on $X_0$ and as $h/|h|$ on $X\backslash(h^{-1}(0) \cup Z)$, is continuous. As $X\backslash((h^{-1}(0)\backslash X_0) \cup Z)$ is a dense open subset of $X$, Corollary 5.2.11 applies, and $g$ has an extension $u$ in $C(X)$. For $p$ in $X\backslash Z$, $u(p)|h(p)| = h(p)$. As $|h| \in \mathcal{S}(X)$, $\mathrm{Re}\, h$ and $\mathrm{Im}\, h$ defined on $X\backslash Z$ have $(\mathrm{Re}\, u) \hat{\cdot} |h|$ and $(\mathrm{Im}\, u) \hat{\cdot} |h|$ as normal extensions. Thus $h = h_1 \hat{+} ih_2$ with $h_1$ and $h_2$ in $\mathcal{S}(X)$. Choosing $A_1$ and $A_2$ in $\mathcal{S}(\mathcal{A})$ such that $\varphi(A_1) = h_1$ and $\varphi(A_2) = h_2$, we have $\varphi(A_1 \hat{+} iA_2) = h_1 \hat{+} ih_2 = h$. Hence $\varphi$ maps $\mathcal{N}(\mathcal{A})$ onto $\mathcal{N}(X)$.

If $f$ in $\mathcal{N}(X)$ is defined on $X\backslash Z$, we have just seen that the real and imaginary parts of $f$ have normal extensions. Denote these extensions by $\mathrm{Re} f$ and $\mathrm{Im} f$. With $g$ in $\mathcal{N}(X)$ defined on $X\backslash Z'$, $f + g$ and $fg$ are defined and continuous on $X\backslash(Z \cup Z')$ and have the normal extensions

$$(\mathrm{Re} f \hat{+} \mathrm{Re}\, g) \hat{+} i(\mathrm{Im} f \hat{+} \mathrm{Im}\, g)\ (= f \hat{+} g)$$

and

$$(\mathrm{Re} f \hat{\cdot} \mathrm{Re}\, g \hat{+} - \mathrm{Im} f \hat{\cdot} \mathrm{Im}\, g) \hat{+} i(\mathrm{Re} f \hat{\cdot} \mathrm{Im}\, g \hat{+} \mathrm{Im} f \hat{\cdot} \mathrm{Re}\, g)\ (= f \hat{\cdot} g).$$

With the indicated operations, $\mathcal{N}(X)$ becomes an associative, commutative algebra with unit 1 and adjoint operation $f \to \bar{f}$. The mapping $\varphi$, as extended from $\mathcal{S}(\mathcal{A})$, is a * isomorphism of $\mathcal{N}(\mathcal{A})$ onto $\mathcal{N}(X)$. If $A \,\eta\, \mathcal{A}$ and $E$ is a bounding projection in $\mathcal{A}$ for $A$, we have, now, $\varphi(AE) = \varphi(A \hat{\cdot} E) = \varphi(A) \hat{\cdot} \varphi(E)$. ■

In the preceding proof, we note that, with $h$ in $\mathcal{N}(X)$, $h^{-1}(0)$, a subset of $X\backslash Z$, is closed in $X$. The same argument establishes that $h^{-1}(\mathcal{K})$, a subset of $X\backslash Z$, is closed, hence compact, in $X$, for each compact subset $\mathcal{K}$ of $\mathbb{C}$. It follows that the range of $h$ intersects $\mathcal{K}$ in a compact subset (since this intersection is $h(h^{-1}(\mathcal{K}))$, the continuous image of a compact set). Thus if $z_0$ is not in the range of $h$ it is at positive distance from that range, and $1/(h - z_0 1)$ is a bounded continuous function on $X\backslash Z$ tending to 0 at each point of $Z$. Hence $z_0$ is not in the range of $h$ if and only if $h - z_0 1$ has an inverse in $\mathcal{N}(X)$ and that inverse is in $C(X)$. This observation suggests the concept of spectrum to us and will play the key role in identifying the "spectrum" of an unbounded operator with the range of its representing function. (See Proposition 5.6.20.)

The *spectrum* sp$(T)$ *of a closed densely defined linear operator* $T$ on a Hilbert space $\mathscr{H}$ is the set of those complex numbers $z$ such that $T - zI$ is not a one-to-one mapping of $\mathscr{D}(T)$ onto $\mathscr{H}$. If $z_0 \notin \text{sp}(T)$, then $T - z_0 I$ is a one-to-one linear mapping of $\mathscr{D}(T)$ onto $\mathscr{H}$ and has a linear inverse $B$ (mapping $\mathscr{H}$ onto $\mathscr{D}(T)$). Since the graph of $T - z_0 I$ is closed, the graph of $B$ is closed. As $B$ is defined on all of $\mathscr{H}$, the closed graph theorem applies and $B$ is bounded. Thus if $z_0 \notin \text{sp}(T)$, $T - z_0 I$ has a *bounded* (everywhere-defined) inverse; and, of course, conversely, $z_0 \notin \text{sp}(T)$ if $T - z_0 I$ has such an inverse. If $T \eta \mathscr{A}$, for some abelian von Neumann algebra $\mathscr{A}$, and $z_0 \notin \text{sp}(T)$, then $B$ just constructed is in $\mathscr{A}$. As $B$ is bounded and $T - z_0 I$ is closed, $(T - z_0 I)B$ is closed. Hence

$$I = (T - z_0 I)B = (T - z_0 I) \hat{\cdot} B = B \hat{\cdot} (T - z_0 I).$$

It follows that $z_0 \notin \text{sp}\, T$ if and only if $T - z_0 I$ has an inverse $B$ in the algebra $\mathscr{N}(\mathscr{A})$ and $B$ lies in $\mathscr{A}$.

5.6.20. Proposition. *If $A$ is a normal operator affiliated with the abelian von Neumann algebra $\mathscr{A}$, then* sp$(A)$ *coincides with the range of $\varphi(A)$, where $\varphi$ is the isomorphism of $\mathscr{N}(\mathscr{A})$ onto $\mathscr{N}(X)$ extending the isomorphism of $\mathscr{A}$ with $C(X)$.*

*Proof.* By definition, $z_0 \notin \text{sp}(A)$ if and only if there is a bounded $B$ inverse to $A - z_0 I$. If $U$ is a unitary operator in $\mathscr{A}'$, then $U^*(A - z_0 I)U = A - z_0 I$, since $A - z_0 I \eta \mathscr{A}$. Hence $U^*BU = B$ for each such $U$; and $B \in \mathscr{A}$. Since there is a $B$ in $\mathscr{A}$ such that $(A - z_0 I) \hat{\cdot} B = I$, equivalently, $(\varphi(A) - z_0 1) \hat{\cdot} \varphi(B) = 1$, if and only if $z_0 \notin \text{sp}(A)$; and there is such a $\varphi(B)$ in $C(X)$ if and only if $z_0$ is not in the range of $\varphi(A)$, sp$(A)$ is the range of $\varphi(A)$. ∎

Is there an interpretation of "spectrum relative to $\mathscr{N}(\mathscr{A})$"? Equivalently, when does $f - z_0 1$ fail to have an inverse in $\mathscr{N}(X)$? This occurs if and only if $f - z_0 1$ vanishes on some non-null clopen subset of $X$. Viewed in $\mathscr{N}(\mathscr{A})$ this amounts to the existence of a non-zero projection $E_0$ in $\mathscr{A}$ such that $(A - z_0 I) \hat{\cdot} E_0 = 0$ or, equivalently, to the existence of a unit vector $x_0$ such that $(A - z_0 I)x_0 = 0$. In this case, we say that $z_0$ is in the *point spectrum* of $A$. Thus the spectrum of $A$ relative to $\mathscr{N}(\mathscr{A})$ is its point spectrum.

If we define a self-adjoint operator $A$ to be *positive* (and write $A \geqslant 0$) when $\langle Ax, x \rangle \geqslant 0$ for each $x$ in $\mathscr{D}(A)$, the question (answered for bounded $A$ in Theorem 4.2.6) of the relation of this condition to the nature of sp$(A)$ arises. It is easily settled with the help of Proposition 5.6.21.

5.6.21. Proposition. *A self-adjoint operator $A$ is positive if and only if $a \geqslant 0$ when $a \in \text{sp}(A)$.*

*Proof.* We may suppose, from Lemma 5.6.7, that $A$ is affiliated with an abelian von Neumann algebra $\mathscr{A}$. From Theorem 5.6.19, there is an isomorphism $\varphi$ of $\mathcal{N}(\mathcal{A})$ onto $\mathcal{N}(X)$ mapping $\mathcal{A}$ onto $C(X)$, where $X$ is an extremely disconnected compact Hausdorff space. If $\varphi(A)$ is defined on $X\backslash Z$ and $\varphi(A)(p) < 0$ for some $p$ in $X\backslash Z$, then there is a non-null clopen set $X_0$ containing $p$ and contained in $X\backslash Z$ such that $\varphi(A)(q) < a < 0$ for each $q$ in $X_0$. If $E_0$ is the projection in $\mathcal{A}$ corresponding to the characteristic function of $X_0$, then $E_0$ is a non-zero bounding projection for $A$ and $AE_0 \leqslant aE_0$. With $x$ a unit vector in the range of $E_0$, $\langle Ax, x\rangle \leqslant a < 0$ and $A$ is not positive. Thus $\mathrm{sp}(A) \geqslant 0$ if $A \geqslant 0$.

On the other hand, if $\varphi(A)$ has range consisting of non-negative real numbers, its (positive) square root $g$ is a normal (self-adjoint) function on $X$. If $\varphi(B)$ (in $\mathcal{S}(X)$) is $g$, then $B^2 = B \hat{\cdot} B = A$ (recall Theorem 5.6.15(iv)); and, with $x$ a unit vector in $\mathcal{D}(A)$, $x \in \mathcal{D}(B)$ and $\langle Ax, x\rangle = \langle Bx, Bx\rangle \geqslant 0$. Thus $A \geqslant 0$ if $\mathrm{sp}(A) \geqslant 0$. ■

It follows now that the set of positive elements in $\mathcal{N}(\mathcal{A})$ (in $\mathcal{S}(\mathcal{A})$) forms a positive cone and that $\mathcal{S}(\mathcal{A})$ is a partially ordered vector space relative to the partial ordering induced by this cone. (See the discussion preceding Definition 3.4.5.) The same is true for $\mathcal{S}(X)$. Of course $I$ is *not* an order unit for $\mathcal{S}(\mathcal{A})$ in the present case. The following lemma, an analogue for $\mathcal{N}(X)$ of Lemma 5.2.10, will form the basis for a proof that $\mathcal{S}(X)$ is a bounded $\sigma$-lattice in the given ordering as well as providing the basis for our Borel function calculus.

5.6.22. LEMMA. *Each Borel function $g$ on an extremely disconnected compact Hausdorff space $X$ agrees with a unique normal function $f$ on the complement of a meager set. The mapping that assigns $f$ to $g$ is a conjugation-preserving homomorphism of the algebra $\mathcal{B}_u(X)$ of Borel functions on $X$ onto $\mathcal{N}(X)$ with kernel consisting of those functions in $\mathcal{B}_u(X)$ vanishing on the complement of a meager set.*

*Proof.* We note first that if $f$ and $g$ are normal functions defined on $X\backslash Z$ and $X\backslash Z'$, respectively, and $f(p) = g(p)$ for $p$ in $X\backslash(Z \cup Z' \cup M)$, where $M$ is a meager subset of $X$, then $Z = Z'$ and $f = g$; for $Z \cup Z' \cup M$ is meager in $X$, so that $X\backslash(Z \cup Z' \cup M)$ is dense in $X$ and Lemma 5.6.6 applies. Thus there can be at most one normal function agreeing with *any* function on the complement of a meager set.

If $g$ is a Borel function on $X$ and $\mathbb{D}_n$ is the closed disk in $\mathbb{C}$ with center 0 and radius $n$, then $g^{-1}(\mathbb{D}_n)$ is a Borel subset $S_n$ of $X$. According to Lemma 5.2.10, there is a clopen set $X_n$ such that $(S_n\backslash X_n) \cup (X_n\backslash S_n)$ is meager in $X$. If $g_n$ is the (Borel) function equal to $g$ on $S_n$ and 0 on $X\backslash S_n$, then $\|g_n\| \leqslant n$. Again from Lemma 5.2.10 there is a (unique) continuous function $f_n$ on $X$ that agrees with $g_n$ on the complement of a meager subset $M_n$ of $X$. Since $g_n$ vanishes on $X\backslash S_n$, $f_n$ vanishes on $X\backslash(S_n \cup M_n)$ which contains $(X\backslash X_n)\backslash(M_n \cup (S_n\backslash X_n) \cup (X_n\backslash S_n))$. As $M_n \cup (S_n\backslash X_n) \cup (X_n\backslash S_n)$ is meager and $f_n$ is continuous, $f_n$ vanishes on $X\backslash X_n$. Note that $S_n \subseteq S_{n+1}$, so that $e_n(p) \leqslant e_{n+1}(p)$ for each $p$ outside a meager set, where $e_j$ is the characteristic function of $X_j$. By continuity, $e_n \leqslant e_{n+1}$ and $X_n \subseteq X_{n+1}$. Again, since $g_{n+1}$ agrees with $g_n$ on $S_n$ and $f_n$ and $f_{n+1}$ are continuous, $f_{n+1}$ agrees with $f_n$ on $X_n$. In the same way, $n \leqslant |f_m(p)|$ if $p \in X_m \backslash X_n$, where $n < m$, since $n \leqslant |g_m(q)|$ if $q \in S_m \backslash S_n$. As $\bigcup_{n=1}^{\infty} S_n = X$, we have

$$(Z =) X \backslash \bigcup_{n=1}^{\infty} X_n \subseteq \bigcup_{n=1}^{\infty} (S_n \backslash X_n) \cup (X_n \backslash S_n).$$

Thus $Z$ is closed and meager, hence, nowhere dense in $X$. If $f(p)$ is defined as $f_n(p)$ when $p \in X_n$, then $f$ is continuous on $X \backslash Z$. For $p$ in $Z$, $X \backslash X_n$ is an open set containing $p$ such that, for $q$ in $X \backslash (Z \cup X_n)$, $n \leqslant |f(q)|$. Thus $f$ is normal. Moreover, $f$ and $g$ agree on the complement of $Z \cup (\bigcup_{n=1}^{\infty} M_n)$, a meager subset of $X$. The mapping assigning $f$ to $g$ is a homomorphism of $\mathscr{B}_u(X)$ onto $\mathscr{N}(X)$ with kernel consisting of those Borel functions on $X$ vanishing on the complement of a meager set. ∎

5.6.23. Proposition. *If $\mathscr{A}$ is an abelian von Neumann algebra and $\{A_n\}$ is an increasing sequence of operators in $\mathscr{S}(\mathscr{A})$ with upper bound $A_0$ in $\mathscr{S}(\mathscr{A})$, then $\{A_n\}$ has a least upper bound in $\mathscr{S}(\mathscr{A})$.*

*Proof.* With $\mathscr{A} \cong C(X)$, let $f_n$ in $\mathscr{N}(X)$ represent $A_n$. If $\{A_n \hat{+} -A_1\}$ has a least upper bound $B$ in $\mathscr{S}(\mathscr{A})$, then $B \hat{+} A_1$ is the least upper bound of $\{A_n\}$. We may assume, without loss of generality, that each $A_n \geqslant 0$. If $f_n$ is defined on $X \backslash Z_n$, then $Z_n = (Z_n)_+$, in this case, and $Z_n \subseteq Z_{n+1}$. If $p$ is not in $\bigcup_{n=0}^{\infty} Z_n$ $(= Z)$, then $f_n(p)$ is defined for all $n$ and $\{f_n(p)\}$ has $f_0(p)$ as an upper bound. Thus $\{f_n(p)\}$ converges to some $g(p)$. If we define $g$ to be 0 on $Z$, then $g$ is a Borel function on $X$. From Lemma 5.6.22, there is an $f$ in $\mathscr{N}(X)$ agreeing with $g$ on $X \backslash M$, where $M$ is a meager subset of $X$. Hence $\{f_n(p)\}$ converges to $f(p)$ on the dense set $X \backslash (M \cup Z)$. It follows that $f$ is an upper bound for $\{f_n\}$ and the *least* upper bound. If $A$ in $\mathscr{S}(\mathscr{A})$ is represented by $f$ in $\mathscr{S}(X)$, then $A$ is the least upper bound of $\{A_n\}$. ∎

5.6.24. Remark. In view of Proposition 5.6.23, we may extend the notion of $\sigma$-normal homomorphism, in the obvious way, to apply to $\mathscr{N}(\mathscr{A})$, $\mathscr{N}(X)$, and to $\mathscr{B}_u$ (the algebra of complex-valued Borel functions on $\mathbb{C}$), $\mathscr{B}_u(X)$. With this extension, we observe that the homomorphism of Lemma 5.6.22 mapping $\mathscr{B}_u(X)$ onto $\mathscr{N}(X)$ is $\sigma$-normal. If $\{g_n\}$ is an increasing sequence of Borel functions on $X$ tending pointwise to the (Borel) function $g_0$ and $f_n$ is the normal function corresponding to $g_n$, then $\{f_n\}$ has $f_0$ as its least upper bound in $\mathscr{N}(X)$. To see this, suppose $f_n$ and $g_n$ agree on the complement of the meager set $M_n$. Then $\{f_n(p)\}$ tends to $f_0(p)$ for each $p$ in $X \backslash \bigcup_{n=0}^{\infty} M_n$, a dense subset of $X$. If $h$ in $\mathscr{N}(X)$ is an upper bound for $\{f_n\}$, then $f_n(p) \leqslant h(p)$ for all $n$ and all $p$ in the complement of some meager set. Thus $f_0(p) \leqslant h(p)$ for all $p$ in the complement of some meager set, and $h \hat{+} -f_0$ takes non-negative values on a dense set. Hence $f_0 \leqslant h$, and $f_0$ is the least upper bound in $\mathscr{N}(X)$ of $\{f_n\}$. ∎

5.6.25. Remark. With the aid of Lemma 5.6.22, we can define $g(A)$ for an arbitrary Borel function $g$ on $\operatorname{sp}(A)$ and an arbitrary normal operator $A$. From Theorem 5.6.18, $A$, $A^*$, and $I$ generate an abelian von Neumann algebra $\mathscr{A}_0$ (and $A \,\eta\, \mathscr{A}_0$). From Theorem 5.2.1, $\mathscr{A}_0$ is isomorphic to $C(X)$, where $X$ is an extremely disconnected compact Hausdorff space. From Theorem 5.6.19, there is an isomorphism $\varphi$ of $\mathscr{N}(\mathscr{A}_0)$ onto $\mathscr{N}(X)$. If $\varphi(A)$ is defined on $X \backslash Z$, then $\tilde{g}$ defined as 0 on $Z$ and $g \circ \varphi(A)$ on $X \backslash Z$ is in $\mathscr{B}_u(X)$. From Lemma 5.6.22, there is a function $h$ in $\mathscr{N}(X)$ agreeing with $\tilde{g}$ on the complement of a meager set in $X$. *We define* $g(A)$ *to be* $\varphi^{-1}(h)$. If $\operatorname{sp}(A)$ is a subset of the Borel set $S$ and $g$ is a Borel function defined on $S$, then $g(A)$ will denote $g_r(A)$, where $g_r$ is the restriction of $g$ to $\operatorname{sp}(A)$. ∎

5.6.26. Theorem. *If $\mathscr{A}_0$ is the abelian von Neumann algebra generated by a normal operator $A$ acting on a Hilbert space $\mathscr{H}$, the mapping $g \to g(A)$ of the algebra $\mathscr{B}_u(\operatorname{sp}(A))$ of Borel functions on $\operatorname{sp}(A)$ into $\mathscr{N}(\mathscr{A}_0)$ is a $\sigma$-normal homomorphism mapping the constant function 1 onto $I$ and the identity transformation $\iota$ on $\operatorname{sp}(A)$ onto $A$. The mapping $S \to E(S)$ of Borel subsets $S$ of $\mathbb{C}$ into $\mathscr{A}_0$ is a projection-valued measure on $\mathbb{C}$, where $E(S) = g(A)$ and $g$ is the characteristic function of $S$. If $h$ is in $\mathscr{B}$, the algebra of bounded Borel functions on $\mathbb{C}$, then*

$$\|h(A)\| \leqslant \sup\{|h(a)| : a \in \mathbb{C}\} \ (= \|h\|). \tag{16}$$

*If $x$ is a vector in $\mathscr{H}$ and $\mu_x(S) = \langle E(S)x, x\rangle$, then, for each $h$ in $\mathscr{B}$,*

$$\langle h(A)x, x\rangle = \int_{\mathbb{C}} h(a)\, d\mu_x(a). \tag{17}$$

*With $f$ in $\mathscr{B}_u$, $x \in \mathscr{D}(f(A))$ if and only if*

$$\int_{\mathbb{C}} |f(a)|^2\, d\mu_x(a) \ (= \|f(A)x\|^2) < \infty; \tag{18}$$

*and (17) is valid with $f$ in place of $h$, in this case. If $\varphi$ is the extension to $\mathscr{N}(\mathscr{A}_0)$ of the isomorphism of $\mathscr{A}_0$ with $C(X)$ and $\nu_x$ is the regular Borel measure on $X$ such that $\langle Bx, x\rangle = \int_X (\varphi(B))(p)\, d\nu_x(p)$ for each $B$ in $\mathscr{A}_0$, then $x \in \mathscr{D}(T)$, with $T$ in $\mathscr{N}(\mathscr{A}_0)$, if and only if*

$$\int_X |(\varphi(T))(p)|^2\, d\nu_x(p) \ (= \|Tx\|^2) < \infty. \tag{18$'$}$$

*If $A$ is a self-adjoint operator, its spectral resolution is $\{E_\lambda\}$, where $E_\lambda = I - E((\lambda, \infty))$; and $x \in \mathscr{D}(f(A))$ if and only if $\int_{-\infty}^{\infty} |f(\lambda)|^2\, d\langle E_\lambda x, x\rangle < \infty$. If $x \in \mathscr{D}(f(A))$, $\langle f(A)x, x\rangle = \int_{-\infty}^{\infty} f(\lambda)\, d\langle E_\lambda x, x\rangle$.*

*Proof.* The mapping $g \to \tilde{g}$ of $\mathscr{B}_u(\mathrm{sp}(A))$ into $\mathscr{B}_u(X)$ defined in Remark 5.6.25 is a $\sigma$-normal homomorphism. From Remark 5.6.24, the mapping assigning $h$ (of Remark 5.6.25) in $\mathscr{N}(X)$ to $\tilde{g}$ is a $\sigma$-normal homomorphism, as is $\varphi$. It follows that the mapping $g \to g(A)$ is a $\sigma$-normal homomorphism of $\mathscr{B}_u$ into $\mathscr{N}(\mathscr{A}_0)$. If $g$ is 1, $h$ is the constant function 1 on $X$ and $g(A)$ is $I$. If $g$ is $\iota$, $h$ is $\varphi(A)$; so that $\iota(A) = A$.

With $g$ the characteristic function of $S$, a Borel subset of $\mathbb{C}$, $g(A)^* = \bar{g}(A) = g(A)$ and $g(A) = g^2(A) = g(A)^2$. Thus $g(A)$ is a projection $E(S)$ in $\mathscr{A}_0$. If $g$ is 0, $g(A) = 0$, so that $E(\varnothing) = 0$. If $g$ is 1, $g(A) = I$; and $E(\mathbb{C}) = I$. If $\{S_j\}$ is a countable disjoint family of Borel subsets of $\mathbb{C}$ and $g_j$ is the characteristic function of $S_j$, then $g_1 + \cdots + g_n$ $(= h_n)$ is an increasing sequence tending pointwise to the characteristic function $h$ of $\bigcup_{j=1}^{\infty} S_j$ $(= S)$. Since the mapping $g \to g(A)$ is a $\sigma$-normal homomorphism $\{\sum_{j=1}^{n} E(S_j)\}$ has least upper bound $E(S)$ – that is $\sum_{j=1}^{\infty} E(S_j) = E(S)$. With this notation

$$\langle h(A)x, x\rangle = \langle E(S)x, x\rangle = \mu_x(S) = \int_{\mathbb{C}} h(a)\, d\mu_x(a);$$

so that (17) is valid for (Borel) step functions $h$. With $h$ a bounded Borel function on $\mathbb{C}$, $\|\tilde{h}\| \leqslant \|h\|$ and the function $f$ in $\mathscr{N}(X)$ corresponding to $\tilde{h}$ lies in $C(X)$. (See Lemma 5.2.10.) Moreover, $\|f\| \leqslant \|\tilde{h}\| \leqslant \|h\|$. Thus

$$\|h(A)\| = \|\varphi^{-1}(f)\| = \|f\| \leqslant \|h\|.$$

As each $h$ in $\mathscr{B}$ is a norm limit of (Borel) step functions, (17) now follows for each such $h$.

Suppose $A$ is self-adjoint and $g$ is the characteristic function of $(\lambda, \infty)$. Then $\tilde{g}$ is the characteristic function of $\varphi(A)^{-1}((\lambda, \infty))$, an open subset of $X \backslash Z$ (hence, of $X$) and the function in $C(X)$ corresponding to $\tilde{g}$ is the characteristic function, $1 - e_\lambda$, of $\varphi(A)^{-1}((\lambda, \infty))^-$. Thus $E((\lambda, \infty))$ $(= g(A))$ in $\mathscr{A}_0$ corresponds to $1 - e_\lambda$; and, from Theorem 5.6.12(v), $E_\lambda = I - E((\lambda, \infty))$. It follows that $\langle (E_{\lambda'} - E_\lambda)x, x\rangle = \mu_x((\lambda, \lambda'])$ when $\lambda \leqslant \lambda'$, and

$$\int_{-\infty}^{\infty} |f(\lambda)|^2\, d\langle E_\lambda x, x\rangle = \int |f(\lambda)|^2\, d\mu_x(\lambda)$$

for each $f$ in $\mathscr{B}_u$. Hence the last assertions of our theorem reduce to (17) and (18).

With $f$ in $\mathscr{B}_u$, let $k_n$ be the characteristic function of $|f|^{-1}([0, n])$ $(= S_n)$ and $F_n$ be $k_n(A)$. Then $f_n(A) = f(A)F_n$, where $f_n = f \cdot k_n$, from the first part of this proof. Thus, with $x$ in $\mathscr{H}$,

$$\|f(A)F_n x\|^2 = \langle |f_n|^2(A)x, x\rangle = \int_{S_n} |f(a)|^2\, d\mu_x(a) \tag{19}$$

and, if $m \leqslant n$,

(20) $$\|f(A)F_nx - f(A)F_mx\|^2 = \int_{S_n \setminus S_m} |f(a)|^2 \, d\mu_x(a).$$

As $\{k_n\}$ is increasing and tends pointwise to 1, $\{F_n\}$ is increasing with least upper bound $I$. Since $F_nf(A) \subseteq f(A)F_n$, $F_nf(A)x = f(A)F_nx$ if $x \in \mathscr{D}(f(A))$. Thus $f(A)F_nx \to f(A)x$ and (18) follows from (19). Conversely, if $\int_{\mathbb{C}} |f(a)|^2 \, d\mu_x(a)$ converges, $\{f(A)F_nx\}$ is a Cauchy sequence from (20), and converges to some vector in $\mathscr{H}$. Since $\{F_nx\}$ tends to $x$ and $f(A)$ is closed, $x \in \mathscr{D}(f(A))$. A completely analogous argument establishes (18′). With $x$ in $\mathscr{D}(f(A))$, $f \in L_2(\mathbb{C}, \mu_x) \subseteq L_1(\mathbb{C}, \mu_x)$ since $\mu_x$ is a finite measure. Thus

$$\langle f(A)x, x \rangle = \lim \langle f(A)F_nx, x \rangle = \lim \int_{\mathbb{C}} f_n(a) \, d\mu_x(a) = \int_{\mathbb{C}} f(a) \, d\mu_x(a). \quad \blacksquare$$

The projection $E(S)$ appearing in the statement of Theorem 5.6.26 is often referred to as *the spectral projection for $A$ corresponding to the Borel subset $S$ of $\mathbb{C}$.*

5.6.27. THEOREM. *If $A$ is a normal operator affiliated with an abelian von Neumann algebra $\mathscr{A}$ acting on a Hilbert space $\mathscr{H}$ and $\psi$ is a $\sigma$-normal homomorphism of $\mathscr{B}_u$, the algebra of Borel functions on $\mathbb{C}$, into $\mathscr{N}(\mathscr{A})$ such that $\psi(1) = I$ and $\psi(\iota) = A$, where $\iota$ is the identity transform on $\mathbb{C}$, then $\psi(f) = f_r(A)$ for each $f$ in $\mathscr{B}_u$, where $f_r$ is the restriction of $f$ to* sp$(A)$.

*Proof.* Since $\psi$ is $\sigma$-normal, $\psi$ is adjoint preserving. Positive elements of $\mathscr{B}_u$ have positive square roots, so that $\psi$ is order preserving. As $\psi(1) = I$, it follows that $\psi$ maps the algebra $\mathscr{B}$ of bounded Borel functions in $\mathscr{B}_u$ into $\mathscr{A}$ and does not increase norm.

Suppose $A$ is bounded and $g_0$ is the characteristic function of $\mathbb{C} \setminus \mathbb{D}_0$, where $\mathbb{D}_0$ is the closed disk in $\mathbb{C}$ with center 0 and radius $2\|A\|$. Then $0 \leqslant (2\|A\|)^n g_0 \leqslant |\iota|^n$ for each positive integer $n$. Now $\psi(|\iota|^n) = |A|^n$, so that $0 \leqslant (2\|A\|)^n \psi(g_0) \leqslant |A|^n$. Thus $\|\psi(g_0)\| \leqslant 2^{-n}$ for each positive integer $n$, and $\psi(g_0) = 0$. If $g_1$ is the characteristic function of $\mathbb{D}_0$, then $\psi(g_1) = I$, so that $\psi(g_1 h) = \psi(h)$ for each $h$ in $\mathscr{B}_u$. Let $h_0$ denote the restriction of $h$ to $\mathbb{D}_0$ and let $\psi_0(h_0)$ be $\psi(h)$. Then $\psi_0$ is a $\sigma$-normal homomorphism of $\mathscr{B}_u(\mathbb{D}_0)$ into $\mathscr{N}(\mathscr{A})$ mapping the constant function 1 on $\mathbb{D}_0$ onto $I$, $\iota_0$ onto $A$, and $\mathscr{B}(\mathbb{D}_0)$ into $\mathscr{A}$. Since $C(\mathbb{D}_0)$ is a $C^*$-algebra whose unit is the constant function 1 on $\mathbb{D}_0$, Proposition 4.4.7 applies and $\psi_0(f) = \psi_0(f(\iota_0)) = f(A)$, for each $f$ in $C(\mathbb{D}_0)$. As noted in the proof of Theorem 5.2.8, the characteristic function $h_1$ of the open subset $\mathbb{D}_0 \setminus \text{sp}(A)$ of $\mathbb{D}_0$ is the pointwise limit of an increasing sequence $\{f_n\}$ of positive continuous functions on $\mathbb{D}_0$. Thus $\psi_0(h_1)$ is the least upper bound in $\mathscr{A}$ of $\psi_0(f_n)$, by $\sigma$-normality of $\psi_0$. But $\psi_0(f_n) = f_n(A)$ and $f_n(A) = 0$, since $f_n$ is continuous and vanishes on sp$(A)$. Thus $\psi_0(h_1) = 0$. If $f_1$ is the characteristic function of sp($A$) (as a subset of $\mathbb{D}_0$), $\psi_0(f_1) = I$ and $\psi_0(f_1 h_0) = \psi_0(h_0)$ for each $h_0$ in $\mathscr{B}_u(\mathbb{D}_0)$. If $\psi_1(k_1) = \psi_0(k)$, where $k_1$ is the restriction of $k$ in $\mathscr{B}_u(\mathbb{D}_0)$ to sp($A$), then $\psi_1$ is a $\sigma$-normal homomorphism of $\mathscr{B}_u(\mathrm{sp}(A))$ into $\mathscr{N}(\mathscr{A})$ mapping the constant function 1 on sp($A$) onto $I$, the identity transform $\iota_1$ on sp($A$) onto $A$, and $\mathscr{B}(\mathrm{sp}(A))$ into $\mathscr{A}$. Theorem 5.2.9 applies to the restriction of $\psi_1$ to $\mathscr{B}(\mathrm{sp}(A))$, and $\psi_1(f) = f(A)$ for each $f$ in $\mathscr{B}(\mathrm{sp}(A))$. Each positive $g$ in $\mathscr{B}_u(\mathrm{sp}(A))$ is the pointwise limit of an increasing sequence of positive functions in $\mathscr{B}(\mathrm{sp}(A))$. Since $\psi_1$ and the homomorphism $f \to f(A)$ of $\mathscr{B}_u(\mathrm{sp}(A))$ into $\mathscr{N}(\mathscr{A})$ are $\sigma$-normal, $\psi_1(g) = g(A)$ for each positive $g$ in $\mathscr{B}_u(\mathrm{sp}(A))$. Each $h$ in $\mathscr{B}_u(\mathrm{sp}(A))$ is a linear combination of four (or fewer) positive functions in $\mathscr{B}_u(\mathrm{sp}(A))$, so that $\psi_1(h) = h(A)$ for each $h$ in $\mathscr{B}_u(\mathrm{sp}(A))$. If $k \in \mathscr{B}_u$ and $k_0$ and $k_{\mathrm{r}}$ are its restrictions to $\mathbb{D}_0$ and sp($A$), respectively, then $\psi(k) = \psi_0(k_0) = \psi_1(k_{\mathrm{r}}) = k_{\mathrm{r}}(A)$.

With $A$ now an arbitrary (normal) operator in $\mathscr{N}(\mathscr{A})$ and $E$ a bounding projection for $A$ in $\mathscr{A}$, the mapping $\varphi$ that assigns $(B \hat{\cdot} E)|E(\mathscr{H})$ (in $\mathscr{N}(\mathscr{A}E)$ acting on $E(\mathscr{H})$) to $B$ in $\mathscr{N}(\mathscr{A})$ is a $\sigma$-normal homomorphism of $\mathscr{N}(\mathscr{A})$ into $\mathscr{N}(\mathscr{A}E)$. Composing $\varphi$ with $\psi$ yields a $\sigma$-normal homomorphism $\psi_2$ of $\mathscr{B}_u$ into $\mathscr{N}(\mathscr{A}E)$ mapping 1 onto $E|E(\mathscr{H})$ and $\iota$ onto $A|E(\mathscr{H})$. At the same time, the composition of $\varphi$ with the mapping $f \to f_{\mathrm{r}}(A)$ of $\mathscr{B}_u$ into $\mathscr{N}(\mathscr{A})$ is another such homomorphism of $\mathscr{B}_u$ into $\mathscr{N}(\mathscr{A}E)$. Since $A|E(\mathscr{H})$ is bounded, the first part of this proof applies and

$$(\psi(f) \hat{\cdot} E)|E(\mathscr{H}) = \psi_2(f) = f_{\mathrm{r}}(A|E(\mathscr{H})) = (f_{\mathrm{r}}(A) \hat{\cdot} E)|E(\mathscr{H}).$$

From Theorem 5.6.15(i), there is a common bounding sequence $\{E_n\}$ for $A$, $\psi(f)$, and $f_{\mathrm{r}}(A)$, in $\mathscr{N}(\mathscr{A})$, where $f$ is a given element of $\mathscr{B}_u$. As

$$\begin{aligned}(\psi(f) \hat{\cdot} E_n)|E_n(\mathscr{H}) &= (\psi(f)E_n)|E_n(\mathscr{H}) = (f_{\mathrm{r}}(A) \hat{\cdot} E_n)|E_n(\mathscr{H}) \\ &= (f_{\mathrm{r}}(A)E_n)|E_n(\mathscr{H}),\end{aligned}$$

$\psi(f)E_n = f_{\mathrm{r}}(A)E_n$ for each $n$. Since $\cup_{n=1}^{\infty} E_n(\mathscr{H})$ is a core for both $\psi(f)$ and $f_{\mathrm{r}}(A)$, $\psi(f) = f_{\mathrm{r}}(A)$. ■

5.6.28. Remark. The process described in Remark 5.6.25 for forming $g(A)$ in $\mathscr{A}_0$ could be applied, equally well, to $\mathscr{N}(Y)$, where $\mathscr{A}$ is another abelian von Neumann algebra with which $A$ is affiliated and $\mathscr{A} \cong C(Y)$. Theorem 5.6.27 assures us that the operator in $\mathscr{N}(\mathscr{A})$ formed in this way is $g(A)$ (and lies in $\mathscr{N}(\mathscr{A}_0)$). ■

5.6.29. Corollary. *If $A$ is a normal operator and $f$ and $g$ are in $\mathscr{B}_u$, then*

$$(f \circ g)(A) = f(g(A)). \tag{21}$$

*Proof.* If $\mathscr{A}_0$ is the abelian von Neumann algebra generated by $A$, $A^*$, and $I$, then $g(A) \in \mathscr{N}(\mathscr{A}_0)$ and $f \to f \circ g$ is a $\sigma$-normal homomorphism $\varphi$ of $\mathscr{B}_u$ into

$\mathscr{B}_u$. Composing $\varphi$ with the $\sigma$-normal homomorphism $h \to h(A)$ of $\mathscr{B}_u$ into $\mathscr{N}(\mathscr{A}_0)$ yields a $\sigma$-normal homomorphism, $f \to (f \circ g)(A)$, of $\mathscr{B}_u$ into $\mathscr{N}(\mathscr{A}_0)$ that maps 1 onto $I$ and $\iota$ onto $g(A)$. From Theorem 5.6.27, (21) follows. ∎

5.6.30. PROPOSITION. *If $\psi$ is a $\sigma$-normal homomorphism of $\mathscr{N}(\mathscr{A}_1)$ into $\mathscr{N}(\mathscr{A}_2)$ such that $\psi(I) = I$, where $\mathscr{A}_1$ and $\mathscr{A}_2$ are abelian von Neumann algebras, then $\psi(f(A)) = f(\psi(A))$ for each $A$ in $\mathscr{N}(\mathscr{A}_1)$ and each $f$ in $\mathscr{B}_u$.*

*Proof.* The mapping $f \to \psi(f(A))$ of $\mathscr{B}_u$ into $\mathscr{N}(\mathscr{A}_2)$ is a $\sigma$-normal homomorphism mapping 1 onto $I$ and $\iota$ onto $\psi(A)$. From Theorem 5.6.27, $\psi(f(A)) = f(\psi(A))$ for each $f$ in $\mathscr{B}_u$. ∎

5.6.31. COROLLARY. *If $\mathscr{A}$ is an abelian von Neumann algebra acting on the Hilbert space $\mathscr{H}$, $E$ is a projection in $\mathscr{A}$, $A \,\eta\, \mathscr{A}$, and $f \in \mathscr{B}_u$, then*

$$f((A \mathbin{\hat{\cdot}} E)|E(\mathscr{H})) = (f(A) \mathbin{\hat{\cdot}} E)|E(\mathscr{H}).$$

*Proof.* The mapping $B \to (B \mathbin{\hat{\cdot}} E)|E(\mathscr{H})$ is a $\sigma$-normal homomorphism $\psi$ of $\mathscr{N}(\mathscr{A})$ onto $\mathscr{N}(\mathscr{A}E|E(\mathscr{H}))$ such that $\psi(I) = E|E(\mathscr{H})$ (and $E|E(\mathscr{H})$ is the identity operator on $E(\mathscr{H})$). Thus, from Proposition 5.6.30,

$$f((A \mathbin{\hat{\cdot}} E)|E(\mathscr{H})) = f(\psi(A)) = \psi(f(A)) = (f(A) \mathbin{\hat{\cdot}} E)|E(\mathscr{H}). \quad ∎$$

5.6.32. REMARK. If $A$ and $B$ are (unbounded) normal operators whose spectra are contained in the domain of a Borel function $g$ and $g$ has a Borel inverse function $f$, then $g(A) = g(B)$ if and only if $A = B$; for if $g(A) = g(B)$, then, from Corollary 5.6.29,

$$A = (f \circ g)(A) = f(g(A)) = f(g(B)) = (f \circ g)(B) = B.$$

As an application of this comment, we note that if $A^2 = B^2$ with $A$ and $B$ positive operators, then $A = B$, so that a positive operator has a unique positive square root.

We illustrate the use of Corollary 5.6.31 with the observation that $f(A)x_0 = f(\lambda)x_0$ when $Ax_0 = \lambda x_0$, where $A$ is a normal operator, $f$ is a Borel function whose domain contains $\text{sp}(A)$, and $x_0$ is a non-zero vector. To see this, let $E$ be the projection with range $\{x: Ax = \lambda x\}$ (which is closed since $A$ is a closed operator); and note that, since $EA \subseteq AE = \lambda E$,

$$\begin{aligned} f(A)x_0 &= [(f(A) \mathbin{\hat{\cdot}} E)|E(\mathscr{H})]x_0 = f((A \mathbin{\hat{\cdot}} E)|E(\mathscr{H}))x_0 \\ &= f(\lambda E|E(\mathscr{H}))x_0 = f(\lambda)x_0. \quad ∎ \end{aligned}$$

As we noted in Theorem 5.6.15, $\mathscr{N}(\mathscr{A})$ becomes a * algebra when the usual operations of operator addition and multiplication are "refined" by passing to closures. In certain instances, the combination of operators in question is closed – for example, $A^*A$ is self-adjoint and, hence, closed for each closed densely defined $A$. In Theorem 5.6.19 we see that $\mathscr{N}(X)$ is an algebra when the usual operations of function addition and multiplication are "refined" by passing to a (unique) normal extension. In certain instances, the combination of functions in question is normal – for example, $f^n$ is normal if $f$ is. Theorem 5.6.19 tells us, as well, that $\mathcal{N}(\mathcal{A})$ and $\mathcal{N}(X)$ are isomorphic when endowed with these operations by an isomorphism that extends the isomorphism of $\mathcal{A}$ with $C(X)$. With all this in view, the temptation is great to assume that taking the closure of an algebraic combination of operators affiliated with $\mathcal{A}$ is unnecessary exactly when the same combination of the corresponding normal functions is normal (without requiring a proper extension). This is not so, as we shall see in the following example. At the same time, this example provides a nice illustration of the isomorphism of $\mathcal{N}(\mathcal{A})$ onto $\mathcal{N}(X)$ and the way we work with it.

5.6.33. EXAMPLE. Let $\mathcal{H}$ be a separable Hilbert space, $\{e_n\}_{n=1,2,\ldots}$ an orthonormal basis for $\mathcal{H}$, and $\mathcal{A}$ the algebra of operators in $\mathcal{B}(\mathcal{H})$ having each $e_n$ as an eigenvector (that is, $\mathcal{A}$ is the algebra of bounded diagonal matrices relative to $\{e_n\}$). In this case, $\mathcal{A} \cong C(X)$ and $X$ is $\beta(\mathbb{N})$, the $\beta$-compactification of the numbers $\{1, 2, \ldots\}$. The points $p_n$ corresponding to the pure states $T \to \langle Te_n, e_n\rangle$ of $\mathcal{A}$, $n = 1, 2, \ldots$ form a dense subset of $X$, for if $0 = \langle Te_n, e_n\rangle$, with $T$ in $\mathcal{A}$, then $T = 0$. Thus each function in $C(X)$ vanishing on $\{p_n\}_{n\in\mathbb{N}}$ is 0, from which the density of $\{p_n\}$ follows. A function 1 at $p_1$ and 0 at $p_n$, $n = 2, 3, \ldots$, is therefore 0 at all points of $X$ other than $p_1$ if it lies in $C(X)$. The projection whose range is generated by $e_1$ lies in $\mathcal{A}$ and corresponds to such a function in $C(X)$. It follows that $\{p_1\}$ is an open subset of $X$ as is each one-point set formed from a $p_n$. Thus $\{p_n\}_{n=1,2,\ldots}$ is an open dense subset of $X$ and its complement $Z$ is a closed nowhere-dense subset of $X$.

The function $h$ defined as $b_n$ at $p_n$, where $|b_n| \to \infty$, is normal (and defined on $X\backslash Z$). Letting $b_n$ be $n$, we have a normal function $f$ corresponding to an operator $A$ affiliated with $\mathcal{A}$ (and $Ae_n = ne_n$). Letting $b_n$ be $n^{1/4} - n$, we have a normal function $g$ corresponding to an operator $B$ affiliated with $\mathcal{A}$ (and $Be_n = (n^{1/4} - n)e_n$). If $x$ is $\sum_{n=1}^{\infty} n^{-1}e_n$, then $x \in \mathcal{H}$ and $\nu_x(\{p_n\}) = n^{-2}$. Thus $\int_X |f(p)|^2\, d\nu_x(p) = \infty$ and $\int_X |(f+g)(p)|^2\, d\nu_x(p) = \sum_{n=1}^{\infty} n^{-3/2} < \infty$. From Theorem 5.6.26, $x \notin \mathcal{D}(A)$ and $x \in \mathcal{D}(A \hat{+} B)$. Note for this that $f + g$ is normal as defined on $X\backslash Z$ since $|(f+g)(p_n)| = n^{1/4} \to \infty$, so that $f + g$ $(= f \hat{+} g)$ corresponds to $A \hat{+} B$. It follows that $A + B \neq A \hat{+} B$.

This same structure provides us with an example of self-adjoint operators $A$ and $C$ affiliated with $\mathcal{A}$ such that $hf$ is normal but $CA$ is not closed. With $A, f$, and $x$ as before, choose $h$ so that $h(p_n) = n^{-3/4}$. Then $C$, corresponding to $h$, is bounded and $hf$ is normal as defined on $X\backslash Z$. Thus $hf$ corresponds to $C \hat{\cdot} A$. Now

$$\int_X |(hf)(p)|^2\, d\nu_x(p) = \sum_{n=1}^{\infty} n^{-3/2} < \infty,$$

so that $x \in \mathscr{D}(C \hat{\cdot} A)$. But $x \notin \mathscr{D}(CA)$ since $x \notin \mathscr{D}(A)$. Thus $CA \neq C \hat{\cdot} A$. At the same time, this provides an example of a bounded operator $C$ and a closed operator $A$ such that $CA$ is not closed. (We noted in the first paragraph of the proof of Theorem 5.6.15 that this product, in the reverse order, is automatically closed.) ■

5.6.34. REMARK. If $A$ and $B$ are positive operators affiliated with the abelian von Neumann algebra $\mathscr{A}$ acting on $\mathscr{H}$ and $\mathscr{A} \cong C(X)$, then $A$ and $B$ correspond to positive normal functions $f$ and $g$ defined on $X \backslash Z$ and $X \backslash Z'$, respectively. Hence $f + g$, defined on $X \backslash (Z \cup Z')$, is normal and corresponds to $A \hat{+} B$. In this case, with $x$ in $\mathscr{D}(A \hat{+} B)$,

$$0 \leqslant \int_X |f(p)|^2 \, dv_x(p) \leqslant \int_X |f(p) + g(p)|^2 \, dv_x(p) < \infty.$$

Thus $x \in \mathscr{D}(A)$ and, similarly, $x \in \mathscr{D}(B)$. It follows that $x \in \mathscr{D}(A + B)$ and that $A + B = A \hat{+} B$.

Again, with this same notation, but no longer assuming that $A$ and $B$ are positive, if $Z \cap Z' = \varnothing$, then there are disjoint open sets $\mathscr{O}_1$ and $\mathscr{O}_2$ containing $Z$ and $Z'$, respectively. Thus $\mathscr{O}_1^- \subseteq X \backslash \mathscr{O}_2$ and there is a clopen set $Y\ (= \mathscr{O}_1^-)$ containing $Z$ such that $X \backslash Y$ contains $Z'$. It follows that $g$ is bounded on $Y$ and $f$ is bounded on $X \backslash Y$. If $E$ is the projection in $\mathscr{A}$ corresponding to the characteristic function of $Y$, then $BE$ and $A(I - E)$ are in $\mathscr{B}(\mathscr{H})$. As $A$ is closed and $E$ is bounded, $AE$ is closed and $AE = A \hat{\cdot} E$. Thus

$$(A \hat{+} B)E = AE \hat{+} BE = AE + BE = (A + B)E,$$

and, similarly, $(A \hat{+} B)(I - E) = (A + B)(I - E)$. If $x \in \mathscr{D}(A \hat{+} B)$, then $Ex$ and $(I - E)x$ are in $\mathscr{D}(A \hat{+} B)$. Thus $Ex$ and $(I - E)x$ are in $\mathscr{D}(A + B)$, and $x \in \mathscr{D}(A + B)$. Hence $A + B = A \hat{+} B$. ■

Polynomials in a single variable provide an important case in which we need not pass to a closure. Thus "$p(A)$" refers to the same operator whether viewed in the customary sense or as a Borel function of $A$, for a polynomial $p$. For the purpose of the statement of the following proposition, "$p(A)$" refers to the operator obtained by forming the Borel function $p$ of $A$.

5.6.35. PROPOSITION. *If $A \,\eta\, \mathscr{A}$, where $\mathscr{A}$ is an abelian von Neumann algebra acting on the Hilbert space $\mathscr{H}$, and $p(z) = a_n z^n + \cdots + a_1 z + a_0$, with $a_n$ not 0, then $a_n A^n + \cdots + a_1 A + a_0 I$ is closed and equal to $p(A)$.*

*Proof.* Suppose $\mathscr{A} \cong C(X)$ and $A$ corresponds to the normal function $f$ defined on $X \backslash Z$. Then $a_n f^n + \cdots + a_1 f + a_0$ is defined on $X \backslash Z$ and normal. Hence it corresponds to $p(A)$. Now $x \in \mathscr{D}(p(A))$ if and only if $\int_X |a_n f^n(p) + \cdots + a_1 f(p) + a_0|^2 \, dv_x(p) < \infty$. Let $X_k$ be $\{p : |f(p)| < k\}^-$.

For $k$ large enough and $p$ in $X\backslash(Z \cup X_k)$,

$$\tfrac{1}{2}|a_n f^n(p)| \leqslant |a_n f^n(p) + \cdots + a_1 f(p) + a_0|,$$

so that

$$\int_{X\backslash(Z\cup X_k)} |f(p)|^{2n}\, dv_x(p) < \infty.$$

Since $f$ is bounded on $X_k$ and $v_x$ is a finite measure on $X$, $f \in L_{2n}(X, v_x)$. Hence $f \in L_k(X, v_x)$ for $1 \leqslant k \leqslant 2n$. As $f^n$ is defined on $X\backslash Z$ and is normal, it represents $\overline{A^n}$. From the foregoing and Theorem 5.6.26 (especially (18′)), $x \in \mathscr{D}(\overline{A^k})$ for $k = 1, \ldots, n$. In particular $x \in \mathscr{D}(\bar{A}) = \mathscr{D}(A)$. We show, by induction, that $\overline{A^k} = A^k$ for all $k$. Assume this for $k = 1, \ldots, n-1$. If $y \in \mathscr{D}(\overline{A^n})$, then, from what we have established to this point, $y \in \mathscr{D}(A)$. Let $\{E_m\}$ be a bounding sequence in $\mathscr{A}$ for $\overline{A^{n-1}}$ and $A$. Then $\overline{A^n} E_m = (\overline{A^{n-1}} \mathbin{\hat{\cdot}} A)E_m = \overline{A^{n-1}} E_m A E_m$. Hence $E_m \overline{A^n} y = \overline{A^n} E_m y = \overline{A^{n-1}} E_m A E_m y = \overline{A^{n-1}} E_m A y$. Now $E_m \overline{A^n} y \to \overline{A^n} y$ and $E_m A y \to Ay$. Since $\overline{A^{n-1}}$ is closed, $Ay \in \mathscr{D}(\overline{A^{n-1}})$ and $\overline{A^{n-1}} Ay = \overline{A^n} y$. Hence $\overline{A^n} \subseteq \overline{A^{n-1}} A$. But $\overline{A^{n-1}} = A^{n-1}$ by inductive assumption. Thus $\overline{A^n} \subseteq A^n$ and $\overline{A^n} = A^n$, completing the induction.

It follows now that $x \in \mathscr{D}(A^k)$ for $k = 1, \ldots, n$, and

$$x \in \mathscr{D}(a_n A^n + \cdots + a_1 A + a_0 I),$$

so that

$$p(A) = a_n A^n + \cdots + a_1 A + a_0 I. \quad \blacksquare$$

We apply (unbounded-)spectral-resolution considerations to an analysis of the unitary representations of $\mathbb{R}$ – the one-parameter unitary groups. We use the notation and definitions in Theorem 4.5.9 and the discussion preceding it.

5.6.36. Theorem (Stone's theorem). *If $H$ is a (possibly unbounded) self-adjoint operator on the Hilbert space $\mathscr{H}$, then $t \to \exp itH$ is a one-parameter unitary group on $\mathscr{H}$. Conversely, if $t \to U_t$ is a one-parameter unitary group on $\mathscr{H}$, there is a (possibly unbounded) self-adjoint operator $H$ on $\mathscr{H}$ such that $U_t = \exp itH$ for each real $t$. The domain of $H$ consists of precisely those vectors $x$ in $\mathscr{H}$ for which $t^{-1}(U_t x - x)$ tends to a limit as $t$ tends to 0, in which case this limit is $iHx$.*

*If $f \in L_1(\mathbb{R})$ and $x$, $y$ are in $\mathscr{H}$, then*

$$\langle \hat{f}(H)x, y\rangle = \int_{\mathbb{R}} f(t)\langle e^{itH}x, y\rangle\, dt.$$

*Proof.* From Theorem 5.6.26 and the properties of the function $\lambda \to \exp it\lambda$ on $\mathbb{R}$, $\exp itH (= U_t)$ is a unitary operator and $U_{t+t'} = U_t U_{t'}$. To see that $t \to U_t$ is strong-operator continuous, it will suffice to show that $\|U_t x - x\| \to 0$ as $t \to 0$ for each vector $x$ in $F_n(\mathscr{H})$, where $F_n = E_n - E_{-n}$ and $\{E_\lambda\}$ is the spectral resolution of $H$. Now $HF_n (= H_n)$ is bounded; and, with $x$ in $F_n(\mathscr{H})$, $(\exp itH)x = (\exp itH_n)x$, from Corollary 5.6.31. Examining the function representation of $\exp itH_n$, we have

$$\|e^{itH_n} - I\| \leqslant |t| \cdot \|H_n\|, \tag{22}$$

so that $\|U_t x - x\| \to 0$ as $t \to 0$ for each $x$ in $F_n(\mathscr{H})$. We see, from (22) as well that $\|U_t - I\| \to 0$ if $H$ is bounded.

Suppose now that $t \to U_t$ is a one-parameter unitary group. The argument we shall give to show that $U_t = \exp itH$ for some self-adjoint operator $H$ and all real $t$ constructs the (unbounded) spectral resolution of $H$. We begin by "extending" the representation $t \to U_t$ of $\mathbb{R}$ to a representation $\varphi$ of $\mathfrak{A}_0(\mathbb{R})$ (as described in Theorem 4.5.9). In the proof of Theorem 3.2.27 (especially, the last paragraph of that proof), we construct a homeomorphism $\Lambda'$ of the pure state space of $\mathfrak{A}_0(\mathbb{R})$ (denoted, there, by $\mathscr{M}(\mathbb{R})$), in its weak* topology, onto the one-point compactification, $\{\hat{\mathbb{R}}, \infty\}$, of $\hat{\mathbb{R}}$. Let $\hat{\varphi}$ be the representation of $C(\{\hat{\mathbb{R}}, \infty\})$ on $\mathscr{H}$ obtained by composing, successively, the isomorphism of $C(\{\hat{\mathbb{R}}, \infty\})$ with $C(\mathscr{M}(\mathbb{R}))$, the isomorphism of $C(\mathscr{M}(\mathbb{R}))$ with $\mathfrak{A}_0(\mathbb{R})$, and $\varphi$. From Theorem 4.5.9, $\varphi$ restricts to an essential representation of $\mathscr{A}_1(\mathbb{R})$, so that $\hat{\varphi}$ is $\mathbb{R}$-essential and gives rise, as in Corollary 5.2.7, to a (possibly unbounded) resolution of the identity $\{E_\lambda\}$. With $f$ in $L_1(\mathbb{R})$, the Fourier transform $\hat{f}$ of $f$ is the image of $L_f$ in $C(\{\hat{\mathbb{R}}, \infty\})$ under the isomorphism of $\mathfrak{A}_0(\mathbb{R})$ onto $C(\{\hat{\mathbb{R}}, \infty\})$. Employing, in succession, the fact that $\varphi$ extends $t \to U_t$, the definition of $\hat{\varphi}$ and the identification of $\hat{f}$ just noted, the choice of $\{E_\lambda\}$ and Corollary 5.2.7, 3.2(3) of Theorem 3.2.26, an interchange in the order of integration, and Theorems 5.6.12, 5.6.26, we have, for some self-adjoint operator $H$ on $\mathscr{H}$,

$$\begin{aligned} \langle \varphi(L_f)x, x\rangle &= \int_{\mathbb{R}} f(t)\langle U_t x, x\rangle\, dt = \langle \hat{\varphi}(\hat{f})x, x\rangle \\ &= \int_{\hat{\mathbb{R}}} \hat{f}(\lambda)\, d\langle E_\lambda x, x\rangle = \int_{\hat{\mathbb{R}}} \left(\int_{\mathbb{R}} f(t) e^{it\lambda}\, dt\right) d\langle E_\lambda x, x\rangle \\ &= \int_{\mathbb{R}} f(t)\left(\int_{\hat{\mathbb{R}}} e^{it\lambda}\, d\langle E_\lambda x, x\rangle\right) dt = \int_{\mathbb{R}} f(t)\langle e^{itH}x, x\rangle\, dt \end{aligned} \tag{23}$$

for each $x$ in $(E_n - E_{-n})(\mathscr{H})$. The interchange in the order of integration is justified by noting that $f \in L_1(\mathbb{R})$, the Borel measure on $\mathbb{R}$ corresponding to $d\langle E_\lambda x, x\rangle$ has support in $[-n, n]$ for the given $x$, and $\lambda \to \exp it\lambda$ is bounded and continuous on $\mathbb{R}$. It follows from (23) that

$$\int_{\mathbb{R}} f(t)(\langle U_t x, x\rangle - \langle e^{itH}x, x\rangle)\, dt = 0,$$

for each $f$ in $L_1(\mathbb{R})$, whence the (continuous) functions $t \to \langle U_t x, x\rangle$ and $t \to \langle(\exp itH)x, x\rangle$ coincide when $x \in (E_n - E_{-n})(\mathscr{H})$. Since

$$\bigcup_{n=1}^{\infty} (E_n - E_{-n})(\mathscr{H})$$

is a dense submanifold of $\mathscr{H}$, $U_t = \exp itH$ for all real $t$. As

$$\int_{\mathbb{R}} \hat{f}(\lambda)\, d\langle E_\lambda x, x\rangle = \langle \hat{f}(H)x, x\rangle,$$

we may read from (23) that, for $f$ in $L_1(\mathbb{R})$,

$$\langle \hat{f}(H)x, y\rangle = \int_{\mathbb{R}} f(t)\langle e^{itH}x, y\rangle\, dt.$$

If $F_n = E_n - E_{-n}$, $H_n = H|F_n(\mathscr{H})$, and $x \in F_n(\mathscr{H})$, then

$$t^{-1}[U_t x - x] = t^{-1}[e^{itH_n} - F_n]x \to iH_n x \qquad (t \to 0); \tag{24}$$

for, employing the function representation of the (commutative) $C^*$-algebra on $F_n(\mathscr{H})$ generated by $H_n$ and $F_n$, we have that $t^{-1}[\exp itH_n - F_n]$ converges in norm to $iH_n$ in $\mathscr{B}(F_n(\mathscr{H}))$. In fact, for small $t$,

$$\|t^{-1}[e^{itH_n} - F_n] - iH_n\| \leqslant |t| \left[ \frac{\|H_n\|^2}{2!} + \frac{\|H_n\|^3}{3!} + \cdots \right] \leqslant |t| e^n.$$

With $x$ arbitrary, if $t^{-1}[U_t x - x]$ tends to $y$ in $\mathscr{H}$ as $t$ tends to 0, then

$$F_n y = \lim_{t \to 0} F_n(t^{-1}[U_t x - x]) = \lim_{t \to 0} t^{-1}[U_t F_n x - F_n x] = iHF_n x.$$

As $F_n x \to x$, $F_n y (= iHF_n x) \to y$, and $H$ is closed; it follows that $x \in \mathscr{D}(H)$ and $y = iHx$.

If $x \in \mathscr{D}(H)$, then, from Theorem 5.6.26,

$$\|Hx\|^2 = \int_{-\infty}^{\infty} \lambda^2\, d\langle E_\lambda x, x\rangle$$

and

$$\begin{aligned} \|t^{-1}[U_t x - x] - iHx\|^2 &= \int_{-\infty}^{\infty} |t^{-1}[e^{it\lambda} - 1] - i\lambda|^2\, d\langle E_\lambda x, x\rangle \\ &= \int_{-\infty}^{\infty} |(t\lambda)^{-1}(1 + it\lambda - e^{it\lambda})|^2 \lambda^2\, d\langle E_\lambda x, x\rangle. \end{aligned}$$

When $t$ and $\lambda$ are different from 0,

$$|(t\lambda)^{-1}(1 + it\lambda - e^{it\lambda})|^2 = \left(\frac{1 - \cos t\lambda}{t\lambda}\right)^2 + \left(\frac{t\lambda - \sin t\lambda}{t\lambda}\right)^2 \leqslant 2.$$

Thus, given a positive $\varepsilon$, for sufficiently large $n$ and all non-zero $t$,

$$\int_{\mathbb{R}\backslash[-n,n]} |t^{-1}[e^{it\lambda} - 1] - i\lambda|^2 \, d\langle E_\lambda x, x\rangle \leqslant 2\int_{\mathbb{R}\backslash[-n,n]} \lambda^2 \, d\langle E_\lambda x, x\rangle \leqslant \frac{\varepsilon}{2}.$$

On the other hand, from (24),

$$\|(t^{-1}[U_t - I] - iH)F_n x\|^2 = \int_{-n}^{n} |t^{-1}[e^{it\lambda} - 1] - i\lambda|^2 \, d\langle E_\lambda x, x\rangle \leqslant \frac{\varepsilon}{2}$$

for all small, non-zero $t$. Hence

$$\|t^{-1}[U_t x - x] - iHx\| \to 0 \qquad (t \to 0),$$

for each $x$ in $\mathscr{D}(H)$. ∎

5.6.37. REMARK. Formal differentiation of $U_t x$ at 0 would lead us to expect that $t^{-1}[U_t x - x]$ tends to $iHx$ for $x$ in $\mathscr{D}(H)$, when we keep in mind the eventual expression of $U_t$ as $\exp itH$. This process of differentiation can be used as an alternative starting point for the construction of $H$ from $U_t$.

The proof we give of Stone's theorem actually starts in Subsection 3.2, *The Banach algebra $L_1(\mathbb{R})$ and Fourier analysis.* It is formulated in such a way that little needs to be added in order to arrive at a "spectral decomposition" of a unitary representation of a general locally compact abelian group. The representation is extended, again, to a representation of an abelian $C^*$-algebra associated with the group. Theorem 5.2.6 applies and yields a projection-valued measure on the pure state space of that algebra (the one-point compactification of the dual group). The entire process illustrates the way in which a representation of a given system can be studied through the corresponding representation of an associated $C^*$-algebra. At the same time, it underscores the strong interrelations among Fourier analysis, representations of abelian groups, and representations of abelian $C^*$-algebras. Broadened to the general (non-abelian) case, this area of study is often referred to as "non-commutative harmonic analysis." ∎

*Bibliography*: [1, 3, 10, 12, 13, 14, 15, 18, 20, 21]

## 5.7. Exercises

5.7.1. Show that the mapping $A \to A^*$ of $\mathscr{B}(\mathscr{H})$ into $\mathscr{B}(\mathscr{H})$ is weak-operator continuous.

5.7.2. Show that the weak-operator topology on $\mathscr{B}(\mathscr{H})$ is *strictly* weaker (coarser) than the strong-operator topology on $\mathscr{B}(\mathscr{H})$.

5.7.3. Let $\mathscr{P}$ denote the set of all projections from a Hilbert space $\mathscr{H}$ onto its closed subspaces, and suppose that $F \in \mathscr{P}$ and $0 \neq F \neq I$. Prove that the mappings

$$E \to E \wedge F, \qquad E \to E \vee F \qquad (\mathscr{P} \to \mathscr{P})$$

are not continuous from $\mathscr{P}$ with the norm topology to $\mathscr{P}$ with the weak-operator topology. (Compare Exercise 2.8.17.)

5.7.4. Let $\mathscr{H}$ be a Hilbert space and $\mathscr{T}_s, \mathscr{T}_w$ be the restrictions of the strong- and weak-operator topologies to the set $\mathscr{P}$ of projections in $\mathscr{B}(\mathscr{H})$. Show that $\mathscr{T}_s = \mathscr{T}_w$. Conclude that if a sequence of projections tends to a *projection* in the weak-operator topology, it tends to that projection in the strong-operator topology (but compare Exercise 5.7.8(ii)).

5.7.5. Let $\mathscr{H}$ be a Hilbert space and $\mathscr{T}'_s, \mathscr{T}'_w$ be the restrictions of the strong- and weak-operator topologies to the set $\mathscr{U}$ of unitary operators in $\mathscr{B}(\mathscr{H})$. Show that $\mathscr{T}'_s = \mathscr{T}'_w$. Conclude that if a sequence of unitary operators tends to a *unitary operator* in the weak-operator topology, it tends to that unitary operator in the strong-operator topology.

5.7.6. Let $\mathscr{H}$ be a Hilbert space, $\{y_a\}_{a\in\mathbb{A}}$ be an orthonormal basis for $\mathscr{H}$, and $\mathscr{S}$ be a bounded subset of $\mathscr{B}(\mathscr{H})$. With $\mathbb{F}$ a finite subset of $\mathbb{A}$, $\varepsilon$ a positive number, and $S_0$ in $\mathscr{S}$,

$$\{S : |\langle (S - S_0)y_a, y_{a'}\rangle| < \varepsilon,\ a, a' \in \mathbb{F},\ S \in \mathscr{S}\}\ (= \mathscr{V}_{\mathbb{F},\varepsilon})$$

is a weak-operator open neighborhood of $S_0$. Show that $\{\mathscr{V}_{\mathbb{F},\varepsilon}\}$ is a base for the weak-operator open neighborhoods of $S_0$ in $\mathscr{S}$.

5.7.7. Let $\mathscr{H}$ be a separable Hilbert space and $\{y_1, y_2, \ldots\}$ be an orthonormal basis for $\mathscr{H}$. Show that the equation

$$d(S, T) = \sum_{n,m=1}^{\infty} 2^{-(n+m)}|\langle (S - T)y_n, y_m\rangle|$$

defines a translation-invariant metric $d$ on $\mathscr{B}(\mathscr{H})$ (that is, $d(S + R, T + R) = d(S, T)$ for each $R$ in $\mathscr{B}(\mathscr{H})$), and the associated metric topology coincides on bounded subsets of $\mathscr{B}(\mathscr{H})$ with the weak-operator topology.

5.7.8. With the notation of Exercise 5.7.4, assume that $\mathscr{H}$ is infinite dimensional.

(i) Show that $\mathscr{P}$ is weak-operator dense in $(\mathscr{B}(\mathscr{H}))_1^+$, the set of positive operators in the unit ball of $\mathscr{B}(\mathscr{H})$. [*Hint.* Note that

$$\begin{bmatrix} A & (A - A^2)^{1/2} \\ (A - A^2)^{1/2} & I - A \end{bmatrix}$$

is a projection in $\mathscr{B}(\mathscr{K} \oplus \mathscr{K})$ when $A \in (\mathscr{B}(\mathscr{K}))_1^+$.]

(ii) Show that $\mathscr{P}$ is strong-operator closed in $\mathscr{B}(\mathscr{H})$, and conclude that there is a sequence of projections tending to an operator in $(\mathscr{B}(\mathscr{H}))_1^+$ in the weak-operator topology, when $\mathscr{H}$ is separable, that does not tend to it in the strong-operator topology. (Compare Exercises 5.7.2 and 5.7.4 and Theorem 5.1.2.)

5.7.9. Let $\mathscr{H}$ be a Hilbert space.

(i) Show that the mapping $A \to AB$ of $\mathscr{B}(\mathscr{H})$ into $\mathscr{B}(\mathscr{H})$ is weak-operator continuous for each $B$ in $\mathscr{B}(\mathscr{H})$.

(ii) Show that the mapping $A \to BA$ of $\mathscr{B}(\mathscr{H})$ into $\mathscr{B}(\mathscr{H})$ is weak-operator continuous for each $B$ in $\mathscr{B}(\mathscr{H})$.

(iii) When $\mathscr{H}$ is infinite dimensional, show that the mapping $(A, B) \to AB$ of $(\mathscr{B}(\mathscr{H}))_1^+ \times (\mathscr{B}(\mathscr{H}))_1^+$ into $\mathscr{B}(\mathscr{H})$ is not weak-operator continuous.

5.7.10. Let $\mathscr{H}$ be a Hilbert space and $(\mathscr{B}(\mathscr{H}))_r$ be $\{T \in \mathscr{B}(\mathscr{H}) : \|T\| \leqslant r\}$.

(i) Show that the mapping

$$(A, B) \to AB: \quad (\mathscr{B}(\mathscr{H}))_r \times \mathscr{B}(\mathscr{H}) \to \mathscr{B}(\mathscr{H})$$

is continuous when $(\mathscr{B}(\mathscr{H}))_r \times \mathscr{B}(\mathscr{H})$ is provided with the product of the weak-operator topology on $(\mathscr{B}(\mathscr{H}))_r$ and the strong-operator topology on $\mathscr{B}(\mathscr{H})$, and the range $\mathscr{B}(\mathscr{H})$ is provided with the weak-operator topology. (Compare Exercise 2.8.33.)

(ii) Let $\mathscr{H}$ be infinite dimensional and separable. Show that the mapping

$$(A, B) \to AB: \quad (\mathscr{B}(\mathscr{H}))_1 \times (\mathscr{B}(\mathscr{H}))_1 \to \mathscr{B}(\mathscr{H})$$

is *not* continuous when the first factor of $(\mathscr{B}(\mathscr{H}))_1 \times (\mathscr{B}(\mathscr{H}))_1$ is provided with the strong-operator topology and the second is provided with the weak-operator topology.

5.7.11. Let $\mathscr{H}$ be a Hilbert space and $\mathscr{R}$ be a von Neumann algebra on $\mathscr{H}$.

(i) Show that $(\mathscr{R})_1$ is weak-operator compact.

(ii) Show that $(\mathscr{R})_1^+$ is weak-operator compact.

(iii) With $\mathscr{H}$ infinite dimensional, show that $(\mathscr{B}(\mathscr{H}))_1$ and $(\mathscr{B}(\mathscr{H}))_1^+$ are not strong-operator compact.

5.7.12. With the notation of Exercise 5.7.5, assume that $\mathscr{H}$ is infinite dimensional.

(i) Show that $\mathscr{U}$ is weak-operator dense in $(\mathscr{B}(\mathscr{H}))_1$. [*Hint*. Follow the pattern of the solution to Exercise 5.7.8(i) and use the facts that each operator on a finite-dimensional Hilbert space has the form $VH$, where $V$ is unitary and $H$ is positive, and that such an operator $R_0$ is unitary if *either* of $R_0^*R_0$ or $R_0R_0^*$ is $I$.]

(ii) Show that the set of isometries in $\mathscr{B}(\mathscr{H})$ is strong-operator closed.

(iii) Conclude that, with $\mathscr{H}$ separable, there is a sequence of unitary operators in $\mathscr{B}(\mathscr{H})$ that tends to an operator in $(\mathscr{B}(\mathscr{H}))_1$ in the weak-operator topology but not in the strong-operator topology.

(iv) Find a sequence of unitary operators that converges in the strong-operator topology to an operator (isometry) that is not a unitary.

5.7.13. Let $\mathscr{H}$ be an infinite-dimensional Hilbert space.

(i) Show that $(\mathscr{H})_1$ is a weakly compact convex subset of $\mathscr{H}$ whose extreme points form a dense subset of it.

(ii) Show that $(\mathscr{B}(\mathscr{H}))_1^+$ is a weak-operator compact convex subset of $\mathscr{B}(\mathscr{H})$ whose extreme points form a dense subset of it.

(iii) Show that $(\mathscr{B}(\mathscr{H}))_1$ is a weak-operator compact convex subset of $\mathscr{B}(\mathscr{H})$ whose extreme points form a dense subset of it.

5.7.14. Let $X$ be an extremely disconnected compact Hausdorff space, and let $\{f_a : a \in \mathbb{A}\}$ be a family of real-valued functions in $C(X)$ bounded above by some constant.

(i) Suppose that each $f_a$ is the characteristic function of some clopen subset $X_a$ of $X$. Show that $[\bigcup_{a\in\mathbb{A}} X_a]^-$ is a clopen set whose characteristic function $\bigvee_{a\in\mathbb{A}} f_a$ is the least upper bound of $\{f_a\}$ and that the interior of $\bigcap_{a\in\mathbb{A}} X_a$ is a clopen set whose characteristic function $\bigwedge_{a\in\mathbb{A}} f_a$ is the greatest lower bound of $\{f_a\}$ in $C(X)$.

(ii) Show that

$$X\setminus\left[\bigcup_{a\in\mathbb{A}}\{x\in X : f_a(x) > \lambda\}\right]^- \quad (= X_\lambda)$$

is a clopen subset of $X$ and that if $Y$ is a clopen subset of $X$ with the property that $f_a(p) \leqslant \lambda$ for all $a$ in $\mathbb{A}$ and all $p$ in $Y$, then $Y \subseteq X_\lambda$.

(iii) Let $e_\lambda$ be the characteristic function of $X_\lambda$. Let $k$ be a constant that bounds $\{f_a\}$ above and such that $-k \leqslant f_{a'}$ for some $a'$ in $\mathbb{A}$. Show that

(1) $e_\lambda = 0$ for $\lambda < -k$ and $e_\lambda = 1$ for $\lambda > k$;

(2) $e_\lambda \leqslant e_{\lambda'}$ if $\lambda \leqslant \lambda'$;

(3) $e_\lambda = \bigwedge_{\lambda'>\lambda} e_{\lambda'}$.

(iv) Show that $\int_{-k}^{k} \lambda\, de_\lambda$ converges in norm (in the sense of approximating Riemann sums) to a function $f$ in $C(X)$ and that $X_\lambda$ is the largest clopen set on which $f$ takes values not exceeding $\lambda$.

(v) Show that $f$ is the least upper bound of $\{f_a\}$, and conclude that $C(X)$ is a boundedly complete lattice.

(vi) Deduce that $C(X)$ is a boundedly complete lattice if and only if $X$ is extremely disconnected.

5.7.15. Let $X$ be a compact Hausdorff space.

(i) Show that $X$ is extremely disconnected if and only if disjoint open subsets have disjoint closures.

(ii) Show that $X$ is extremely disconnected if and only if it satisfies the following two conditions:

(a) $X$ is totally disconnected;

(b) the family $\mathscr{C}$ of clopen subsets of $X$ partially ordered by inclusion is a complete lattice.

5.7.16. With the notation of Exercises 3.5.4, 3.5.5, and 3.5.6:

(i) show that $l_\infty$ is isometrically isomorphic to the maximal abelian algebra $\mathscr{A}$ of multiplication operators $M_f$ ($f \in l_\infty$) on $l_2(\mathbb{N}, \mathbb{C})$ via the mapping $f \to M_f$;

(ii) show that the pure state space of the quotient $C^*$-algebra $\mathscr{A}/\mathscr{C}_0$ (see Exercise 4.6.60) is (naturally homeomorphic to) $\beta(\mathbb{N})\backslash\mathbb{N}$, where $\mathscr{C}_0$ is the image of $c_0$ under the mapping $f \to M_f$.

5.7.17. With the notation of Exercise 5.7.16, let $E_0$ be a projection in the quotient $C^*$-algebra $\mathscr{A}/\mathscr{C}_0$.

(i) Show that there is a projection $E$ in $\mathscr{A}$ such that $E$ maps onto $E_0$ under the quotient mapping.

(ii) Let $Y_0$ be a subset of $\beta(\mathbb{N})\backslash\mathbb{N}$ clopen in the relative topology on $\beta(\mathbb{N})\backslash\mathbb{N}$. Show that there is a clopen subset $Y$ of $\beta(\mathbb{N})$ such that $Y \cap (\beta(\mathbb{N})\backslash\mathbb{N}) = Y_0$.

5.7.18. With the notation of Exercise 3.5.5:

(i) show that $\mathcal{O}^- = (\mathcal{O} \cap \mathbb{N})^-$ for each open subset $\mathcal{O}$ of $\beta(\mathbb{N})$;

(ii) show that a subset $Y_0$ of $\beta(\mathbb{N})\backslash\mathbb{N}$ is clopen in $\beta(\mathbb{N})\backslash\mathbb{N}$ if and only if it has the form $\mathbb{N}_0^- \cap (\beta(\mathbb{N})\backslash\mathbb{N})$ for some subset $\mathbb{N}_0$ of $\mathbb{N}$, and that $Y_0$ is non-empty if and only if $\mathbb{N}_0$ is infinite.

5.7.19. Let $\varphi$ be a one-to-one mapping of the set of rational numbers onto $\mathbb{N}$. For each real number $t$, choose a sequence $\{r_1, r_2, \ldots\}$ of distinct rational numbers tending to $t$. With the notation of Exercise 5.7.16, let $\mathbb{N}_t$ be $\{\varphi(r_1), \varphi(r_2), \ldots\}$ and let $Y_t$ be $\mathbb{N}_t^- \cap (\beta(\mathbb{N})\backslash\mathbb{N})$.

(i) Show that $Y_t$ and $Y_s$ are disjoint, non-empty, clopen subsets of $\beta(\mathbb{N})\backslash\mathbb{N}$ when $s \neq t$.

(ii) Let $S$ be a subset of $\mathbb{R}$ and let $Y_S$ be the closure of $\cup_{s\in S} Y_s$. Show that if $t \notin S$, then $Y_t \cap Y_S = \varnothing$.

(iii) Show that $Y_S$ is not a clopen subset of $\beta(\mathbb{N})\backslash\mathbb{N}$ for some subset $S$ of $\mathbb{R}$. [*Hint*. "Count" subsets of $\mathbb{R}$ and of $\mathbb{N}$ and use Exercise 5.7.18(ii).]

(iv) Deduce that $\beta(\mathbb{N})\backslash\mathbb{N}$ is totally disconnected but not extremely disconnected and that $\mathscr{A}/\mathscr{C}_0$ is not a boundedly complete lattice.

5.7.20. Let $X$ be a complete metric space. We say that an open subset of $X$ is *regular* when it coincides with the interior of its closure.

(i) Show that the interiors of closed (hence of the closures and complements of open) sets in $X$ are regular.

(ii) Show that each open subset of $X$ differs from a regular open subset on a meager set.

(iii) Show that each Borel subset of $X$ differs from a regular open subset on a meager (Borel) set. [*Hint*. Follow the pattern of the argument of the first paragraph of the proof of Lemma 5.2.10.]

(iv) Show that there is a *unique* regular open subset of $X$ that differs from a given Borel set on a meager (Borel) set.

(v) Let $\mathscr{F}_0$ be the family of regular open subsets of $X$ partially ordered by inclusion. Show that $\mathscr{F}_0$ is a complete lattice.

(vi) Let $\mathscr{F}$ be the family of Borel subsets of $X$ and $\mathscr{M}$ the $\sigma$-ideal of meager Borel subsets of $X$ (a countable union of sets in $\mathscr{M}$ is in $\mathscr{M}$ and the intersection of a set of $\mathscr{M}$ with any set of $\mathscr{F}$ is in $\mathscr{M}$). Let $\mathscr{F}/\mathscr{M}$ be the family of equivalence classes of sets in $\mathscr{F}$ under the relation $S \sim S'$ when $S$ and $S'$ differ by a meager set. With $\mathscr{S}$ and $\mathscr{S}'$ in $\mathscr{F}/\mathscr{M}$, define $\mathscr{S} \precsim \mathscr{S}'$ when $S \subseteq S'$ for some $S$ in $\mathscr{S}$ and $S'$ in $\mathscr{S}'$. Show that $\precsim$ is a partial ordering of $\mathscr{F}/\mathscr{M}$ (the "quotient" of "inclusion" on $\mathscr{F}$ by the ideal $\mathscr{M}$), that each $\mathscr{S}$ in $\mathscr{F}/\mathscr{M}$ contains precisely one regular open set, and that the mapping that assigns to each $\mathscr{S}$ in $\mathscr{F}/\mathscr{M}$ the regular open set it contains is an order isomorphism of $\mathscr{F}/\mathscr{M}$ onto $\mathscr{F}_0$. Conclude that $\mathscr{F}/\mathscr{M}$ is a complete lattice.

(vii) Show that the algebra $\mathscr{B}(X)$ of bounded Borel functions on $X$ is a commutative $C^*$-algebra and that the family $\mathscr{M}_0$ of functions in $\mathscr{B}(X)$ that vanish on the complement of a meager Borel set is a closed ideal in $\mathscr{B}(X)$. Conclude that $\mathscr{B}(X)/\mathscr{M}_0$ is a commutative $C^*$-algebra.

(viii) Let $Y$ be the compact Hausdorff space such that $\mathscr{B}(X)/\mathscr{M}_0 \cong C(Y)$. Show that $Y$ is totally disconnected and that the family of clopen subsets of $Y$, partially ordered by inclusion, form a complete lattice. Conclude that $Y$ is extremely disconnected and that $C(Y)$ (and $\mathscr{B}(X)/\mathscr{M}_0$) are boundedly complete lattices.

5.7.21. With the notation of Exercise 5.7.20, assume that $X$ is $[0, 1]$ and let $\rho$ be a state of $C(Y)$.

(i) Suppose $\rho(\bigvee_{n=1}^{\infty} e_n) = \sum_{n=1}^{\infty} \rho(e_n)$ whenever $\{e_n\}$ is a countable family of idempotents in $C(Y)$ such that $e_n \cdot e_{n'} = 0$ unless $n = n'$. (We say that $\rho$ is a *normal* state in this case.) Show that $\rho(\bigvee_{n=1}^{\infty} f_n) \leqslant \sum_{n=1}^{\infty} \rho(f_n)$ for each countable set $\{f_n\}$ of idempotents $f_n$ in $C(Y)$ (where "$a \leqslant +\infty$" is envisaged in the inequality of this assertion).

(ii) Enumerate the open intervals in $[0, 1]$ ($= X$) with rational endpoints and let $f_1, f_2, \ldots$ be the idempotents in $C(Y)$ that are the images of their characteristic functions (in $\mathscr{B}(X)$) under the composition of the quotient mapping of $\mathscr{B}(X)$ onto $\mathscr{B}(X)/\mathscr{M}_0$ and the isomorphism of $\mathscr{B}(X)/\mathscr{M}_0$ with $C(Y)$. For each $j$ in $\{1, 2, \ldots\}$, let $e_j$ be an idempotent in $C(Y)$ such that $0 < e_j \leqslant f_j$. Show that $\bigvee_{j=1}^{\infty} e_j = 1$.

(iii) With the notation of (ii) and given a positive $\varepsilon$, show that $e_j$ can be chosen such that $\rho(e_j) \leqslant 2^{-j}\varepsilon$. Conclude that $C(Y)$ has no normal states.

(iv) Deduce that $C(Y)$ is isomorphic to no abelian von Neumann algebra although $Y$ is extremely disconnected.

5.7.22. Let $\mathscr{A}$ be an abelian von Neumann algebra acting on a Hilbert space $\mathscr{H}$, and let $A$ be an operator in $\mathscr{A}$. Suppose $\mathscr{A} \cong C(X)$, where $X$ is an extremely disconnected compact Hausdorff space, and $f$ in $C(X)$ represents $A$. Show that $Ax = \lambda x$ for some unit vector $x$ in $\mathscr{H}$ and some complex number $\lambda$ if and only if $f^{-1}(\lambda)$ contains a non-empty clopen subset of $X$.

5.7.23. Let $\mathscr{R}$ be a von Neumann algebra and $\mathscr{P}$ be its family of projections. Show that, with $\rho_0$ in $\mathscr{R}^{\#}$, the family of all sets,

$$\{\rho \in \mathscr{R}^{\#} : |(\rho - \rho_0)(E_j)| < \varepsilon, j \in \{1, \ldots, n\}\} \; (= \mathscr{V}(E_1, \ldots, E_n, \varepsilon)),$$

where $\varepsilon > 0$ and $\{E_1, \ldots, E_n\} \subseteq \mathscr{P}$ is a base for the open neighborhoods of $\rho_0$ in the weak* topology on a bounded subset of $\mathscr{R}^{\#}$.

5.7.24. Let $\mathscr{R}$ be a von Neumann algebra acting on a Hilbert space $\mathscr{H}$.

(i) Let $\varphi$ be a representation of a $C^*$-algebra $\mathfrak{A}$ with image $\mathscr{R}$, and let $V$ be a unitary operator in $\mathscr{R}$. Show that there is a unitary operator $U$ in $\mathfrak{A}$ such that $\varphi(U) = V$.

(ii) Show that the unitary group $\mathscr{R}_u$ of $\mathscr{R}$ is (pathwise) connected (in its norm topology).

5.7.25. Let $S$ be a locally compact topological space, $\mathscr{S}$ the $\sigma$-algebra of Borel sets, and $m$ a $\sigma$-finite regular Borel measure on $S$. Let $\mathfrak{A}$ be the algebra of multiplications by bounded continuous functions on $L_2(S, m)$ and $\mathscr{A}$ be its weak-operator closure. Show that $\mathscr{A}$ is the algebra of multiplications by essentially bounded measurable functions on $S$.

5.7.26. Let $(S, \mathscr{S}, m)$ be a $\sigma$-finite measure space, $f$ be a measurable function on $S$, and $A$ be a bounded operator on $L_2(S, m)$ such that $f \cdot g = Ag$ almost everywhere for each essentially bounded measurable function $g$ in $L_2(S, m)$. Show that $f$ is essentially bounded and that $M_f = A$.

5.7.27. Let $(S, \mathscr{S}, m)$ be a $\sigma$-finite measure space and $g$ be an essentially bounded measurable function on $S$.

(i) Show that $m(g^{-1}(D)) = 0$ for each closed disk $D$ contained in $\mathbb{C}\backslash\text{sp}(g)$, where $\text{sp}(g)$ is the essential range of $g$ (defined and studied in Example 3.2.16).

(ii) Note that each open subset of $\mathbb{C}$ is the union of a countable family of closed disks and conclude that $m(g^{-1}(\mathbb{C}\backslash\text{sp}(g))) = 0$.

(iii) Let $\lambda$ be some point in $\text{sp}(g)$ and define $g_0(s)$ to be $g(s)$ for $s$ in $g^{-1}(\text{sp}(g))$ and $\lambda$ for $s$ in $g^{-1}(\mathbb{C}\backslash\text{sp}(g))$. Show that $\text{sp}(g_0) = \text{sp}(g)$.

(iv) Let $f$ be a bounded Borel function on $\text{sp}(g)$. Show that $f(M_g) = f(M_{g_0}) = M_{f \circ g_0}$. [*Hint*. Use uniqueness of the Borel function calculus.]

5.7.28. Let $\mathscr{R}$ be a von Neumann algebra acting on a Hilbert space $\mathscr{H}$, and let $A$ be a self-adjoint operator in $\mathscr{B}(\mathscr{H})$ such that $UA + AU \leqslant 2A$ for each self-adjoint unitary operator $U$ in $\mathscr{R}$. Show that $A \in \mathscr{R}'$.

5.7.29. Let $\mathscr{S}$ and $\mathscr{T}$ be two families of bounded operators on a Hilbert space $\mathscr{H}$, and suppose that $\mathscr{S} \subseteq \mathscr{T}$.

(i) Show that $\mathscr{T}' \subseteq \mathscr{S}'$.

(ii) Show that $\mathscr{S}' = (\mathscr{S}')''$ $(= \mathscr{S}''')$. (Compare Theorem 5.3.1.)

5.7.30. Let $\mathscr{H}$ be a Hilbert space of dimension greater than 1. Find a weak-operator closed subalgebra $\mathscr{B}$ of $\mathscr{B}(\mathscr{H})$ such that $\mathscr{B} \neq \mathscr{B}''$.

5.7.31. Let $\mathscr{H}$ be a Hilbert space and $\mathfrak{A}_0$ be a self-adjoint subalgebra of $\mathscr{B}(\mathscr{H})$. Assume that $\mathfrak{A}_0(\mathscr{H})$ is dense in $\mathscr{H}$ but *not* that $I \in \mathfrak{A}_0$. Show that $\mathfrak{A}_0''$ is the strong-operator closure of $\mathfrak{A}_0$ (and that $I \in \mathfrak{A}_0''$).

5.7.32. Use the double commutant theorem to re-prove (compare Proposition 5.1.8), in the special case of a von Neumann algebra $\mathscr{R}$ (so $\mathscr{R}$ is assumed to contain $I$), that the union and intersection of each family of projections in $\mathscr{R}$ lie in $\mathscr{R}$.

5.7.33. Let $\mathfrak{A}$ be a simple $C^*$-algebra (that is, $\mathfrak{A}$ has no proper two-sided ideals) acting on a Hilbert space $\mathscr{H}$.

(i) Show that the center of $\mathfrak{A}$ is $\{aI : a \in \mathbb{C}\}$.

(ii) Suppose $\mathfrak{A}$ contains a maximal abelian subalgebra of $\mathscr{B}(\mathscr{H})$. Show that $\mathfrak{A}$ acts irreducibly on $\mathscr{H}$.

5.7.34. Find a von Neumann algebra $\mathscr{R}$ and a strong-operator dense, self-adjoint subalgebra $\mathfrak{A}_0$ (containing $I$) such that *no* unitary operator in $\mathscr{R}$ other than a scalar is the strong-operator limit of unitary operators in $\mathfrak{A}_0$. [*Hint.* Consider polynomials on $[0, 1]$.)

5.7.35. Let $\mathscr{H}$ be a Hilbert space, $S$ a closed subset of $\mathbb{R}$, $\mathscr{S}$ the set of self-adjoint operators on $\mathscr{H}$ with spectrum in $S$, and $h$ a real-valued, bounded, continuous function defined on $S$. With $A_0$ in $\mathscr{S}$, let $k$ be a continuous function on $\mathbb{R}$ that takes the value 1 at each point of $\mathrm{sp}(A_0)$ and vanishes outside of $[-(\|A_0\| + 1), \|A_0\| + 1]$. Let $p$ be $hk$ and $q$ be $1 - k + p$.

(i) Show that $p(A_0) = q(A_0) = h(A_0)$, $h = (1 - h)p + hq$, and

$$h(A) - h(A_0) = (I - h(A))(p(A) - p(A_0)) + h(A)(q(A) - q(A_0)).$$

(ii) Show that the mapping $A \to h(A)$ of $\mathscr{S}$ into (the set of normal operators in) $\mathscr{B}(\mathscr{H})$ is strong-operator continuous.

5.7.36. Let $\mathscr{H}$ be a Hilbert space, $S$ a closed subset of $\mathbb{R}$, and $\mathscr{S}$ the set of self-adjoint operators in $\mathscr{B}(\mathscr{H})$ with spectrum in $S$.

(i) Show that the mapping $A \to |A|$ is strong-operator continuous on $\mathscr{S}$.

(ii) Let $h$ be a real-valued, continuous function on $S$ (so $h$ is bounded on bounded subsets of $S$). Suppose $S_0$ is a bounded subset of $S$ such that $g$ is bounded on $S \backslash S_0$, where $g(t) = h(t)/|t|$ for $t$ in $S \backslash S_0$. Show that the mapping $A \to h(A)$ is strong-operator continuous on $\mathscr{S}$.

(iii) Deduce that $A \to A^{1/n}$ is a strong-operator continuous mapping on $\mathscr{B}(\mathscr{H})^+$ for each positive integer $n$.

5.7.37. Let $S$ be a subset of $\mathbb{R}$. Suppose $f$ is a function defined on $S$ such that the mapping $A \to f(A)$ is strong-operator continuous on $\mathscr{S}$ for each Hilbert space $\mathscr{H}$, where $\mathscr{S}$ is the set of self-adjoint operators in $\mathscr{B}(\mathscr{H})$ with spectrum in $S$. Show that $f$ is continuous on $S$, bounded on bounded subsets of $S$, and that there is a bounded subset $S_0$ of $S$ such that $\{f(t)/|t| : t \in S \backslash S_0\}$ is bounded.

5.7.38. Let $\mathscr{R}$ be a von Neumann algebra acting on a Hilbert space $\mathscr{H}$, and let $E$ be a projection in $\mathscr{R}$. Find the central carrier of a projection $F$ in $E\mathscr{R}E$ relative to $E\mathscr{R}E$ in terms of its central carrier $C_F$ relative to $\mathscr{R}$.

5.7.39. Let $\mathscr{R}$ be a von Neumann algebra of infinite linear dimension. Show that there is an orthogonal infinite family of non-zero projections with sum $I$.

5.7.40. Let $\mathfrak{A}$ be a $C^*$-algebra acting on a Hilbert space $\mathscr{H}$, and let $\{E'_n\}$ be an orthogonal family of non-zero projections in the commutant $\mathfrak{A}'$ of $\mathfrak{A}$ with sum $I$. Suppose $x_0$ is a generating vector for $\mathfrak{A}$ and $E'_n x_0 = a_n x_n$, where $|a_n| = \|E'_n x_0\|$ and $\|x_n\| = 1$.

(i) Show that $a_n \neq 0$ and that $x_0 = \sum_n a_n x_n$.

(ii) Suppose $\{E'_n\}$ is a (countably) infinite family (indexed by positive integers). Choose $n(j)$ $(> n(j-1))$ such that $|a_{n(j)}| < j^{-2}$ for $j$ in $\mathbb{N}$. Let $x'$ be $\sum_{j=1}^{\infty} j^{-1} x_{n(j)}$. Show that for each $A$ in the strong-operator closure $\mathfrak{A}^-$ of $\mathfrak{A}$, $Ax_0 \neq x'$.

(iii) Conclude that $\mathfrak{A}'$ is finite dimensional if there is a vector $x_0$ in $\mathscr{H}$ such that $\{Ax_0 : A \in \mathfrak{A}^-\} = \mathscr{H}$.

5.7.41. Let $\mathfrak{A}$ be a $C^*$-algebra that acts topologically irreducibly on a Hilbert space $\mathscr{H}$, and let $\{x_1, \ldots, x_n\}$ and $\{y_1, \ldots, y_n\}$ be sets of vectors in $\mathscr{H}$.

(i) Let $H$ be a self-adjoint operator in $\mathscr{B}(\mathscr{H})$ such that $Hx_j = y_j$ for $j$ in $\{1, \ldots, n\}$. Show that there is a self-adjoint operator $K$ in $\mathfrak{A}$ such that $Kx_j = y_j$ for $j$ in $\{1, \ldots, n\}$ and $\|K\| \leqslant \|H\|$. [*Hint*. Use a diagonalizing orthonormal basis for $EHE$ restricted to $[x_1, \ldots, x_n, y_1, \ldots, y_n]$, where $E$ is the projection in $\mathscr{B}(\mathscr{H})$ with this subspace as range, and apply Remark 5.6.32.]

(ii) Let $B$ be an operator in $\mathscr{B}(\mathscr{H})$ such that $Bx_j = y_j$ for $j$ in $\{1, \ldots, n\}$. Show that there is an operator $A$ in $\mathfrak{A}$ such that $Ax_j = y_j$ for $j$ in $\{1, \ldots, n\}$ and $\|A\| \leqslant \|B\|$. [*Hint*. With $E$ as in the hint to (i), use the fact that $EBE | E(\mathscr{H})$ has the form $VH$ with $V$ a unitary operator and $H$ a positive operator on $E(\mathscr{H})$.]

5.7.42. Let $\mathscr{H}_0$ be a Hilbert space and $\mathscr{H}$ be the direct sum $\sum_n \oplus \mathscr{H}_n$ of countably many copies $\mathscr{H}_n$ of $\mathscr{H}_0$.

(i) With the notation of Subsection 2.6, *Matrix representations*, let $\mathscr{R}$ be the subalgebra of $\mathscr{B}(\mathscr{H})$ consisting of operators whose matrix has the same element of $\mathscr{B}(\mathscr{H}_0)$ at each diagonal entry and 0 at each off-diagonal entry. (With $\mathscr{H}$ viewed as a tensor product $\mathscr{H}_0 \otimes \mathscr{K}$ of $\mathscr{H}_0$ with a separable, infinite-dimensional Hilbert space $\mathscr{K}$, $\mathscr{R}$ is $\{T \otimes I : T \in \mathscr{B}(\mathscr{H}_0)\}$.) Show that $\mathscr{R}'$ consists of those operators whose matrix representations have scalar multiples of $I$ at each entry. (In tensor product form, $\mathscr{R}' = \{I \otimes S : S \in \mathscr{B}(\mathscr{K})\}$.) Show that $\mathscr{R}'' = \mathscr{R}$ and conclude that $\mathscr{R}$ is a von Neumann algebra (as well as $\mathscr{R}'$).

(ii) Let $x_1, x_2, \ldots$ be a sequence of vectors in $\mathscr{H}_0$. Call this sequence $l_2$-*independent* when $\sum_{j=1}^{\infty} \|x_j\|^2 < \infty$ and $\sum_{j=1}^{\infty} a_j x_j = 0$ for a sequence $\{a_j\}$ in $l_2(\mathbb{N}, \mathbb{C})$ only if $a_j = 0$ for all $j$. Note that an $l_2$-independent sequence is linearly independent and find a linearly independent sequence that is not $l_2$-independent when $\mathscr{H}_0$ is infinite dimensional.

(iii) Show that $\{x_1, x_2, \ldots\}$ is a generating vector for $\mathscr{R}$ if and only if $x_1, x_2, \ldots$ is $l_2$-independent.

(iv) Show that $\{x_1, x_2, \ldots\}$ is separating for $\mathscr{R}$ if and only if $[x_1, x_2, \ldots] = \mathscr{H}_0$.

5.7.43. Let $\mathscr{H}$ be $L_2([0,1])$ relative to Lebesgue measure. With $\mathscr{A}$ the multiplication algebra of $\mathscr{H}$ and $\mathfrak{A}$ the $C^*$-algebra of multiplications by continuous functions, find a unit vector $u$ that is separating for $\mathfrak{A}$ (if $A \in \mathfrak{A}$ and $Au = 0$, then $A = 0$) but not for $\mathscr{A}$. [*Hint.* Use the "Cantor process" to find an open dense subset of $[0,1]$ that has measure $\frac{1}{2}$.]

5.7.44. Let $\mathscr{H}$ be $L_2([0,1])$ relative to Lebesgue measure and $\mathscr{A}$ be the multiplication algebra of $\mathscr{H}$.

(i) Describe the vectors in $\mathscr{H}$ that are generating for $\mathscr{A}$.
(ii) Show that the set of generating vectors for $\mathscr{A}$ is dense in $\mathscr{H}$.
(iii) Deduce that a norm limit of generating vectors need not be a generating vector.

5.7.45. Let $\mathscr{R}$ be a von Neumann algebra and $\{E_1, E_2, \ldots\}$ be a countable family of countably decomposable projections in $\mathscr{R}$. Show that $\bigvee_{n=1}^{\infty} E_n$ is a countably decomposable projection in $\mathscr{R}$.

5.7.46. Let $\mathscr{R}$ be a countably decomposable von Neumann algebra acting on a Hilbert space $\mathscr{H}$. Define a metric on $\mathscr{R}$ with the property that its associated metric topology coincides with the strong-operator topology on bounded subsets of $\mathscr{R}$.

5.7.47. Let $(S, \mathscr{S}, m)$ be a $\sigma$-finite measure space, $\mathscr{H}$ be $L_2(S, m)$, $\mathscr{A}$ be the multiplication algebra, and $f$ and $g$ be measurable functions on $S$ finite almost everywhere. Show that:

(i) $M_f = M_g$ if and only if $f = g$ almost everywhere;
(ii) $M_{af+g} = aM_f \hat{+} M_g$ for each scalar $a$;
(iii) $M_{f \cdot g} = M_f \hat{\cdot} M_g$;
(iv) $M_f \geqslant 0$ if and only if $f \geqslant 0$ almost everywhere.

5.7.48. Let $(S, \mathscr{S}, m)$ be a $\sigma$-finite measure space, $\mathscr{H}$ be $L_2(S, m)$, $\mathscr{A}$ be the multiplication algebra, $g$ be a measurable function on $S$ finite almost everywhere, and $f$ be a Borel function on $\mathrm{sp}(M_g)$.

(i) Define a concept of "essential range" $\mathrm{sp}(g)$ analogous to that of Example 3.2.16 and show that $\mathrm{sp}(g) = \mathrm{sp}(M_g)$.
(ii) Show that there is a measurable $g_0$ on $S$ equal to $g$ almost everywhere such that the range of $g_0$ is contained in $\mathrm{sp}(g)$.
(iii) With the notation of (ii), show that $f(M_g) = M_{f \circ g_0}$.

5.7.49. Let $\mathscr{H}$ be $L_2(\mathbb{R})$ relative to Lebesgue measure and $A$ be the (unbounded) multiplication operator corresponding to the identity transform $\iota$ (the function $t \to t$) on $\mathbb{R}$ with domain $\mathscr{D}$ consisting of those $f$ in $L_2(\mathbb{R})$ such that $\iota \cdot f \in L_2(\mathbb{R})$.

(i) Note that $A$ is self-adjoint and that the spectral resolution for $A$ is $\{E_\lambda\}$, where $E_\lambda$ is the multiplication operator corresponding to the characteristic function of $(-\infty, \lambda]$.

(ii) Let $\mathscr{D}_0$ be the set of continuously differentiable functions on $\mathbb{R}$ that vanish outside a finite interval. Note that $\mathscr{D}_0$ is a dense linear submanifold of $\mathscr{H}$. Let $D_0$ be the operator with domain $\mathscr{D}_0$ that assigns $if'$ $(= i\,df/dt)$ to $f$. With $T$ the unitary operator defined in Theorem 3.2.31, show that $T^{-1}ATf = D_0 f$ for each $f$ in $\mathscr{D}_0$.

(iii) Conclude that $T^{-1}AT$ $(= D)$ with domain $T^{-1}(\mathscr{D})$ is a self-adjoint extension of $D_0$.

(iv) Show that $\exp(itD)$ is the unitary operator $U_t$, where $(U_t f)(p) = f(p - t)$. How does this relate to Stone's theorem?

5.7.50. Let $f$ be a jointly continuous function of two complex variables defined on $S_1 \times S_2$, where $S_1$ and $S_2$ are subsets of $\mathbb{C}$. Let $A$ be a normal operator such that $\operatorname{sp} A \subseteq S_2$ acting on a Hilbert space $\mathscr{H}$. For each $z_1$ in $S_1$, the mapping $z \to f(z_1, z)$ is a continuous (hence, Borel) function defined on $\operatorname{sp} A$ so that $f(z_1, A)$ is a normal operator on $\mathscr{H}$. Define an operator-valued function $g$ on $S_1$ by $g(z) = f(z, A)$.

(i) Suppose $A$ is bounded and $S_0$ is a compact subset of $S_1$. Show that the restriction of $g$ to $S_0$ is norm continuous (that is, the mapping $z \to g(z)$ is continuous from $S_0$ to $\mathscr{B}(\mathscr{H})$ with its norm topology).

(ii) Suppose $S_1$ is closed and $f$ is bounded (but no longer that $A$ is bounded). Show that $g$ is strong-operator continuous.

(iii) Let $H$ be a positive operator on $\mathscr{H}$. Show that $\exp(-izH)$ $(= U_z)$ is defined for each $z$ in the closed lower half plane $\mathbb{C}_-$ $(= \{z : \operatorname{Im} z \leqslant 0\})$, that $\|U_z\| \leqslant 1$, and that the mapping $z \to U_z$ is strong-operator continuous.

5.7.51. (i) With the notation and hypotheses of Exercise 5.7.50(ii), make the following additional assumptions about $f$:

(1) for each $z_2$ in $S_2$, $z \to f(z, z_2)$ is differentiable at each point $z_0$ of the interior $S_1^0$ of $S_1$ with derivative $f_1(z_0, z_2)$,

(2) given $z_0$ in $S_1^0$, a bounded subset $S_2'$ of $S_2$, and a positive $\varepsilon$, there are a positive $\delta$, a closed disk $D$ with center $z_0$ in $S_1^0$, and a positive $C$, such that for all $z_2$ in $S_2'$

$$|[f(z, z_2) - f(z_0, z_2)](z - z_0)^{-1} - f_1(z_0, z_2)| < \varepsilon,$$

provided $0 < |z - z_0| < \delta$ and $z \in S_1^0$ (that is, $z \to f(z, z_2)$ is differentiable on $S_1^0$ *uniformly* on bounded subsets of $S_2$) and such that for all $z$ in $D \backslash \{z_0\}$ and $z'$ in $S_2$

$$|[f(z,z') - f(z_0,z')](z - z_0)^{-1}| \leqslant C.$$

(3) $z \to f_1(z_0, z)$ is continuous on $S_2$ for each $z_0$ in $S_1^0$.

Show that for each pair of vectors $x$, $y$ in $\mathscr{H}$, $z \to \langle g(z)x, y\rangle$ is analytic on $S_1^0$ with derivative $\langle f_1(z_0, A)x, y\rangle$ at each $z_0$ in $S_1^0$. [*Hint*. With $z_0$ in $S_1^0$, let $h(z_1, z_2)$ be $[f(z_1, z_2) - f(z_0, z_2)](z_1 - z_0)^{-1}$ when $(z_1, z_2) \in S_1^0 \times S_2$ and $z_1 \neq z_0$, and let $h(z_0, z_2)$ be $f_1(z_0, z_2)$. Use Exercise 5.7.50(ii) to establish strong-operator continuity of $z \to h(z, A)$.]

(ii) With the notation and assumptions of Exercise 5.7.50(iii), show that the function $z \to \langle U_z x, y\rangle$ is analytic in the open lower half-plane $\mathbb{C}_-^0$ $(=\{z : \operatorname{Im} z < 0\})$ for each pair of vectors $x$, $y$ in $\mathscr{H}$.

(iii) Show that $z \to \langle U_z x, y\rangle$ is entire for each pair of vectors $x$ and $y$ in $\mathscr{H}$ when $H$ is bounded. Re-prove the results of (ii) by using a bounding sequence $\{E_n\}$ of projections for $H$ and considering first the case where $x, y \in E_n(\mathscr{H})$.

5.7.52. Let $\mathscr{R}$ be a von Neumann algebra acting on a Hilbert space $\mathscr{H}$ and $t \to \exp(-itH)$ $(= U_t)$ be a one-parameter unitary group on $\mathscr{H}$, where $H$ is a self-adjoint operator on $\mathscr{H}$ with domain $\mathscr{D}$. Let $x_0$ be a unit vector in $\mathscr{D}$. Make the following assumptions about $H$, $U_t$, $\mathscr{R}$, and $x_0$:

(1) $U_t A U_{-t} \in \mathscr{R}$ for each $A$ in $\mathscr{R}$ and each $t$ in $\mathbb{R}$;
(2) $H \geqslant 0$;
(3) $Hx_0 = 0$;
(4) $x_0$ is generating for $\mathscr{R}$.

(i) Define $U_z$ as in Exercise 5.7.50(iii) when $z \in \mathbb{C}_-$, and show that $U_z x_0 = x_0$ for each $z$ in $\mathbb{C}_-$.

With $A$ and $A'$ self-adjoint operators in $\mathscr{R}$ and $\mathscr{R}'$, respectively, define $f(z)$ to be $\langle U_z A x_0, A' x_0\rangle$ for $z$ in $\mathbb{C}_-$.

(ii) Show that $f(t)$ is a real number for real $t$, and that $f$ is continuous on $\mathbb{C}_-$, analytic on $\mathbb{C}_-^0$, and bounded on $\mathbb{C}_-$.

(iii) Show that $U_t \in \mathscr{R}$ for each real $t$.

5.7.53. Let $H$ be a self-adjoint operator acting on a Hilbert space $\mathscr{H}$, $\mathscr{A}$ be the von Neumann algebra generated by $H$, and $\mathscr{A}_0$ be the von Neumann algebra generated by $\{U_t : t \in \mathbb{R}\}$, where $U_t = \exp(-itH)$.

(i) Assume $H$ is bounded and show that the $C^*$-algebras $\mathfrak{A}$ and $\mathfrak{A}_0$ generated by $H$ (and $I$) and by $\{U_t : t \in \mathbb{R}\}$, respectively, coincide.

(ii) Show that $\mathscr{A} = \mathscr{A}_0$.

(iii) With the notation and assumptions of Exercise 5.7.52, show that $H \eta \mathscr{R}$.

5.7.54. Let $A$ be a normal operator acting on a Hilbert space $\mathscr{H}$.

(i) Show that $\exp iA = I$ if $\mathrm{sp}\, A \subseteq \{2\pi n : n \in \mathbb{Z}\}$ and only if $A$ is self-adjoint and $\mathrm{sp}\, A \subseteq \{2\pi n : n \in \mathbb{Z}\}$.

(ii) Show that $\exp itA = I$ for all $t$ in $\mathbb{R}$ if and only if $A = 0$.

(iii) Let $A$ and $B$ be self-adjoint operators on $\mathscr{H}$ such that $\exp itB = \exp itA$ for each real $t$. Show that $A = B$.

(iv) Let $t \to U_t$ be a one-parameter unitary group acting on $\mathscr{H}$. Show that there is a *unique* self-adjoint operator $H$ on $\mathscr{H}$ such that $U_t = \exp itH$ for all real $t$.

5.7.55. With the notation of Exercise 5.7.52, assume conditions (1), (2) and (3), and in place of (4) assume that $x_0$ is separating for the center $\mathscr{C}$ of $\mathscr{R}$. Show that there is a positive self-adjoint operator $K$ on $\mathscr{H}$ such that $K \eta \mathscr{R}$ and $W_t A W_{-t} = U_t A U_{-t}$ for each $A$ in $\mathscr{R}$ and all real $t$, where $W_t = \exp(-itK)$ $(\in \mathscr{R})$, and such that $Kx_0 = 0$. [*Hint*. Consider the projection $E'$ with range $[\mathscr{R}x_0]$ and the von Neumann algebra $\mathscr{R}E'$ acting on $E'(\mathscr{H})$.]

# BIBLIOGRAPHY


## General references

[H] P. R. Halmos, "Measure Theory." D. Van Nostrand, Princeton, New Jersey, 1950; reprinted, Springer-Verlag, New York, 1974.

[K] J. L. Kelley, "General Topology." D. Van Nostrand, Princeton, New Jersey, 1955; reprinted, Springer-Verlag, New York, 1975.

[R] W. Rudin, "Real and Complex Analysis," 2nd ed. McGraw-Hill, New York, 1974.

## References

[1] W. Ambrose, Spectral resolution of groups of unitary operators, *Duke Math. J.* **11** (1944), 589–595.

[2] J. Dixmier, "Les $C^*$-Algèbres et Leurs Représentations." Gauthier-Villars, Paris, 1964. [*English translation:* "$C^*$-Algebras." North-Holland Mathematical Library, Vol. 15. North-Holland Publ., Amsterdam, 1977.]

[3] J. M. G. Fell and J. L. Kelley, An algebra of unbounded operators, *Proc. Nat. Acad. Sci. U.S.A.* **38** (1952), 592–598.

[4] I. M. Gelfand and M. A. Neumark, On the imbedding of normed rings into the ring of operators in Hilbert space, *Mat. Sb.* **12** (1943), 197–213.

[5] J. G. Glimm and R. V. Kadison, Unitary operators in $C^*$-algebras, *Pacific J. Math.* **10** (1960), 547–556.

[6] F. Hansen and G. K. Pedersen, Jensen's inequality for operators and Löwner's theorem. *Math. Ann.* **258** (1982), 229–241.

[7] D. Hilbert, Grundzüge einer allgemeinen Theorie der linearen Integralgleichungen IV, *Nachr. Akad. Wiss. Göttingen Math.-Phys. Kl.* 1904, 49–91.

[8] R. V. Kadison, Irreducible operator algebras, *Proc. Nat. Acad. Sci. U.S.A.* **43** (1957), 273–276.

[9] I. Kaplansky, A theorem on rings of operators, *Pacific J. Math.* **1** (1951), 227–232.

[10] J. L. Kelley, Commutative operator algebras, *Proc. Nat. Acad. Sci. U.S.A.* **38** (1952), 598–605.

[11] J. von Neumann, Zur Algebra der Funktionaloperationen und Theorie der normalen Operatoren, *Math. Ann.* **102** (1930), 370–427.

[12] J. von Neumann, Allgemeine Eigenwerttheorie Hermitescher Funktionaloperatoren, *Math. Ann.* **102** (1930), 49–131.

[13] J. von Neumann, Über Funktionen von Funktionaloperatoren, *Ann. of Math.* **32** (1931), 191–226.

[14] J. von Neumann, Über adjungierte Funktionaloperatoren, *Ann. of Math.* **33** (1932), 294–310.

[15] M. Neumark, Positive definite operator functions on a commutative group (in Russian, English summary), *Bull. Acad. Sci. URSS Sér. Math.* [*Iz. Akad. Nauk SSSR Ser. Mat.*] **7** (1943), 237–244.

[16] G. K. Pedersen, "$C^*$-Algebras and Their Automorphism Groups," London Mathematical Society Monographs, Vol. 14. Academic Press, London, 1979.

[17] F. Riesz, "Les Systèmes d'Équations Linéaires à une Infinité d'Inconnues." Gauthier-Villars, Paris, 1913.

[18] F. Riesz, Über die linearen Transformationen des komplexen Hilbertschen Raumes, *Acta Sci. Math.* (*Szeged*) **5** (1930–1932), 23–54.

[19] I. E. Segal, Irreducible representations of operator algebras, *Bull. Amer. Math. Soc.* **53** (1947), 73–88.

[20] M. H. Stone, On one-parameter unitary groups in Hilbert space, *Ann. of Math.* **33** (1932), 643–648.

[21] M. H. Stone, "Linear Transformations in Hilbert Space and Their Applications to Analysis". American Mathematical Society Colloquium Publications, Vol. 15. Amer. Math. Soc., New York, 1932.

[22] M. H. Stone, The generalized Weierstrass approximation theorem, *Math. Mag.* **21** (1948), 167–183, 237–254.

[23] M. H. Stone, Boundedness properties in function-lattices, *Canad. J. Math.* **1** (1949), 176–186.

[24] S. Strătilă and L. Zsidó, "Lectures on von Neumann Algebras." Abacus Press, Tunbridge Wells, 1979.

[25] M. Takesaki, "Theory of Operator Algebras I." Springer-Verlag, Heidelberg, 1979.

# INDEX OF NOTATION

**Algebras and related matters**

**Direct sums**



**Inner products and norms**



**Linear operators**

**Linear spaces**



**Linear topological spaces, Banach spaces, Hilbert spaces**

**Sets and mappings**



**Special Banach spaces**



**Tensor products**

# INDEX

## D



## E

**T**

# Pure and Applied Mathematics

A Series of Monographs and Textbooks



RECENT TITLES

SIGURDUR HELGASON. Differential Geometry, Lie Groups, and Symmetric Spaces

ROBERT B. BURCKEL. An Introduction to Classical Complex Analysis: Volume 1

JOSEPH J. ROTMAN. An Introduction to Homological Algebra

C. TRUESDELL AND R. G. MUNCASTER. Fundamentals of Maxwell's Kinetic Theory of a Simple Monatomic Gas: Treated as a Branch of Rational Mechanics

BARRY SIMON. Functional Integration and Quantum Physics

GRZEGORZ ROZENBERG AND ARTO SALOMAA. The Mathematical Theory of L Systems.

DAVID KINDERLEHRER and GUIDO STAMPACCHIA. An Introduction to Variational Inequalities and Their Applications.

H. SEIFERT AND W. THRELFALL. A Textbook of Topology; H. SEIFERT. Topology of 3-Dimensional Fibered Spaces

LOUIS HALLE ROWEN. Polynominal Identities in Ring Theory

DONALD W. KAHN. Introduction to Global Analysis

DRAGOS M. CVETKOVIC, MICHAEL DOOB, AND HORST SACHS. Spectra of Graphs

ROBERT M. YOUNG. An Introduction to Nonharmonic Fourier Series

MICHAEL C. IRWIN. Smooth Dynamical Systems

JOHN B. GARNETT. Bounded Analytic Functions

EDUARD PRUGOVEČKI. Quantum Mechanics in Hilbert Space, Second Edition

M. SCOTT OSBORNE AND GARTH WARNER. The Theory of Eisenstein Systems

K. A. ZHEVLAKOV, A. M. SLIN'KO, I. P. SHESTAKOV, AND A. I. SHIRSHOV. Translated by HARRY SMITH. Rings That Are Nearly Associative

JEAN DIEUDONNÉ. A Panorama of Pure Mathematics; Translated by I. Macdonald

JOSEPH G. ROSENSTEIN. Linear Orderings

AVRAHAM FEINTUCH AND RICHARD SAEKS. System Theory: A Hilbert Space Approach

ULF GRENANDER. Mathematical Experiments on the Computer

HOWARD OSBORN. Vector Bundles: Volume 1, Foundations and Stiefel-Whitney Classes

BHASKARA RAO AND BHASKARA RAO. Theory of Charges

RICHARD V. KADISON AND JOHN R. RINGROSE. Fundamentals of the Theory of Operator Algebras, Volume I

IN PREPARATION

ROBERT B. BURCKEL. An Introduction to Classical Complex Analysis: Volume 2

RICHARD V. KADISON AND JOHN R. RINGROSE. Fundamentals of the Theory of Operator Algebras, Volume II

BARRETT O'NEILL. Semi-Riemannian Geometry: With Applications to Relativity

EDWARD B. MANOUKIAN. Renormalization

E. J. MCSHANE. Unified Integration

A. P. MORSE. A Theory of Sets, Revised and Enlarged Edition